T0348741

INFRARED SPECTROSCOPY FOR FOOD QUALITY ANALYSIS AND CONTROL

Infrared Spectroscopy for Food Quality Analysis and Control

Edited by
Da-Wen Sun
Director of the Food Refrigeration and
Computerised Food Technology Research Group
National University of Ireland, Dublin
(University College Dublin)
Agriculture and Food Science Centre
Belfield
Dublin 4
Ireland

AMSTERDAM • BOSTON • HEIDELBERG • LONDON • NEW YORK • OXFORD
PARIS • SAN DIEGO • SAN FRANCISCO • SINGAPORE • SYDNEY • TOKYO
Academic Press is an imprint of Elsevier

Academic Press is an imprint of Elsevier
30 Corporate Drive, Suite 400, Burlington, MA 01803, USA
525 B Street, Suite 1900, San Diego, CA 92101-4495, USA
32 Jamestown Road, London, NW1 7BY, UK
360 Park Avenue South, New York, NY 10010-1710, USA

First edition 2009

Library of Congress Cataloging-in-Publication Data
A catalog record for this book is available from the Library of Congress

British Library Cataloguing in Publication Data
A catalogue record for this book is available from the British Library

ISBN: 978-0-12-374136-3

For information on all Academic Press publications
visit our website at www.elsevierdirect.com

Typeset by Charon Tec Ltd., A Macmillan Company
www.macmillansolutions.com

Contents

About the Editor

Born in Southern China, Professor Da-Wen Sun is a world authority in food engineering research and education. His main research activities include cooling, drying and refrigeration processes and systems, quality and safety of food products, bioprocess simulation and optimisation, and computer vision technology. Especially, his innovative studies on vacuum cooling of cooked meats, pizza quality inspection by computer vision, and edible films for shelf-life extension of fruit and vegetables have been widely reported in national and international media. Results of his work have been published in over 200 peer reviewed journal papers and more than 200 conference papers.

He received a first class BSc Honours and MSc in Mechanical Engineering, and a PhD in Chemical Engineering in China before working in various universities in Europe. He became the first Chinese national to be permanently employed in an Irish University when he was appointed College Lecturer at National University of Ireland, Dublin (University College Dublin) in 1995, and was then continuously promoted in the shortest possible time to Senior Lecturer, Associate Professor and Full Professor. Dr Sun is now Professor of Food and Biosystems Engineering and Director of the Food Refrigeration and Computerised Food Technology Research Group at University College Dublin.

As a leading educator in food engineering, Professor Sun has significantly contributed to the field of food engineering. He has trained many PhD students, who have made their own contributions to the industry and academia. He has also given lectures on advances in food engineering on a regular basis in academic institutions internationally and delivered keynote speeches at international conferences. As a recognised authority in food engineering, he has been conferred adjunct/ visiting/ consulting professorships from ten top universities in China including Zhejiang University, Shanghai

Jiaotong University, Harbin Institute of Technology, China Agricultural University, South China University of Technology, Jiangnan University and so on. In recognition of his significant contribution to Food Engineering worldwide and for his outstanding leadership in the field, the International Commission of Agricultural Engineering (CIGR) awarded him the CIGR Merit Award in 2000 and again in 2006, the Institution of Mechanical Engineers (IMechE) based in the UK named him "Food Engineer of the Year 2004", in 2008 he was awarded CIGR Recognition Award in honour of his distinguished achievements as the top one percent of Agricultural Engineering scientists in the world.

He is a Fellow of the Institution of Agricultural Engineers and a Fellow of Engineers Ireland. He has also received numerous awards for teaching and research excellence, including the President's Research Fellowship, and has twice received the President's Research Award of University College Dublin. He is a Member of CIGR Executive Board and Honorary Vice-President of CIGR, Editor-in-Chief of *Food and Bioprocess Technology – an International Journal* (Springer), Series Editor of "Contemporary Food Engineering" book series (CRC Press / Taylor & Francis), former Editor of *Journal of Food Engineering* (Elsevier), and Editorial Board Member for *Journal of Food Engineering* (Elsevier), *Journal of Food Process Engineering* (Blackwell), *Sensing and Instrumentation for Food Quality and Safety* (Springer) and *Czech Journal of Food Sciences*. He is also a Chartered Engineer.

Contributors

Manel Alcalà Bernàrdez (Ch. 3) Grup de Quimiometria Aplicada, Departament de Química (Unitat Analítica), Facultat de Ciències, Universitat Autònoma de Barcelona, 01893 Bellaterra, Spain

Murad A Al-Holy (Ch. 6) Department of Clinical Nutrition and Dietetics, Faculty of Allied Health Sciences, Hashemite University, Zarqa-Jordan

Josse De Baerdemaeker (Ch. 15) Division of Mechatronics, Biostatistics and Sensors (MeBioS), Department of Biosystems, K.U. Leuven, Kasteelpark Arenberg 30, B-3001, Leuven, Belgium

Davide Ballabio (Ch. 4) Milano Chemometrics and QSAR Research Group, Department of Environmental Sciences, University of Milano-Bicocca, P.zza della Scienza 1, 20126 Milano, Italy

Søren Balling Engelsen (Ch. 2) Faculty of Life Sciences, Chemometrics and Spectroscopy Group, Quality and Technology, Department of Food Science, University of Copenhagen, Rolighedsvej 30, DK-1958 Frederiksberg C, Denmark

Flip Bamelis (Ch. 15) Division of Mechatronics, Biostatistics and Sensors (MeBioS), Department of Biosystems, K.U. Leuven, Kasteelpark Arenberg 30, B-3001, Leuven, Belgium

Malgorzata Baranska (Ch. 12) Jagiellonian University, Department of Chemistry, Chemical Physics Division, 3 Ingardena str., 30–060 Krakow, Poland

Marcelo Blanco Romía (Ch. 3) Grup de Quimiometria Aplicada, Departament de Química (Unitat Analítica), Facultat de Ciències, Universitat Autònoma de Barcelona, 01893 Bellaterra, Spain

Rasmus Bro (Ch. 2) Faculty of Life Sciences, Chemometrics and Spectroscopy Group, Quality and Technology, Department of Food Science, University of Copenhagen, Rolighedsvej 30, DK-1958 Frederiksberg C, Denmark

Anna G Cavinato (Ch. 6, 13) Department of Chemistry and Biochemistry, Eastern Oregon University, One University Blvd., La Grande, OR 97850, USA

Daniel Cozzolino (Ch. 14) Australian Wine Res. Inst., Waite Rd., Urrbrae, PO Box 197, Adelaide, SA 5064, Australia

Robert G Dambergs (Ch. 14) The Australian Wine Research Institute. Waite Road, Urrbrae. PO Box 197, Adelaide, SA 5064 Australia

Bart De Ketelaere (Ch. 15) Division of Mechatronics, Biostatistics and Sensors (MeBioS), Department of Biosystems, K.U. Leuven, Kasteelpark Arenberg 30, B-3001, Leuven, Belgium

Éric Dufour (Ch. 1) UR Typicite des Produits Alimentaires, ENITA de Clermont Ferrand, Site de Marmilhat BP 35, F-63370 Lempdes, France

Colette C Fagan (Ch. 10) Biosystems Engineering, School of Agriculture, Food Science and Veterinary Medicine, University College Dublin, Belfield, Dublin 4, Ireland

Rahul Reddy Gangidi (Ch. 8) Spectroscopist Research Scientist, Thermo Fisher Scientific Inc., Process Instruments Division, 2026 Brewster Street, #5, St. Paul, MN 55108, USA

Yiqun Huang (Ch. 13) Department of Family, Nutrition, and Exercise Sciences, Queens College, City University of New York, 65–30 Kissena Boulevard, Flushing, NY 11367, USA

Birthe Møller Jespersen (Ch. 11) Faculty of Life Sciences, Chemometrics and Spectroscopy Group, Quality and Technology, Department of Food Science, University of Copenhagen, Rolighedsvej 30, DK-1958 Frederiksberg C, Denmark

Romdhane Karoui (Ch. 15) Unité de Recherche Typicité des Produits Alimentaires, Enita Clermont, Site de Marmilhat—BP 35, F-63370 Lempdes, France

Bart Kemps (Ch. 15) Division of Mechatronics, Biostatistics and Sensors (MeBioS), Department of Biosystems, K.U. Leuven, Kasteelpark Arenberg 30, B-3001, Leuven, Belgium

Mengshi Lin (Ch. 6) Food Science Program, Division of Food Systems & Bioengineering, 256 Stringer Wing, University of Missouri, Columbia 65211, USA

Kristof Mertens (Ch. 15) Division of Mechatronics, Biostatistics and Sensors (MeBioS), Department of Biosystems, K.U. Leuven, Kasteelpark Arenberg 30, B-3001, Leuven, Belgium

Lars Munck (Ch. 11) Faculty of Life Sciences, Chemometrics and Spectroscopy Group, Quality and Technology, Department of Food Science, University of Copenhagen, Rolighedsvej 30, DK-1958 Frederiksberg C, Denmark

Lars Nørgaard (Ch. 2) Faculty of Life Sciences, Chemometrics and Spectroscopy Group, Quality and Technology, Department of Food Science, University of Copenhagen, Rolighedsvej 30, DK-1958 Frederiksberg C, Denmark

Emiko Okazaki (Ch. 9) Food Quality Section, Division of Food Technology and Biochemistry, National Research Institute of Fisheries Science, 2–12–4 Fukuura, Kanazawa, Yokohama 236–8648, Japan

CP O'Donnell (Ch. 10) Biosystems Engineering, School of Agriculture, Food Science and Veterinary Medicine, University College Dublin, Belfield, Dublin 4, Ireland

Andrew Proctor (Ch. 8) Department of Food Science, 2650 N Young Ave., University of Arkansas, Fayetteville, AR 72704, USA

Barbara A Rasco (Ch. 6, 13) Department of Food Science & Human Nutrition, Washington State University, P.O. Box 646376, Pullman, WA 99164-6376, USA

Åsmund Rinnan (Ch. 2, 5) Faculty of Life Sciences, Chemometrics and Spectroscopy Group, Quality and Technology, Department of Food Science, University of Copenhagen, Rolighedsvej 30, DK-1958 Frederiksberg C, Denmark

Luis Rodriguez-Saona (Ch. 7) Department of Food Science and Technology, 110 Parker Food Science & Technology Building, Columbus OH 43210, USA

Hartwig Schulz (Ch. 12) Federal Research Centre for Cultivated Plants, Institute for Ecological Chemistry, Plant Analysis and Stored Product Protection, Erwin-Baur-Strasse 27, D-06484 Quedlinburg, Germany

Anand Subramanian (Ch. 7) Department of Food Science and Technology, The Ohio State University, 2015 Fyffe Ct, Columbus, OH 43210, USA

L Rudzik (Ch. 10) Institutsleiter, Institut für Lebensmittelqualität, LUFA Nord-West, Ammerländer Heerstr.115–117, 26129 Oldenburg, Germany

Jonas Thygesen (Ch. 2) The Royal Veterinary and Agricultural University, Chemometrics Group, Quality & Technology, Department of Food Science, Rolighedsvej 30, DK-1958 Frederiksberg C, Denmark

Roberto Todeschini (Ch. 4) Milano Chemometrics and QSAR Research Group, Department of Environmental Sciences, University of Milano-Bicocca, P.zza della Scienza 1, 20126 Milano, Italy

Musleh Uddin (Ch. 9) Food Quality Section, Division of Food Technology and Biochemistry, National Research Institute of Fisheries Science, 2–12–4 Fukuura, Kanazawa, Yokohama 236–8648, Japan

Frans van den Berg (Ch. 2, 5) Faculty of Life Sciences, Chemometrics and Spectroscopy Group, Quality and Technology, Department of Food Science, University of Copenhagen, Rolighedsvej 30, DK-1958 Frederiksberg C, Denmark

E Wüst (Ch. 10) University of Applied Scienes and Arts Hannover, Faculty of Mechanical and Bioprocess Engineering, Heisterbergallee 12, D-30453 Hannover, Germany

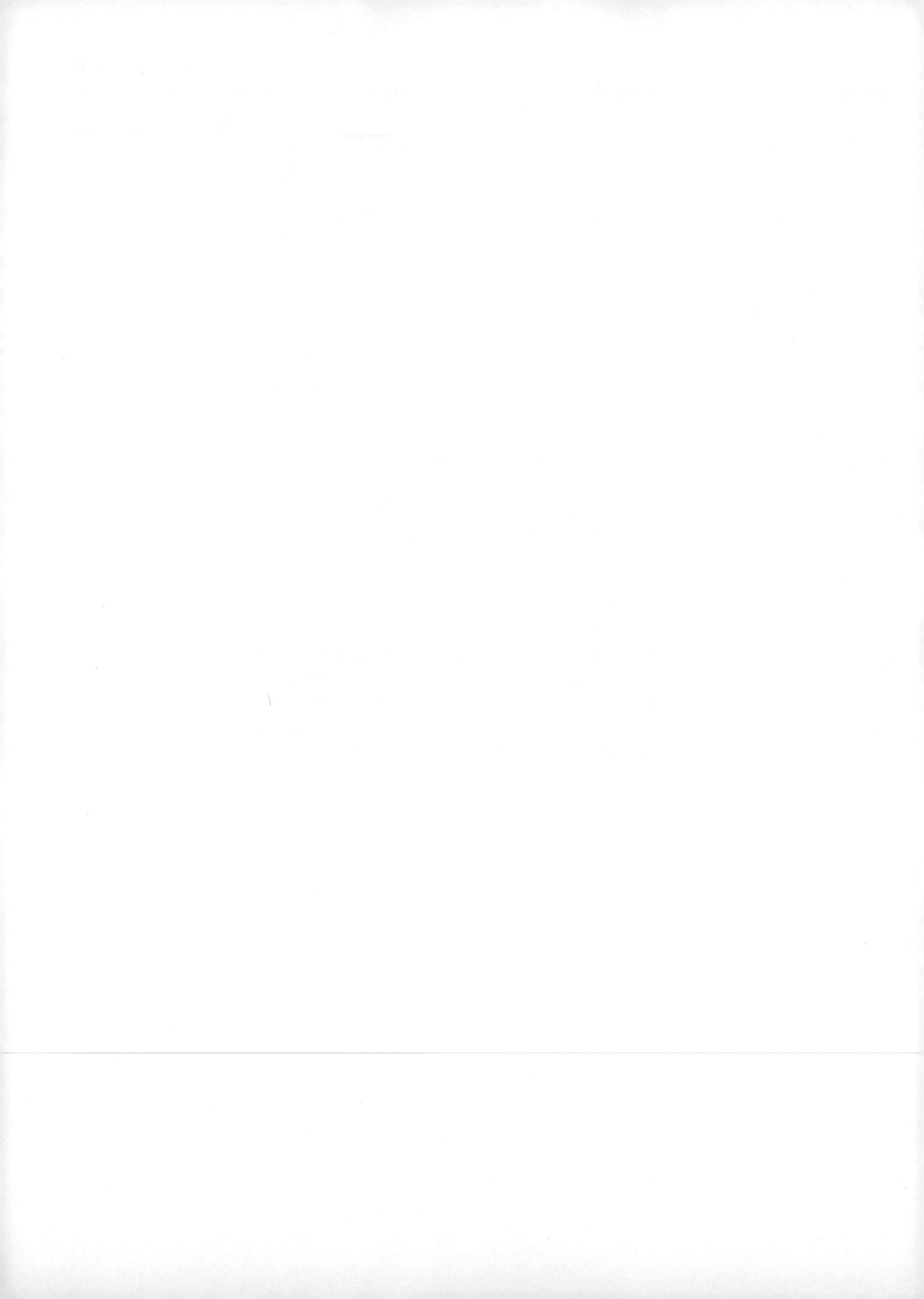

Preface

Infrared (IR) spectroscopy deals with the infrared part of the electromagnetic spectrum, it measures the absorption of different IR frequencies by a sample positioned in the path of an IR beam. Currently, infrared spectroscopy is one of the most common spectroscopic techniques used by the industry. With the rapid development in infrared spectroscopic instrumentation software and hardware, the application of this technique has expanded into many areas of food research. Infrared spectroscopy has become a powerful, fast and non-destructive tool for food quality analysis and control.

In order to reflect this trend of rapid technology development, it is appropriate to publish *Infrared Spectroscopy for Food Quality Analysis and Control*. The book is divided into two parts. Part I deals with principles and instruments including theory, data treatment techniques and infrared spectroscopy instruments. Part II covers its applications in quality analysis and control for various foods, for example, meat and meat products, fish and related products, vegetables, fruits, dairy products and cereals.

Infrared Spectroscopy for Food Quality Analysis and Control is written by international peers who have both academic and professional credentials, highlighting the truly international nature of the work. It aims to provide the engineer and technologist working in research, development, and operations in the food industry with critical and readily accessible information on the art and science of infrared spectroscopy technology. The book should also serve as an essential reference source to undergraduate and postgraduate students and researchers in universities and research institutions.

PART I

Fundamentals and Instruments

Principles of Infrared Spectroscopy

1

Éric Dufour

Introduction

The development of rapid analytical methods for food products relies mainly upon two approaches: the use of physical properties of substrates as an information supply and the automation of chemical methods. Most rapid analytical methods based on the physical properties of food products are spectroscopic methods. Spectroscopy can be split into two large groups (Wilson, 1994): photonic spectroscopy, which is based on the study of the interaction of an electromagnetic wave with matter, and particle spectroscopy. The first group comprises spectroscopic methods exhibiting an analytical potential for rapid control. The second group is represented by mass spectrometry and derived methods.

All the spectroscopic methods, except mass spectrometry, can be classified according to the energy involved during measurement. Electromagnetic radiation, of which visible light forms a tiny part, exists as waves that are propagated from a source and move in a straight line if they are not reflected or refracted. The undulatory phenomenon is a magnetic field associated with an electric one. The speed of the electromagnetic wave

Infrared Spectroscopy for Food Quality Analysis and Control
ISBN: 978-0-12-374136-3

is a universal constant "c," equal to 3×10^8 m/s. This wave can be represented as a sinusoidal function of time:

$$y = A \sin w.t \tag{1.1}$$

where A is signal amplitude, w is the pulsation expressed in radians per second (rad/s), and t is the time in seconds. In a second, the shape of the wave is repeated $w/2\pi$ times. This value is the frequency, υ, in cycles per second (s^{-1}, or Hertz, for which the symbol is Hz). The above equation represents a wave as a temporal phenomenon. A wave can also be represented as a function of the covered distance, x, expressed by the following equation, which takes into account the relation between time and distance:

$$x = c.t \tag{1.2}$$

Combining equations (1.1) and (1.2) gives:

$$y = A \sin 2\pi\upsilon t = A \sin (2\pi\upsilon x)/c \tag{1.3}$$

Wave can then be characterized by another value, the wavelength, which is the distance covered by light during a full cycle. Considering that the speed of the wave is "c" meters per second and that there are "υ" cycles per second, we get the following relation:

$$\lambda = c/\upsilon \tag{1.4}$$

In spectroscopy, the wavelengths are expressed using different units, aiming to avoid the manipulation of large number in the considered spectral region. Usually centimeter, millimeter, micrometer ($1\,\mu m = 10^{-6}$ m), nanometer ($1\,nm = 10^{-9}$ m), angström ($1\,Å = 10^{-10}$ m) are used. Another unit is generally used in the mid-infrared spectral region, the wavenumber, $\bar{\upsilon}$. Wavenumber is defined as the inverse of the wavelength expressed in centimeters:

$$\bar{\upsilon} = 1/\lambda \tag{1.5}$$

As the wavenumber is proportional to the frequency:

$$\upsilon = c \cdot \bar{\upsilon} \tag{1.6}$$

The conversion relationship is $\bar{\upsilon}$ (cm^{-1}) $= 10^7/\lambda$, with λ expressed in nanometers; and λ (nm) $= 10^7/\bar{\upsilon}$, with $\bar{\upsilon}$ expressed in centimeters^{-1}.

In this chapter, wavelength expressed in nanometers will be used for the near-infrared spectral region and wavenumber for the mid-infrared spectral region. Spectral regions, several of them being of interest for analytical purposes, can be defined as a function of wavelength (Figure 1.1):

- X-ray region (wavelengths between 0.5 and 10 nm) is involved in energy changes of electrons of the internal layers of atoms and molecules.
- Far-ultraviolet region (10–200 nm) is the zone corresponding to electronic emission from valence orbitals. In the near-UV region (200–350 nm), electronic transitions of the energetic levels of valence orbitals are observed. This spectral region is characterized by the absorption of peptidic bonds in proteins and of molecules presenting conjugated double bonds such as aromatic amino acids

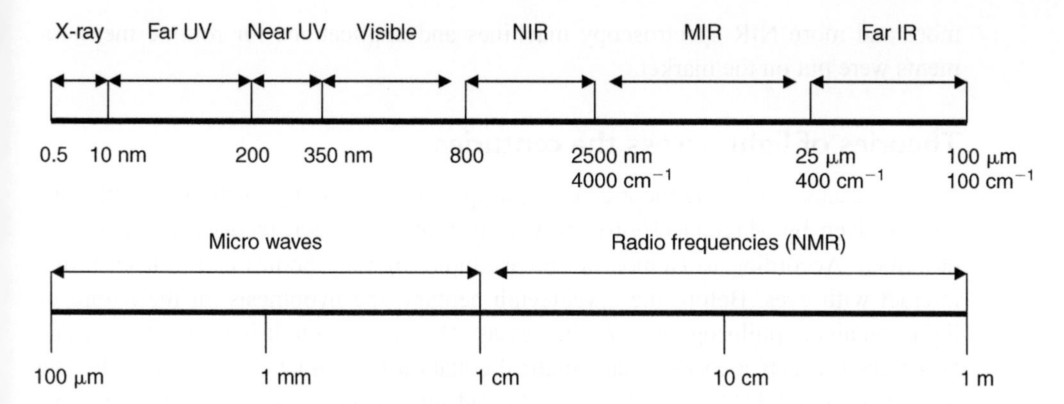

Figure 1.1 Spectral regions of interest for analytical purposes.

of proteins or vitamins such as vitamins A and E. In this wavelength range, luminescence (fluorescence and phosphorescence) may also be observed.

- The visible region (350–800 nm) is another zone where electronic transitions occur. Molecules exhibiting a large number of conjugated double bonds such as carotenoids, chlorophylls, and porphyrins absorb energy in this region. And their absorption properties may be used to evaluate the color of food products.
- The near-infrared (NIR) region (800–2500 nm or 12 500–4000 cm^{-1}) is the first spectral region exhibiting absorption bands related to molecule vibrations. This region is characterized by harmonics and combination bands and is widely used for composition analyses of food products.
- The mid-infrared (MIR) region (2500–25 000 nm or 4000–400 cm^{-1}) is the main region of vibrational spectroscopy. This region retains information, allowing organic molecules to be identified and the structure and conformation of molecules such as proteins, polysaccharides, and lipids to be characterized. In general, the absorption of an infrared radiation corresponds to an energy change ranging between 2 and 10 kcal mol^{-1}.
- In the microwave region (100 μm–1 cm), absorbed energy is related to molecule rotation. The radiofrequency region (1 cm–10 m) is the region investigated by nuclear magnetic resonance (NMR) and electron spin resonance.

History of the analytical development of infrared spectroscopy

Before the beginning of the twentieth century, infrared spectroscopy and theoretical studies of light evolved in parallel. In the first part of the twentieth century, spectroscopy developed at a fundamental level. The development of analytical methods based on NIR spectroscopy starts in the 1960s with the work of Karl H Norris (Norris, 1992). During the 1970s, the number of publications on the application of NIR spectroscopy in agriculture and food sectors increased tremendously, and from this date, more and more NIR spectroscopy machines and applications for routine measurements were put on the market.

Theories of light across the centuries

A good discussion of the theories of light up to the beginning of the twentieth century is given by Massain (1966). In Ancient Greece, several theories of light were described. According to Democrite, for example, lighting bodies emit particles that interact with eyes. Before the seventeenth century, the hypothesis on the nature of light remained philosophical. In his essay *Dioptrique*, published in 1637, René Descartes presents a correct mathematical equation for refraction, known as the law of sine. In about 1666, Isaac Newton showed interest in white light and its decomposition by a prism. Newton was a supporter of the corpuscular theory of light, which was popular for a long time, despite several experimental results disagreeing with this theory. In his famous book published in 1690, C Huygens stated that light originating from a point is a vibration that spherically propagates in a milieu called "ether." From this date, the undulatory theory and the corpuscular theory of light were in competition and gave different results for the speed of light in a refringent medium. However, the measurement of the speed of light became possible much more later.

In about 1800, T Young showed that light is an undulatory phenomena and, in 1862, L Foucault measured the speed of light in air and water. He obtained a good estimation of the speed of light, finding it to be 2.98×10^8 m/s. L Foucault also demonstrated that light travels faster in air than in water, confirming the undulatory nature of light.

The modern theory of light based on the undulation of an electromagnetic field was developed by JC Maxwell (1831–1879). If the undulatory theory of light was established in about 1900, the development of quantum physics at the beginning of the twentieth century strongly suggests the dual nature—undulatory and corpuscular—of light. Two fundamental studies performed by Planck on black-body radiation and Einstein on the photoelectric effect show the quantum nature of light energy. From the hypothesis of Planck, an oscillator of frequency, υ, can give or receive energy only by quanta of amplitude $E = h\upsilon$, where "h" is a new fundamental constant. The value of Planck's constant, h, is: 6.6268×10^{-34} joule seconds. In 1905, Einstein gave a simple explanation of this phenomenon in relation to Planck's hypothesis. The energy of a light beam is formed of quanta with energy equal to $h\upsilon$.

Theoretical bases of spectroscopy

Spectroscopy can be defined as the study of the interaction of an electromagnetic wave with matter. The first spectroscopic studies dealt with emission spectra or atomic absorption. In 1885, JJ Balmer investigated the spectrum of the hydrogen atom. He observed four light lines in the spectrum located at 656, 486, 434, and 410 nm. These wavelengths are related by the following equation:

$$1/\lambda = R((1/2^2) - (1/n^2)) \tag{1.7}$$

where $n = 3, 4, \ldots$ with Rydberg constant $R = 1.097 \times 10^7 \, m^{-1}$.

In fact, it was shown later that the Balmer series extends in the ultraviolet region to the wavelength of 365 nm. Other experiments have shown that additional series of light lines exist in the ultraviolet and infrared regions.

The discrete nature of the wavelengths of the atomic spectrum suggests that the quantum nature of light energy observed by Planck and Einstein is a universal law that applies to electrons and atoms. In about 1912, N Bohr hypothesized that electrons in atoms cannot lose or gain energy according to the continuous law, but only by quantum jumps. Bohr postulated that the electrons move around the atomic nucleus according to circular orbits, but only discrete orbits are allowed.

However, it was soon found that the Bohr model had some theoretical limits and could not be applied to atoms with several electrons. Several years after Bohr proposed his model, E Schrödinger and W Heisenberg separately proposed a new theory, quantum mechanics. This corresponds to a new approach in physics in which newtonian determinism is replaced by a probabilistic approach. A quantum object (photon, electron, as an example) is totally described by a time and space function, the wave function, Ψ, Quantum mechanics is the basis of modern physics and is considered to be the most satisfactory theory at the present time (Feynman and Hibbs, 1965).

Development of spectroscopic techniques

In 1800, an astronomer, W Herschel, demonstrated the existence of infrared radiation for the first time. Later, in 1882, W Abney and ER Festing took pictures of the absorption spectra on 53 compounds and showed correlations between absorption bands and the presence of some chemical groups in the studied molecules. In about 1890, WH Julius investigated the spectra of 20 organic molecules using a sodium chloride prism. He found that the methyl group absorbs at 3450 nm. The first modern investigations were done by WW Coblentz in 1905. He recorded the spectra of 19 compounds between 800 and 2800 nm with a spectrometer equipped with a quartz prism and a home-made radiometer. The motions of the radiometer were measured with a telescope located in a contiguous room. Most of the recorded spectra showed bands of low intensities between 840 and 1200 nm and an intense band at 1700 nm. Coblentz hypothesized that the bands between 840 and 1700 nm are harmonics of a series that goes to 13 700 nm and that the observed bands are related to CH group.

The first experiments showed that each compound has a unique spectrum and that a given chemical group present in different molecules exhibited absorption bands grossly located at the same wavelength.

In 1922, JW Ellis investigated organic liquids using an NIR spectrometer. Most of these liquids showed bands located at 750, 820, 900, 1000, 1200, 1400, 1700, and 2200 nm and Ellis assigned these band to CH bonds. Later, in 1927, he hypothesized that the band at 3400 nm (2940 cm^{-1}) is a fundamental band, whereas bands at 1700 and 1200 nm are first and second harmonics, respectively. Then, the bands of primary and secondary amines at about 1000, 1500, and 2000 nm are identified as harmonics and combination of the two fundamentals located at 3000 and 6200 nm (3330 and 1610 cm^{-1}). In 1928, FS Brackett arrived to split the broad band located at about 1200 nm into three absorption bands at 1190, 1220, and 1230 nm assigned to CH_3, CH_2, and CH groups, respectively.

The first detectors with PbS sensitive in the NIR region were discovered in the early 1950s. As more and more spectrometers were developed, infrared spectroscopy became a common method in the field of chemistry, used mainly to identify organic molecules and to assess the purity of synthesized organic molecules. At the same time, several researchers also investigated the structure of polymers such as proteins found in food products. Elliot and Ambrose (1950) were the first to demonstrate the correlation between the shape of the amide I and II bands and the structure of polypeptides. In the 1960s, Miyazawa and Blout (1961) showed from a detailed evaluation of amide I band that each type of secondary structure (helix, sheet, and random) is associated with one or more characteristic frequencies.

For a long time the low sensitivity of diffraction grating spectrometers and the difficulty of removing the water band in the amide I region limited the investigation of protein solutions. Then in the mid-1960s the first spectrometers including an interferometer and using Fourier transform were launched on the market. The higher sensitivity of these spectrometers made it possible to popularize the technique.

The development of analytical applications of infrared spectroscopy started in 1949 when the US Department of Agriculture launched a research project to evaluate the quality of eggs (Norris, 1992). The first study related to the quantitative analysis of a compound was published by Hart *et al.* (1962). In this study, the authors describe an analytical method based on infrared spectroscopy for the determination of seed moisture.

Over the past 20 years, the development of analytical methods has been strongly linked with the advance of computer technology and the progress in chemometrics. The history of the development of analytical applications based on infrared spectroscopy has been reviewed by Smith (1979), Butler and Burns (1983), Whetsel (1991), Burns and Margoshes (1992) and Bertrand (2006).

Vibrational spectroscopy

Vibrational movements of molecules induce absorption in the infrared region. These absorption bands have been used for quantitative and qualitative analyses of numerous molecules, and the identification and the attribution of these bands to specific chemical groups give specific information on the investigated product.

Infrared radiation can also excite rotational movements of molecules, giving rotation bands. These are generally superimposed on the vibration bands. They can be observed with high-resolution spectrometers and for gaseous molecules exhibiting sharp bands.

It appears important to model the vibrations of a given molecule, starting from a simple diatomic molecule. The simplest model corresponds to the harmonic oscillator. A slightly more complex model is the anharmonic oscillator. The approach based on classical mechanics is a good starting point to study vibrational spectroscopy. It allows the potential energy of this simple molecular system to be calculated and a Hamilton function built. These models can be improved by the introduction of quantum mechanics, allowing the rough calculation of the position of the absorption bands.

Development of vibrational models

As a rough estimate, the vibrational movements of two atoms of a diatomic molecule can be considered to be like the compression and extension movements of a spring—the atoms can attract or push away.

Harmonic oscillator

The simplest model corresponds to a mass, m, bound to a spring with no mass. This model is defined by the strength constant, k, measuring spring tightness, the displacement of the molecule, $q = r - r_0$, and the moving of the molecule from its equilibrium position, r_0. If the spring responds to Hooke's law, the strength, f, applied to the particle is proportional to the molecule movement according to:

$$f = -k(r - r_0) = -k.q \tag{1.8}$$

From this equation, it is possible to calculate the potential energy, and using Newton's equations, the vibration frequency, υ, can be determined since we have:

$$k = 4\pi^2.\upsilon^2.m \tag{1.9}$$

It can be seen that the frequency is only dependent on k and m. This simple model can be improved by using quantum mechanics. Vibrational energy, like all the energies of the molecule, is quantified and can be calculated from the Schrödinger equation (Herzberg, 1950; Szymanski, 1964; Colthup *et al.*, 1990).

The harmonic model can be used to calculate the stretching vibration, $\bar{\upsilon}$, observed in a spectrum. Indeed, Hooke's law can be transformed as follows:

$$\bar{\upsilon} = (1/2\pi.c) \times \sqrt{(k/\mu)} \tag{1.10}$$

where $\mu = ((m_1.m_2)/(m_1 + m_2)) \times (1/6.02 \times 10^{26})$, and μ = reduced mass. For example, the strength constant is 480 $N.m^{-1}$ and μ equals 0.98 for isotope 35 of the HCl molecule. From the above equation, the calculated stretching vibration is $\bar{\upsilon} = 2900\,cm^{-1}$.

Anharmonic oscillator

The harmonic model is the simplest one and it can provide a rough idea about the location of the fundamental bands of very small model molecules. However it is not relevant for real molecules. The anharmonic model is much more complex and it will not be described here. For interested readers, this subject has been addressed by Lachenal (2006), Diem (1993), and Duncan (1991).

Polyatomic molecules

Considering a molecule with N atoms, each atom can be located by three coordinates: x, y, and z. The molecule consequently has $3N$ characteristic coordinates or $3N$ degrees of freedom or $3N$ fundamental vibrations or $3N$ vibration modes. If the values of these coordinates were constants, the molecule would be "frozen" and the bond lengths and values for the stretching angles would be constant. However a molecule can move and deform in the space at room temperature.

The degrees of freedom are split in three groups corresponding to translation, vibration and rotation. A translation movement requires three degrees of freedom among the $3N$ ones, allowing $3N-3$ degrees. If the molecule is non-linear, three additional degrees of freedom, associated with the three orthogonal axes, are necessary to describe rotation movements, leading to $3N-6$ degrees or fundamental vibrations.

A normal mode of vibration of a polyatomic molecule can be defined as a state of vibration where each atom has a simple harmonic movement around its equilibrium position. Each atom of the molecule exhibits the same oscillation frequency and in general, the oscillations are in phase. Figure 1.2 shows the vibration modes for a non-linear molecule—water. And the vibrations of a CH_2 group are shown in Figure 1.3.

A molecule may exhibit one (or more) plane of symmetry (see Szymanski, 1964 and Colthup *et al.*, 1990, for more information on this subject). Water molecules present an axis of symmetry, C_2, and two planes of symmetry (Figure 1.4). A consequence of the plane of symmetry is the existence of symmetric and antisymmetric vibrations (Figures 1.2 and 1.3). By convention, the vibrations are classified according to the wavenumber and as a function of their degrees of symmetry. In that way, the symmetric stretching vibration of water exhibiting the highest frequency (3652 cm^{-1}) is called ν_1. The symmetric bending vibration observed at 1590 cm^{-1} is named ν_2, and the antisymmetric bending vibration at 3755 cm^{-1} is called ν_3. These three frequencies, found in the infrared spectrum of water, are fundamental frequencies.

In general, the bonds between light atoms vibrate at higher frequencies than the bonds between heavy atoms. It is observed for carbon atom bound to another

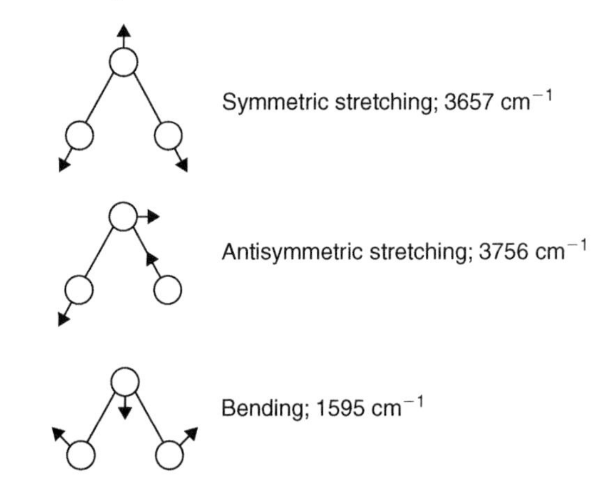

Figure 1.2 Normal vibration modes for a water molecule.

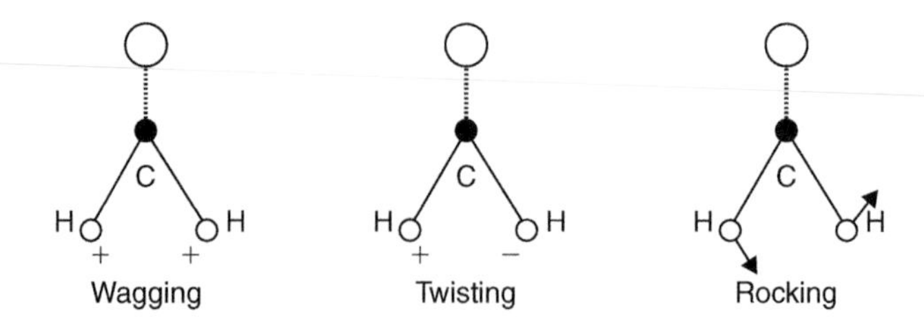

Figure 1.3 Wagging, twisting, and rocking vibrations of the CH_2 group.

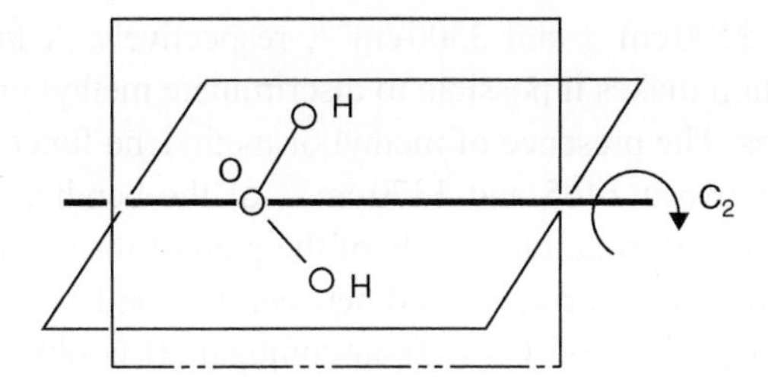

Figure 1.4 Axes of symmetry and planes for water molecule.

atom: when the reduced mass, μ, increases, the frequency decreases. The frequencies of C–H, C–D, C–O, C–Cl, and C–Br bonds are 3000, 2280, 1100, 800, and 550 cm^{-1}, respectively. However the strength constant, *k*, of the bond also has to be taken into account. For example, due to a higher strength constant, the H–F bond vibrates at a higher frequency than the C–H one. The strength constant also changes as a function of the type of bond: the value of the strength constant for the C=C bond is about twice that of the C–C one. As a consequence, the vibration frequency of C=C is located at 1650 cm^{-1}, compared with 1200 cm^{-1} for C–C. It has also been demonstrated that bending movements are less energetic than stretching ones. In that way, the bending frequency of C–H bond is close to 1340 cm^{-1}, whereas its stretching frequency is observed at about 3000 cm^{-1}.

The intensity of the bands is related to the nature and polarity of the bond. Indeed, the C=O bond, formed by different atoms and highly polarized, strongly absorbs in the MIR region, while C=C bond absorbance in the MIR region is much weaker.

Assignment of spectral bands in near- and mid-infrared regions

Major food components are generally complex molecules resulting from the polymerization of monomers such as amino acids or carbohydrates. These monomers exhibit specific chemical groups such as carboxylic and amine functions in amino acids. As each chemical group may absorb in the infrared region, it appears useful in a first step to clearly identify the characteristic absorption bands of these groups in the near- and mid-infrared regions. For further information on this subject, the reader may refer to Robert and Dufour (2006), Osborne and Fearn (1986), Pavia *et al.* (1979), Williams and Norris (1987) and Wojtkowiak and Chabanel (1977).

General rules of assignment

Aliphatic chain

C–H bonds, which are found in large quantities in organic molecules, show stretching vibrations between 2750 and 3320 cm^{-1} in the MIR region. The location of these bands is related to carbon hybridization. As saturated aliphatic molecules are characterized by absorption bands at about 3000 cm^{-1}, vinylic and acetylenic groups present absorption bands at 3100 cm^{-1} and 3300 cm^{-1}, respectively. A fine investigation of stretching band location makes it possible to discriminate methyl groups from methylene and methyne ones. The presence of methyl or methylene function can be assessed by the observation between 1465 and 1370 cm^{-1} of the bending vibrations of C–H bonds. For alcenes, the deformation outside of the plan of the C–H bond is characterized by a relatively intense absorption band between 650 and 1000 cm^{-1}. The stretching vibration of the double bond C=C (non-conjugated) is observed between 1640 and 1666 cm^{-1}.

In the NIR region, the first and second harmonics for C–H stretching vibrations are observed at about 1700 nm and 1200 nm, respectively. Combination bands involving stretching and bending of the C–H bond may be identified between 2000 and 2500 nm and, with a lower intensity, between 1300 and 1440 nm.

The spectra of hexane and dodecane in the MIR region are shown in Figure 1.5. Stretching vibrations observed between 2850 and 2962 cm^{-1} are characteristic of sp^3 carbons and allow the identification of methyl and methylene groups. For methyl groups, the asymmetric vibration $\nu_a CH_3$ is located at 2962 cm^{-1} for hexane and at 2956 cm^{-1} for dodecane, whereas symmetric vibrations $\nu_s CH_3$ are observed at 2876 and 2872 cm^{-1}, respectively. The hexane methylene group shows an asymmetric stretching vibration $\nu_a CH_2$ at 2926 cm^{-1} (2922 cm^{-1} for dodecane), as well as a symmetric vibration $\nu_s CH_2$ at 2864 cm^{-1} (2852 cm^{-1} for dodecane). Dodecane, exhibiting a larger number of methylene than hexane, presents a lower absorbance ratio, A_{CH3}/A_{CH2}, than hexane. The asymmetric bending vibrations $\delta_a CH_3$ of methyl groups, as well as the bending vibration of methylene group δCH_2 are located at about 1466 cm^{-1}. An identification of methyl groups can be performed by the analysis of the symmetric bending band at 1378 cm^{-1} ($\delta_s CH_3$). For a number of CH_2 groups equal to or larger than 4, such as in hexane and dodecane, the methylene bending vibration in the plan, $\delta_r CH_2$, shows an intense absorption band at about 720 cm^{-1}.

In the NIR region (Figure 1.5), the second harmonics assigned to the stretching of C–H bonds give a weak absorption band at about 1200 nm. In this region, hexane shows two bands at 1186 and 1208 nm, whereas dodecane is characterized by a band at 1208 nm and a shoulder at 1186 nm. The stretching and bending combination bands of C–H groups are observed between 2250 and 2500 nm. In the 1400 nm region, weak combination bands are assigned to 2νC–H + δC–H.

The MIR spectrum of 1-hexene shows characteristic absorption bands of C=C double bond and of the terminal methylene group at 3084, 1642, 992, and 908 cm^{-1}. As the stretching vibrations ν = C–H are observed at 3084 cm^{-1}, out of plan bending vibration (δ_{op}) of this chemical group are characterized by bands at 992 and 908 cm^{-1}. In the near-infrared region, the first harmonic located at 1628 nm is assigned to =C–H vibrations. In the region of combination bands, three absorption bands at 2112, 2168, and 2228 nm involved stretching vibration of the C=C double bond. While the band at 2112 nm is assigned to ν = CH_2 + νC=C, the ones at 2168 and 2228 nm originate from ν_aCH + νC=C and ν_sCH + νC=C, respectively.

Hydroxyl group

This chemical group, found in molecules such as alcohols, organic acids, or water, exhibits in the MIR region a strong absorption band between 3200 and 3600 cm^{-1}.

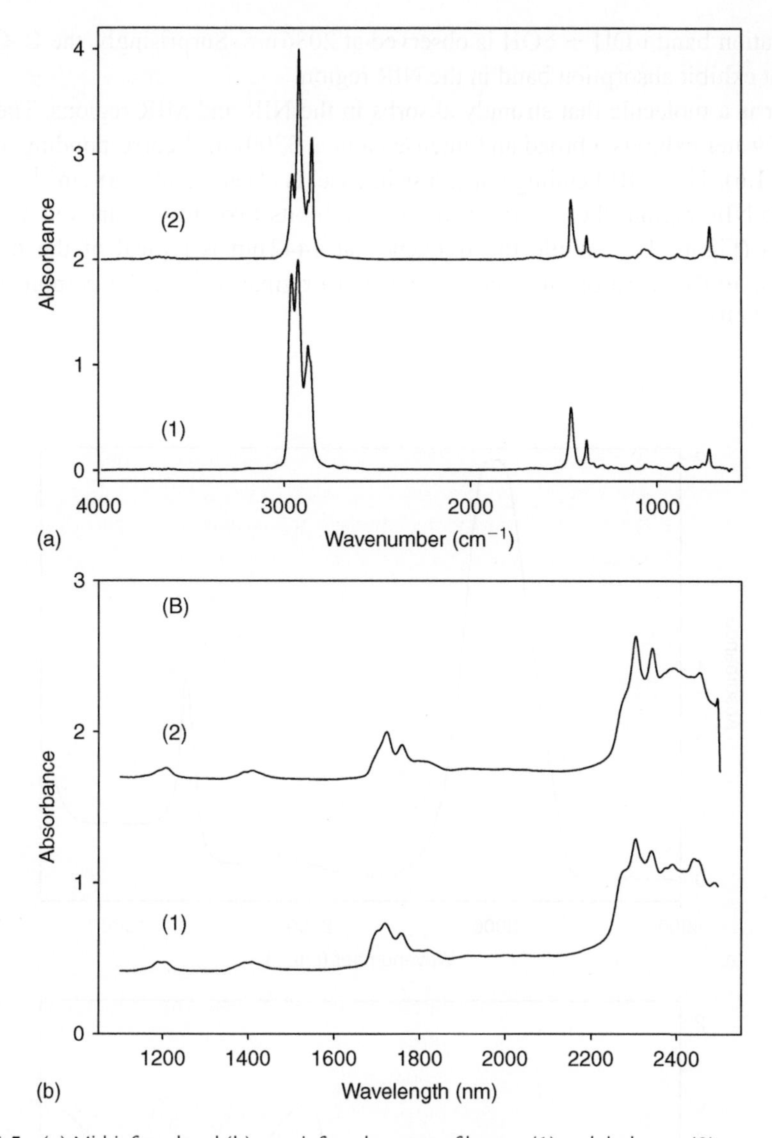

Figure 1.5 (a) Mid-infrared and (b) near-infrared spectra of hexane (1) and dodecane (2).

When this chemical group is involved in hydrogen bonds with other molecules, a broad absorption band centered at about 3300 cm^{-1} is observed. The O–H groups without hydrogen bonding are characterized by a sharp band at about 3600 cm^{-1}. In addition, the position of this stretching band depends on temperature. In the NIR region, the first harmonic of stretching vibration, νOH, is located between 1400 and 1500 nm.

The spectrum of 1-hexanol is characterized by a broad band νOH at 3314 cm^{-1} and by the asymmetric stretching of the C–O bond at 1056 cm^{-1}. This ν_aCO frequency is typical of primary alcohol. In the NIR, the first harmonic of the stretching vibration νOH exhibits a broad band centered at 1500 nm. More specifically, the combination band νOH + δOH is observed at 2086 nm. Surprisingly, the C–O bond does not exhibit absorption band in the NIR region.

Water is a molecule that strongly absorbs in the NIR and MIR regions. The spectrum of water exhibits a broad and intense band at 3300 cm^{-1} corresponding to νOH (Figure 1.6). The δOH bending band, less intense, is observed at 1638 cm^{-1}.

In the NIR region, the spectrum of water exhibits two strong bands at 1442 and 1932 nm (Figure 1.6). While the frequency at 1442 nm is typical of the first harmonic of νOH vibration, the one at 1932 nm originates from the combination of νOH + δOH.

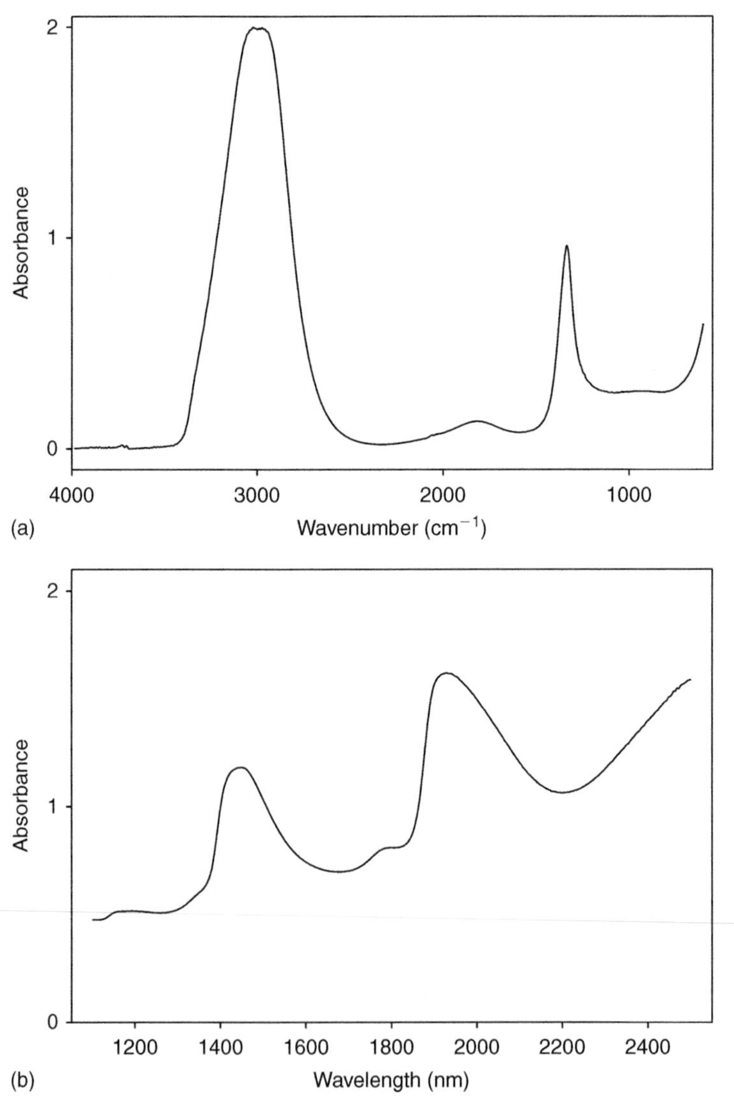

Figure 1.6 (a) Mid-infrared and (b) near-infrared spectra of water.

Carbonyl group

The carbonyl group, found in aldehydes, ketones, acids, esters, and amides, strongly absorbs in the MIR between 1650 and 1850 cm^{-1}. The precise location of the stretching vibration $\nu C{=}O$ depends on resonance effects and hydrogen bonding. In the NIR, harmonics associated with the carbonyl group are expected at about 1160, 1450, and 1950 nm. Even if they have been observed for several molecules, the absorbance of these harmonics is generally so weak that it cannot be used for analytical purposes.

Hexanal MIR spectrum shows a stretching vibration of the carbonyl group at 1724 cm^{-1} and the C–H bond of the aldehyde group is characterized by two bands (2820 and 2716 cm^{-1}) resulting from Fermi resonance between νCH and $2\delta_{op}$CH. In the NIR, the aldehyde group shows two combination bands $\nu CH + \nu C{=}O$ at 2200 and 2246 nm. The shoulder at 2130 nm is assigned to the combination $\nu CH + \nu C{=}O$ for C–H groups which do not belong to the aldehyde group.

The aliphatic ketones, particularly 2-hexanone, show in the MIR a stretching vibration at 1714 cm^{-1} corresponding to the carbonyl group. The ketones absorb at a lower frequency than aldehydes since they incorporate a second acyl group donor of electrons. The absorption band at 1358 cm^{-1}, relatively intense, corresponds to the symmetric bending of the methyl group adjacent to the carbonyl group. Moreover, stretching and bending vibrations coupling of C–CO–C accounts for the absorption band at 1168 cm^{-1}. The NIR spectrum of 2-hexanone is similar to the hexanal spectrum, except for the combination bands specific to the aldehyde group. 2-Hexanone shows absorption bands at 1906, 1960, 2112, and 2150 nm.

The MIR spectrum of hexylacetate is characterized by two intense bands at 1738 and 1232 cm^{-1}. The 1738 cm^{-1} vibration corresponds to $\nu C{=}O$ stretching, whereas ν_aC–O–C is observed at 1232 cm^{-1}. In addition, the symmetric stretching vibration (ν_s) of C–O–C shows a weak band at 1034 cm^{-1}. The NIR spectrum of the ester is similar to the hexane one, except for two weak absorption bands at 1926 nm and 2126 nm corresponding to C=O second harmonic and $\nu CH + \nu C{=}O$ combination, respectively.

The hexanoamide carbonyl group absorbs in the MIR at about 1658 cm^{-1} ($\nu C{=}O$). This absorption band is also found in peptide and protein spectra and is called amide I. In the NIR, the $\nu C{=}O$ second harmonic is predicted at about 2010 nm. However, the band observed at this wavelength is assigned to a combination vibration involving N–H stretching vibration. Nevertheless, the band at 2210 nm corresponds to $\nu CH + \nu C{=}O$ combination.

Nitrogen group

The bands of N–H stretching vibrations, located between 3300 and 3500 cm^{-1}, are generally weaker and sharper than the O–H ones. Whereas the primary amines are characterized by asymmetric and symmetric vibrations at about 3400 and 3300 cm^{-1}, the secondary amines show only one band. The tertiary amines do not absorb in this spectral region. The bending vibration δNH is expected between 1560 and 1640 cm^{-1}, whereas the out-of-plan bending δ_{op}NH shows a broad band at about 800 cm^{-1}. Finally, the stretching vibration of the C–N bond is observed between 1000 and 1350 cm^{-1}. In the NIR, the first harmonic associated with NH groups shows an absorption band between 1500 and 1550 nm. Combination bands involving NH groups are also observed at 2000 nm.

The stretching vibrations, $\nu_a NH_2$ and $\nu_s NH_2$, in the spectrum of 1-hexylamine are observed at 3370 and 3288 cm^{-1}, respectively. While the bending δ_{op}NH gives a broad band at 800 cm^{-1}, the bending vibration δNH absorbs at 1604 cm^{-1}. Finally, the stretching vibration of the C–N bond is observed at 1070 cm^{-1}. In the NIR, the primary amine is particularly characterized by the first harmonic of ν_sNH at 1524 nm, as well as by a combination band νNH + δNH at 2018 nm. Two other combination bands are observed at 2108 and 2136 nm.

The NH_2 group of hexanoamide is characterized in the MIR by asymmetric (3354 cm^{-1}) and symmetric (3186 cm^{-1}) stretching, as well as by out-of-plan bending (634 cm^{-1}). The stretching associated with the C–N bond is observed at 1414 cm^{-1}. In the NIR, combination bands involving N–H bond are located at 2010 and 2074 nm. Considering the first harmonics of νN–H vibration, broad bands are observed between 1500 and 1600 nm.

The assignments performed on pure organic compounds are mostly transposable to the major components (protein, lipid, and carbohydrate) of food products. The following sections investigate these assignments.

Protein, lipid and carbohydrate absorption bands in the infrared region

With regard to food components such as triacylglycerides and proteins, the acyl chain of fatty acids is mainly responsible for the absorption observed between 3000 and 2800 cm^{-1} (Figure 1.7), whereas the peptidic bound C–NH is mainly responsible for the absorption occurring between 1700 and 1500 cm^{-1}. Most of the absorption bands in the MIR region, but not in the NIR region, have been identified and attributed to chemical groups. The triacylglycerols ester linkage C–O ($\sim$1175 cm^{-1}), C=O ($\sim$1750 cm^{-1}) group, and acyl chain C–H (3000–2800 cm^{-1}) stretch wavenumbers are commonly used to determine fat (Table 1.1). The infrared bands appearing in the 3000–2800 cm^{-1} region are particularly useful because they are sensitive to the conformation and the packing of the phospholipid acyl chains (Unemera *et al.*, 1980; Casal and Mantsch, 1984; Mendelsohn and Mantsch, 1986). For example, the phase transition of phospholipids (sol to gel state transition) can be followed by MIR spectroscopy: increasing temperature results in a shift of the bands associated with C–H ($\sim$2850, 2880, 2935, and 2960 cm^{-1}) and carbonyl stretching mode of the phospholipids. Table 1.2 presents the main bands of lipids in the NIR region.

The development of Fourier transform infrared (FTIR) spectroscopy in recent years also affords the possibility of obtaining unique information about protein structure and protein–protein and protein–lipid interactions without introducing perturbing probe molecules (Casal and Mantsch, 1984). The amide I, II, and III bands (1700–1500 cm^{-1}) are known to be sensitive to the conformation adopted by the

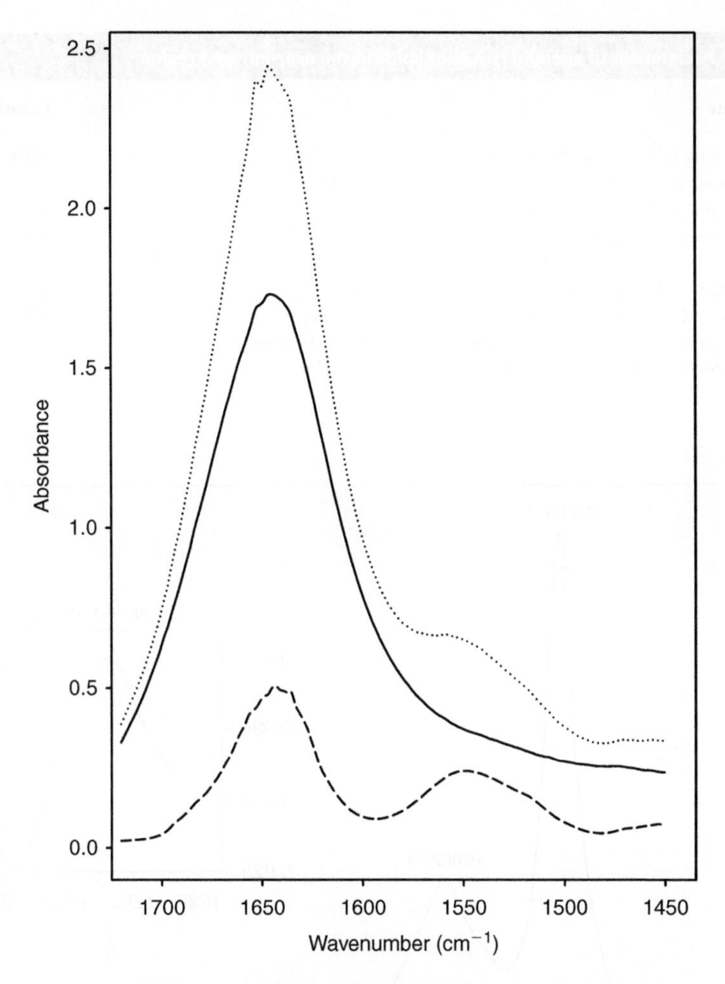

Figure 1.7 Spectra in the amides I and II region of Tris 50 mM (solid line), pH 7, buffer, of β-lactoglobulin dissolved in this buffer (dotted line) and of the protein (dashed line) after buffer subtraction.

Table 1.1 Assignment of spectral bands of stearic acid methyl ester in the mid-infrared region

Assignment	Location (cm^{-1})
CH_3 asymmetric stretching	2961
CH_2 asymmetric stretching	2935
CH_3 symmetric stretching	2880
CH_2 symmetric stretching	2863
C=O ester stretching	1754
CH_2 scissoring	1463
CH_3 symmetric scissoring	1381
C-O ester symmetric stretching	1176
CH_3 wagging	1123
CH_2 wagging	727

Table 1.2 Assignment of spectral bands of lipids in the near-infrared region

Assignment	Location (nm)
C-H stretching of CH_2, 2nd harmonic	1208
Combination: 2 × C-H stretching + C-H bending of CH_2	1416
C-H stretching of CH_2 group, 1st harmonic	1724
C-H stretching of CH_2 group, 1st harmonic	1760
Combination: =C-H stretching + C=C stretching	2144
Combination: CH_2 asymmetric stretching + C=C stretching	2190
Combination: C-H stretching + C-H bending of CH_2 group	2304
Combination: C-H symmetric stretching of CH_2 + $=CH_2$ bending	2348
Combination: C-H stretching of CH_2 group + C-C stretching	2380

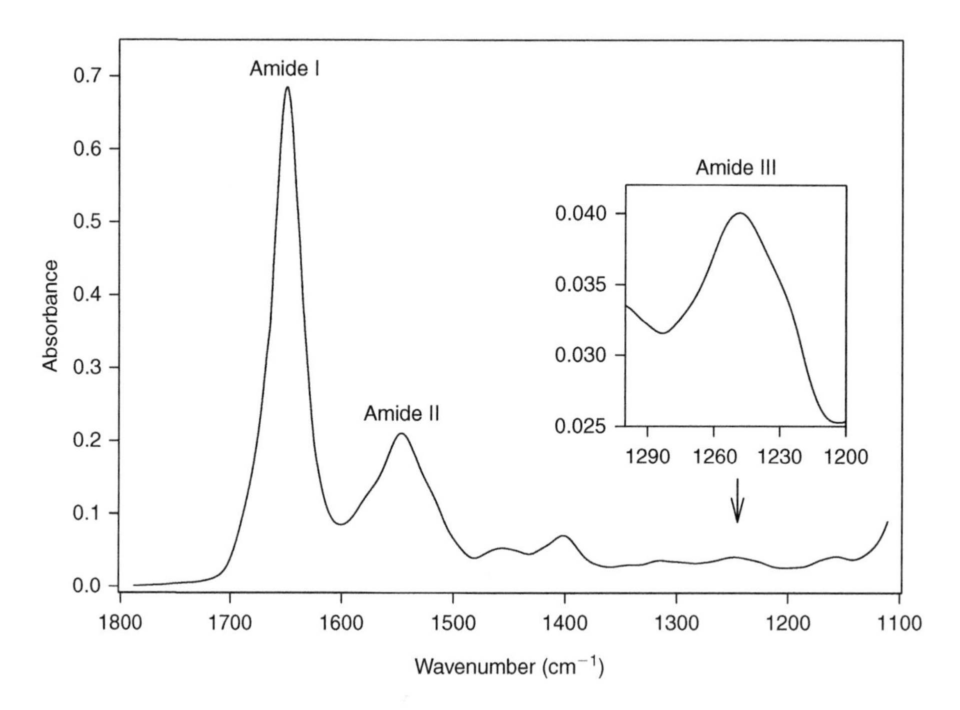

Figure 1.8 Mid-infrared spectrum of α-lactalbumin—amide I, II, and III bands.

protein backbone (Figure 1.8, Table 1.3). Figure 1.9 shows the spectra of β-lactoglobulin, α-lactalbumin, and β-casein, which are known to exhibit mainly β-sheet, α-helix, and unordered secondary structures, respectively.

The secondary structures of proteins can be deduced from their FTIR spectra since there are good correlations between the amide I band (1700–1600 cm^{-1}) (Fox, 1989) and the levels of α-helix, β-sheet and unordered structure in proteins (Dousseau and Pézolet, 1990).

Although the peptide bonds are essentially responsible for the absorbance of proteins in the 1700–1500 cm^{-1} region, the side-chains of several amino acids (glutamic

Mode	Frequency (cm^{-1})		Potential energy distribution[a] (%)
Amide I	~1655	Stretching C-O	83
		Stretching C-N	15
		Bending C-C-N	11
Amide II	~1560	Wagging N-H	49
		Stretching C-N	33
		Wagging C-O	12
		Stretching C-C	10
		Stretching N-C_α	9
Amide III	~1300	Wagging N-H	52
		Stretching C-C	18
		Stretching C-N	14
		Wagging C-O	11
Amide A	~3300	Stretching N-H	95
Amide V	~660	Twisting C-N	60
		Wagging N-H	30

Table 1.3 Amides bands of proteins in the mid-infrared region

[a]Only contributions >5% are mentioned.

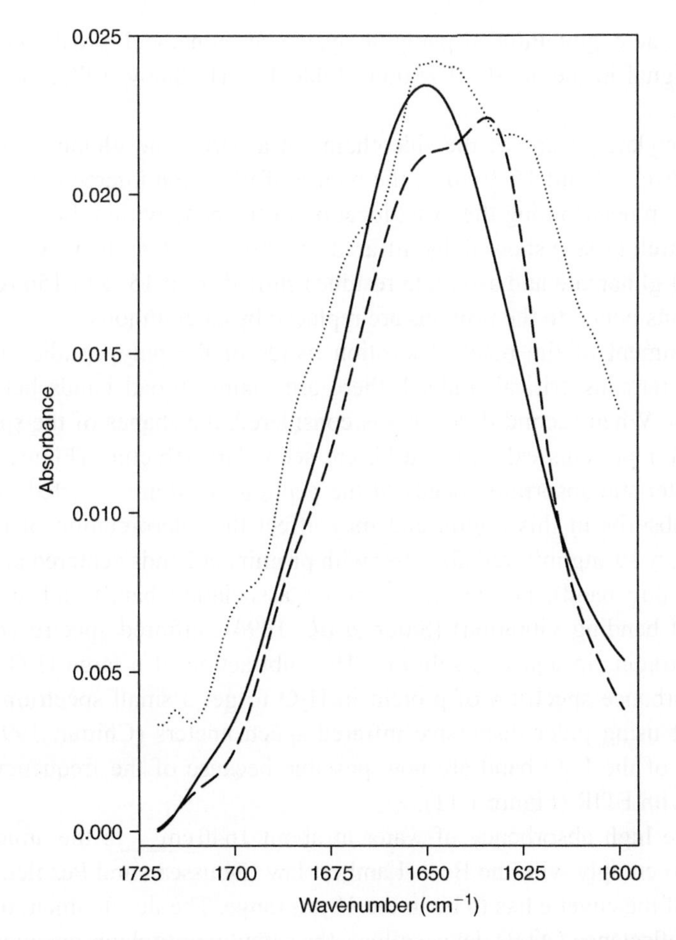

Figure 1.9 Mid-infrared spectra of β-lactoglobulin (dashed line), α-lactalbumin (solid line), and casein β (dotted line).

Table 1.4 Characteristic bands between 1800 and 1400 cm^{-1} and molar extinction coefficients of amino acid side-chains in D_2O

Amino acid		Frequency (cm^{-1})	Absorbance (L/mol/cm)
Asp	-COO$^-$	1584	820
	-COOH	1713	290
Glu	-COO$^-$	1567	830
	-COOH	1706	280
Asn	-C=O	1648	570
Gln	-C=O	1635	550
Arg	-CN3+H5	1608	500
		1586	460
Tyr	-OH	1615	160
		1515	500
	-O$^-$	1603	350
		1500	650

acid, aspartic acid, glutamine, asparagine, lysine, arginine, and tyrosine) can contribute to the signal in the amide II region (Table 1.4) (Bellamy, 1975; Goormaghtigh *et al.*, 1990).

The carboxylate groups of the side-chains of aspartic and glutamic acids absorb between 1580 cm^{-1} and 1520 cm^{-1}, the nature of the anion interacting with the carboxylate group determining the exact location of the band within this interval. Thus, Byler and Farell (1989) showed, by infrared spectroscopy, that the O–C–O stretching vibrations of glutamate and aspartate residues shifted from 1575 to 1565 cm^{-1} when potassium ions bound to the proteins are replaced by calcium ions.

The assignment of the main absorption bands of the polypeptidic chain in the NIR region remains critical. Indeed, there are mainly broad bands between 1100 and 2500 nm. When second derivative is considered, the shapes of the spectra show differences for proteins exhibiting different secondary structure (Figure 1.10). The main characteristic absorption bands of the peptidic bond are reported in Table 1.5. The water absorbs in this region and may affect the interpretation of the spectra. Water is a very strong infrared absorber with prominent bands centered at 3360 cm^{-1} (H–O stretching band), at 2130 cm^{-1} (water association band) and at 1640 cm^{-1} (the H–O–H bending vibration) (Safar *et al.*, 1994). Infrared spectroscopy can be used with proteins in aqueous solution. The subtraction of a large H_2O band from a large absorbance spectrum of protein in H_2O to get a small spectrum of protein was difficult using older dispersive infrared spectrometers (Chittur, 1999). Precise subtractions of the H_2O band are now possible because of the frequency precision achievable with FTIR (Figure 1.11).

Due to the high absorbency of water at about 1640 cm^{-1} in the amide I and II region and to comply with the Beer–Lambert law (Dousseau and Pézolet, 1990), the pathlength of the cuvette has to be in the 10 μm range. The development of the attenuated total reflectance (ATR) device allows the sampling problems encountered when collecting spectra from opaque and viscous samples to be overcome. This simple and

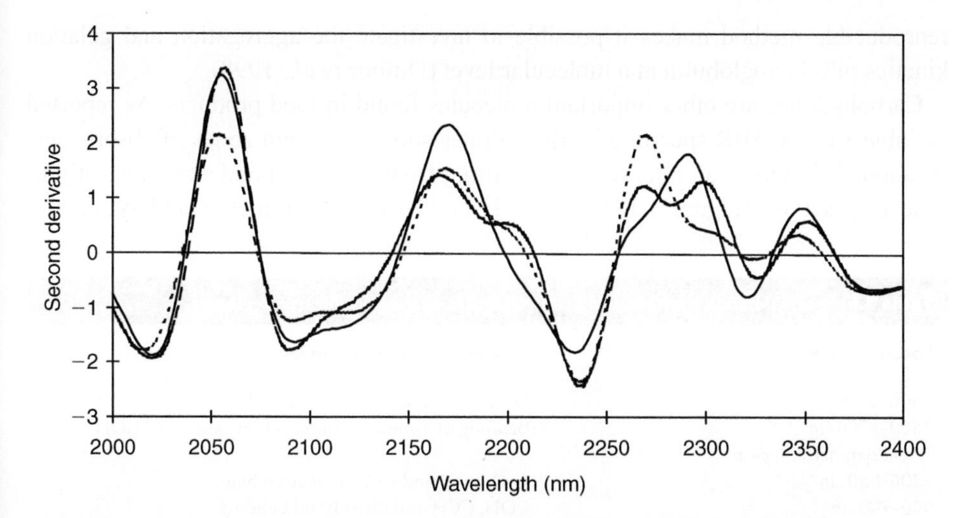

Figure 1.10 Second derivative near-infrared spectra of β-lactoglobulin (dashed line), α-lactalbumin (solid line) and casein β (dotted line).

Table 1.5 Bands of polypeptidic skeleton of proteins in the near-infrared region

Harmonic/combination band	Wavelength (nm)
Amide A, 1st harmonic	1523
Free NH stretching/amide II (1st harmonic), combination	1600
Amide A/amide I, combination	2050
Amide A:amide III, combination, or	
Amide I (1st harmonic)/amide III, combination	2180

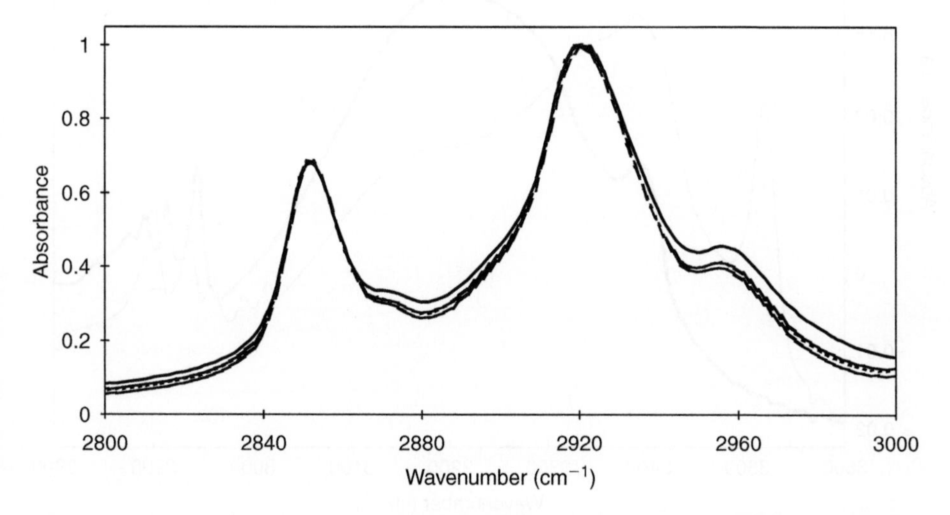

Figure 1.11 Mid-infrared spectra (3000–2800 cm^{-1}) of cheeses recorded at four different times during ripening: 1 day (—), 21 days (— —), 51 days (-—-) and 81 days (-·-·).

reproducible method makes it possible to investigate the aggregation and gelation kinetics of β-lactoglobulin at a molecular level (Dufour *et al.*, 1998).

Carbohydrates are other important molecules found in food products. As reported in Table 1.6, the MIR spectra of carbohydrates show four main zones of absorbance. At about 3220 cm^{-1}, an intense band resulting from the O–H bond stretching of glucose is observed (Figure 1.12). The C–H bond shows asymmetric and symmetric

Table 1.6 Spectral bands characterizing carbohydrates in the mid-infrared region

Spectral region	Observed vibrational modes
3600–2800 cm^{-1}	O–H bond and C–H bond stretching
1500–1200 cm^{-1} "local symmetric region"	Bending of symmetric bond (HCH) and of CH_2OH
1200–950 cm^{-1}	C–O bond and C–C bond stretching
950–700 cm^{-1} "fingerprint region"	COH, CCH and OCH bond bending Anomeric bands between 930 and 840 cm^{-1} C–C bond stretching
Below 700 cm^{-1} "carbon squeletal vibration"	Exocyclic bending of CCO bond "crystalline region" (700 and 500 cm^{-1}) Endocyclic bending of CCO and CCC bonds (<700 cm^{-1})

From Mathlouli and Koenig (1986).

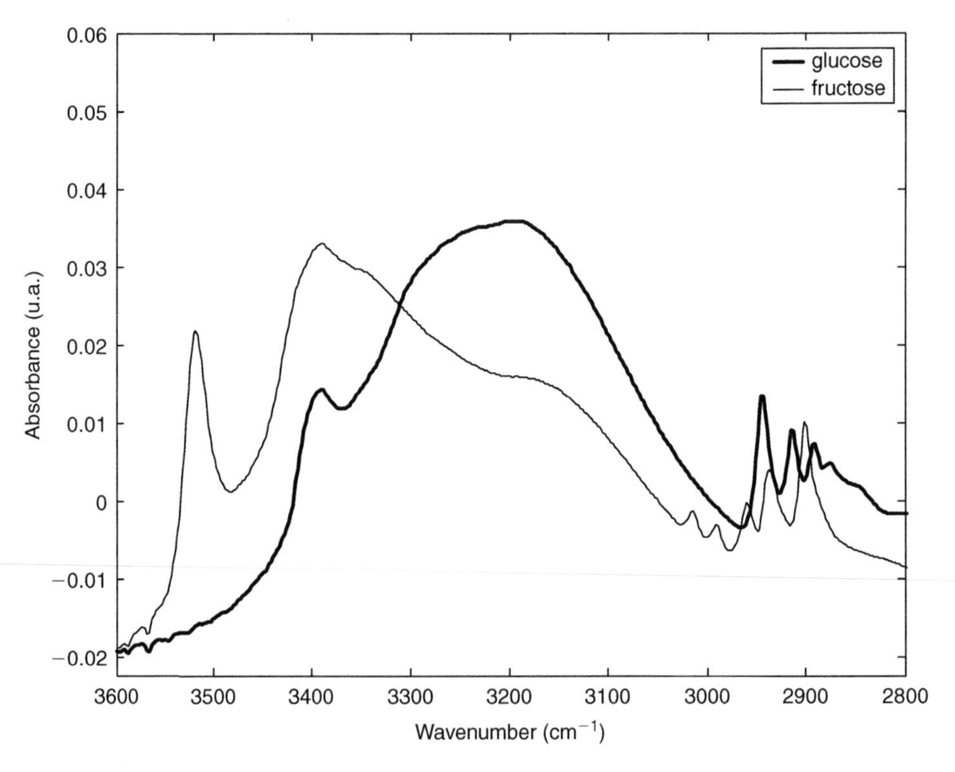

Figure 1.12 Mid-infrared spectra of glucose and fructose.

Table 1.7 Vibration frequencies of C–H bond in solid glucose

Assignment	Location (cm^{-1})
CH_2	1457
CH_2	1337
CH_2	1219
CH_2	1011
CH_2	1404
C_1–H	1360
C_1–H	1250
C_1–H	1076
C_1–H	1047
C_1–H	911
C_1–H	836

From Vasco *et al.* (1971).

bending bands at 1470 cm^{-1} and 1380 cm^{-1} (Table 1.7), respectively. Bands assigned to C–O and C–C vibrations are observed at about 1100 cm^{-1}. In the region close to 920 cm^{-1} two vibrations of C–O–C asymmetric stretching corresponding to α and β anomers are observed. The main bands observed in the MIR region for monosaccharides, oligosaccharides, and polysaccharides are presented in Table 1.8.

Conclusions

For a long time, infrared spectroscopy was considered to be a method for fundamental research and characterization of the chemical structure of purified molecules. In addition, due to the broad absorbance bands, the NIR region was considered to have poor potential until the early 1960s. With the development of electronics and computing, and the relative simplicity of NIR spectrometers, numerous applications based on NIR spectra have been developed over the last 30 years. The development of analytical methods has been made possible by the development of chemometrics. Indeed, the first methods were based on multilinear regressions. By the 1980s, the global predictive methods became popular as PCs developed. They allow the identification of absorption bands involved in the prediction. Predictive methods such as principal component regression and partial least squares are now widely used in the development of analytical methods for the prediction of foodstuff composition and quality.

Moreover, chemometrics makes it possible to extract relevant information from spectral databases related to the molecular structure of carbohydrates, proteins, and fats in food products, and to address the relation between their structure and texture (Dufour *et al.*, 2000; Karoui and Dufour, 2006). In the coming years, the development of spectroscopic methods coupled with chemometrics should increase our understanding of the determinants of food texture and may allow us to devise the structural engineering of foodstuffs.

Table 1.8 Observed bands in the 1200–500 cm^{-1} region for different carbohydrates

Monosaccharides		Oligosaccharides		Polysaccharides	
Arabinose	1130	Lactose	1167	Starch	1148
	1102		1139		1103
	1089		1116		1077
	1064		1093		1013
	1047		1083		992
	992		1070		
			1058	Inulin	1197
Fructose	1176		1031		1162
	1148		1018		1119
	1133		1005		1023
	1093		989		985
	1075				932
	1047	Maltose	1150		
	1029		1134		
	975		1101		
			1070		
Galactose	1151		1033		
	1138		1025		
	1102		1005		
	1077		989		
	1051				
	1043	Melibiose	1186		
	995		1160		
	975		1153		
	955		1132		
			1122		
Glucose	1145		1107		
	1105		1083		
	1077		1070		
	1047		1049		
	1017		1010		
	992		971		
Mannose	1166	Saccharose	1170		
	1118		1162		
	1084		1125		
	1070		1115		
	1058		1103		
	1033		1064		
	1006		1047		
			1037		
Xylose	1190		1011		
	1147		1003		
	1125		989		
	1081				
	1053	Raffinose	1192		
	1033		1152		
	1016		1127		
			1105		
			1077		
			1044		
			1028		
			1014		
			995		
			965		
			935		

Nomenclature

δ_{op} out of plan bending vibration
δ_s symmetric bending
h Planck's constant
λ wavelength
R Rydberg constant
υ frequency
$\bar{\upsilon}$ wavenumber
ν_a asymmetric stretching vibration
ν_s symmetric stretching vibration
Ψ wave function
μ reduced mass

References

Bellamy LJ (1975) *The infrared spectra of complex molecules.* New York: John Wiley and Sons, Inc.

Bertrand D (2006) Les méthodes d'analyse rapide dans les industries agroalimentaires. In: *La spectroscopie infrarouge et ses applications analytiques* (Bertrand D, Dufour E, eds). Paris: Lavoisier.

Burns DA, Margoshes M (1992) Historical development. In: *Handbook of Near Infrared Analysis* (Burns DA, Ciurczak EW, eds). New York: Marcel Dekker, pp. 1–11.

Butler LA (1983) The history and background of NIR. *Cereal Food World*, **3**, 238–240.

Byler DM, Farell HM (1989) Infrared spectroscopic evidence for calcium ion interaction with carboxylate groups of casein. *Journal of Dairy Science*, **72**, 1719–1723.

Casal HL, Mantsch HH (1984) Polymorphic phase behavior of phopholipid membranes studied by infrared spectroscopy. *Biochimica et Biophysica Acta*, **779**, 382–401.

Chittur KK (1999) FTIR and protein structure at interfaces. *Bulletin BMES*, **23**, 3–9.

Colthup NB, Daly LH, Wiberley SF (1990) *Introduction to Infrared and Raman Spectroscopy.* Boston: Academic Press.

Diem M (1993) *Introduction to Modern Vibrational Spectroscopy.* New York: John Wiley and Sons.

Dousseau F, Pézolet M (1990) Determination of the secondary structure contents of proteins in aqueous solutions from their amide I and amide II infrared bands. Comparison between classical and partial least squares methods. *Biochemistry*, **29**, 8771–8779.

Dufour E, Robert P, Renard D, Lamas G (1998) Investigation of β-lactoglobulin gelation in water/ethanol solutions. *International Dairy Journal*, **8**, 87–93.

Dufour E, Mazerolles G, Devaux MF, Duboz G, Duployer MH, Mouhous Riou N (2000) Phase transition of triglycerides during semi-hard cheese ripening. *Journal of Dairy Science*, **10**, 81–93.

Duncan JL (1991) The determination of vibrational anharmonicity in the molecules from spectroscopic observations. *Spectrochimica Acta*, **47**, 1–27.

Elliot A, Ambrose EJ (1950) Structure of synthetic polypeptides. *Nature*, **165**, 921–922.

Feynman RP, Hibbs AR (1965) *Quantum Mechanics and Path Integrals.* New York: McGraw-Hill.

Fox PF (1989) *Development in Dairy Chemistry.* New York: Elsevier.

Goormaghtigh E, Cabiaux V, Ruysschaert JM (1990) Secondary structure and dosage of soluble and membrane proteins by attenuated total reflection Fourier-transform infrared spectroscopy on hydrated films. *European Journal of Biochemistry*, **193**, 409–420.

Hart JR, Norris KH, Golumbic C (1962) Determination of the moisture content of seeds by near infrared spectrophotometry of their methanol extracts. *Cereal Chemistry*, **39**, 94–99.

Herzberg G (1950) *Spectra of Diatomic Molecules.* Princeton, NJ: Van Nostrand.

Karoui R, Dufour É (2006) Prediction of the rheology parameters of ripened semi-hard cheeses using fluorescence spectra in the UV and visible ranges recorded at a young stage. *International Dairy Journal*, **16**, 1490–1497.

Lachenal G (2006) Introduction à la spectroscopie infrarouge. In: *La spectroscopie infrarouge et ses applications analytiques* (Bertrand D, Dufour E, eds). Paris: Lavoisier.

Massain R (1966) Les théories de la lumière jusqu'à Fresnel. In: *Physique et physicians* (Massain R, ed.). Paris: Magnard, pp. 270–319.

Mathlouthi M, Koenig JL (1986) Vibrational spectra of carbohydrates. *Advances in Carbohydrate Chemistry and Biochemistry*, **44**, 7–89.

Mendelsohn R, Mantsch HH (1986) Fourier transform infrared studies of lipid-protein interactions. In: *Progress in Protein–Lipid Interactions*, Vol. 2 (Watts A, Pont JJH, eds). Amsterdam: Elsevier Sciences Publishers, pp. 103–145.

Miyazawa T, Blout ER (1961) The infrared spectra of polypeptide in various conformations. *Journal of the American Chemical Society*, **83**, 712–719.

Norris KH (1992) Early history of near infrared for agricultural applications. *NIR News*, **3**, 12–13.

Osborne BG, Fearn T (1986) *Near Infrared Spectroscopy in Food Analysis.* London: Longman Scientific & Technical.

Pavia DL, Lampman GM, Kriz GS (1979) *Introduction to Spectroscopy: A Guide for Students of Organic Chemistry.* Philadelphia, PA: Saunders Golden Sunburst Series.

Robert P, Dufour E (2006) Règles générales d'attribution des bandes spectrales. In: *La spectroscopie infrarouge et ses applications analytiques* (Bertrand D, Dufour E, eds). Paris: Lavoisier.

Safar M, Bertrand D, Robert P, Devaux MF, Genot C (1994) Characterization of edible oils, butters, and margarines by Fourier transform infrared spectroscopy with attenuated total reflectance. *Journal of the American Oil Chemists' Society*, **71**, 371–377.

Smith AL (1979) *Applied Infrared Spectroscopy.* New York: John Wiley and Sons.

Szymanski HA (1964) *IR Theory and Practice of Infrared Spectroscopy.* New York: Plenum Press.

Unemera JDG, Cameron DG, Mantsch HH (1980) A Fourier transform infrared spectroscopic study of the molecular interaction of cholesterol with 1, 2 dipalmitoyl-sn-glycero-3-phosphocholine. *Biochimica et Biophysica Acta*, **602**, 32–44.

Vasko PD, Blackwell J, Koenig JL (1971) Infrared and Raman spectroscopy of carbohydrates. Part 1. Identification of O–H and C–H related vibrational modes for

D-glucose, maltose, cellobiose and dextran by deuterium-substitution methods. *Carbohydrate Research*, **19**, 297–310.

Whetsel KB (1991) The first fifty years of near-infrared spectroscopy in America. *NIR News*, **2**, 4–5.

Williams P, Norris K (1987) *Near Infrared Technology in the Agricultural and Food Industries.* St. Paul, MN: American Association of Cereal Chemists.

Wilson RH (1994) *Spectroscopic Techniques for Food Analysis.* New York: VCH Publishers.

Wojtkowiak B, Chabanel M (1977) *Spectrochimie Moléculaire.* Paris: Technique & Documentation.

Data Pre-processing

2

Åsmund Rinnan, Lars Nørgaard, Frans van den Berg, Jonas Thygesen, Rasmus Bro, and Søren Balling Engelsen

Introduction

Pre-processing of spectral data is often of vital importance if reasonable results are to be obtained whether the analysis is concerned with exploratory data mining, classification or building a good and robust prediction model. This chapter will focus on the pre-processing of near-infrared (NIR) data, but all the methods presented are applicable to mid-infrared (MIR) and infrared (IR) spectra as well. It should be noted, however, that the scatter phenomena mentioned regarding NIR spectroscopy are seldom observed in IR spectra of liquids; while in IR spectra of solid samples (e.g. using an ATR-IR probe) some of the same scatter effects might be present.

The main goal of pre-processing is to transform data in such a way that the (multivariate) signals will better adhere to Beer's law, which states that absorbance and concentration are linearly correlated:

$$A = \varepsilon \, l \, c \tag{2.1}$$

Infrared Spectroscopy for Food Quality Analysis and Control
ISBN: 978-0-12-374136-3

where ε is the molar absorptivity, l is the (effective) path length, and c is the concentration of the constituent of interest. The estimation of ε or the correct value of l is not important; what is aimed for is that the collective term $\varepsilon \times l$ is constant for the data set, thus making the relationship between A and c linear.

Many physical and chemical phenomena can cause a deviation from this linear relationship, including scatter from particulates, interferents, molecular interactions, changes in refractive index at high concentrations, shifts in chemical equilibrium as a function of concentration, stray light, changes in sample size/path length, etc. By focusing on reference-independent methods (see below), two typical features of the spectra which have to be corrected for can be identified. This is illustrated in Figure 2.1 showing the constant baseline offset and a curved/linear baseline.

Pre-processing techniques are designed to compensate for these deviations from linear relationships and thus to improve the linear relationship between the spectral signals and analyte concentrations. Pre-processing techniques can be divided into two major groups: those which directly use available reference values for the pre-processing operation, and those that do not. The latter group is thus a reference-independent pre-processing group, and as such provides more general tools suitable for studies such as exploratory studies, for example, where often no reference value is available. The reference-independent techniques can further be divided into two subgroups: scatter correction methods and derivation methods. Scatter correction methods include: multiplicative signal correction (MSC), also known as multiplicative scatter correction (including extended MSC, inverse MSC, inverse extended MSC and de-trending), standard normal variate (SNV) scaling, normalization and baseline correction. Derivation includes the following techniques: finite difference, Savitzky–Golay, and Norris–Williams. Only Savitzky–Golay derivatives will be discussed further in this work, as this is proven to be the most applied option. The

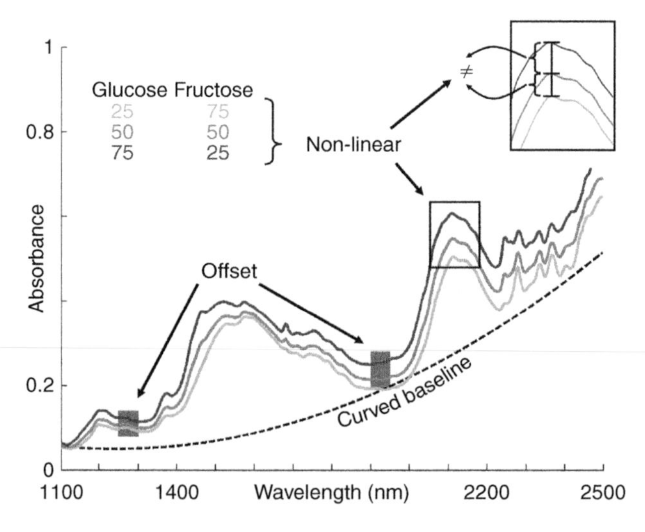

Figure 2.1 The non-linearity in the spectra is in general caused by two scatter effects: offset and curved baseline.

reference-dependent pre-processing methods comprise primarily those techniques that orthogonalize the data with respect to a reference of interest (Karstang and Manne, 1992; Goicoechea and Olivieri, 2001; Westerhuis *et al.*, 2001). Such methods are not generally applicable and will require special attention as the response variables are used actively in the modeling. This chapter will therefore focus on the generally applicable reference-independent techniques.

In the following sections, data presented in Figure 2.1 will be used as examples. The data were collected from a mixture of three sugars: fructose, glucose, and sucrose. There are a total of 231 samples, where the concentration of the three sugars ranges from 0% to 100% (w/w). The crystalline samples were all measured with an NIRS 6500 II instrument (Foss A/S, Denmark) over the VIS/NIR range 400–2498 nm with a 2 nm sampling interval. A small sample cup was used for the measurements. The data were obtained from a project conducted at the University of Copenhagen, Denmark, and have not been published before.

Overview of pre-processing techniques

As mentioned earlier, the pre-processing techniques can be divided into two main categories: reference-independent techniques and reference-dependent techniques. The reference-independent techniques will be discussed thoroughly, including figures of their application on the same data as shown in Figure 2.1. The reference-dependent techniques will be discussed only briefly, including the necessary references appropriate to explore them further.

The multiplicative signal correction family

The multiplicative signal correction (MSC) technique was introduced by Martens *et al.* (1983) and further elaborated on by Geladi *et al.* (1985). Originally MSC meant "multiplicative scatter correction" but the abbreviation has changed meaning over the years because it has been found to be useful for types of multiplicative problems other than those arising from scatter only. The basic concept of the MSC is to remove non-linearities in the data caused by scatter from particulates in the samples. The MSC operation is divided into two steps: estimation of the correction coefficients,

$$x_{org} = b_0 + b_{ref,1}\,\mathbf{x}_{ref} + \mathbf{e} \tag{2.2}$$

and correction of the spectra

$$\mathbf{x}_{corr} = \frac{\mathbf{x}_{org} - b_0}{b_{ref,1}} \tag{2.3}$$

where the *b*'s are the correction coefficients, $\mathbf{e}$ is the unmodeled part, and $\mathbf{x}_{org}$, $\mathbf{x}_{ref}$, and $\mathbf{x}_{corr}$ are the original, reference, and corrected spectra, respectively. This can be visualized in Figure 2.2, in which the slope of the curve is $b_{\text{ref},1}$, while the intercept with the ordinate axis is the b_0.

In the original work (Martens *et al.*, 1983; Geladi *et al.* 1985) it was suggested that the correction coefficients could be estimated from a smaller spectral range containing no chemical information. This is indeed a useful approach, but it can be difficult, especially in NIR spectroscopy, to find such an "empty" spectral range. In practise the b-coefficients are often found using the entire spectrum in equation (2.2). Using MSC this way no subjective evaluation of the spectrum is required and any two users will get the same results from the MSC pre-processing. Another important aspect to be considered in MSC is the definition of the reference spectrum $\mathbf{x}_{\text{ref}}$. This can either be an a priori defined reference or the average spectra over a set of samples (e.g. the calibration set). Since it can be difficult to select one appropriate spectrum as the reference spectrum, the average over a set of samples is typically used.

Martens and Stark (1991) later expanded MSC to include wavelength corrections and corrections for known spectra representing interferents or desired constituents. The first expansion, inclusion of wavelength dependency, can be seen as a merging of the de-trending technique—first introduced by Barnes *et al.* in 1989—with the MSC. The latter expansion is a novel idea, but with some resemblance to methods such as OSC, O-PLS, etc. The inclusion of the wavelength dependency in the MSC is called extended MSC (EMSC) in the literature, while the addition of the known spectra is named spectral interference subtraction (SIS) (Martens and Stark, 1991). However, these two expansions can be readily included into one equation, thus determining all the correction coefficients simultaneously:

$$\mathbf{x}_{\text{org}} = \begin{bmatrix} 1 & \mathbf{x}_{\text{ref}} & \mathbf{x}_{\text{ref}}^2 & \lambda & \lambda^2 & \mathbf{x}_{\text{known},1} & \mathbf{x}_{\text{known},2} & \cdots \end{bmatrix} \mathbf{b} + \mathbf{e} \tag{2.4}$$

$$\mathbf{b} = \mathbf{x}_{\text{org}} \mathbf{X}_{\text{c}}^{+} \tag{2.5}$$

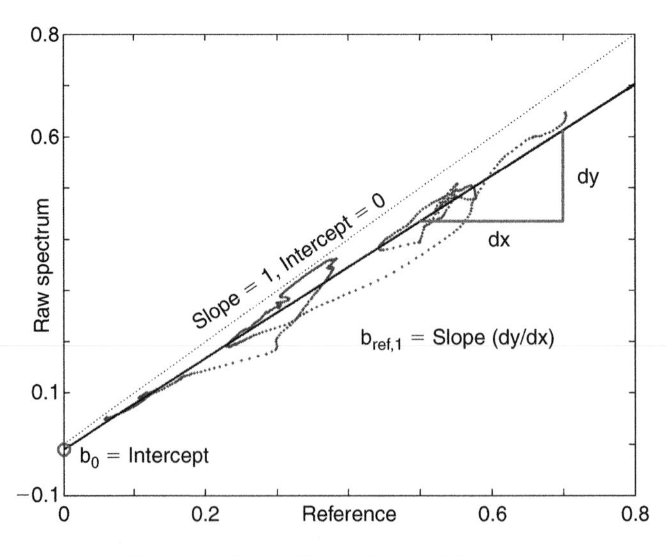

Figure 2.2 The estimation of the correction coefficients for multiplicative signal correction.

$$\mathbf{b} = \begin{bmatrix} b_0 & b_{ref,1} & b_{ref,2} & b_{\lambda,1} & b_{\lambda,2} & b_{k,1} & b_{k,2} & \cdots \end{bmatrix} \quad (2.6)$$

where the vector **b** holds the EMSC correction coefficients, **1** is a vector only consisting of ones, λ is the wavelength axis and the $\mathbf{x}_{\mathbf{known}}$'s are the known spectra. $\mathbf{X_c}$ in equation (2.5) denotes the entire correction matrix enclosed by brackets in equation (2.4), while the "+" indicates that the so-called Moore–Penrose inverse is used to solve the system. Equation (2.6) indicates that the correction vector is built up from one coefficient per correction term in equation (2.4). As mentioned by Martens and Stark (1991) the different $\mathbf{x}_{\mathbf{known}}$ vectors should typically be orthogonalized to make the computation of the matrix inverse in equation (2.5) numerically stable. In addition to the above-mentioned examples, it is also possible to add higher order wavelength dependencies or higher order reference dependencies. However, orders higher than the first lead to two or more possible solutions for the correction (resulting from multiple roots for the polynomial expansion). The correct solution can be found as only one of the solutions gives physical/spectroscopic meaningful results. This can be seen from Figure 2.3, where the two solutions to the second-order equation are shown for the sugar data. It becomes clear by looking at this figure that only the positive root gives physical/spectroscopic sensible result. Solving the equation for second- and third-order reference dependencies can be done mathematically, while higher order correction only can be solved numerically.

From equation (2.4) it can be seen that by removing everything but the two first terms EMSC turns into the original MSC, while removing the reference $\mathbf{x}_{\mathbf{ref}}$ and the constituent correction $\mathbf{x}_{\mathbf{known}}$ the equation turns into the standard spectral dependent de-trending equation. Thus equation (2.4) summarizes a variety of pre-processing techniques, and for ease of use, it will be denoted simply the "MSC equation." In some cases it is not so simple to obtain a known spectrum of interferences to use in the correction, as the samples are complex (especially in food and feed analysis by NIR spectroscopy). Therefore the last part of equation (2.4) will not be available for many practical applications of MSC.

The effect of three variations of MSC is illustrated in Figure 2.4, where regular MSC (linear reference dependency), EMSC (linear reference plus first- and second-order wavelength dependency) and de-trending (first- and second-order wavelength dependency) are compared.

As becomes apparent from Figure 2.4, inclusion of the reference spectra ($\mathbf{x}_{\mathbf{ref}}$) in the correction leads to only minor changes compared with the raw spectra. On the other hand, the addition of wavelength correction changes the pre-processed spectra appearance considerably; spectra are positioned along the abscissa in contrast to along a diagonal. As can be seen for both MSC (Figure 2.4b) and EMSC (Figure 2.4c) the pre-processing has led to a decrease in the offset between the three samples, and the data have become visibly more linear dependent on the concentration of the sugars. The peak around 1600 nm varies linearly with the concentration of fructose, while the smaller peak at 1500 nm is related linearly to the concentration of glucose. In the de-trending case (Figure 2.4d), the general slope in the spectra is removed, giving a nice flat spectrum, but the non-linearity between absorbance and concentration is still apparent. The offset however is somewhat reduced.

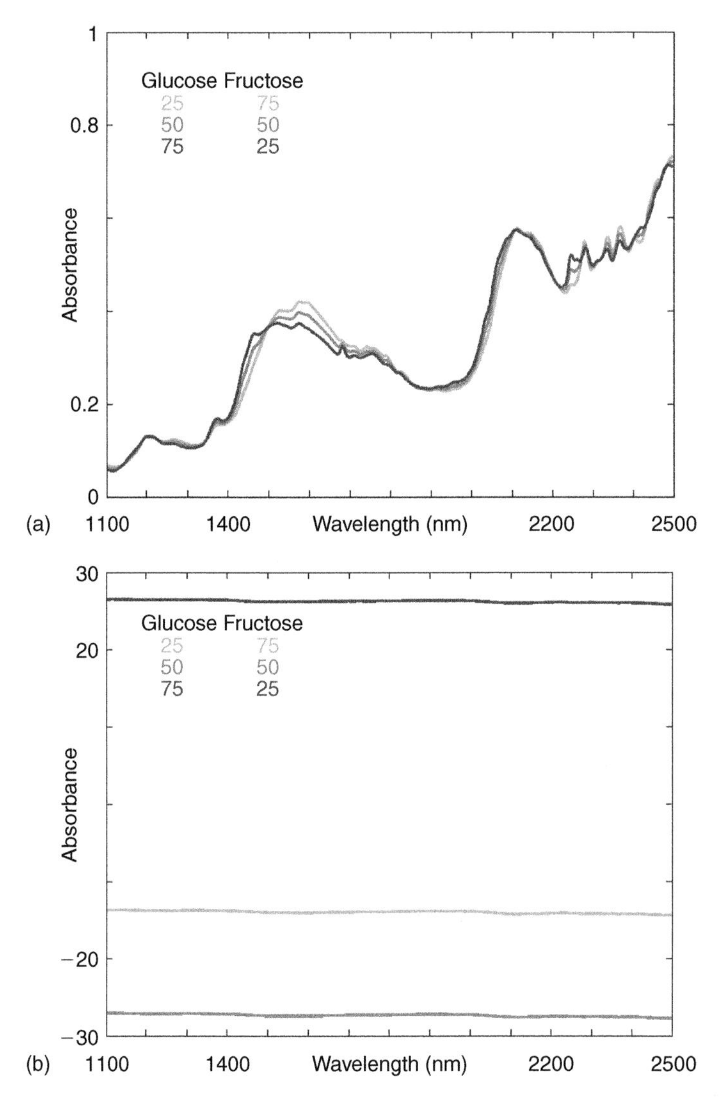

Figure 2.3 Solutions for the second-order reference correction. (a) Solution from the positive root and (b) solution from the negative root of the polynomial expansion.

Inverse versions of both MSC and EMSC have also been presented in literature, named ISC (Helland *et al.*, 1995) and EISC (Pedersen *et al.*, 2002) respectively. The main difference between the "normal" signal correction and the inverse is that in the inverse the reference is regressed on the measured spectrum, while for the normal versions this is the opposite (i.e. the measured spectrum is regressed on the reference spectrum). Performing the inverse operation makes the scatter correction equations appear simpler and better conditioned as shown in the mentioned references (Helland *et al.*, 1995; Pedersen *et al.*, 2002). Apart from this the use of the inverse approach is only justified if the reference spectrum contains more noise than the measured spectrum. This is normally not the case since the reference spectrum is often taken as the

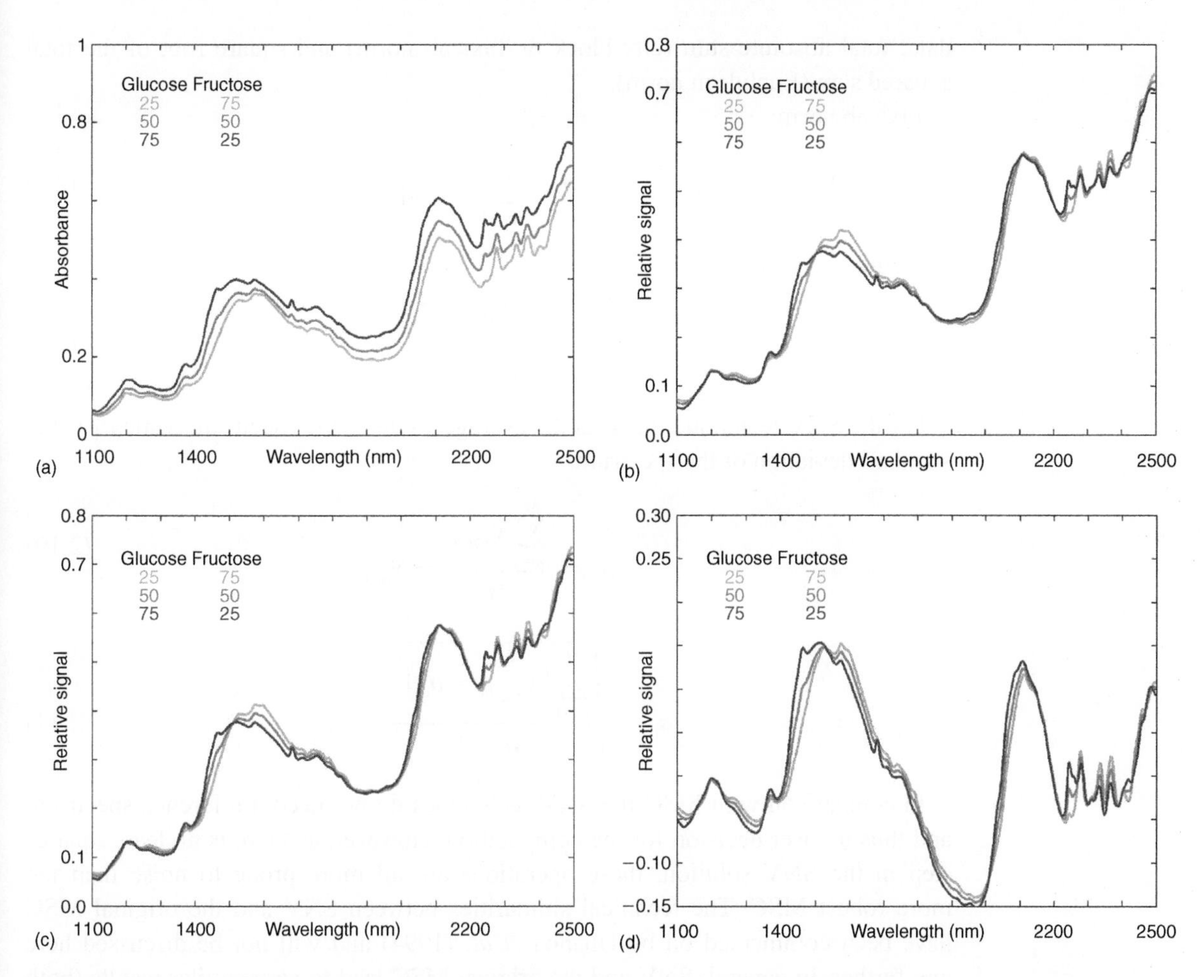

Figure 2.4 The effect of three variations of multiplicative signal correction (MSC) on the sugar-mixture spectra. (a) The raw data, (b) the MSC data: [**1** $\mathbf{x}_{ref}$], (c) the extended MSC data: [**1** $\mathbf{x}_{ref}$ $\boldsymbol{\lambda}$ $\boldsymbol{\lambda}^2$], and (d) the de-trended data: [**1** $\boldsymbol{\lambda}$ $\boldsymbol{\lambda}^2$]. (The matrix brackets indicate which parts of equation (2.4) is included.)

average over a set of N samples, indicating that the noise level of the reference spectrum is of a magnitude $\sqrt{N}$ smaller than for the individual samples. Thus normally the original MSC should be preferred over the inverse versions.

Standard normal variate and normalization

The basic equation for standard normal variate (SNV) correction (Barnes *et al.*, 1989) and normalization has the same form as equation (2.2) for MSC:

$$\mathbf{x}_{\mathbf{corr}} = \frac{\mathbf{x}_{\mathbf{org}} - a_0}{a_1} \tag{2.7}$$

For normalization, a_0 will always equal zero, while a_1 depends on the type of normalization performed. There are two typical types of normalization used on spectral data: total absolute sum (city-block or Taxicab norm) and square root of the total squared sum (Euclidean norm).

Taxicab norm:

$$a_1 = \sum_{m=1}^{M} \left| x_{org,m} \right| \tag{2.8}$$

Euclidean norm:

$$a_1 = \sqrt{\sum_{m=1}^{M} x_{org,m}{}^2} \tag{2.9}$$

For the SNV technique the a_0 is the average value of the spectrum, while a_1 is the standard deviation of the spectrum:

$$a_0 = \frac{\sum_{m=1}^{M} x_{org,m}}{M} = \bar{x}_{org} \tag{2.10}$$

$$a_1 = \sqrt{\frac{\sum_{m=1}^{M} \left(x_{org,m} - a_0\right)^2}{M-1}} = s_{x_{org}} \tag{2.11}$$

In comparison with MSC the SNV techniques do not need a reference spectrum, and thus no user decision for the computation. However, as there is no least squares step in the SNV solution, these operations are all more prone to noise than the more robust MSC. The technical similarities between SNV and the original MSC have been commented on by Dhanoa *et al.* (1994) and will not be discussed here any further. In general, SNV and the original MSC lead to very similar results (both spectrally and regression-wise).

Figure 2.5 illustrates the application of SNV to the sugar data set. By comparison with Figure 2.4, it can be seen that the MSC and SNV corrected spectra are very similar, with the main difference being the ordinate axis scale. As in the case for MSC and EMSC, SNV has transformed the data in such a way that the corrected data generally have a more linear relationship between signal and concentration. The Euclidean base normalization often gives results similar to SNV (as is also clear in Figure 2.5), since it is based on the square root of a squared sum as in SNV.

Baseline correction

The most common baseline correction technique used in NIR (and IR) is the de-trending technique introduced by Barnes *et al.* (1989). However, more sophisticated methods to perform this operation exist, such as wavelets (Schulze *et al.*, 2005; Davis *et al.*, 2007) and iterative polynomial baseline fitting (IPBF) (Lieber and Mahadevan-Jansen, 2003) to mention just two.

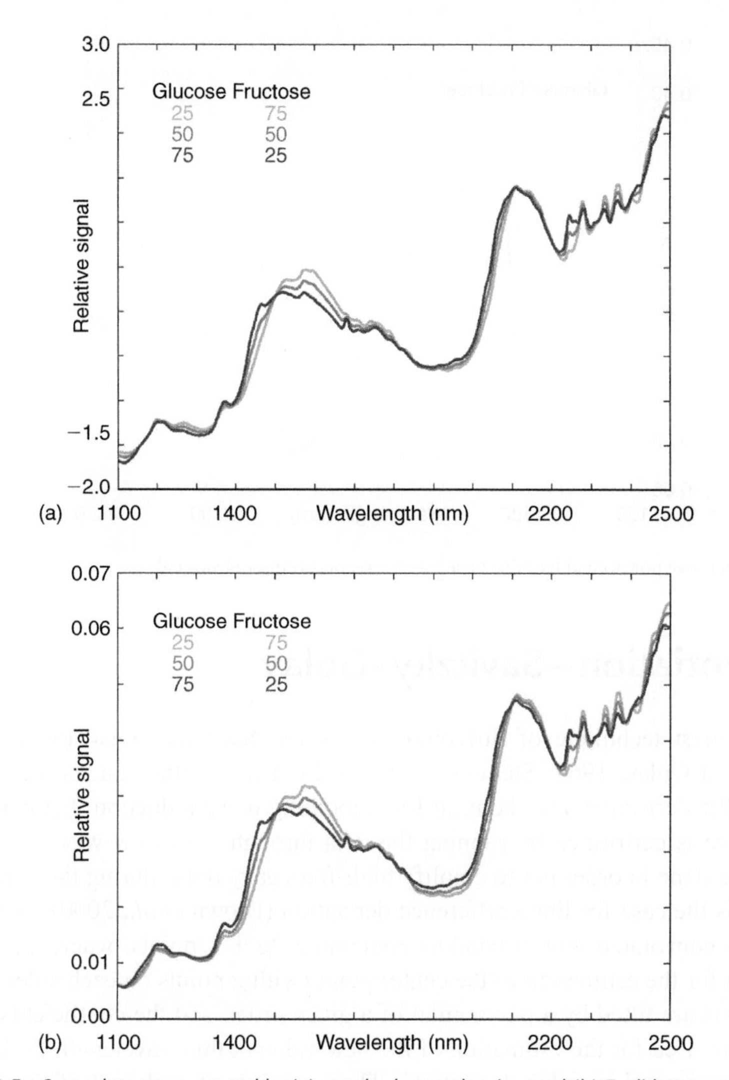

Figure 2.5 Sugar data pre-processed by (a) standard normal variate and (b) Euclidean norm.

IPBF uses the same principle as de-trending and the basic difference is that in IPBF the fitting of the baseline is done iteratively. First a baseline of a chosen polynomial order is fitted to the sample spectrum. Then the measurement points lying above the estimated baseline are replaced by the predictions from the fitted baseline. This new (artificial) baseline spectrum is then fitted again with the same polynomial order, and this procedure is repeated until no new sample points are replaced. IPBF assumes that the baseline of the spectrum is given by the lowest points along the spectrum, which is not necessarily the case for NIR. However, for IR absorbance spectra this assumption is generally valid wherefore IR spectra could benefit from IPBF.

Comparing IPBF in Figure 2.6 with de-trending in Figure 2.4, it is quite clear that these two methods perform similarly. As for de-trending, IPBF has changed the spectra, forcing them to have an almost zero baseline. The offset and spectral trend has been removed, but it is difficult to find parts of the spectra that obey Beer's law.

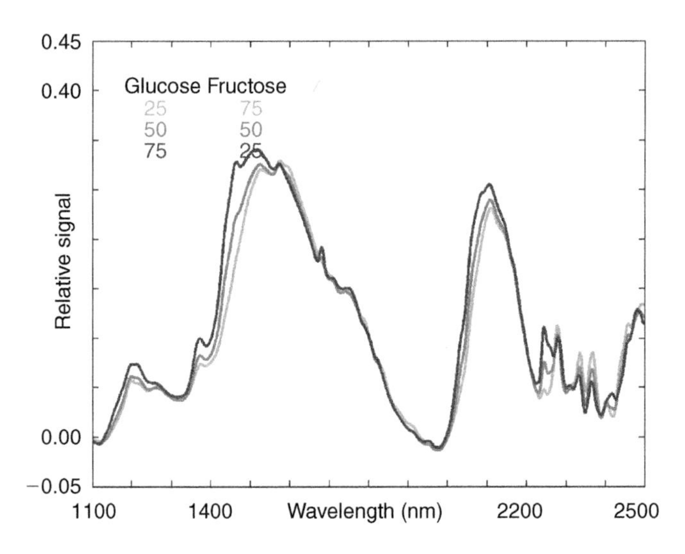

Figure 2.6 Iterative polynomial baseline fitting with a second-order polynomial.

Differentiation—Savitzky–Golay

The commonest technique of differentiation is the Savitzky–Golay (SG) routine (Savitzky and Golay, 1964; Steinier *et al.*, 1972). This routine can, in addition to estimating the derivative, also be used for smoothing/noise reduction. Estimation of the derivative is performed by running the data through a window-wise symmetric filter. This is done in order not to amplify high-frequency noise during the derivation process as is the case for finite difference derivation (Brown *et al.*, 2000). In SG the spectrum is convoluted with a window containing $2g + 1$ points, where each window is used for the estimation of the center point (with g points on each side). These $2g + 1$ points are fitted by a polynomial of a given order, and the coefficients found by this fit are used for the estimation of the new value at this wavelength (either just smoothing or smoothing plus derivation). Thus g points at each end of the spectra will be lost. Gorry (1990) proposed a method to circumvent this problem of losing points using asymmetric windows. However, this method will probably lose fine structure at the ends of the spectra unless the polynomial fit is of a high order. Unless the number of wavelengths is very limited and the loss of $2g$ points thus will be detrimental to subsequent analysis, this technique of circumventing the loss of points is not recommended.

The polynomial coefficient estimation is in principle only done once, as the coefficients will not change from one window to the next when the wavelengths are placed equidistant on the axis, meaning that the same vector of coefficients can be used throughout the spectra (minus the two sets of g points at the spectral ends), making the computation efficient. In the original article by Savitzky and Golay (1964), tables with the coefficients for various derivations and smoothing for many different polynomial fittings were presented. However, there are errors in these tables, as pointed out and corrected by Steinier *et al.* (1972). Nowadays, there is no need for these tables as the coefficients can be found easily through an inversion of a small matrix **C**:

$$\mathbf{C} = \begin{bmatrix} 1 & -g & \dots & (-g)^i \\ 1 & -(g-1) & \dots & [-(g-1)]^i \\ \dots & \dots & 0 & \dots \\ 1 & g-1 & \dots & (g-1)^i \\ 1 & g & \dots & g^i \end{bmatrix} \tag{2.12}$$

where i is the polynomial order to which the points are fitted. The size of this matrix will thus be $(2g + 1) \times (i + 1)$, and as such simple to compute.

As noted by Wentzell and Brown (2000), for example, matching pairs of polynomials give the same estimate for the derivatives due to redundancy in the estimation of the coefficients (i.e. linear and quadratic fitting gives the same estimates for the first derivative, while quadratic and cubic give the same estimate for the second derivative, etc.).

As can be seen for both of the derivative estimates in Figure 2.7 the amount of noise, especially at the end of the spectra has increased. This can be countered by selecting a larger smoothing window, at the cost of a possible loss of important information at the spectral edges. Normally, using 7–11 points for smoothing and a second or fourth degree polynomial for the fitting procedure is sufficient for typical high-resolution spectra data. For the estimation of the second derivative, a higher number of smoothing points should in general be used, as this estimation is more prone to the amplification of noise than the first derivative, reducing the overall signal-to-noise ratio. The higher the degree of the derivative, the higher the number of points in the smoothing window is required. Even though the estimation of the second derivative looks quite noisy, there are areas of linearity in both the derivative estimations. The baseline effect has been removed in the second derivative, and the offset has been removed by both techniques as expected.

Alternative methods

Many alternatives to the pre-processing methods mentioned here exist, especially for normalization, baseline correction, and derivation. For normalization any pseudo-norm could be used for correction such as the spectral median, interquartile range, and/or median average deviation (MAD). Baseline correction can not only be done by de-trending (inside the MSC) or IPBF, but also via a wavelet transformation or a spline-based correction, to mention just some alternatives. For the derivative there are two main obvious alternative methods: finite differences and Norris–Williams derivation (NW) (Norris, 1983; Norris and Williams, 1984). The first method calculates a simple difference spectrum between two adjacent points for the first derivative and is therefore sensitive to high-frequency noise, making it inadequate for most spectroscopic data. NW is based on a simple moving average (Massart *et al.*, 1997) over points, followed by calculating the finite difference on this smoothed spectrum, but

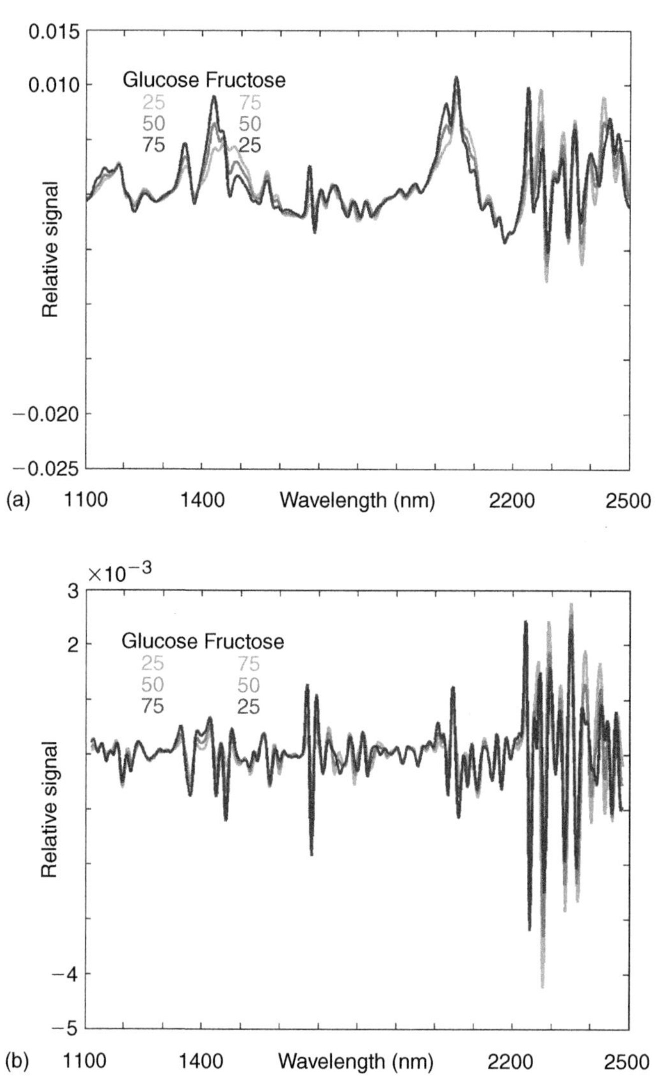

Figure 2.7 Savitzky–Golay (a) first and (b) second derivative estimations of the sugar spectra. Both have been estimated by a 9-point second-order polynomial smoothing.

with a gap between the points used in the estimation of the derivative. The use of the smoothing and a gap-size makes NW less prone to high-frequency noise, as is the case for Savitzky–Golay derivatives, but it is difficult to defend the use of a gap on spectroscopic data (assuming that the spectra are not presented in the time domain). However, using the right settings, NW gives similar estimates of the derivative as the SG.

Example of reference-independent methods

Until now only observations on the effect of reference-independent pre-processing techniques on the raw spectra have been made. However, the user is often more interested

Table 2.1 Prediction results for nine different pre-processing techniques

Method	No. of PLS factors	RMSEP (% w/w)
Raw	8	1.03 (0.08)
MSC A[a]: $[\mathbf{1}\ \boldsymbol{\lambda}\ \boldsymbol{\lambda}^2]$	7	1.06 (0.09)
MSC B[a]: $[\mathbf{1}\ \mathbf{x}_{\mathbf{ref}}]$	6	0.90 (0.07)
MSC C[a]: $[\mathbf{1}\ \mathbf{x}_{\mathbf{ref}}\ \boldsymbol{\lambda}\ \boldsymbol{\lambda}^2]$	6	0.89 (0.05)
SNV	6	0.90 (0.09)
Normalization—Euclidean norm	6	1.25 (0.12)
IPBF—second-order polynomial	7	1.48 (0.10)
SG1—first derivative, 9 point window, second-order polynomial	8	0.80 (0.04)
SG2—second derivative, 9 point window, second-order polynomial	5	1.25 (0.05)

RMSEP is root mean squared error of prediction reported with uncertainty intervals.

[a]Indicates which part of equation (2.4) used in the correction.

PLS, partial least squares; MSC, multiplicative signal correction; SNV, standard normal variate; IPBF, iterative polynomial baseline fitting; SG, Savitzky-Golay.

in seeing how the pre-processing affects, for example, the performance and number of factors required in regression modeling. To illustrate this point Table 2.1 presents different pre-processing schemes in combination with partial least squares (PLS) regression to the analyte concentration in the sugar mixture design data set. In order to get a good indication of the uncertainty, 1000 bootstrap samples have been drawn, and the remaining samples have been predicted. Thus the root mean squared error of prediction (RMSEP) values are an average over these bootstrap drawings (one standard deviation).

Four pre-processing methods all give similar, optimal RMSEP values, with MSC B and SNV giving near identical results and MSC C and SG1 showing even lower values. SG2 gives the lowest number of factors but suffers from a high RMSEP value. The SG2 is prone to the effect of high-frequency noise, and this starts to influence the model already after five factors indicated by overfitting. Of the remaining pre-processing techniques, the MSC A, Raw, Normalization by Euclidean norm and especially IPBF give worse predictions than the best pre-processing methods. This does not come as a surprise as the latter pre-processing only showed a minor linearization effect on the spectra. The fact that MSC B and SNV perform similarly is in accordance with observations made by Dhanoa *et al.* (1994).

It is possible to further evaluate the linearity in the data prior to and after pre-processing by subtracting the sample with the average concentration for glucose and fructose from all the three spectra used in the figures so far. As can be seen from Figure 2.8, MSC B correction of the spectra succeeds in linearizing the data (the distance from the high glucose sample to the mean is close to equal to the distance from the mean to the low glucose sample), while this is not the case for the raw spectra. Overall, the MSC corrected multivariate signals adhere much closer to Beer's law in equation (2.1).

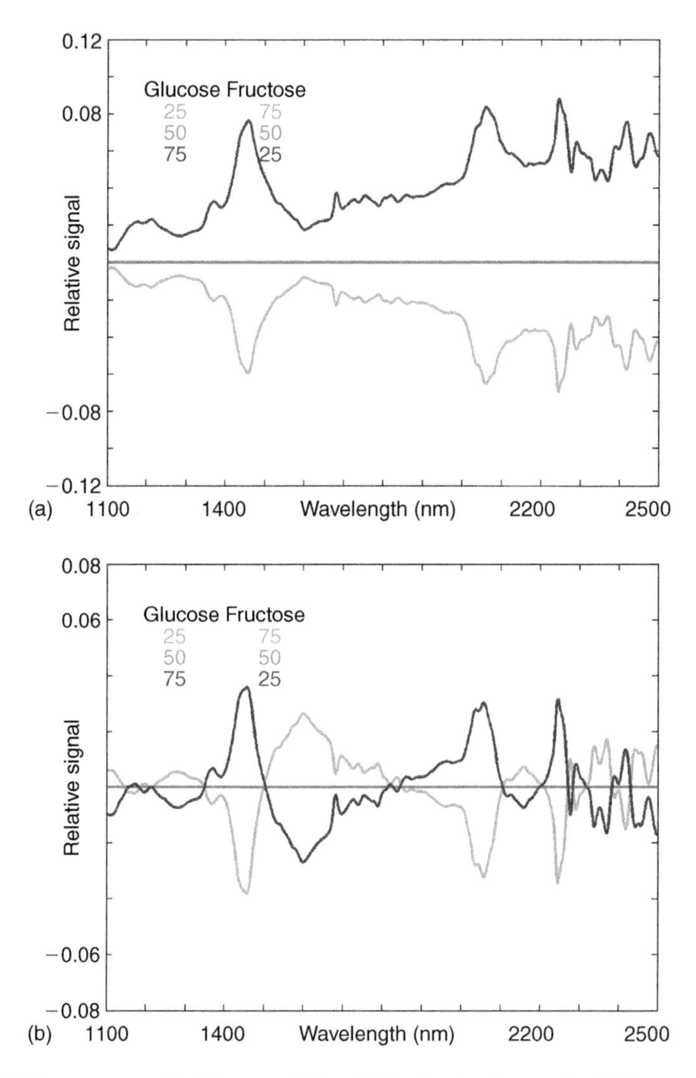

Figure 2.8 Difference spectra for (a) raw and (b) multiplicative signal correction B (the spectra are subtracted by the green spectrum).

Reference-dependent techniques

There are several reference-dependent techniques, but here we only focus on three of them: orthogonalization, optimized scaling, and net analyte pre-processing.

Orthogonalization is a group of methods and algorithms rather than one single pre-processing technique. The goal of all the orthogonalization methods is to remove variability in the spectra which does not correlate to the reference value. The first of these methods—orthogonal signal correction (OSC)—was introduced by Wold *et al.* (1998), which was quickly followed by several suggestions for improvement: Sjöblom *et al.* (1998), Wise (1998), Andersson (1999), and Fearn (2000). Westerhuis *et al.* (2001) provided an excellent summary on how the different methods perform. A slightly different way of approaching the problem was suggested by Trygg and Wold (2002).

Optimized scaling (OS) was introduced by Karstang and Manne (1992), but has received limited attention. Stordrange *et al.* (2002) showed that it outperforms other pre-processing methods such as MSC, OSC, normalization, and derivation. Optimized scaling gives each calibration sample a different scaling in order to correct for non-linearities which especially are present in closed data systems where the molar absorptivity of the different constituents of the sample may vary significantly.

Net analyte pre-processing (NAP) was developed from the net analyte signal (NAS), a figure of merit for multivariate calibration, leading to prediction uncertainty estimation, etc. Goicoechea and Olivieri (2001) showed that it can also be used for pre-processing of data, much in the same way as OSC orthogonalization does.

Case study

Using a case study, we will demonstrate how the different pre-processing techniques affect the analysis of a real data set. The data are available from the Faculty of Life Sciences, University of Copenhagen (http://models.life.ku.dk/), and have previously been reported by Pedersen *et al.* (2002). In this publication the data were analyzed with special regard to the effect of EISC pre-processing on the prediction of protein in single wheat kernels. Using this near-infrared transmittance (NIT) data set which includes a large number of samples measured over two periods we will demonstrate how the different pre-processing techniques perform.

Data

The data consists of 523 single wheat kernels measured by NIT in the wavelength range 850–1048 nm, measured every other nm. The protein content was provided by a reference method measured in the laboratory. It varied from 6.8 to 17.0% protein content in dry matter.

Methods

The PLS predictions are performed several times by randomly extracting bootstrap drawings from the total pool of 523 wheat samples. Each of the drawings contains 471 measurements (90% of all data), and these samples make up the calibration set, while the remaining samples are used as the prediction set. Since each bootstrap drawing will contain more than one copy of a number of individual samples, the size of the prediction set varies from draw to draw. On average the prediction set contained 320 calibration independent measurements. A total of 1000 bootstrap draws were made, giving a good indication of the uncertainty in future predictions and the variation of the RMSEP value.

The pre-processing techniques tested are: raw spectra (no pre-processing), SNV, normalization by the Euclidean norm, SG with first derivative, SG with second derivative, five types of MSC and IPBF with a second-degree polynomial fitting. The polynomial fitted for the SG was a second-degree polynomial for both first and second derivatives. The smoothing window in SG was set to 7, 11, or 15 in order to evaluate the effect of this parameter on the estimation of the derivative. The five MSC variations are: linear reference, quadratic wavelength dependency, linear reference and up to quadratic wavelength dependency, up to quadratic reference, and quadratic reference and quadratic wavelength dependency. Mean-centering is performed in all cases after the selected pre-processing. The number of factors in a model is based on the average RMSEP values from the bootstrap re-drawing, the standard deviation of the RMSEP, the shape of the regression coefficients (the vector should be relatively smooth), and the correlation between the dependent reference and the independent spectral scores in the PLS model.

Results and discussion

First it is of interest to view how the different pre-processing techniques affect the original spectra (Figure 2.9). A Savitzky–Golay estimate of the first derivative with 7 points gave small oscillations at the end of the spectra, therefore a larger amount of points was required in the smoothing, in Figure 2.9d shown with 11 points. To further appreciate this difference one should look at Figure 2.9e and Figure 2.9f showing two different estimates of the second derivative by SG using 7 and 15 points in the smoothing. As shown in Figure 2.9e, using only 7 points in the smoothing creates artifacts in the spectra, as oscillations are especially apparent at high wavelengths. By investigating the raw spectra, there is nothing indicating that such a fluctuation is real. Thus the 15-point smoothing, showing a smooth second derivative, is probably a more accurate estimate of the true second derivative. This also indicates that it is of great importance to (visually) inspect the pre-processed spectra, since artifacts may pop up.

Figure 2.9 shows the effect all the 11 selected pre-processing techniques have on the spectra. The three samples shown are selected so that the relative distance between the spectra should be similar if Beer's law is applicable to the protein content for these three multivariate signals. It is difficult to fully evaluate this by simple inspection of Figure 2.9, and even by zooming in on the MSC corrections, for example, no obvious linear areas appear. This suggests that the number of factors should be quite high in this complex system.

These pre-processed samples were used in PLS models, and 1000 bootstrap drawings were run as explained earlier. The RMSEP values given in Table 2.2 are the mean bootstrap RMSEP and the standard deviation of the RMSEP values—an indication of how precise the models are.

By examining data in Table 2.2, it becomes apparent that the quadratic wavelength dependency correction is important in order to push the RMSEP below 0.5. A minimal improvement can be achieved by adding the quadratic reference correction. The remaining nine methods all behave similarly, giving RMSEP values from 0.53 to 0.58.

The above case study shows that when dealing with real spectra, it is important not to look at the model's predictive performance (RMSEP or RMSECV) only, but also to evaluate the appropriate character of the pre-processing in relation to the data recorded and the artifacts they contain. Inadequate pre-processing settings and/or techniques may lead to artifacts in the spectra, giving the potential of suboptimal correlation in the subsequent analysis.

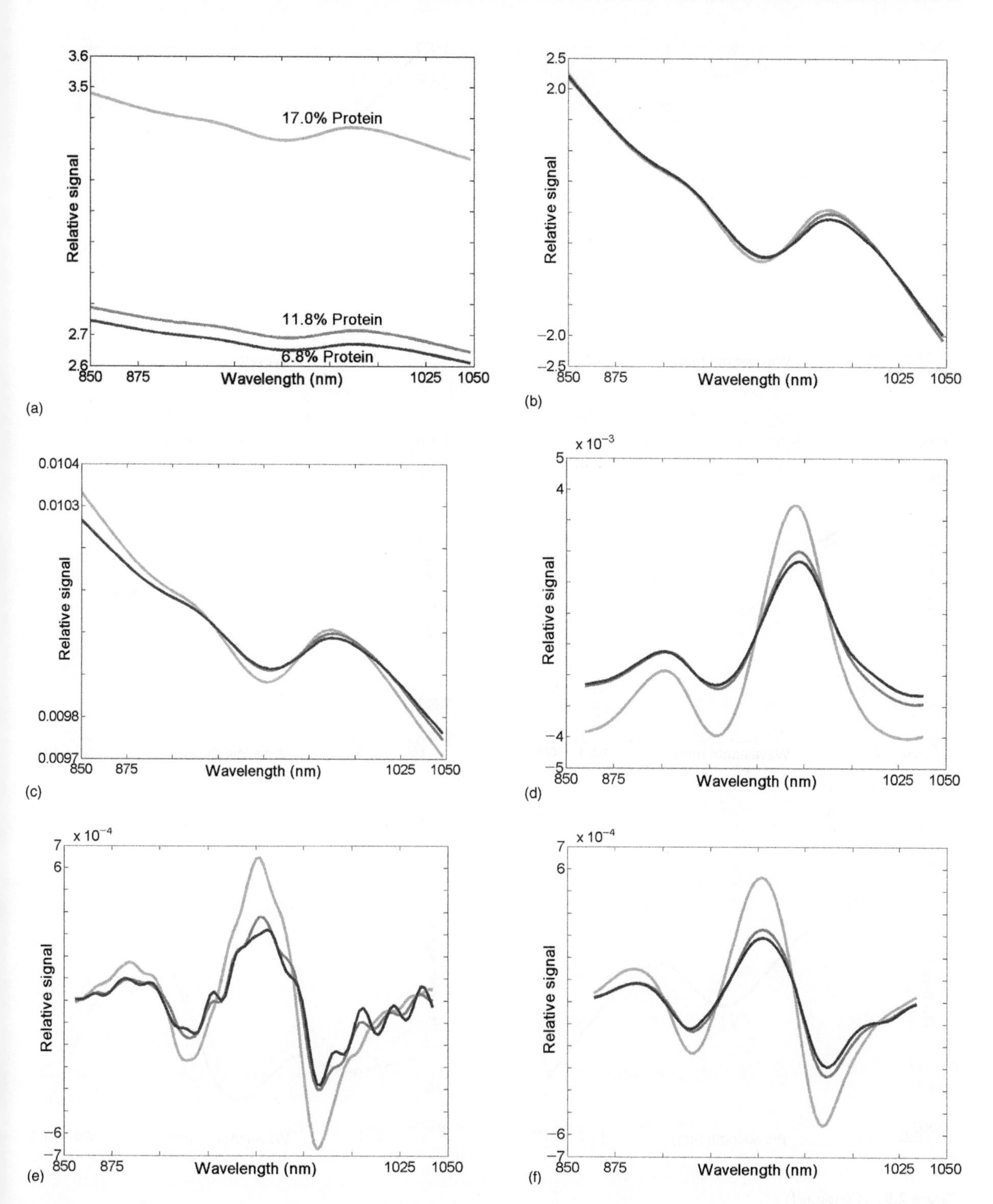

Figure 2.9 The effect the 11 pre-processing techniques on three selected spectra from the raw data. The protein contents for the three spectra are shown in the figure of the raw data. (a) Raw data, (b) standard normal variate, (c) Euclidean norm, (d) Savitzky-Golay (SG) first derivative with 11 points smoothing, (e) SG second derivative with 7 points smoothing, (f) SG second derivative with 15 points smoothing, (g) multiplicative signal correction (MSC): $[\mathbf{1}\ \boldsymbol{\lambda}\ \boldsymbol{\lambda}^2]$, (h) MSC: $[\mathbf{1}\ \mathbf{x}_{\mathbf{ref}}]$, (i) MSC: $[\mathbf{1}\ \mathbf{x}_{\mathbf{ref}}\ \boldsymbol{\lambda}\ \boldsymbol{\lambda}^2]$, (j) $[\mathbf{1}\ \mathbf{x}_{\mathbf{ref}}\ \mathbf{x}_{\mathbf{ref}}^2]$, (k) $[\mathbf{1}\ \mathbf{x}_{\mathbf{ref}}\ \mathbf{x}_{\mathbf{ref}}^2\ \boldsymbol{\lambda}\ \boldsymbol{\lambda}^2]$ and (l) IPBF with second-order polynomial fitting. (The matrix brackets indicate which parts of equation (2.4) are included.)

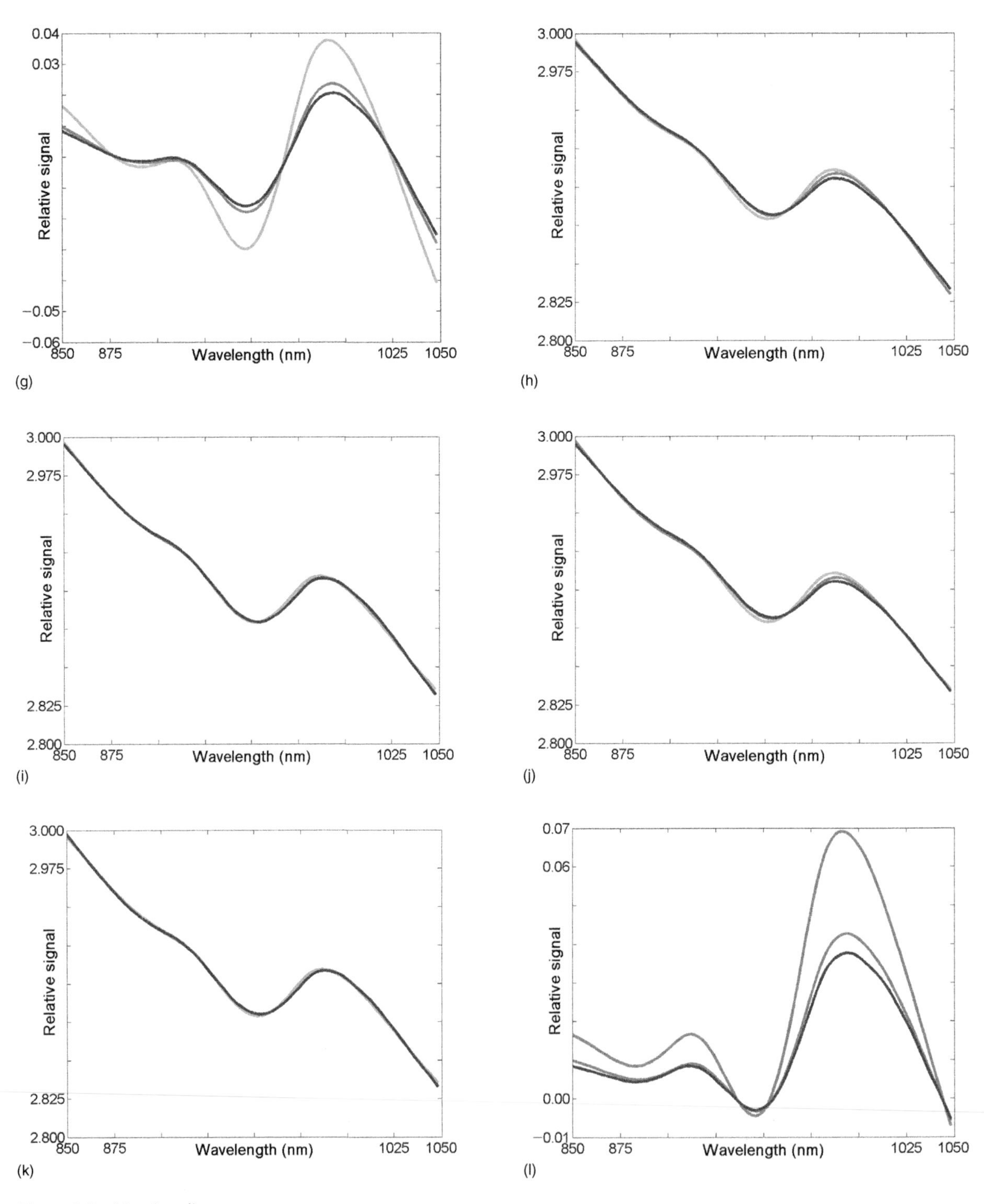

Figure 2.9 (Continued)

Table 2.2 The accuracy of the partial least squares (PLS) models calculated on the wheat kernels with different pre-processing techniques

Method	No. of PLS factors	RMSEP (% protein)
Raw	10	0.54 (0.01)
SNV	8	0.53 (0.01)
Euclidean norm	8	0.53 (0.01)
SG–first derivative, 11 point window, second-order polynomial	8	0.56 (0.01)
SG–second derivative, 15 point window, second-order polynomial	7	0.54 (0.01)
MSC[a]:$[\mathbf{1}\ \lambda\ \lambda^2]$	6	0.56 (0.01)
MSC[a]: $[\mathbf{1}\ \mathbf{x}_{ref}]$	7	0.58 (0.01)
MSC[a]: $[\mathbf{1}\ \mathbf{x}_{ref}\ \lambda\ \lambda^2]$	5	0.47 (0.01)
MSC[a]: $[\mathbf{1}\ \mathbf{x}_{ref}\ \mathbf{x}_{ref}^2]$	9	0.54 (0.01)
MSC[a]: $[\mathbf{1}\ \mathbf{x}_{ref}\ \mathbf{x}_{ref}\ \lambda\ \lambda^2]$	7	0.42 (0.01)
IPBF: second-order polynomial	9	0.56 (0.01)

[a]Indicates which part of equation (2.4) used in the correction.

RMSEP, root mean squared error of prediction; MSC, multiplicative signal correction; SNV, standard normal variate; IPBF, iterative polynomial baseline fitting; SG, Savitzky-Golay.

Conclusions

As shown and discussed previously, it is important to select the correct pre-processing technique for the data. In general, using MSC with linear reference and quadratic wavelength correction leads to satisfying results. On the other hand, the use of baseline correction in general gives bad results. However, as mentioned above, this is a relevant technique for MIR and IR. This is of no surprise since these techniques correct for effects that are not normally observed in NIR/NIT spectroscopy; the baseline does not typically "move" with the spectra.

Although combining different pre-processing techniques is common in research and industry, it will rarely lead to significant improvements to model simplicity or quantitative regression performance. On the other hand, from a pragmatic point of view the optimal pre-processing method for a given data analytical problem is the combination or single transformation that gives the best regression performance after rigorous validation for all possible variations in the data.

Nomenclature

Variables given as italic small letters (e.g. *a*) signify scalars, bold small letters (e.g. **a**) are vectors, and bold capital letters (e.g. **A**) are matrices.

1	Vector of ones
a	Correction coefficients for SNV and normalization
b	Correction coefficient for MSC

$\mathbf{b}$	Vector of correction coefficients for MSC
$\mathbf{C}$	Smoothing filter for Savitzky–Golay
$\mathbf{E}$	Unmodeled part
g	Size of smoothing window for Savitzky–Golay
i	Polynomial order
$\boldsymbol{\lambda}$	Vector for the wavelength axis
m	Variable index
M	Number of variables (or wavelengths)
n	Sample index
N	Number of samples
s_x	Estimated standard deviation of the vector $\mathbf{x}$
$\bar{x}$	Estimated average of the vector $\mathbf{x}$
$\mathbf{x}_{corr}$	Corrected sample spectrum
$\mathbf{x}_{known}$	A priori known constituent spectra (interferent or desired constituent)
$\mathbf{x}_{org}$	Original sample spectrum
$\mathbf{x}_{ref}$	Reference spectrum
$\mathbf{X}_C$	Correction matrix for MSC
x	Absolute value of x

References

Andersson CA (1999) Direct orthogonalization. *Chemometrics and Intelligent Laboratory Systems*, **47**, 51–63.

Barnes RJ, Dhanoa MS, Lister SJ (1989) Standard normal variate transformation and de-trending of near-infrared diffuse reflectance spectra. *Applied Spectroscopy*, **43**, 772–777.

Brown CD, Vega-Montoto L, Wentzell PD (2000) Derivative preprocessing and optimal corrections for baseline drift in multivariate calibration. *Applied Spectroscopy*, **54**(7), 1055–1068.

Davis RA, Charlton AJ, Godward J, Jones SA, Harrison M, Wilson JC (2007) Adaptive binning: An improved binning method for metabolomics data using the undecimated wavelet transform. *Chemometrics and Intelligent Laboratory Systems*, **85**, 144–154.

Dhanoa MS, Lister SJ, Sanderson R, Barnes RJ (1994) The link between Multiplicative Scatter Correction (MSC) and Standard Normal Variate (SNV) transformations of NIR spectra. *Journal of Near Infrared Spectroscopy*, **2**, 43–47.

Fearn T (2000) On orthogonal signal correction. *Chemometrics and Intelligent Laboratory Systems*, **50**, 47–52.

Geladi P, MacDougal D, Martens H (1985) Linearization and scatter correction for near-infrared reflectance spectra of meat. *Applied Spectroscopy*, **39**, 491–500.

Goicoechea HC, Olivieri AC (2001) A comparison of orthogonal signal correction-next term and net analyte preprocessing methods. *Chemometrics and Intelligent Laboratory Systems*, **56**(2), 73–81.

Gorry PA (1990) General least-squares smoothing and differentiation by the convolution (Savitzky–Golay) method. *Analytical Chemistry*, **62**, 570–573.

Helland IS, Næs T, Isaksson T (1995) Related versions of the multiplicative scatter correction method for preprocessing spectroscopic data. *Chemometrics and Intelligent Laboratory Systems*, **29**, 233–241.

Karstang TV, Manne R (1992) Optimized scaling—A novel approach to linear calibration with closed data sets. *Chemometrics and Intelligent Laboratory Systems*, **14**(1–3), 165–173.

Lieber CA, Mahadevan-Jansen A (2003) Automated method for subtraction of fluorescence from biological raman spectra. *Applied Spectroscopy*, **57**(11), 1363–1367.

Martens H, Stark E (1991) Extended multiplicative signal correction and spectral interference subtraction: new preprocessing methods for near infrared spectroscopy. *Journal of Pharmaceutical and Biomedicinal Analysis*, **9**, 625–635.

Martens H, Jensen SA, Geladi P (1983) Multivariate linearity transformations for near infrared reflectance spectroscopy. In: *Proceedings of the Nordic Symposium on Applied Statistcs* (Christie OHJ, ed.). Stavanger, Norway: Stokkland Forlag, pp. 205–234.

Massart DL, Vandeginste BGM, Buydens LMC, de Jong S, Lewi PJ, Smeyers–Verbeke J (1997) *Handbook of Chemometrics and Qualimetrics Part A*. Amsterdam: Elsevier, p. 162.

Norris KH (1983) Extraction information from spectrophotometric curves. Predicting chemical composition from visible and near-infrared spectra. In: *Food Research and Data Analysis* (Martens H, Russwurm H Jr, eds). London: Applied Science, pp. 95–114.

Norris KH, Williams PC (1984) Optimization of mathematical treatments of raw near-infrared signal in the measurement of protein in hard red spring wheat: I. Influence of particle size. *Cereal Chemistry*, **61**, 158–165.

Pedersen D, Martens H, Nielsen JP, Engelsen SB (2002) Near-infrared absorption and scattering separated by extended inverse signal correction (EISC): Analysis of near-infrared transmittance spectra of single wheat seeds. *Applied Spectroscopy*, **56**, 1206–1214.

Savitzky A, Golay MJE (1964) Smoothing and differentiation of data by simplified least squares procedures. *Analytical Chemistry*, **36**(8), 1627–1639.

Schulze G, Jirasek A, Yu MML, Lim A, Turner RFB, Blades MLW (2005) Investigation of selected baseline removal techniques as candidates for automated implementation. *Applied Spectroscopy*, **59**, 545–574.

Sjöblom J, Svensson O, Josefson M, Kullberg H, Wold S (1998) An evaluation of orthogonal signal correction applied to calibration transfer of near infrared spectra. *Chemometrics and Intelligent Laboratory Systems*, **44**(1–2), 229–244.

Steinier J, Termonia Y, Deltour J (1972) Comments on smoothing and differentiation of data by simplified least squares procedure. *Analytical Chemistry*, **44**(11), 1906–1909.

Stordrange L, Libnau FL, Malthe-Sorenssen D, Kvalheim OM (2002) Feasibility study of NIR for surveillance of a pharmaceutical process, including a study of different preprocessing techniques. *Journal of Chemometrics*, **16**(8–10), 529–541.

Trygg J, Wold S (2002) Orthogonal projections to latent structures (O-PLS). *Journal of Chemometrics*, **16**(3), 119–128.

Wentzell PD, Brown CD (2000) Signal processing in analytical chemistry. In: *Encyclopedia of Analytical Chemistry*, Volume 11 (Meyers RA, ed.). Chichester, UK: Wiley, pp. 9764–9800.

Westerhuis JA, de Jong S, Smilde AK (2001) Direct orthogonal signal correction. *Chemometrics and Intelligent Laboratory Systems*, **56**, 13–25.

Wise B (1998) Orthogonal signal correction. In: *Functions in the MATLAB User Area*. Eigenvector Research, Wenatchee, WA 98801, USA. Available at http://www.eigenvector.com/MATLAB/OSC.html

Wold S, Antti H, Lindgren F, Öhman J (1998) Orthogonal signal correction of near-infrared spectra. *Chemometrics and Intelligent Laboratory Systems*, **44**, 175–185.

3

Multivariate Calibration for Quantitative Analysis

Marcelo Blanco Romía and Manel Alcalà Bernàrdez

Introduction

Calibration is the process by which the mathematical relationship between the values provided by a measuring instrument or system and those known for the measured

Infrared Spectroscopy for Food Quality Analysis and Control
ISBN: 978-0-12-374136-3

material object is established. The mathematical expression relating analytical responses or signals to concentrations is known as the calibration equation (Mark and Workman, 2003). Most analytical techniques use a straight line for calibration on account of its straightforward equation and its ability to illustrate a direct relationship between measured signals and concentrations (univariate calibration). However, a linear calibration model can only be useful for quantitation purposes if the analytical signal depends exclusively on the concentration of the specific analyte for which the model has been developed. Such exclusive dependence is the exception rather than the rule in the analysis of complex samples by spectroscopic techniques such as infrared and near-infrared (NIR) spectroscopies. Spectroscopic measurements are used to establish a linear relation between the absorbance (i.e. the inverse logarithm of the transmittance) or apparent absorbance (viz. the inverse logarithm of the reflectance) and concentration via the Beer–Lambert law (Meehan, 1981). Calibration models for these techniques are usually constructed by least squares regression (LSR) of the absorbance (or apparent absorbance) values for a set of standards against their concentrations.

Most often, the analytical signal comprises contributions from several analytes or even the sample matrix, which precludes the use of models other than LSR (e.g. multivariate calibration) in order to accurately predict analyte concentrations. Multiple linear regression (MLR), which is an extension of linear regression, involves using more than one variable in order to predict the concentration of one or more analytes (Mark, 1991). Mathematically, MLR can be formulated as

$$\mathbf{y} = \mathbf{X} \cdot \mathbf{b} + \mathbf{b} \cdot \mathbf{1} + \mathbf{f} \tag{3.1}$$

where $\mathbf{X}$ is the matrix containing the responses of the different variables considered and $\mathbf{b}$ the vector containing the regression coefficients for the variables. $\mathbf{X}$ and $\mathbf{b}$ can be expanded by including an offset term for the MLR coefficients. Such coefficients can be calculated by using least-squares methodology in the form

$$\hat{\mathbf{b}} = (\mathbf{X}^{\mathrm{T}}\mathbf{X})^{-1}\mathbf{X}^{\mathrm{T}}\mathbf{y} \tag{3.2}$$

Multiple linear regression is subject to two major restrictions. One is the dimension of matrix $\mathbf{X}$; thus, the number of variables used cannot exceed that of samples. The other is that no two X variables should be mutually related; otherwise, the matrix $(\mathbf{X}^{\mathrm{T}}\mathbf{X})$ cannot be inverted. In real-world applications, where data are typically noisy, variables are highly unlikely to be fully correlated; however, a substantial degree of correlation between variables can lead to an unstable inverted matrix.

Multivariate calibration methods

A wide variety of multivariate calibration methods have now been reported, and can be classified according to whether or not they possess a given property. Thus, an immediate distinction can be made between *linear methods* and *non-linear methods*. Linear methods are formulated mathematically as

$$Y = b_0 + \sum_{k=1}^{K} b_k x_k \tag{3.3}$$

where b_0 and b_k are the target parameters, y is the dependent variable and x_k denotes the independent variables. Therefore, linearity here refers to the relationship between the dependent variable and regression parameters rather than to that between variables. Non-linear methods use a non-linear relationship between parameters. Thus, in

$$y = \alpha x^{\beta} \tag{3.4}$$

α and β are the target parameters. Although linear methods are more widely used, a number of determinations require non-linear methods.

Another classification distinguishes between *direct methods* and *indirect methods*. The former calculate the calibration parameters from the individually recorded signal for each analyte, whereas the latter use the analytical signals for mixtures of the analytes to obtain such parameters.

Depending on the specific quantity used as independent variable, calibration methods can be of the *classical* or *inverse* type. Classical calibration relies on a criterion directly related to the Beer–Lambert law and uses the analytical signal as a concentration-dependent variable. Inverse calibration uses a more mathematical concept and, because the ultimate aim is to calculate a concentration, it uses concentration as a dependent variable and the analytical signal as the independent variable.

Yet another classification of multivariate calibration methods distinguishes between *rigid methods* and *flexible methods*. The former use a preset number of terms in the regression equation, whereas the latter use the optimum number established by the method itself. Those methods using analytical information contained in a large, unrestricted number of variables are designated as *full-spectrum methods*, whereas those reducing an initially large number of variables to a much smaller one without losing relevant analytical information are known as *variable-compression methods*. Table 3.1 compares the characteristics of various calibration methods.

In Table 3.1, the first four methods are all linear (i.e. they use a calibration model based on an MLR equation), whereas the fifth, artificial neural networks (ANNs), can be applied to both linear and non-linear systems.

Advantages of multivariate calibration methods

Although multivariate calibration methods can obviously be applied to any analytical technique, the ease at which multi-parameter signals (e.g. the absorbance at several wavelengths) can be obtained in practice has facilitated its preferential expansion among spectroscopic techniques. The discussion that follows is therefore conducted in spectrophotometric terms, but can be extended to non-spectroscopic techniques.

Table 3.1 Comparison of various calibration methods

Calibration method	Characteristics
Classical least squares (CLS)	Rigid, full-spectrum method compatible with direct and indirect calibration
Inverse least squares (ILS)	Flexible method indirect calibration only and unrestricted as regards the number of variables
Principal component regression (PCR)	Flexible, full-spectrum, variable-compression method compatible with inverse and indirect calibration
Partial least squares regression (PLSR)	Flexible, full-spectrum, variable-compression method compatible with inverse and indirect calibration
Artificial neural networks (ANNs)	Flexible, compression method using a restricted number of input variables and compatible with inverse and indirect calibration

In conventional analytical methodology, analyte responses are made selective by using specific sensors or by physically isolating the analyte from its interferents. In many cases, however, isolating the analyte can be expensive or extremely difficult—or even illogical if the interferents are to be determined as well. Multivariate calibration solves many determination problems arising from interferences or matrix effects.

Traditional methods for obtaining linear signals fail sometimes. For example, the analysis of highly absorbing samples with classical methodology requires their dilution or, alternatively, using cells of a shorter light path. Such seemingly simple operations, however, can be very difficult to perform or introduce new problems. Non-linear responses can be of instrumental, physical and/or chemical origin, and result in a curved response–concentration line by effect of a non-linear detector response, straight light generated at high optical densities, baseline drift (e.g. that caused by physical scattering of the light by solid particles), and shifts in the positions of bands or changes in their widths by effect of changes in temperature or the nature of the solvent.

Some spectral pre-treatments are efficient in reducing, but not completely suppressing, non-linearity. Also, occasionally, no theoretical ground for obtaining the signal–concentration relationship exists with which an appropriate prelinearizing treatment can be chosen.

Some slight deviations from linearity can be modeled with multivariate calibration methods at the expense of an increased complexity (e.g. by using principal component regression (PCR) or partial least squares regression (PLSR) with additional principal components). Complex non-linear systems can be resolved by using non-linear calibration methods including some PCR and PLSR variants, or intrinsically non-linear methods such as ANNs.

One major advantage of multivariate calibration methods is the reliability of their results for unknown samples. With univariate calibration, the presence of an uncontrolled interference in a sample will go undetected and invariably introduce an error in the results. With multivariate calibration, however, provided that a large number of variables are used for the analytical signal, examining the residuals can expose whether a given sample is "different" from those used for calibration and the result that it provides is thus unreliable.

Stepwise multiple linear regression

The primary purpose of using a regression technique is constructing models allowing the value of the dependent variable, Y, which is usually a concentration, to be predicted from experimental data (absorbance in our case) represented by the independent variable, X (Kramer, 1998). The need to use q independent variables, X_q, in order to explain the results entails using multiple linear regression models where each value of the dependent variable is expressed as a combination of polynomial terms:

$$y_i = a_0 + a_1 x_{i1} + a_2 x_{i2} + a_3 x_{i3} + \cdots + a_q x_{iq} = a_0 + \sum a_q x_{iq} + \varepsilon_i \tag{3.5}$$

This general expression is substantially simpler in matrix form:

$$\mathbf{Y} = \mathbf{Xa} + \varepsilon \tag{3.6}$$

Usually, the target parameters are estimated by minimizing the summation of the squares of the errors ($\Sigma\varepsilon_i^2$).

Classical least squares

This multivariate method was developed for processing spectroscopic data. As implied by its designation, it assumes fulfillment of Beer's law by each individual component of a mixture throughout the working range and additivity of individual absorbances in the mixture (Coello and Maspoch, 2007). Errors in CLS models are assumed to be due to the spectral data used.

The absorbance at wavelength j of a mixture of P components can be expressed as

$$x_j = k_{j1}c_1 + k_{j2}c_2 + k_{j3}c_3 + \cdots + k_{jP}c_P + e_j \tag{3.7}$$

where e_j is the random error in the measurement, c_i the concentration of component i, and k_{ji} the product of the light path length (which will be constant at any wavelength) by the absorptivity coefficient of component i at wavelength j.

By making measurements at K different wavelengths such that $K \geq P$, one can obtain a system of K equations that can be expressed in matrix form as

$$\mathbf{X} = \mathbf{CK} + \mathbf{E} \tag{3.8}$$

where $\mathbf{X}$ ($K \times N$) contains the spectra for N samples, $\mathbf{C}$ ($K \times P$) the concentrations for $\mathbf{P}$ analytes in the N samples, $\mathbf{K}$ ($P \times N$) the pure spectra for the P analytes and $\mathbf{E}$ the error of the model.

The spectrum for any sample will simply be a linear combination of the spectra for the pure components and the error come from the spectroscopic data rather than the concentrations. Application of this model entails the prior knowledge of the total number of analytes present in the samples.

Depending on the way matrix $\mathbf{K}$ is calculated, calibration will be of the direct or indirect type. In the former case, $\mathbf{K}$ will be obtained by recording the spectra for the pure components in the sample; in this way, if a constant light path length is used, then each column in **K** will represent the spectrum for the standard pure component concerned at unity concentration.

Indirect calibration is done by using mixtures of all the components. With M calibration standards of known concentration for the **P** analytes one can obtain an absorbance matrix **A** ($K \times M$) and a concentration matrix **C** ($P \times M$) allowing the previous equation to be rewritten as

$$\mathbf{A_{st}} = \mathbf{KC} + \mathbf{E} \tag{3.9}$$

where **E** ($K \times M$) is the absorbance residuals matrix.

The values of matrix **K** ($K \times P$) are estimated by minimizing the summation of the squares of the spectral errors using LSR. The solution thus obtained is

$$\hat{\mathbf{K}} = (\mathbf{CC}^{\mathrm{T}})^{-1}\mathbf{C}^{\mathrm{T}}\mathbf{A_{st}} \tag{3.10}$$

where the hat symbol denotes an estimated value of **K** and the superscript T the transpose of the matrix.

The estimated pure component spectra $\hat{K}$ (or the measured spectra for the pure compounds, **K**) constitute the parameters of the CLS model.

The concentration of each analyte in an unknown sample can be determined by applying the following equation to its spectrum:

$$\hat{\mathbf{c}} = (\hat{\mathbf{K}}^{\mathrm{T}}\hat{\mathbf{K}})^{-1}\hat{\mathbf{K}}^{\mathrm{T}}\mathbf{x} \tag{3.11}$$

where **x** is the spectrum for the unknown sample.

Although indirect calibration has some advantages over direct calibration, it is much more time-consuming and less widely used as a result. On the other hand, the simplicity of calibration with direct CLS, which only requires recording the spectra for the pure components, has made it a widespread choice, especially in UV-Vis spectrophotometry. For direct CLS to be applicable, the following two requirements must be met: (a) the components contributing to the analytical signal should all be known; and (b) the Beer–Lambert law should be obeyed (i.e. the absorbance of each component at any wavelength should be a linear function of its concentration and no interaction between analytes or with the sample matrix should exist).

Because it requires the knowledge of every possible contribution to the signal, direct CLS can only be used with samples of accurately known composition. Also, the spectra for the pure components should all be recordable and not altered by the presence of other components, which in practice is only the case with very dilute solutions. Many UV-Vis spectrophotometers come with control software including a *multicomponent analysis* routine based on this procedure.

Direct CLS has a number of advantages such as the following:

(a) It allows multi-determinations to be accomplished in a simple manner.
(b) All analyte responses are modeled simultaneously; as a result, predictions are independent of the light path length used and of multiplicative changes in the sample spectrum.

(c) The use of the estimated spectra for the pure components, $\hat{K}$, allows qualitative spectral information for the analytes or their mutual interactions to be extracted.

(d) Because it is a full-spectrum method, CLS can provide increased precision relative to other methods using a limited number of variables.

However, some of these advantages depend on the precision in the calculated concentrations, which is obviously dictated by that in the measured absorbance, and also on the accuracy of the chosen model. Also influential is the "quality" of matrix **K**, which is a function of spectral noise and the degree of similarity between spectra. Thus, very similar spectra can introduce high collinearity in the matrix columns (i.e. near-linear relationships between the absorbance at different wavelengths) and lead to spurious results in calculating the inverse of ($\mathbf{K}^T\mathbf{K}$).

Increasing the precision of the results requires reducing spectral collinearity, and can be accomplished by appropriate selection of the variables or spectral mode (absorbance or *n*th derivative mode). In fact, the selection process is the most crucial step in developing a CLS-based analytical procedure. The use of derivative spectra provides some advantages such as reduced spectral overlap and also reduced baseline shift and drift; however, it also reduces the signal-to-noise ratio, which can detract from precision. There are no universal rules for using derivative spectra in preference over absorbance spectra, so the choice is always necessarily empirical.

The analytical precision can be substantially improved by appropriate selection of variables. As a rule, the precision can be maximized by using a number of variables only slightly greater than that of analytes. However, using too few variables detracts from one of the principal advantages of multivariate calibration as applied to control analyses: the ability to detect samples departing from those used for calibration (i.e. samples containing uncontrolled interferences, which can be detected by examining the standard deviation of the spectral fitting) (Frans and Harris, 1985).

The CLS method performs poorly when the detector response or strong spectral interactions between analytes cause substantial deviations from linearity. Also, the method is rigid and limited in scope. In any case, it is used here as the background for discussion of other methods in the following sections.

Inverse least squares

When MLR is used to construct a predictive model based on signals from a multi-analyzer (e.g. wavelengths) as inputs and a property of interest (e.g. the concentration of a component) as output, the method is referred to as *inverse linear regression* (ILR) (Barnett and Bartoli, 1960). The word "inverse" here signifies that the model uses the inverse form of Beer's law, where the concentration is expressed as a function of the absorbance:

$$Y_i = b_0 + b_1 x_1 + \cdots + b_{K-1} x_{K-1} + e_i \tag{3.12}$$

This equation can be expressed in matrix form as

$$\mathbf{Y} = \mathbf{XB} + \mathbf{E} \tag{3.13}$$

where matrix $\mathbf{Y}$ ($M \times P$) contains concentrations and matrix $\mathbf{X}$ ($M \times K$) the body of spectroscopic data selected for calibration, $\mathbf{E}$ ($M \times P$) is the matrix of the random residuals of the concentrations, and $\mathbf{B}$ ($K \times P$) is the unknown matrix of regressors and can be calculated during calibration.

This model has the advantage that it requires no prior knowledge of all absorbing species present in the sample in order to quantify the P target analytes. However, those components not included in the quantitation process should be present in all samples and are implicitly modeled. This has made the ILS method widely popular, especially for IR spectroscopy; recently, however, it has been gradually superseded by variable-compression methods such as PCR and PLSR.

The regression coefficients can be calculated by least squares regression in the form

$$\hat{\mathbf{B}} = (\mathbf{X}^{\mathrm{T}}\mathbf{X})^{-1}\mathbf{X}^{\mathrm{T}}\mathbf{Y} \tag{3.14}$$

Once matrix $\hat{\mathbf{B}}$ has been obtained, property Y (the analyte concentration) for a new, unknown sample can be estimated from the spectral data for the sample, $\mathbf{x}$, as given below.

$$\hat{Y} = \mathbf{x}^{\mathrm{T}}\hat{\mathbf{B}} \tag{3.15}$$

The most salient implication of the ILS method is that it assumes the error of the model to be present in the data matrix $\mathbf{Y}$, and the method minimizes the square of the errors in the concentrations. Although this is not strictly true in practice (where the variables X inevitably contain some noise), it does not preclude analytical application of the method.

The ILS method is probably the simplest of all analytically useful multivariate calibration methods and simplicity is a valuable asset when automation and reliability over time are sought. Its principal use is in the determination of a single component in a complex mixture. Although it could also be effective for the simultaneous determination of several components, the need to use an increased number of variables would detract from performance. Calibration with ILS is always of the indirect type and conducted with samples where the target parameter has previously been determined by using a reference method.

The greatest disadvantages of ILS are that it requires using a number of samples exceeding that of variables and that selecting the target variables is no easy task. In fact, ensuring acceptable results with ILS entails careful selection of the target wavelengths and using only a fairly small number—otherwise, the number of calibration samples to be analyzed with the reference method can go beyond practical limits.

In addition, using too many wavelengths can give rise to the typical collinearity problems of CLS and detract from precision. Therefore, ILS cannot benefit from increased precision and the use of full spectra.

The ILS method has been the foundation for quantitative analyses with filter near-infrared (NIR) photometers. The development of scanning instruments has facilitated the use of calibration techniques such as principal component regression (PCR) and partial least squares regression (PLRS), which are more robust and avoid the need for careful selection of the target wavelengths.

The procedure used to select wavelengths depends on the number of variables available. Today's computers are fast enough to calculate every possible combination of up to six variables—using a greater number is discouraged as it considerably increases the risk of overfitting. The aim is to select the calibration equation providing the closest fitting. This is impossible with a large number of variables. In this situation, variables are selected by using the *stepwise ascending method*, which compares all equations involving a single variable in order to select the most suitable among them. After the first variable has been set, all binary combinations of the other variables are assayed; this is followed by ternary combinations and so on until the inclusion of a new term results in no significantly improved fitting (Stenberg *et al.*, 1960).

Calibration methods based on variable reduction

These methods combine the advantages of the two above-described least-squares methods. Because they are inverse, indirect methods, they allow individual analytes in mixtures to be quantified without the need to know the other components. Also, they use the information contained in the whole spectrum, and "compress" it into a small number of variables. This avoids the need to select variables and facilitates detection of interferences and outliers.

The number of variables of a system can be reduced in various manners including the use of wavelets or Fourier coefficients, among others. Only the two most widely used for multivariate calibration in analytical chemistry are discussed here, namely: principal component regression (PCR) and partial least squares regression (PLSR). Both follow the principles of principal component regression.

Both PCR and PLSR assume that the information contained in the measured variables can be concentrated into a smaller number of variables in order to reduce signal to noise without losing relevant information. The two methods regress the new variables rather than the measured responses, and hence simplify construction of the calibration model and interpretation of the results. Their importance lies in the ability to solve—at least partly—typical problems such as the following:

(a) *Poor selectivity*. Measurements of the variables X can be influenced by the presence of interferents accompanying the analyte. This requires using x variables to predict y.

(b) *Collinearity*. The information contained in X can be redundant or even correlated.

(c) The lack of an accurate knowledge of the influence of Y on X. One may not know all sample components influencing X or their mutual interactions. Also, the instrument response may be impossible to linearize in full.

In addition to the typical features of variable reduction methods, PCR and PLSR use an orthogonal space for the regression, thereby avoiding the problems derived from collinearity between variables. Interested readers can find a comprehensive description of their principles, and a comparative study of their properties and results, elsewhere (Beebe and Kowalski, 1987; Haaland and Thomas, 1988; Martens and Naes, 1989; Thomas and Haaland, 1990; Thomas, 1994).

Briefly, principal component regression (PCR) involves a principal component analysis (PCA) of the data matrix, **X**, followed by least squares regression between the scores of the selected calibration samples and the reference values of the parameter to be modeled. The regression model is constructed similarly as in ILS except that PCR uses PCA scores rather than the original variables (viz. absorbance values at the target wavelengths). This entails previously subjecting the body of calibration spectra to PCA. Although Chapter 4 provides a more detailed description of PCA, the following section provides a brief introduction to the variable reduction process.

Principal component analysis

Principal component analysis is a variable compression method that reduces the data set of matrix **X** ($K \times N$) to a much smaller number of A variables called principal components (PCs). The corresponding mathematical model is constructed from the expression

$$\mathbf{X} = \mathbf{T}\mathbf{P}^{\mathrm{T}} + \mathbf{E} \tag{3.16}$$

where **T** ($N \times A$) is a matrix containing the A scores for the PCs, **P** ($K \times A$) that containing the A loadings for the PCs and **E** ($K \times N$) the residuals matrix of the model. The scores are the intensities of the new A variables for the samples and the loadings the new variables obtained from the original ones. The PCs are orthogonal to each another, so both vectors are completely uncorrelated. One major consequence of the orthogonality in the PC vectors is that correlation is completely eliminated by using the new variables instead of the original **X**.

The aim of PCA is to identify the directions, allowing the original data matrix to be reduced to a simpler one while deleting useless information. The mathematical algorithm used simply calculates the eigenvectors and eigenvalues of a matrix; as can be easily demonstrated, if the variables X are centered, then the vectors of the loadings p_a (with $a = 1, 2, \ldots, A$ PCs) are the eigenvectors of the matrix ($\mathbf{X}^{\mathrm{T}}\mathbf{X}$) and those of the scores t_a the eigenvalues of the matrix ($\mathbf{X}\mathbf{X}^{\mathrm{T}}$). The most common among the computational algorithms available for this purpose calculate PCs in a sequential manner via an iterative least-squares process followed by subtraction of the contribution of each component. Each PC is determined in such a way that it will account for the residual variance in the data matrix **X** and the process is allowed to progress until the PCs equal the original variables in number and account for 100% of the variance in the data.

The algorithm initially determines the direction of maximum variability in the objects, then the next in significance and so on. To this end, it uses any vector in the K-dimensional space and rotates it about the origin in order to reach the position best fitting the principal direction of the data. Points are projected onto the new, P_1 axis, and their variance is calculated. An angle exists where the variance is maximal; in that position, vector P_1 will be the first PC (viz. the vector best describing the potential principal direction of variability in the points). The director cosines of the vector will be the loadings and reflect the position of the new axis in the K-dimensional space. By projecting the points in the space of K dimensions (the objects) onto this vector, one can calculate the coordinates of the points with respect to the first PC (i.e. the scores). A simple PC computational example is described by Massart *et al.* (1988).

Multiplying the scores, t_1, by the loadings, p_1^T, allows the original matrix to be converted into a new matrix $\mathbf{X}_1$ such that

$$\mathbf{X}_1 = \mathbf{t}_1\mathbf{p}_1^{\mathrm{T}} \tag{3.17}$$

$\mathbf{X}_1$ will be different from the original matrix, $\mathbf{X}$, but constitute its best possible reproduction. A second PC can then be calculated from a loadings vector and its corresponding scores vector:

$$\mathbf{X}_1 = \mathbf{X} - \mathbf{t}_1\mathbf{p}_1^{\mathrm{T}} = \mathbf{t}_2\mathbf{p}_2^{\mathrm{T}} \tag{3.18}$$

The process can be continued in this way until the whole original matrix has been resolved.

The product $\mathbf{TP}^{\mathrm{T}}$ provides a better approximation to $\mathbf{X}$ than does that obtained in the previous step. Accurately reproducing $\mathbf{X}$ would in principle require using K loadings vectors and scores vectors; however, the original matrix can also be accurately represented by using a number of vectors A smaller than K:

$$\mathbf{X} = \mathbf{t}_1\mathbf{p}_1^{\mathrm{T}} + \mathbf{t}_2\mathbf{p}_2^{\mathrm{T}} + \cdots + \mathbf{t}_{\mathrm{A}}\mathbf{p}_{\mathrm{A}}^{\mathrm{T}} + \mathbf{E} \tag{3.19}$$

since the relevant information is contained in the first A components and the others only describe noise-related variations. In physical terms, the number of loadings vectors will be the same as that of sources of systematic variation in the data.

However, the data are not compressed until the user chooses the number of PCs, which is much smaller than that of original variables ($A << K$). In practice, the choice is subjective and should rely on a balanced compromise between the need to explain the variance in the original data and that to avoid overfitting.

Principal component regression

Principal component regression (PCR) performs multiple inverse regression (ILS) of the predictor variables against the scores rather than the original data. To this end, it uses the ability of PCA to resolve variability sources since the scores of matrix $\mathbf{X}$ will contain the same information as the matrix itself.

The first step in the process involves resolving matrix $\mathbf{X}$ into its PCs as described below:

$$\mathbf{X} = \mathbf{TP}^{\mathrm{T}} + \mathbf{E} = \sum_{a=1}^{A} t_a p_a^{\mathrm{T}} + \mathbf{E} \tag{3.20}$$

Once the optimum number A of PCs describing the original matrix has been chosen, **X** can be represented by its scores matrix, **T**:

$$\mathbf{T} = \mathbf{XP} \tag{3.21}$$

where the scores matrix, **T**, and the loadings matrix, **P**, are both obtained from the data matrix, **X**.

Matrix **Y** can be calculated by regressing **Y** against **T**:

$$\mathbf{Y} = \mathbf{TB} + \mathbf{E} \tag{3.22}$$

where **B** is the regressors matrix and as in ILS can be calculated by least squares regression provided the values of **Y** in the calibration are known:

$$\hat{\mathbf{B}} = (\mathbf{T}^{\mathrm{T}}\mathbf{T})^{-1}\mathbf{T}^{\mathrm{T}}\mathbf{Y} \tag{3.23}$$

$\mathbf{T}^{\mathrm{T}}\mathbf{T}$ can be easily inverted since, unlike with the original data, the scores are orthogonal.

Once a seemingly accurate model is developed, predicting the results for a set of new samples involves performing calculations similar to those used in its construction. First, the matrix containing the spectroscopic data for the samples in the prediction set (or the vector if only one sample is to be predicted), $\mathbf{X}^{*}$, is centered or autoscaled by using values calculated from the data matrix **X** used for calibration (the superscripted asterisk denotes new samples for prediction). The loadings matrix (with A optimum components) obtained in the calibration process is used to calculate the scores of the new samples:

$$\mathbf{T}^{*} = \mathbf{X}^{*}\mathbf{P} \tag{3.24}$$

and the regressors matrix, also obtained during calibration, is used in combination with the scores for the new samples to calculate the corresponding concentrations:

$$\mathbf{Y} = \mathbf{T}^{*}\mathbf{B} \tag{3.25}$$

The greatest problem with PCR is that the principal components best describing the matrix of spectroscopic data, **X**, may not be the optimum PCs for predicting new concentrations. Fairly often, the first few PCs account for variations in the analytical signal that bears no relationship to the target analyte. This problem has been addressed by introducing an intermediate step involving estimating the correlation coefficient between the analyte concentrations and the scores for each component and using only those components exhibiting significant correlation in the subsequent regression.

The most salient advantage of PCR over inverse MLR is that, because the former considers the variance between the different variables X, it circumvents the potential problems arising in the mathematical calculations (viz. inverting the matrix $\mathbf{T}^{\mathrm{T}}\mathbf{T}$) and avoids the need to select variables sufficiently independent of one another. However, one should always avoid the risk of *overfitting* by using more factors than needed as this can lead to a model vulnerable to unexpected distortions. Figure 3.1 depicts the calibration and prediction process as conducted by using a PCR algorithm.

Overfitting PCR models can be avoided by using a number of techniques. Among them, one of the most widespread used involves choosing the number of PCs causing a

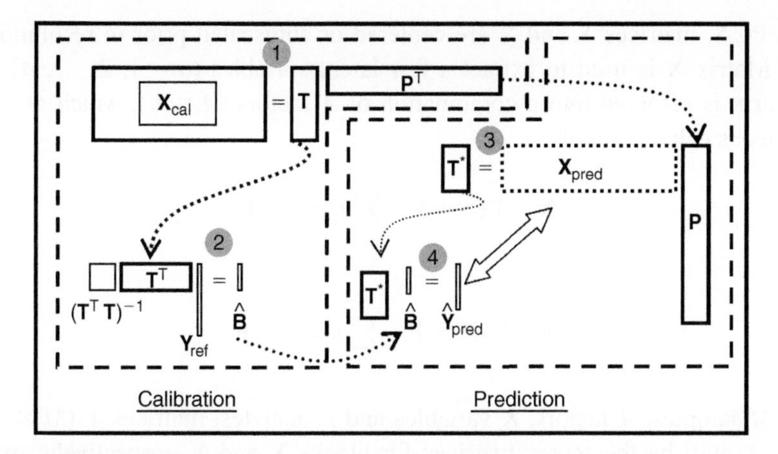

Figure 3.1 Schema for the calibration of a PCR model and the prediction of new samples. Steps: (1) Decomposition of calibration data matrix in scores and loadings matrix. (2) Calculation of regression vector. (3) Projection of prediction spectra into calibration space. (4) Prediction of the property of the new samples.

change in the graph of per cent explained variance in the data against the number of PCs. It should be noted that PCR produces especially stable, user-friendly methods; however, the way PCs are obtained may not be the best with a view to constructing an appropriate calibration model. In fact, the variance in *X* data that is relevant for predicting may constitute a small contribution to the total variance in **X** and could be lost if "weak" PCs are deleted.

Partial least squares regression

The PLSR method, which was introduced by Wold (1975), has been used as an alternative to ordinary least squares regression for solving problems involving high collinearity or the need to calculate correlated *Y* variables. Since its original formulation, PLSR has been associated with other mathematical methods and algorithms. The algorithms most widely used to implement PLSR are non-linear iterative partial least squares (NIPALS) and SIMPLS. Some recent ones, however, depart from the classical iterative procedure and facilitate more global and faster regression. In any case, the variables y are related to the variables x via auxiliary variables called latent variables, or PLSR factors or components, which are linear combinations of the variables $x_1, x_2, \ldots, x_K$, and highly similar to the PCs calculated by PCA and used for PCR. A detailed description of this calibration method can be found elsewhere (Geladi and Kowalski, 1986).

The difference from PCR is that PLSR aims at ensuring that the first few latent variables will contain as much information of predictive use as possible. For this purpose, the PLSR algorithm uses the information contained in both the spectroscopic data matrix, **X**, and the concentration matrix, **Y**, during calibration and compresses data in such a way that the most variance in both **X** and **Y** is explained. In this way, PLSR reduces the potential impact of large, though irrelevant, variations in **X** during calibration. In PLSR, each component is obtained by maximizing the covariance between y and every possible linear function of **X**.

As in PCA, matrices **X** and **Y** are centered or autoscaled prior to resolution into factors. Matrix **X** is used to extract a few latent variables ($a = 1, 2, \ldots, A$). Thus, each matrix is resolved into a combination of A factors ($A \leq K$), which allows the simultaneous calculation of

$$\mathbf{X} = \mathbf{T}\mathbf{P}^{\mathrm{T}} + \mathbf{E} = \sum_{a=1}^{A} t_a p_a{}^{\mathrm{T}} + \mathbf{E} \tag{3.26}$$

$$\mathbf{Y} = \mathbf{U}\mathbf{Q}^{\mathrm{T}} + \mathbf{F} = \sum_{a=1}^{A} u_a q_a{}^{\mathrm{T}} + \mathbf{F} \tag{3.27}$$

With M samples, A factors, K variables and P analytes, matrices **T** ($M \times A$) and **U** ($M \times A$) will be the scores matrices for blocks **X** and **Y**, respectively; matrices $\mathbf{P}^{\mathrm{T}}$ ($A \times K$) and $\mathbf{Q}^{\mathrm{T}}$ ($A \times P$) the loadings matrices for blocks **X** and **Y**, respectively; and **E** and **F** the residuals matrices for blocks **X** and **Y**, respectively (Kramer, 1998; Martens and Martens, 2001; Naes *et al.*, 2002).

The process is started by calculating a small, though adequate, number of latent variables, $W(X)$ (loading weights), which are extracted from the variables in matrix **X**; the desired number of latent variables are stored in a scores matrix **T** which is used to iteratively model the variables in **X** and **Y** until convergence is reached. Similarly, the variables in **Y** can be modeled from those in **X** via the matrix of regression coefficients **B**. The coefficients in **B** can be estimated as a function of the loadings of **X** and **Y**, and **P** and **Q**, respectively, in addition to $W(X)$:

$$\hat{\mathbf{B}} = \mathbf{W}(\mathbf{P}^{\mathrm{T}}\mathbf{W})^{-1}\mathbf{Q}^{\mathrm{T}} \tag{3.28}$$

Unlike PCA, the loadings do not coincide fully with the direction of maximum variation since they have been corrected in order to maximize the predictive ability of matrix **Y**.

If only the concentration of one of the components in **Y** is to be determined, even if all others are known, then the algorithm, PLS1, is a simplified version of the complete algorithm and is designated as PLS2.

For calibration, the regressors matrix $\hat{\mathbf{B}}$ allowing a sample to be predicted without the need to resolve it into scores and loadings matrices is calculated. Thus, if the spectrum for a given sample is defined by vector x_i, the concentrations of the analytes y can be calculated from

$$\hat{y}_i = x_i{}^{\mathrm{T}}\hat{\mathbf{B}} \tag{3.29}$$

Figure 3.2 depicts the calibration and prediction processes.

Artificial neural networks

Occasionally, the mathematical relationship between spectra and the variable to be modeled is non-linear. Non-linearity in a model can result from spectrum-related

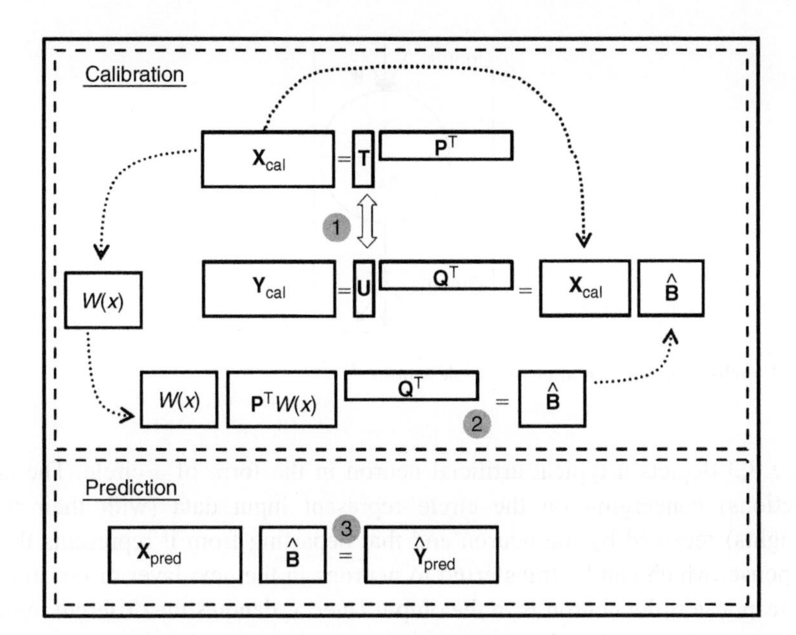

Figure 3.2 Schema for the calibration of a PLSR model and the prediction of new samples. Steps: (1) Simultaneous decomposition of **X** and **Y** matrix and calculation of weights matrix. (2) Calculation of regression matrix. (3) Prediction of new samples.

factors (e.g. deviations from the Beer–Lambert law at high analyte concentrations, a non-linear detector response). These deviations from linearity can be corrected by using an appropriate mathematical treatment. On the other hand, non-linearity arising from intrinsic characteristics of the target parameter can only be corrected by using non-linear calibration methods. A large number of methods for developing non-linear calibration models into especially prominents have been reported, among which are those based on *artificial neural networks* (ANNs). ANNs mimic the parallel processing capabilities of the human brain: a series of processing units (neurons) are used to convert input variable responses into a concentration (or property) output. Neural networks span a very wide range of techniques that are also used for a wide range of applications. An ANN can be defined as an iterative computational method intended to reproduce, in a simple manner, the network connecting neurons in the human brain (Zupan and Gasteiger, 1993).

Most quantitative applications of ANNs in chemical analysis rely on so-called *multilayer feed-forward networks* (MLFs), which use a back-propagation algorithm for learning. This type of algorithm is highly effective as it affords supervised learning (i.e. it can use a data set providing known responses to predict the responses of another data set).

The idea behind MLFs is very simple: a sample set for which the target parameter is known (calibration samples) is used to model the parameter as a function of the product of the measured variables by an appropriate statistical weight. Such a weight is iteratively calculated as the value minimizing the summation of the squares of the differences between the estimated and reference values.

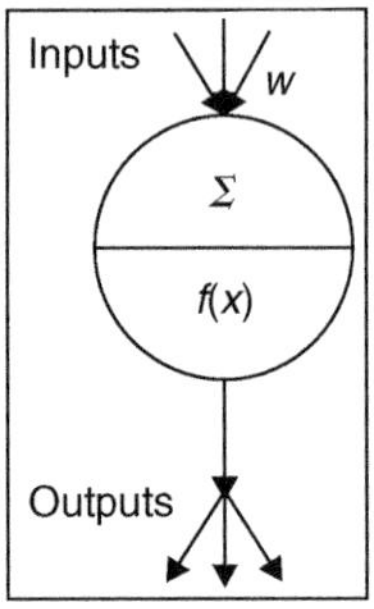

Figure 3.3 Schematic representation of a single neuron.

Figure 3.3 depicts a typical artificial neuron in the form of a circle. The arrows (connections) converging on the circle represent input data (with their respective weights) received by the neuron and that departing from it represents the neuron response, which can be transferred to neurons in the next layer or constitute the sought response if the neuron is in the output layer. x_i denotes the different inputs for a neuron j, w_{ij} the weight of the connection through which signal x_i enters neuron j and o_j the neuron's output. Because the neuron performs mathematical operations in two steps, it is often depicted with a horizontal line splitting it into two halves. In the first step, the neuron evaluates the combination of all weighted signals reaching it in order to calculate a net summation:

$$Net_j = \sum_i w_{ij} x_i + \theta_j \tag{3.30}$$

where parameter θ_j (bias) is a non-zero constant that is dealt with as another weight. Such a weight is also adjusted during the learning process and has a unity value for input data. The second step involves calculating the neuron output value, o_j, which need not coincide with Net_j and is related to it through a transfer function.

Architecture of neural networks

Neurons in an ANN are grouped into layers. The J neurons present in each layer receive an identical number of inputs I and hence possess an identical number of weights w_{ij} (with $i = 1, 2, \ldots, I$ and $j = 1, 2, \ldots, J$). All neurons in each layer receive a signal X ($x_1, x_2, \ldots, x_I$) from a series of I neurons in the previous layer (the input layer receives the input data for the network and therefore contains one neuron per datum to be processed). The Jo_j outputs of each layer are calculated simultaneously. In a multilayer architecture, the outputs are transferred to the next layer, which will contain K neurons each receiving J entries; therefore, the layer will have $J \times K$ weights and provide K outputs (as many as neurons it contains) for transfer to the next layer. The output values from the last layer constitute the network output. The process is illustrated in Figure 3.4, which depicts an artificial neuron comprising an input layer, a hidden layer and an output layer. The input layer contains three neurons, which are depicted as squares rather than circles as they perform no computations,

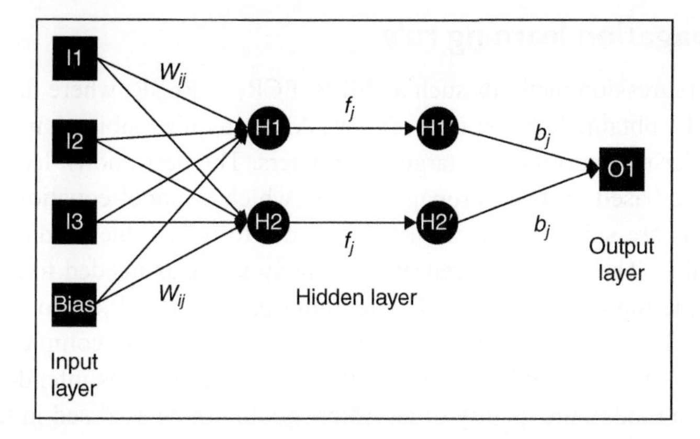

Figure 3.4 Schema for a typical network with three input, two hidden, and one output layers (3, 2, 1).

the two hidden layers and the output layer a single one. In order to avoid confusion with the weights for the hidden layer, w_{ij}, those for the output layer are denoted by b_j. The optimum number of layers for network and that of neurons for each layer are dictated by the particular application.

A network architecture can be represented by (i, hl, o), with i being the number of neurons in the input layer, hl that in the first hidden layer, and o that in the output layer. The network in Figure 3.4 can thus be represented as (3, 2, 1). Some authors use an alternative description including the number of neurons in the hidden and output layers, and stating the particular types of transfer functions used in both.

Transfer functions

The transfer functions used in ANNs can be linear or non-linear. Most often, they are of the sigmoidal (or logistic) type:

$$f(x) = 1/(1+e^{-x}) \tag{3.31}$$

However, sine, hyperbolic tangent, and simple linear functions are also widely used. The only conditions to be met by the function of choice are as follows: (a) the function should be distinguishable at every point in its domain; and (b) it should grow monotonically (i.e. it should only increase or decrease throughout its domain). The use of non-linear transfer functions in a multilayer neural network enables the modeling of non-linear relationships; this makes them especially attractive for modeling signals, whose relationships with the target parameter (whether linear, logarithmic, sinusoidal) are unknown. Sigmoidal and hyperbolic-tangent functions, which allow non-linear relationships to be fitted, are among the most widely used. The transfer functions applied to the output layer can also be linear or non-linear; however, the former are to be preferred as they possess a wider scope.

Back-propagation learning rule

Unlike other regression methods such as MLR, PCR, or PLSR, where the calibration equation can be obtained almost immediately, ANNs require subjection to a learning process in order to determine the target parameters. The best-known learning procedure is the one based on back-propagation, by which a model equation is obtained and its weights are subsequently adjusted in order to reduce the prediction error in the output value. The process is repeated as many times as needed to achieve convergence on the model parameters. The learning process is subject to specific rules intended to reduce the time needed to optimize the ANN, its complexity and the resulting error (Wythoff, 1983). Although the learning process is intended to reduce the error of the model, the quality of an ANN should not be assessed in terms of the calibration error obtained from the training samples, but rather by analysing a set of samples not used in the training step (a validation set).

The calculated structure of an ANN tends to stabilize during the learning process; in fact, the calculated error for the model tends to decrease insignificantly, which can result in overfitting (or overtraining) of the ANN. It is therefore essential to accurately identify the point in the learning process where the error is a minimum and stop it in order to avoid overfitting.

The learning process is intended to produce the known output vector Y from an input signal vector X. This entails using a series of input–output pairs (X_h, Y_h), where X_h can be a spectrum and Y_h the concentration of one analyte or several, and subscript h denotes a specific sample. While learning, the network calculates an Y'_h value for each X_h and compares it with the known value, Y_h; based on this comparison, the weights are modified in order to improve the consistency between Y_h and Y'_h. Each iteration in the process involves processing all (X, Y) pairs once and is repeated as many times as needed to ensure acceptable consistency of all (X_h, Y_h) pairs with the corresponding Y'_h outputs.

Weight correction can be done in various ways, but is usually based on the generalized delta rule: the correction ΔW to be introduced is assumed to be proportional to a parameter δ, which is in turn proportional to the error, and to the input vector X. Once the weight vector has been corrected, the output signal for X should be closer to the correct (known) value. The delta rule is defined mathematically as

$$\Delta W = \eta \delta X \tag{3.32}$$

where η is a proportionality constant known as the learning rate and dictates the rate at which weights are to be adjusted, and δ is the sought correction factor.

With the delta rule, a randomly chosen series of weights are used to calculate the outputs of the hidden layer from an input vector X (x_1, x_2, … , x_l) by using the following expression

$$o_j = f(Net_j) \tag{3.33}$$

where f can be a function of the sigmoidal or another type. Each o_j value thus obtained is used as an input for the neurons in the next layer in order to calculate the output for such a layer and so on until the output for the last layer in the network has been obtained. The output from the last layer, o_j^{last}, will be the network response. Once it is obtained, the error δ is calculated as the difference between the output (o_j^{last}) and the known value (y_j). If a linear transfer function has been used—as is usually the case for the output layer—then the error will be given by

$$\delta_j^{last} = y_j - o_j^{last} \tag{3.34}$$

If the transfer function is of the sigmoidal type, then the error will be the result of multiplying the previous one by the derivative of the sigmoidal function, $o_j^{last}(1 - o_j^{last})$:

$$\delta_j^{last} = (y_j - o_j^{last})o_j^{last}(1 - o_j^{last}) \tag{3.35}$$

With a non-sigmoidal function, the derivative will obviously be different (Wythoff, 1983).

Once the errors for all neurons in the last layer have been calculated, its weights, w_{ij}^{last}, are calculated from

$$\Delta W_{ij}^{last} = \eta \delta_j^{last} o_j^{last-1} + \mu \Delta W_{ij}^{last(prev)} \tag{3.36}$$

The above equation includes the output value for the last layer (o_j^{last}, which is included in δ_j^{last}) and that for the previous layer (o_i^{last-1}, which is the input value for the last layer). This equation results from expanding the delta rule with a second term containing constant μ (momentum) multiplied by the weight correction calculated in the previous iteration (with the subscript prev in brackets); this is intended to prevent abrupt changes in the direction of the corrections and the model from stopping at a local minimum as a result.

The errors δ_{ij}^{l} for each hidden layer from 1 = (last − 1) to 1 = 1 are then calculated, using the expression below:

$$\delta_j^l = \left(\sum_{k=1}^{r} \delta_k^{l+1} w_{jk}^{l+1} \right) o_j^l \left(1 - o_j^l\right) \tag{3.37}$$

where r denotes the number of neurons in the (1 + 1) layer. At that point, all weights w_{ij}^{l} in layer 1 are corrected by using

$$\Delta W_{ij}^{l} = \eta \delta_j^l o_j^{l-1} \mu \Delta w_{ij}^{l(prev)} \tag{3.38}$$

The previous procedure is then repeated with each new (X, Y) pair. Interested readers are referred to papers by Zupan and Gasteiger (1993), Despagne and Massart (1998), Naes *et al.* (1993), and López (1997) for a detailed description of the use of the delta rule.

The learning rate constant is very important as it dictates the rate at which weights are to be adjusted; the greater the constant is, the more often will weights have to be changed in each iteration. However, if weights vary too quickly, the process may stop at a local minimum or the prediction error oscillates or grows. On the other hand, too small changes can result in too slow learning by the network. No universal optimum value for this parameter exists as it depends on the particular network architecture and number of samples, among other factors. Nor is there an optimum value for the constant as it depends on that of the learning rate.

Although the ANN is a powerful tool for developing quantitative models, it is more susceptible to overfitting through the use of too many nodes in the hidden layer. Cross-validation techniques can be used to optimize the number of hidden nodes but this procedure is cumbersome because separate ANN models, with different number of hidden nodes, must be developed separately.

Finally, for the parameters of the ANN model there is very little or no interpretative value that eliminates any useful means for improving the confidence of a predictive model. These facts may explain why there are very few software packages for developing and implementing an ANN model.

Constructing multivariate calibration models

Because NIR spectroscopy is a relative technique, the samples used for calibration must be previously analyzed with adequate accuracy and precision. This entails using an instrument capable of remaining operational for a long time and a simple, robust enough model capable of retaining its predictive ability for new samples over long periods.

Constructing a multivariate calibration model is a complex, time-consuming process that requires careful selection of variables in order to ensure accurate prediction of unknown samples. This requires knowledge not only of the target samples, but also of chemometric techniques in order to obtain a model retaining its predictive ability over time and amenable to easy updating. Because the model will usually be applied by unskilled operators, it should deliver analytical information in an easily interpreted manner.

The process of obtaining a robust model involves the following steps: choosing the samples for inclusion in the calibration set, determining the property to be predicted by using an appropriate method to measure such samples, obtaining the analytical spectral signal, constructing the model, validating it and, finally, using it to predict unknown samples. Below is described in detail each step involved in the modeling of analytical data.

Selection of calibration samples

This is one of the most important steps in constructing a calibration model and involves choosing a series of samples, which ideally should encompass all possible sources of physical and chemical variability in the samples to be subsequently predicted. The model will only operate accurately if both the calibration samples and the prediction samples belong to the same population. Usually, the body of available samples is split into two subsets, one of which is the training or calibration set and is used to construct the model; and the other is the validation set and is used for validation. Variability in the samples used to construct the model is due to the body of factors affecting some property of the samples in such a way as to reflect in their spectra. Variability sources can be of diverse nature and origin. In any case, the samples included in the training set should be representative of the whole population and exhibit values of the target parameter uniformly spanning its potential range of variation. New samples will predict by interpolation within the model limits as no accurate prediction can be ensured by extrapolation.

One other key consideration in selecting samples to develop a calibration model is the potential presence of collinearity between the values of the parameters to be determined. Models established from collinear data are scarcely robust and can produce mathematical artifacts leading to spurious predictions (Deming and Morgan, 1987). Non-uniformity in some physical properties of the samples such as particle size or particle distribution may be an undesirable source of variability that can be corrected—but not suppressed—by spectral treatment.

Each sample is associated with two types of variables: independent (spectra) and dependent (the target parameter). The samples included in the calibration set should span the whole variability in both; thus, the selected samples should be uniformly distributed throughout the calibration range in the multidimensional space defined by spectral variability. One simple method for selecting samples based on spectral variability uses a scatter plot obtained from a PCA applied to the whole set of available spectra. Inspecting the most salient PCs in the graph allows one to clearly envisage the distribution of the sample spectra; those to be included in the calibration set are chosen from both the extremes and the middle of the score maps obtained and simultaneously checked to uniformly encompass the range spanned by the quantity to be determined. This method is effective when the first two or three PCs contain a high proportion of the total variance (Deming and Morgan, 1987).

A number of chemometric algorithms are currently available for selecting calibration samples in an efficient manner in accordance with the previous criteria. Such algorithms include the D-optimal (Ferré and Rius, 1997), Duplex (Fearn, 1997), OptiSim (Clark, 1997), Næs *et al.* (1990) and Kennard and Stone (1969) algorithms.

The Kennard–Stone algorithm, which aims at maximizing the Euclidian distance between the sample spectra, is probably the most popular one. The process is started by selecting a sample from the available set and then that exhibiting the greatest Euclidian distance from it. The process is repeated by using the second sample to calculate the distances from all others. The sample selected in each iteration should be at the maximum possible distance from those selected in all previous iterations.

The D-optimal algorithm selects samples in accordance with the particular calibration model to be used. To this end, it minimizes the variance of the regression coefficients. The optimum set will be that minimizing the determinant of matrix $(\mathbf{X}^T\mathbf{X})^{-1}$. Unlike the Kennard–Stone algorithm, where samples are selected at random and the choice of the initial sample determines the composition of the final set, the D-optimal algorithm selects samples in accordance with the equation of the particular model. However, this algorithm is much more complex in computational terms and a compromise between the amount of information sought and its cost must inevitably be made.

The Duplex algorithm is a modified version of the Kennard–Stone algorithm that allows both the calibration set and the prediction set to be established during the sample selection process.

OptiSim also uses a threshold for the minimum distance between two samples in an iterative process. Finally, the Næs–Isaksson algorithm performs a cluster analysis on the scores provided by a PCA for the spectra and selects as many samples as groups are defined in the clustering step.

The Kennard–Stone algorithm is among the most widely used and has been incorporated into some commercial software packages.

Reference methods and obtaining the analytical signal

Constructing a calibration model involves performing a multiple linear regression between the spectral target variables and those to be predicted, the value of which must be determined by using an appropriate reference method. The reference method used should provide accurate, precise values if the multivariate model finally developed is to be accurate as well. However, the precision of the model may be better than that of the reference values since regression averages random errors.

The analytical signal (viz. the body of spectra for the samples used to construct the model) should be obtained with the same instrument and under identical conditions as those subsequently used for routine analysis in order to ensure that all spectral will contain the same sources of instrumental variability. To this end, if the target model is to be implemented in two or more different instruments, then one can record the spectra for the samples in each instrument and use all to construct the model.

The essential condition for spectra to be useful for constructing calibration models is that they should contain the information to be modeled, which is not always the case. Thus, the samples used to determine ash in flour, total salts in aqueous solutions, or metal ions in water typically possess properties that are not necessarily reflected in their IR spectra; this precludes constructing accurate calibration models from IR spectral parameters. One should bear in mind that chemometrics can extract information present in a data set, but not create it from scratch. Also, the amount of information contained in the set should be large enough to allow the development of models with an adequate predictive ability for the target parameters.

Calculation of the calibration model

Calculating a calibration model involves processing the analytical signal in order to establish its most simple possible relationship with the target parameter (whether an analyte concentration or some physical property of the sample) (Martens and Martens, 2001; Naes *et al.*, 2002).

The aim of calibrating is to calculate the parameters in an equation allowing a property in future, unknown samples to be accurately determined (i.e. with as small as possible a departure from the actual values). The quality of calibration models can be assessed via some statistical parameters, of which those allowing the mean error for the whole population rather than a single sample are to be preferred. The statistics typically used to assess the quality of calibration models calculate the error of prediction as the summation of the squares of the residuals,

$$\sum_{i=1}^{n}(\hat{y}_i - y_i)^2$$

which is usually designated as the predicted residual error sum of squares (PRESS), or its mean value, which is obtained by dividing PRESS by the number of samples (n) to obtain the mean square error (MSE):

$$\mathrm{MSE} = \frac{\sum_{i=1}^{n}(\hat{y}_i - y_i)^2}{n} \tag{3.39}$$

Some authors use the square root of MSE, which is called the root mean square error (RMSE), which is defined as

$$\mathrm{RMSE} = \sqrt{\frac{\sum_{i=1}^{n}(\hat{y}_i - y_i)^2}{n}} \tag{3.40}$$

or the relative standard error (RSE), defined as

$$\mathrm{RSE} = \sqrt{\frac{\sum_{i=1}^{n}(\hat{y}_i - y_i)^2}{\sum_{i=1}^{n}(y_i^2)}} \cdot 100 \tag{3.41}$$

Calibration models are usually constructed from two sample sets (viz. a calibration set and a validation or test set) that are used to calculate MSEP or RMSEP for each principal component, or, alternatively, the equivalent parameters for the calibration set (MSEC and RMSEC). In some cases, MSE is calculated by dividing PRESS by the actual number of degrees of freedom, $n - 1 - a$ (where a is the number of PCs for which the target parameter is calculated), rather than the number of samples in the calibration set.

Spectral scaling and pre-treatments

Pre-treating the analytical signal is intended to suppress the effect on contributions not associated with the information sought from the spectra in order to increase the accuracy and precision of the results. Although spectral signal pre-treatments reduce the contribution of noise, their efficiency depends on the nature of the noise and the specific treatment used. Typical examples of signal processing methods include spectral filtering by use of the Fourier transform and baseline correction methods. These pre-treatments are commonly used in IR spectroscopy, and are required in order to obtain simple, robust models with an acceptable predictive ability; however, the best choice in each situation must be chosen in an empirical manner, using a trial-and-error approach, which is a major disadvantage.

The data-processing treatments used in this context are essentially of two types: spectral pre-treatments (derivative spectra, standard normal variate, multiplicative scattering correction), which are applied to individual spectra; and spectral scaling (mean-centering, autoscaling), which are applied to each individual variable in all selected samples. The data to be used in order to construct a calibration model must almost inevitably be scaled; however, no universal arguments in favor of one specific treatment over the others have been formulated, so the choice is usually dictated by the quality of the resulting model and its predictive ability.

Below are briefly outlined the spectral pre-treatments in widest use at present. A more detailed description is provided in Chapter 2.

Mean centering

This involves subtracting the average response from each individual response for the variable concerned:

$$X_{mc} = X - X_{avg} \tag{3.42}$$

This allows variables to be centered and constant effects suppressed as a result.

Autoscaling

This includes initial centering of the data and subsequent division of each centered variable by the standard deviation for the responses of the variable concerned:

$$X_{auto} = \frac{X - X_{avg}}{X_{sd}} \tag{3.43}$$

Autoscaled variables are centered in zero and possess a unity standard deviation.

Derivation

Derivation of spectra exposes the variation of the absorbance with wavelength. This treatment suppresses constant offset and slope variations in spectra. The Savitzky–Golay algorithm (Savitzky and Golay, 1964), which is the most usual choice for this purpose, usually involves smoothing data by fitting a polynomial expression to the data within the moving window selected for derivation. This entails previously defining the number of points to be included in the window, and the order of the convolution polymer and that of the derivative to be obtained. The first and second spectral derivatives reduce differences in baseline and slope, respectively, between spectra. Higher order derivatives are not recommended, however, as they increase noise in the signal and reduce its magnitude.

Multiplicative scattering correction (MSC)

MSC is used to minimize the additive and multiplicative effects of scatter arising mainly from differences in particle size between samples. This is done by calculating the slope, a, and offset, b, of the regression between each individual spectrum and a reference spectrum (usually, the average spectrum for the calibration set):

$$X = aX_{\text{ref}} + b \tag{3.44}$$

Coefficients a and b are used to correct each spectrum by using the expression below:

$$X_{corr} = \frac{X - b}{a} \tag{3.45}$$

Standard normal variate (SNV)

SNV treatment autoscales each spectrum by calculating the mean and standard deviation between the absorbances for the spectrum:

$$X_{SNV} = \frac{X - X_{avg}}{X_{sd}} \tag{3.46}$$

This pre-treatment also reduces the additive and multiplicative effects of scattering. The SNV and MSC treatments are linearly related and provide similar results. According to Dhanoa *et al.* (1994), the two are equivalent. However, while the output of MSC is influenced by the set of samples used for correction (i.e. it is set-dependent), that of SNV is not (i.e. it is set-independent).

Selection of the calibration method

Once the pre-treatment of choice has been applied, the calibration model is constructed on the grounds of the particular relationship between the analytical signal and the property to be determined (e.g. absorbance via the Beer–Lambert law) or of an empirical relationship. As stated above, a variety of mathematical algorithms are available for constructing models and a wide range of statistical techniques exist for their assessment and optimization.

Current spectrophotometers allow spectra containing thousands of variables for each sample to be obtained; this, together with the fact that samples are usually available in hundreds, makes variable-reduction calibration methods (PCR, PLSR, ANN) the best choices, not only because they reduce variables, but also because they suppress noise. However, filtering instruments that can record spectra at only a few wavelengths continue to be in use for which a calibration method involving no variable reduction (e.g. MLR) may be more effective.

Selection of the spectral range

Choosing the most suitable spectral range for developing a calibration model is not an easy task and frequently it involves an endless sequence of trial-and-error runs until an adequate predictive ability is achieved. When the spectra for the target analyte and its potential interferents in pure form are available, one can choose the range where the analyte exhibits substantial bands and exclude those where the interferents absorb. However, this approach is useless in the NIR region, where bands are typically narrow and strongly overlapped, and the analyte signal is therefore easily concealed by the signal for the sample matrix. In this situation, one can simply calculate the correlation between the absorbance at each wavelength and the target quantity in order to plot the resulting vector against the independent variable. Those intervals exhibiting the strongest correlation can be of help with a view to selecting an appropriate range to develop the model. This approach is used by some software packages to construct MLR models from continuous spectra (Centner *et al.*, 1996; Xu and Schechter, 1996).

Some algorithms allow the most suitable spectral variables for modeling the independent variable to be identified. Specially effective for this purpose is the jack-knife algorithm (Martens and Martens, 2000), which conducts successive computations in alternately suppressed and incorporated spectral subranges. The algorithm calculates the error of the model after each step and the spectral subrange used in each step is adopted or rejected depending on whether it substantially increases or decreases the quality of the model.

Selection of the number of factors (PCs or latent variables)

Choosing the optimum number of factors, principal components (PCR) or latent variables (PLSR) for defining a model is a key step in any calibration process involving variable reduction. Such a number can be selected in various ways, most of which rely on the error of prediction obtained with variable numbers. Using a smaller than optimal value leads to underfitting of the model and hence to large errors, whereas the opposite results in overfitting and leads to increased noise and large errors as well. With small numbers of samples, the method of choice is usually cross-validation, which uses samples from the calibration set to check the goodness of fit of the model. Thus, the calibration set is split into several blocks or segments and the model is constructed as many times as segments are established, using a segment as data block in order to check the results and the other segments to develop the model; in this way, one segment is left out each time. The process is performed for each individual factor in order to calculate MSE for each segment and the output is accumulated in order to obtain a reliable estimate of the predictive ability of the calibration samples. When the number of segments used equals that of calibration samples, the procedure is known as the "leave-one-out method." In each run, one sample is left out and all others are used to construct the model, the process being repeated as many times as samples are in the calibration set. The resulting mean square error of prediction by cross-validation (MSECV) is calculated from

$$\mathrm{MSECV} = \frac{\sum_{i=1}^{n} (\hat{y}_i - y_i)^2}{n} \tag{3.47}$$

One simple way of selecting the optimum number of factors involves plotting MSECV (or indeed any other error descriptor) against the number of factors and choosing the one corresponding to the minimum of the curve. This entails assuming that the error decreases with increasing number of factors up to a point where further factors contribute mainly to noise and MSECV rises by effect of overfitting. Although in theory it seems reasonable, using a limited number of samples inevitably introduces some error in the resulting model.

Figure 3.5 shows the evolution of the calibration error and two different cases of prediction error according to the number of factors used in the model. As can be seen, the calibration error decreases continuously as the number of factors increases. Case no. 1 of prediction error reveals that no minimum can be observed, however, the use of five factors for case no. 2 is clearly justified due to the presence of a minimum of prediction error (i.e. the best predictive ability situation).

In the absence of a minimum, the optimum number of components can be chosen by using the criterion of Haaland and Thomas (1988), which involves selecting the number of factors, with which MSECV is not significantly different from the minimum value for the model. The minimum MSECV will be produced by a number of components a^*, with the MSECV value obtained with a number of factors smaller than a^* being compared with the minimum value via an F-test. This involves calculating

$$F(a) = \frac{\text{MSECV}(a)}{\text{MSECV}(a^*)} \tag{3.48}$$

for each component $a = 1, 2, \ldots, a^*$ and choosing as optimal the smallest number resulting in $F(a) < F_{crit}$, where F_{crit} is the value at a significance level $\alpha = 0.25$ for a and a^* degrees of freedom.

Another recommended procedure, which is in fact the best if a large enough number of samples are available, involves using a set of validation samples not used to construct the model (i.e. test set) and predicting their values with models differing in the number of factors used. The number of choice will be that resulting in the lowest RMSEP.

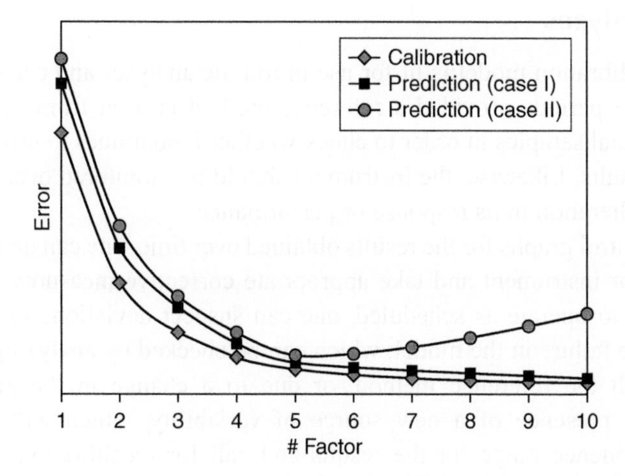

Figure 3.5 Calibration and prediction errors versus number of factors for two current situations: case I, no minimum is observed in prediction error; case II, a clear minimum in prediction error.

Outliers

An outlier can be defined as any observation not fitting the model. One may encounter three different types of outliers in developing a multivariate calibration model, namely: (a) *X*-sample outliers (viz. samples, for which the spectra depart markedly from those for the others); (b) *Y*-sample outliers (viz. samples, for which the model provides a target value considerably different from the actual value); and (c) *X*-variable outliers (viz. spectral variables that behave markedly differently from the others).

It should be noted that the word "outlier" need not be synonymous with "incorrect"; however, one should always ascertain whether an outlier is the result of an actual phenomenon or an artifact arising from some error while constructing the calibration model. In fact, identifying and suppressing outliers is of utmost importance since their presence can adversely affect the robustness and predictive ability of the resulting model. Outliers can be detected by using a number of available methods, and a detailed description is beyond the scope of this chapter (Martens and Naes, 1989; Naes *et al.*, 2002).

Validation of the model

Once the calibration model has been developed, its ability to predict unknown samples not present in the calibration set (i.e. not used to construct the model) should be assessed. This involves applying the model to a limited number of samples not included in the calibration set, for which the target property to be predicted by the model is previously known. The results provided by the model are directly compared with the reference values; if the two are essentially identical, the model will afford accurate predictions and be useful for determining the target property in future (i.e. unknown samples). The quality of the model can be assessed via the above-described statistics. Usually, a model is deemed accurate if it provides an RMSEP value not exceeding 1.4 times the standard error of the laboratory (SEL).

Routine analyses

A validated calibration model is fit for use in routine analyses and can be used unaltered over long periods provided a reference method is used from time to time to analyze anecdotal samples in order to check whether it continues to produce accurate and precise results. Likewise, the instrument should be monitored over time in order to detect any alteration in its response or performance.

By using control graphs for the results obtained over time, one can detect deviations in the model or instrument and take appropriate corrective measures. If the instrument is found to operate as scheduled, one can suspect deviations in the results to be due to some failure in the model, which can be checked by analyzing the samples concerned with the reference method, or due to a change in the target samples caused by the presence of a new source of variability, which will reflect in an expanded confidence range for the results and call for recalibration of the model. Recalibration can be done by expanding the calibration set with the samples used to expose the deviations.

Conclusions

IR spectroscopy, and particularly the NIR, presents doubtless advantages in relation to other spectroscopic techniques due to the ability to obtain the spectral data without the need of sample pre-treatment, and therefore provides fast and reproducible analytical methods. The spectrum not only includes chemical information of the sample components, but also its molecular interactions are revealed by physical properties of the sample; this characteristic does not allow to select individual wavelengths related to the parameters to be quantified. Therefore the use of multivariate calibration techniques is needed to extract the relevant chemical/physical information and obtain proper quantitative models. The multivariate calibration methods are more complex and demand a greater effort for their development in order to obtain chemometric models with a suitable predictive ability; nevertheless this difficulty can be partly compensated by the doubtless advantages in the quality and reliability of the results that they provide.

A wide variety of mathematical algorithms for the development of calibration models are available, with different requirements of application but with also different capacity to model the independent variables, for the obtaining of models with a good predictive ability of external samples. The modern instrumentation incorporates software adapted for the construction of these models and its application to new samples through their spectra. The most important characteristics of these algorithms have been described in this chapter.

In relation to the characteristics of the obtained models, it is important to emphasize that they should be robust, so that they can be used during long periods of time and to make the important effort done profitable. Therefore it is necessary to provide easy tools to update the models periodically (or when they show deviations) in a simple way, with the same samples used in the analytical control. Although it is difficult to establish precise rules for the development of calibration models, the most important aspect to consider in their construction is the detection and elimination of outliers by their negative effect in the predictive capacity of the constructed models. A complete validation of the model is necessary to assure the quality before coming to its application in routine.

Nomenclature

^	means calculated term
T	superscript symbol means transposed
−1	superscript symbol means inversed
*	superscript symbol means prediction set (independent of calibration set)
X	matrix of the independent variables (e.g. spectra)
x	vector of the independent variables (e.g. spectrum)
Y and **C**	matrices of the dependent variables (e.g. concentrations)
y and *c*	scalars of the dependent variable (e.g. concentration)

B	matrix of regression coefficients
b	vector of regression coefficients
T and **U**	scores matrices
t and **u**	score vectors
P and **Q**	loadings matrices
p and **q**	loading vectors
E and **F**	residual matrices
e and **f**	residual vectors
w	weights of neuron connection
θ	bias
ΔW	delta rule
δ	correction factor
o	output dependent value
μ	momentum
k	number of wavelengths
a	number of principal components and/or latent variables
n	number of samples
P	number of mixture components
$i, j, k, 1, \ldots$	running indices

References

Barnett HA, Bartoli A (1960) Least-squares treatment of spectrophotometric data. *Analytical Chemistry*, **32**, 1153.

Beebe KR, Kowalski BR (1987) Introduction to multivariate calibration and analysis. *Analytical Chemistry*, **59**, 1007A.

Centner V, Massart DL, de Noord OE, de Jong S, Vandeginste BM, Sterna C (1996) Elimination of uninformative variables for multivariate calibration. *Analytical Chemistry*, **68**, 3851.

Clark RD (1997) Optisim: an extended dissimilarity selection method for finding diverse representative subsets. *Journal of Chemometrics*, **37**, 1181–1188.

Coello J, Maspoch S (2007) *Calibración Multivariable Cap 6 in Temas Avanzados de Quimiometria.* Spain: Ed. U. Illes Balears.

Deming SN, Morgan SL (1987) *Experimental Design: A Chemometrics Approach.* New York: Elsevier, p. 147.

Despagne F, Massart DL (1998) Neural networks in multivariate calibration. *Analyst*, **123**, 157R.

Dhanoa DJMS, Lister SJ, Sanderson R, Barnes DJ (1994) The link between Multiplicative Scatter Correction (MSC) and Standard Normal Variate (SNV transformations of NIR spectra. *Journal of Near Infrared Spectroscopy*, **2**, 43.

Fearn T (1997) Validation. *NIR News*, **8**, 7–8.

Ferré J, Rius FX (1997) Constructing D-optimal designs from a list of candidate samples. *Trends in Analytical Chemistry*, **16**, 70–73.

Frans SD, Harris JM (1985) Selection of analytical wavelengths for multicomponent spectrophotometric determinations. *Analytical Chemistry*, **57**, 2680.

Geladi P, Kowalski BR (1986) Partial least-squares regression: a tutorial. *Analytica Chimica Acta*, **185**, 1.

Haaland DM, Thomas EV (1988) Partial least-squares methods for spectral analysis. 1. Relation to other quantitative calibration methods and the extraction of qualitative information. *Analytical Chemistry*, **60**, 1193.

Kennard RW, Stone LA (1969) Computer aided design of experiments. *Technometrics,* **11**, 137–148.

Kramer R (1998) *Chemometric Techniques for Quantitative Analysis*. Basel, Switzerland: Marcel Dekker.

López VM (1997) Análisis de componentes principales no lineales mediante redes neuronales artificiales de propagación hacia atrás: aplicaciones del modelo de Kramer. PhD thesis, Universitat Ramón Llull, Barcelona.

Mark H (1991) *Principles and Practice of Spectroscopic Calibration.* Toronto: John Wiley and Sons.

Mark H, Workman J (2003) *Statistics in Spectroscopy*, 2nd edn. London: Elsevier.

Martens H, Martens M (2000) Modified jack-knife estimation of parameter uncertainty in bilinear modelling by partial least squares regression. *Food Quality and Preference*, **11**, 5–16.

Martens H, Martens M (2001) *Multivaried Analysis of Quality. An Introduction.* Chichester: John Wiley & Sons.

Martens H, Naes T (1989) *Multivariate Calibration.* Chichester: John Wiley & Sons.

Massart DL, Vandegiste BGM, Deming SN, Michotte Y, Kaufman L (1988) *Chemometrics: A Textbook.* Amsterdam: Elsevier.

Meehan EJ (1981) In: *Treatise on Analytical Chemistry*, part 1, Volume 7 (Elving PJ, Meehan EJ, Kolthoff IM, eds). New York: Wiley, chapter 2.

Naes T, Kvaal K, Isaksson T, Miller C (1993) Artificial neural networks in multivariate calibration. *Journal of Near Infrared Spectroscopy*, **1**, 126.

Naes T, Isaksson T, Kowalski BR (1990) Locally weighted regression and scatter correction for near-infrared reflectance data. *Analytical Chemistry*, **62**, 664–673.

Naes T, Isaksson T, Fearn T, Davis T (2002) *Multivaried Calibration and Classification.* Chichester: NIR Publications.

Savitzky A, Golay MJE (1964) Smoothing and differentiation of data by simplified least-squares procedures. *Analytical Chemistry*, **36**, 1627–1639.

Stenberg JC, Stillo HS, Schwedeman RH (1960) Spectrophotometric analysis of multicomponent systems using the least squares method in matrix form. *Analytical Chemistry*, **32**, 84.

Thomas EV (1994) A primer on multivariate calibration. *Analytical Chemistry*, **66**, 795A.

Thomas EV, Haaland DM (1990) Comparison of multivariate calibration methods for quantitative spectral analysis. *Analytical Chemistry*, **62**, 1091.

Wold H (1975) Soft modeling by latent variables. In: *The Nonlinear Iterative Partial Least Squares Approach* (Gani J, ed.). Perspectives in Probability and Statistics. London: Academic Press, pp. 520–540.

Wythoff BJ (1983) Backpropagation neural networks. A tutorial. *Chemometrics and Intellingent Laboratory Systems*, **18**, 115.

Xu L, Schechter I (1996) Wavelength selection for simultaneous spectroscopic analysis. experimental and theoretical study. *Analytical Chemistry*, **68**, 2392.

Zupan J, Gasteiger J (1993) *Neural Networks for Chemists; An Introduction.* Weinheim: VCH.

Multivariate Classification for Qualitative Analysis

4

Davide Ballabio and Roberto Todeschini

Introduction

Classification methods are fundamental chemometric techniques designed to find mathematical models able to recognize the membership of each object to its proper class on the basis of a set of measurements. Once a classification model has been obtained, the membership of unknown objects to one of the defined classes can be predicted. While regression methods model quantitative responses on the base of a set of explanatory variables, classification techniques (classifiers) are quantitative methods for the modeling of qualitative responses. In other words, classification methods find mathematical relationships between a set of descriptive variables (e.g. chemical measurements) and a qualitative variable (i.e. the membership to a defined category).

Classification methods (also called supervised pattern recognition methods) are increasingly used in several fields, such as chemistry, process monitoring, medical

Infrared Spectroscopy for Food Quality Analysis and Control
ISBN: 978-0-12-374136-3

sciences, pharmaceutical chemistry, social and economic sciences. Of course, classification is acquiring higher and higher importance in food science too; quality control of production systems and tipicity of products are of increasing interest in the food industry since they represent recent requirements needed to compete in the present-day market. There is a need in the food industry to rationalize and improve quality and process controls; modern production systems require rapid and automatic on-line monitoring, which should be able to extract the maximum amount of available information, in order to assure optimal system functioning. On the other hand, food products acquire a higher value when their tipicity is protected, controlled and assured. As a consequence, it is clear how the development of reliable methods for assuring authenticity is becoming very important and several efforts have been made to authenticate the origin of food products, with different chemical and physical parameters and on several food matrices.

Classification methods appear as optimal tools for facing these purposes, where a qualitative response is studied and modelled. For example, consider a process where different chemical parameters are monitored in order to check the final product quality. Each product can be defined as acceptable or not acceptable on the basis of its chemical properties, i.e. each product (object) can be associated to a qualitative binary response (yes/no). A classification model would be the best way to assign the process outcome to one of the defined classes (acceptable or not acceptable) by using the monitored parameters. Furthermore, consider a consortium that wants to characterize a high-quality food product on the basis of different chemical and physical parameters, in order to assure geographical origin and uniqueness to the product. As before, a classification model can be used in order to distinguish the considered food product from products belonging to other geographical areas. In this model, each object can be associated to a class on the basis of its provenience; when the model will be applied on unknown samples, each new object will be assigned to one of the considered geographical groups. Given these premises, in the following sections the best-known classification techniques will be described, together with some elucidations on the evaluation of classification results.

Principles of classification

The classes

Consider n objects, each described by p variables and divided into G categories (classes); in order to build classification models, these data must be collected in a matrix $\mathbf{X}$, composed of n rows (the objects), and p columns (the explanatory variables). Each entry x_{ij} represents the value of the j-th variable for the i-th object. The additional information concerning the class is collected into an n-dimensional vector $\mathbf{c}$, constituted by G different labels or integers, each representing a class. In most cases, classification methods directly use the class information collected in the $\mathbf{c}$ vector; however, in order to apply certain classification methods, such as partial least squares discriminant analysis and some artificial neural network (ANN) methods, the class vector **c** must be unfolded into a matrix **C**, with n rows (the objects) and G columns (the unfolded class information); each entry c_{ig} of **C** represents the membership of the i-th object to the g-th class expressed with a binary code (0 or 1). Basically, the class unfolding is a procedure transforming an n-dimensional class vector representing G classes into a matrix constituted by n rows and G columns; an example of class unfolding is shown in Table 4.1.

Finally, note that the simplest representation of a single class is its centroid, which is a p-dimensional vector defined as the point whose variables are the mean of the variables of all the objects belonging to the considered class.

Main categories of classification methods

Statisticians and chemometricians have proposed several classifiers, with different characteristics and properties. First, distinctions can be made among the different classification techniques on the basis of the mathematical form of the decision boundary, i.e. on the basis of the ability of the method to detect linear or non-linear boundaries between classes. If a linear classification method is used, the model calculates the best linear boundary for class discrimination, while non-linear classification methods find the best curve (non-linear boundary) for separating the classes.

Moreover, classification techniques can be probabilistic, if they are based on estimates of probability distributions, i.e. a specific underlying probability distribution in the data is assumed. Among probabilistic techniques, parametric and non-parametric methods can be distinguished, when probability distributions are characterized by location and dispersion parameters (e.g. mean, variance, covariance). Classification methods can also be defined as distance-based, if they require the calculation of distances between objects or between objects and models.

Another important distinction can be made between pure classification and class-modeling methods. Pure classification techniques separate the hyperspace in as many regions as the number of available classes. Each object is classified as belonging to the category corresponding to the region of hyperspace where the object is placed. In this way, objects are always assigned to one of the defined classes. For example, in order to discriminate Italian and French wines on the basis of chemical spectra, a pure classification method can be used to predict the origin of unknown wines.

Table 4.1 Example of class unfolding

Object	Class	Class unfolding				
		Class 1	Class 2	Class 3	...	Class G
1	1	1	0	0	...	0
2	1	1	0	0	...	0
3	2	0	1	0	...	0
4	2	0	1	0	...	0
...	...	...	...	...	...	...
n	G	0	0	0	...	1

These new samples will be always recognized as Italian or French, even if they belong to other countries. As a consequence, when pure classification techniques are applied, it is important to assure that the unknown objects to be predicted belong to one of the classes used in the model calculation. On the other hand, class-modeling techniques represent a different approach to classification, since they focus on modeling the analogies among the objects of a class, defining a boundary to separate a specific class from the rest of the hyperspace. Each class is modeled separately and objects fitting the class model are considered element of the class, while objects that do not fit are recognized as non-members of that class. As a consequence, a particular portion of the data hyperspace can be enclosed within the boundaries of more than one class or of none of the classes and three different situations can be encountered: objects can be assigned to a class, to more than one class or to none of the considered classes.

In Figure 4.1, an example of both pure classification and class modeling is shown on a data set including 60 objects described by two variables and grouped into three classes (Circle, Diamond, and Square). When a pure classification technique is applied (Figure 4.1a), the whole data space is divided into three regions, each of them representing the space of a defined category. Consider now three new unknown objects (T1, T2, and T3) that must be classified by means of this model. To do so, these objects are projected into the data space and assigned to the category corresponding to the region of hyperspace where they are placed. T1 and T2 will be assigned to class Circle, even if T1 is far from the Circle samples, while T3 will be recognized as a Diamond object, although it is equally distant from the centroids of the classes. In contrast, if a class-modeling method is applied, each class space is separated by a specific boundary from the rest of the data space, as shown in Figure 4.1b. Of course, the classification results will be different with respect to the previous model: the unknown object T2 will be assigned to class Circle (as before); T1 will not be assigned at all, since it is not placed in a specific class space; T3 can be considered a confused object, since it can be assigned to more than one class (Diamond and Square).

With respect to pure classification techniques, class-modeling methods have some advantages: it is possible to recognize objects that do not fall in any of the considered class spaces and consequently identify members of new classes not considered during the model calculation. Furthermore, as each class is modeled separately, any additional class can be added without recalculating the existing class models.

Finally, it should be noted that unsupervised pattern recognition methods, such as principal components analysis (PCA) (Jolliffe, 1986) and cluster analysis (Massart and Kaufman, 1983), must not be confounded with classification methods (supervised pattern recognition). PCA is a well-known multivariate technique for exploratory data analysis, which projects the data in a reduced hyperspace, defined by the principal components. These are linear combinations of the original variables, with the first principal component having the largest variance, the second principal component having the second-largest variance, and so on. Cluster analysis differs from PCA in that the goal is to detect similarities between objects and find groups in the data on the basis of calculated distances, whereas PCA does not focus on how many groups will be found. Consequently, both PCA and cluster analysis do not use information related to predefined classes of objects. On the other hand, supervised pattern

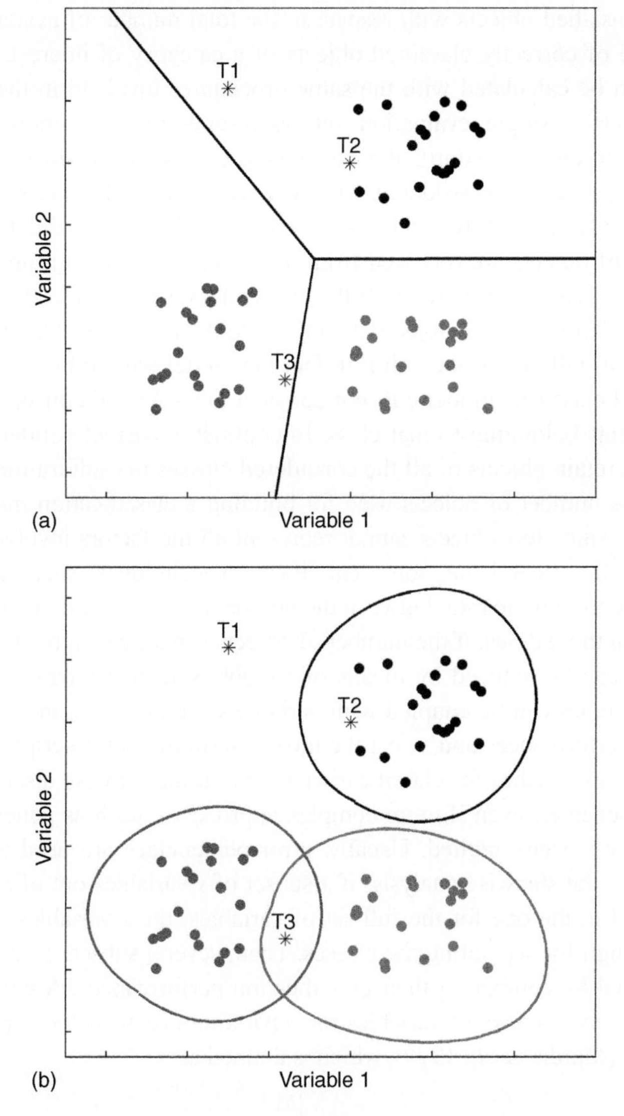

Figure 4.1 Example of both pure classification (a) and class modeling (b) on a data set including 60 objects described by two variables and grouped into three classes (Circle, Diamond, and Square).

recognition requires a priori information on the set of samples that is used for classification purposes.

Validation and variable selection procedures

As well as for regression models, classifiers require cross-validation procedures to analyze the predictive classification capabilities on unknown objects. Obviously, the prediction ability estimation of classification models is performed on different parameters with respect to regression methods, since the modeled response here is qualitative and not quantitative. In any case, several parameters can be used, e.g. the percentage of correctly classified objects with respect to the total number of available objects or the percentage of correctly classified objects of a category of interest. Even if these parameters can be calculated with the same procedures involved in the validation of regression models (single evaluation set, leave-one-out, leave-more-out, repeated training/test splitting, bootstrap), the percentage of objects retained in each cross-validation group has to be considered, when classification models are validated.

Consider a data set with two classes (A and B) and a cross-validation procedure, where groups of objects are removed from the training set, one group at a time, and used to test the classification model. If the entire class A is removed from the data set during the validation (all the objects belonging to A are used to test the model), the validation result will be unsuccessful; in fact the model will be built without objects of the removed class (the model will not consider class A) and consequently will not recognize objects belonging to that class. In contrast, a correct validation procedure should at least retain objects of all the considered classes in each training group.

However, the number of objects used for building a classification model is usually a critical issue, since few objects cannot represent all the factors involved in the class variability. On the other hand, some classification techniques, such as discriminant analysis, can be used if the ratio between the number of objects and the number of variables is high. In these cases, if the number of objects cannot be augmented, the number of descriptors can be reduced by means of variable selection methods. In fact, classification techniques can be coupled with variable selection tools, in order to improve classification performances and select the most discriminating descriptors. The majority of selection approaches for classification are based on stepwise discriminant analysis or similar schemes, even if more complex approaches, such as genetic algorithms, can be (and have been) applied. Usually, error percentages are used as an informal stopping rule in the stepwise analysis; if a subset of s variables out of p gives a lower error compared to the one for the full set of variables, the s variables can be considered to be enough for separating the classes. Then, several subsets of decreasing sizes can be evaluated by comparing their classification performances. A common strategy for selecting the best subset of variables for separating groups is the application of the Wilks' lambda (Mardia *et al.*, 1979), which is defined as:

$$\Lambda = \frac{|\mathbf{W}|}{|\mathbf{W} + \mathbf{B}|} \tag{4.1}$$

where **W** and **B** are the within and between sum of squares, respectively. Wilks' lambda ranges between 0 and 1, where values close to 0 indicate that the group means are different. Consequently, the variables with the lowest Wilks' lambda values can be retained in the classification model.

Evaluation of classification performances

As explained before, several parameters can be used for the quality estimation of classification models, both for fitting and validation purposes (Frank and Todeschini, 1994). Of course, these parameters are related to the presence of errors in the results

Table 4.2 General representation of a confusion matrix

		Assigned class				
		1	2	3	...	G
True class	1	n_{11}	n_{12}	n_{13}	...	n_{1G}
	2	n_{21}	n_{22}	n_{23}	...	n_{2G}
	3	n_{31}	n_{32}	n_{33}	...	n_{3G}
	...	...	...	...	...	...
	G	n_{G1}	n_{G2}	n_{G3}	...	n_{GG}

(objects assigned to the wrong classes), even if errors can be considered with different weights on the basis of the classification aims. All the classification indices can be derived from the *confusion matrix*, which is a square matrix with dimensions $G \times G$, where G is the number of classes. A general representation of a confusion matrix is given in Table 4.2, where each entry n_{gk} represents the number of objects belonging to class g and assigned to class k. Consequently, the diagonal elements n_{gg} represent the correctly classified objects, while the off-diagonal elements represent the objects erroneously classified. Note that the confusion matrix is generally asymmetric since n_{gk} is different from n_{kg}, i.e. the number of objects belonging to class g and assigned to class k is not usually equal to the number of objects belonging to k and assigned to g.

By looking at the confusion matrix (built on fitting or validated outcomes), we can have an idea on how a classification model is performing; of course, some more informative indices can be derived in order to synthesize this information. First, the *non-error rate* (NER) can be defined as follows:

$$\text{NER} = \frac{\sum_{g=1}^{G} n_{gg}}{n} \tag{4.2}$$

where n is the total number of objects. The non-error rate (also called *accuracy* or *classification rate*) is the simplest measure of the quality of a classification model and represents the percentage of correctly assigned objects. The NER complementary index is called the *error rate* (ER); it is the percentage of wrongly assigned objects and is defined as:

$$\text{ER} = \frac{n - \sum_{g=1}^{G} n_{gg}}{n} = 1 - \text{NER} \tag{4.3}$$

NER and ER can simply describe the performance of a model, but the result of a classification tool should be considered suitable in a statistic point of view when the classification ability is significantly larger than that obtained by random assignments to the classes. Thus, the model efficiency can be evaluated by comparing ER with the *no-model error rate* (NOMER), which represents the error rate obtained by assigning all the objects to the largest class and can be calculated as follows:

$$\text{NOMER} = \frac{n - n_M}{n} \tag{4.4}$$

where n_M is the number of objects belonging to the largest class. On the other hand, the error rate can also be compared with the error obtained with a random assignation to one of the defined classes:

$$\text{Random ER} = \frac{\sum_{g=1}^{G} \left(\frac{n - n_g}{n} \right) \cdot n_g}{n} \tag{4.5}$$

where n_g is the number of objects belonging to the g-th class:

$$n_g = \sum_{k=1}^{G} n_{gk} \tag{4.6}$$

Moreover, a different weight can be assigned to each kind of error. Consider for example the quality control step of a generic food process, where acceptable and non-acceptable products are recognized by means of classification and it is preferable to classify acceptable products as non-acceptable rather than the opposite. In this case, a penalty matrix, called *loss matrix* **L**, can be defined. The loss matrix is a $G \times G$ matrix, with diagonal elements being equal to zero and off-diagonal elements representing the user-defined costs of classification errors. Therefore, the *misclassification risk* (MR) can be defined as an estimate of the misclassification probability that takes into account the error costs defined by the user:

$$\text{MR} = \sum_{g=1}^{G} \frac{\left(\sum_{k=1}^{G} L_{gk} n_{gk} \right) \cdot P_g}{n_g} \tag{4.7}$$

where P_g is the *prior class probability*, usually defined as $P_g = 1/G$ or $P_g = n_g/n$.

There are also indices related to the classification quality of a single class. The *sensitivity* (Sn_g) describes the model ability to correctly recognize objects belonging to the g-th class and is defined as:

$$Sn_g = \frac{n_{gg}}{n_g} \tag{4.8}$$

If all the objects belonging to the g-th class are correctly assigned ($n_{gg} = n_g$), Sn_g is equal to 1. The *specificity* (Sp_g) characterizes the ability of the g-th class to reject the objects of all the other classes and is defined as:

$$Sp_g = \frac{\sum_{k=1}^{G} (n'_k - n_{gk})}{n - n_g} \quad \text{for} \quad k \neq g \tag{4.9}$$

Table 4.3 Example of confusion matrix

		Assigned class			
		A	B	C	
True class	A	9	1	0	10
	B	2	8	2	12
	C	1	2	5	8
		12	11	7	n = 30

Table 4.4 Classification parameters calculated on the example of Table 4.3

NER	0.73
ER	0.27
NOMER	0.60
Random ER	0.66
Sn(A)	0.90
Sn(B)	0.67
Sn(C)	0.63
Sp(A)	0.85
Sp(B)	0.83
Sp(C)	0.91
Pr(A)	0.75
Pr(B)	0.73
Pr(C)	0.71

where n'_k is the total number of objects assigned to the k-th class:

$$n'_k = \sum_{g=1}^{G} n_{gk} \tag{4.10}$$

If the objects not belonging to class g are never assigned to g, Sp_g is equal to 1. Finally, the class *precision* (Pr_g) represents the capability of a classification model not to include objects of other classes in the considered class. It can be measured as the ratio between the objects of the g-th class correctly classified and the total number of objects assigned to that class:

$$Pr_g = \frac{n_{gg}}{n'_g} \tag{4.11}$$

If all the objects assigned to class g correspond to the objects belonging to class g, Pr_g is maximum and is equal to 1. In Table 4.3 an example of confusion matrix is shown, and Table 4.4 shows the classification parameters calculated on the example of Table 4.3. Objects are grouped in three classes (10 samples in class A, 12 in class B and 8 in class C).

The parameters used for the evaluation of classification models with G classes have been defined in the previous part of this section. However, it is common to find with classification tasks that a given set of objects is divided into two categories (binary classification) on the basis of whether they cover some property or not. Common binary classification tasks are quality monitoring, to establish if a new product is good enough to be placed on the market or not, and process monitoring, where an outcome can be labeled as acceptable or not acceptable on the basis of defined standards. Binary classification thus takes into consideration only two classes, which can be labeled either as positive (P) or negative (N). Consequently there are four possible outcomes: true positives (TP) are the outcomes effectively recognized as positive, while if the outcome is N and the true value is P, then the outcome is called false negative (FN); true negatives (TN) are the outcomes that occur when both the assigned class and the true class are N, and false positive (FP) when the outcome is P and the true value is N. The four outcomes can be arranged in a 2×2 confusion matrix (or contingency table), as shown in Table 4.5.

In the case of binary classification, the previously described parameters can be defined as follows:

$$\mathrm{NER} = \frac{\mathrm{TP} + \mathrm{TN}}{n} \tag{4.12}$$

$$\mathrm{ER} = \frac{\mathrm{FN} + \mathrm{FP}}{n} \tag{4.13}$$

$$\mathrm{TPR} = Sn = \frac{\mathrm{TP}}{\mathrm{TP} + \mathrm{FN}} \tag{4.14}$$

$$Sp = \frac{\mathrm{TN}}{\mathrm{FP} + \mathrm{TN}} \tag{4.15}$$

$$\mathrm{PPV} = Pr = \frac{\mathrm{TP}}{\mathrm{TP} + \mathrm{FP}} \tag{4.16}$$

Sensitivity and precision are also called true positive rate (TPR) and positive predictive value (PPV), respectively. Moreover, the false positive rate (FPR) can be derived:

$$\mathrm{FPR} = \frac{\mathrm{FP}}{\mathrm{FP} + \mathrm{TN}} = 1 - Sp \tag{4.17}$$

as well as the phi correlation coefficient (phi):

$$\mathrm{phi} = \frac{\mathrm{TP} \cdot \mathrm{TN} - \mathrm{FP} \cdot \mathrm{FN}}{\sqrt{(\mathrm{TP} + \mathrm{FN}) \cdot (\mathrm{TN} + \mathrm{FP}) \cdot (\mathrm{TP} + \mathrm{FP}) \cdot (\mathrm{TN} + \mathrm{FN})}} \tag{4.18}$$

Table 4.5 Confusion matrix for binary classification (contingency table)

		Assigned class	
		P	**N**
True	P	TP	FN
class	N	FP	TN

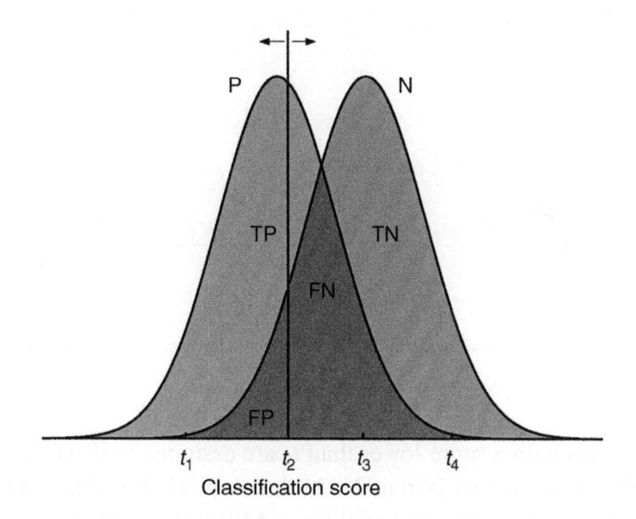

Figure 4.2 Example of binary classification: normal distribution of the classes (P and N) along a classification score. The objects with a score lower than the threshold are assigned to P.

which takes values between -1 and $+1$, where 1 indicates perfect classification, 0 random prediction and values below 0 a classification worse than random prediction.

Starting from a contingency table (Table 4.5), graphical tools (such as receiver operating characteristics) for the analysis of classification results and the selection of optimal models can be built. A receiver operating characteristic, or simply ROC curve, is a graphical plot of FPR and TPR as x and y axes respectively, for a binary classification system as its discrimination threshold is changed.

A single value of FPR and TPR can be calculated from a contingency table and consequently each contingency table represents a single point in the ROC space. Some classification methods produce probability values representing the degree to which class the objects belong. In this case, a threshold value should be defined to determine a classification rule. For each threshold value, a classification rule is calculated and the respective contingency table is obtained. Consequently, by looking at the ROC curve, the optimal threshold value (i.e. the optimal classification model) can be defined. The best possible classification method would yield a point in the upper left corner of the ROC space, representing maximum sensitivity and specificity, while a random classification give points along the diagonal line from the left bottom to the top right corners. An example on the use of ROC curves is shown in Figures 4.2 and 4.3. Consider a binary classification task, where the classes (P and N) are normally distributed along a classification score (Figure 4.2). A threshold value (t_2)

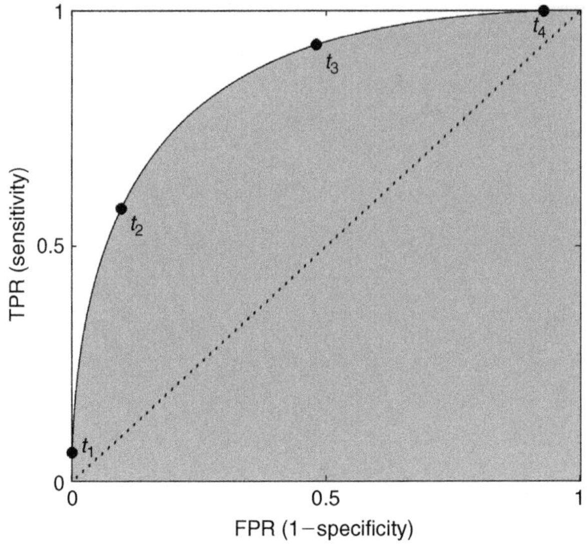

Figure 4.3 Example of binary classification: ROC curve relative to class distribution and threshold values of Figure 4.2.

is set: all the objects with a score lower than t_2 are assigned to P, while objects with a score greater than t_2 are recognized as N. At this step, TP, FP, TN, and FN are calculated, giving TPR (sensitivity) equal to 0.58 and FPR (1 − specificity) equal to 0.1; the point representing this result is placed in the ROC space (Figure 4.3) with these coordinates. Then, the threshold value can be decreased to t_1: in this case another point in the ROC space (TPR equal to 0.06, FPR equal to 0) will be obtained, as well as for threshold values of t_3 and t_4. The complete ROC curve explains how the model is working: in this case, the classification model is performing better than a random classifier, since the ROC curve is higher than the diagonal line; on the other hand, it is far from the best possible model, since the upper left corner of the ROC space is not reached. Finally, on the basis of the classification aim, we can decide which is the optimal balance of sensitivity and specificity and consequently set the best threshold value.

Classification methods

Nearest mean classifier and *K*-nearest neighbors

The nearest mean classifier (NMC) is the simplest classification method; it just considers the centroid of each class and classifies objects with the label of the nearest class centroid, where the centroid of a class is defined as the point whose parameter values are the mean of the parameter values of all the objects belonging to the considered class. NMC is a parametric, unbiased and probabilistic method; it is robust, since generally it has a high error on the training and test sets, but the error on the training data is a good prediction of the error on the test data.

As well as the nearest mean classifier, the K-nearest neighbor (KNN) classification rule (Cover and Hart, 1967) is conceptually quite simple: an object is classified according to the classes of the K closest objects, i.e. it is classified according to the majority of its K-nearest neighbors in the data space. In case of ties, the closer neighbors can acquire a greater weight. In a computational point of view, all that is necessary is to calculate and analyze a distance matrix. The distance of each object from all the other objects is computed, and the objects are then sorted according to this distance. KNN has other advantages: it does not assume a form of the underlying probability density functions (it is a non-parametric classification method) and can handle multiclass problems. Another important advantage is that KNN is a non-linear classification method, since the Euclidean distance between two objects in the data space is a non-linear function of the variables. Because of these characteristics, KNN has been suggested as a standard comparative method for more sophisticated classification techniques (Kowalski and Bender, 1972), while, on the other hand, KNN can be considered very sensitive to the applied distance metric and scaling procedures (Todeschini, 1989).

Of course, when applying KNN, the optimal value of K must be searched for. Even if the selection of the optimal K value can be based on a risk function, there are some practical aids for deciding the number of neighbors to be considered. First of all, distant neighbors (i.e. great values of K) are not useful for classification, while the best empirical rule to follow is to use $K = 1$, if there is not considerable overlap between classes. However, the best way of selecting K is by means of cross-validation procedures, i.e. by testing a set of K values (e.g. from 1 to 10); then, the K giving the lowest classification error can be selected as the optimal one.

Discriminant analysis

Among traditional classifiers, discriminant analysis is probably the most known method (Fisher, 1936; McLachlan, 1992) and can be considered the first multivariate classification technique. Nowadays, several statistical software packages include procedures referred to by various names such as linear discriminant analysis and canonical variate analysis.

Canononical variate analysis (CVA) separates objects into classes by minimizing the within-class variance and maximizing the between-class variance. So, with respect to principal component analysis, the aim of CVA is to find directions (i.e. linear combinations of the original variables) in the data space that maximize the ratio of the between-class to within-class variance, rather than maximizing the between-object variance without taking into account any information on the classes, as PCA does. These directions are called discriminant functions or canonical variates and are in number equal to the number of categories minus 1. Then, an object ($\mathbf{x}$) is assigned to the class with the minimum discriminant score $d_g(\mathbf{x})$:

$$d_g(\mathbf{x}) = (\mathbf{x} - \bar{\mathbf{x}}_g)^T \mathbf{S}_g^{-1} (\mathbf{x} - \bar{\mathbf{x}}_g) + \ln\left|\mathbf{S}_g\right| - 2\ln(P_g) \qquad (4.19)$$

where P_g is the prior class probability (usually defined as $P_g = 1/G$ or $P_g = n_g/n$), $\bar{\mathbf{x}}_g$ and $\mathbf{S}_g$ are the centroid and the covariance matrix of the g-th class, respectively.

The quantity $d_g(\mathbf{x}) + 2\ln(P_g)$ is referred to as discriminant function, while $(\mathbf{x} - \bar{\mathbf{x}}_g)^T \mathbf{S}_g^{-1} (\mathbf{x} - \bar{\mathbf{x}}_g)$ is the Mahalanobis distance between $\mathbf{x}$ and $\bar{\mathbf{x}}_g$.

Quadratic discriminant analysis (QDA) is a probabilistic parametric classification technique and is based on the classification rule described above; basically, it separates the class regions by quadratic boundaries and makes the assumption that each class has a multivariate normal distribution, while the dispersion (represented by the class covariance matrices, $\mathbf{S}_g$) is different in the classes.

A special case, referred to as linear discriminant analysis (LDA), occurs if all the class covariance matrices are assumed to be identical

$$\mathbf{S}_g = \mathbf{S}_p \qquad 1 \leq g \leq G \tag{4.20}$$

where $\mathbf{S}_p$ is the pooled covariance matrix, defined as:

$$\mathbf{S}_p = \frac{\sum_{g=1}^{G} (n_g - 1)\mathbf{S}_g}{n - G} \tag{4.21}$$

where n is the total number of objects, G the number of classes, n_g the number of objects belonging to the g-th class and $\mathbf{S}_g$ is the covariance matrix of the g-th class.

As well as QDA, LDA is a probabilistic parametric classification technique and assumes that each class has a multivariate normal distribution, while the dispersion (covariance matrix) is the same for all the classes.

Consequently, both QDA and LDA are expected to work well if the class conditional densities are approximately normal (i.e. data are multinormally distributed). In addition, when the class object sizes are small compared to the dimension of the measurement space (the number of variables), the inversion of covariance matrices became difficult. So, when applying LDA, the number of objects must be significantly greater than the number of variables, while QDA requires a larger number of objects than LDA, since covariance matrices are calculated for each class. Moreover, when variables are highly correlated among them, i.e. in presence of multicollinearity, discriminant analysis runs the risk of overfitting (Hand, 1997). In order to overcome these problems, a first approach is simply the reduction of the number of variables, by means of variable selection techniques: Stepwise discriminant analysis (SWDA) has been proposed with this aim (Jennrich, 1977). A second approach can be based on PCA: the classification model is performed on the significant scores calculated by means of PCA (i.e. in a reduced hyperspace). Another solution can be the use of alternatives to the usual estimates for the covariance matrices, as proposed by Friedman for regularized discriminant analysis (RDA) (Friedman, 1989).

As explained before, discriminant analysis classification rules always assign objects to classes (i.e. discriminant analysis is not a class modeling technique). A modeling version of QDA has been proposed, known as UNEQ (unequal class modeling) (Derde and Massart, 1986), which is based on the assumption of multivariate normality for each class population, as well as QDA; on the other hand, in UNEQ each class model is represented by the class centroid and the class space is defined on the basis of the Mahalanobis distance from this centroid.

Partial least squares-discriminant analysis (PLS-DA)

Partial least squares (PLS) was originally designed as a tool for statistical regression and nowadays is one of the most commonly used regression techniques in chemistry (Wold, 1966). It is a biased method and its algorithm can be considered as an evolution of the non-linear iterative partial least squares (NIPALS) algorithm. The PLS algorithm has been modified for classification purposes and widely applied in several fields, such as medical, environmental, social, and food sciences. Recently Barker and Rayens (2003) showed that partial least squares-discriminant analysis (PLS-DA) corresponds to the inverse-least-squares approach to LDA and produces essentially the same results but with the noise reduction and variable selection advantages of PLS. Therefore, if PLS is somehow related to LDA, it should be applied for dimension reduction aimed at discrimination of classes, instead of PCA.

The theory of PLS algorithms (PLS1 when dealing with one dependent *Y* variable and PLS2 in presence of several dependent *Y* variables) has been extensively studied and explained in the literature: PLS-DA is essentially based on the PLS2 algorithm that searches for latent variables with a maximum covariance with the *Y* variables. Of course, the main difference is related to the dependent variables, since these represent qualitative (and not quantitative) values, when dealing with classification. In PLS-DA the *Y*-block describes which objects are in the classes of interest. In a binary classification problem, the *Y* variable can be easily defined by setting its values to 1 if the objects are in the class and 0 if not. Then, the model will give a calculated *Y*, in the same way as for a regression approach; the calculated *Y* will not have either 1 or 0 values perfectly, so a threshold (equal to 0.5, for example) can be defined to decide if an object is assigned to the class (calculated *Y* greater than 0.5) or not (calculated *Y* lower than 0.5). When dealing with multiclass problems, the same approach cannot be used: if *Y* is defined with the class numbers (1, 2, 3, … , *G*) this would mean that a mathematical relationship between the classes exists (for example, that class g is somehow in-between class $g - 1$ and class $g + 1$). The solution to this is to unfold the class vector and apply the PLS2 algorithm for multivariate qualitative responses (PLS-DA). For each object, PLS-DA will return the prediction as a vector of size *G*, with values in-between 0 and 1: a g-th value closer to zero indicates that the object does not belong to the g-th class, while a value closer to one the opposite. Since predicted vectors will not have the form (0, 0, … , 1, … , 0) but real values in the range between 0 and 1, a classification rule must be applied; the object can be assigned to the class with the maximum value in the *Y* vector or, alternatively, a threshold between zero and one can be determined for each class. In this case, ROC curves can be used to assess and optimize the class specificity and sensitivity with different thresholds.

Soft independent modeling of class analogy (SIMCA)

As explained before, PCA is not useful for differentiating defined classes, since the class information is not used in the construction of the model and PCA just describes the overall variation in the data. However PCA can be coupled with the class information in order to give classification models by means of soft independent modeling of class analogy (SIMCA) (Wold, 1976). SIMCA was the first class modeling technique introduced in chemistry and nowadays is one the best-known modeling classification methods; it is defined "soft" since no hypothesis on the distribution of variables is made, and "independent" since the classes are modeled one at a time (i.e. each class model is developed independently).

Basically, a SIMCA model consists of a collection of G PCA models, one for each of G defined classes. Therefore, PCA is separately calculated on the objects of each class; since the number of significant components can be different for each category, cross-validation has been proposed as a way of choosing the number of retained components of each class model. In this way, SIMCA defines G subspaces (class models); then, a new object is projected in each subspace and compared to it in order to assess its distance from the class. Finally, the object assignation is obtained by comparing the distances of the object from the class models.

Even if SIMCA is often a useful classification method, it has also some disadvantages. Primarily, the class models in SIMCA are calculated with the aim of describing variation within each class: when PCA is applied on each category, it finds the directions of maximum variance in the class space. Consequently, no attempt is made to find directions that separate classes, on the opposite of, for example, PLS-DA, which directly models the classes on the basis of the descriptors.

Artificial neural networks

Artificial neural networks (ANNs) are increasing in uses related to several chemical applications and nowadays can be considered as one of the most important emerging tools in chemometrics. One of the reasons of their success can be related to the ability of solving both supervised and unsupervised problems, such as clustering and modeling of both qualitative and quantitative responses. Consequently, we have to initially consider the nature of the problem and then look for the best ANN strategy to solve it, since different ANN architectures and different ANN learning strategies have been proposed in literature (Zupan, 1994). Basically, ANN is supposed to mimic the action of a biological network of neurons, where each neuron accepts different signals from neighboring neurons and processes them. Consequently, depending on the outcome of this processing and on the nature of the network, each neuron can give an output signal. The function which calculates the output vector from the input vector is composed of two parts: the first part evaluates the net input and is a linear combination of the input variables, multiplied by coefficients called weights, while the second part transfers the net input in a non-linear manner to the output vector.

Artificial neural networks can be composed of different numbers of neurons; moreover, these neurons can be placed into one or more layers. In chemical applications, the number of neurons changes on the basis of the analyzed data and can range from tens of thousands to as few as less than ten (Zupan and Gasteiger, 1993).

The Kohonen and counterpropagation neural networks are two of the most popular ANN learning strategies (Hecht-Nielsen, 1987; Kohonen, 1988; Zupan *et al.*, 1997). ANNs based on the Kohonen approach (Kohonen maps) are self-organizing

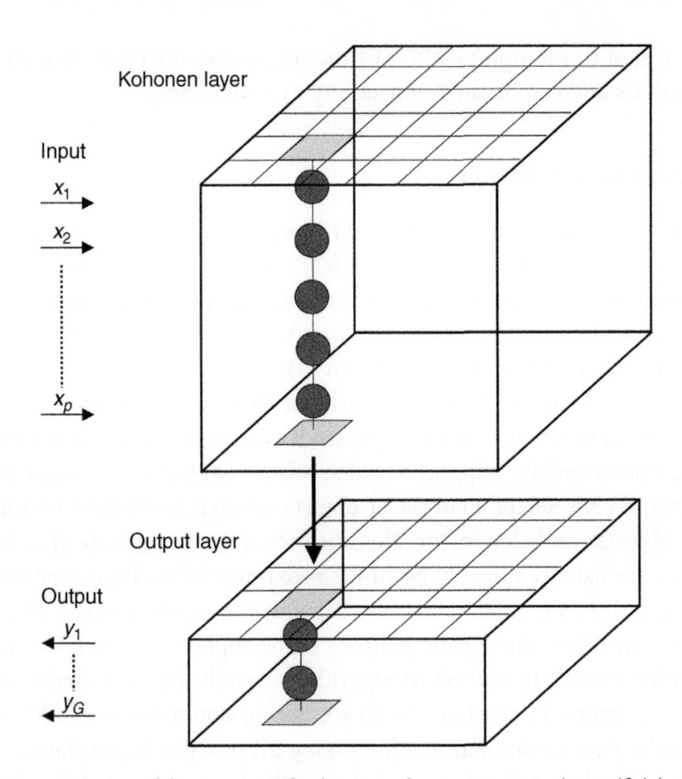

Figure 4.4 Representation of the structure of Kohonen and counterpropagation artificial neural network for a generic data set constituted by p variables and G classes.

systems which are capable of solving unsupervised rather than supervised problems. In Kohonen maps similar input objects are linked to topologically close neurons in the network (i.e. neurons that are located close to each other have similar reactions to similar inputs), while the neurons that are far apart have different reactions to similar inputs. In the Kohonen approach the neurons learn to identify the location in the ANN that is most similar to the input vectors.

Counterpropagation ANN is very similar to the Kohonen maps and is essentially based on the Kohonen approach, but it combines characteristics from both supervised and unsupervised learning. In fact, an output layer is added to the Kohonen map whose neurons have as many weights as the number of responses in the target vectors (the classes). The neuron of the output layer to be corrected is chosen on the basis of the neuron in the Kohonen layer that is more similar to the input vector; then, the weights of the output layer are adapted to the target values. In Figure 4.4, a representation of Kohonen and counterpropagation ANN is shown for a generic data set constituted by p variables and G classes.

Regarding classification, ANNs work better if they deal with non-linear dependence between input and output vectors and generally are efficacious methods for modeling classes separated with non-linear boundaries. In general, since neural networks are non-parametric tools which have adaptable parameters (such as number of neurons, layers, and epochs), most learning schemes require the use of a test set to optimize the structure of the model; in fact one of the major disadvantages of ANNs is probably related to the optimization of the net, since this procedure suffers from some arbitrariness and can be time-consuming in some cases.

Support vector machines

Support vector machines (SVMs) work on binary classification problems, even if they can be extended on multiclass problems, and have gained considerable attention due to their success in classification problems in the last years. SVMs define a function that describes the decision boundary that optimally separates two classes by maximizing the distance between them (Burges, 1998; Vapnik, 1999). Since SVMs are linear classifiers in high-dimensional spaces, the decision boundary can be described as a hyperplane and is expressed in terms of a linear combination of functions parameterized by support vectors, which consist of a subset of training objects. In fact, SVMs select a subset of objects (known as support vectors) among the training objects, and derive the classification rule using only this fraction of objects, which are usually those lying in the proximity of the boundary between the classes. Consequently, the final solution is dependent on only a subset of objects and the removal of any other object (not included in the support vectors) does not change the classification model. In fact SVM algorithms search for the support vectors that give the best separating hyperplane; to do so, during optimization, SVMs search the decision boundary with maximal margin among all possible hyperplanes, where the margin can be intended as the distance between the hyperplane and the closest point for both classes.

With regard to the determination of the parameters of the separating hyperplane, a major advantage of SVMs over other classifiers is that this optimization is a determinate operation where there is only one minimum solution and no local minima can be found. As explained before, SVMs are linear classifiers, but when non-linearly separable classes are present, it is impossible to find a linear boundary that separates all the objects. In this case, a trade-off between maximizing the margin between the classes and minimizing the number of misclassified objects can be defined. On the other hand, it is also possible to improve SVMs by integrating non-linear kernel functions for defining non-linear separations.

Even if SVMs work on binary classification problems, multiclass approaches can be solved combining binary classification functions (e.g. by considering one class at a time and searching a classifier for each class that separates the considered class from all the other classes). Then, an object is assigned to the nearest class, where the distance from each class can be formulated by means of a decision function.

Classification and regression trees

Tree-based approaches have become increasingly popular in recent decades and their application has arisen in several fields. These methods consist of algorithms based on rule induction, which is a way of partitioning the data space into different class subspaces. Basically, the data set is recursively split into smaller subsets where each subset contains objects belonging to as few categories as possible. The purity of each subset can be measured by means of entropy: a subset consisting of objects from one single class has the highest possible purity (the lowest entropy), while the most impure subset is the one where classes are equally represented. Consequently, in each split (node), the partitioning is performed in such a way to reduce entropy (maximize purity) of the new subsets and the final classification model consists of a collection of nodes (tree) that define the classification rule.

Univariate and multivariate strategies for finding the best split can be distinguished; in the univariate approach the algorithm searches at each binary partitioning the single variable that gives the purest subsets; the partitioning can be formulated as a binary rule like "is $x_{ij} < t_k$," where x_{ij} is the value of the j-th variable for the i-th object and t_k is the threshold calculated in the k-th node. All the objects that satisfy the rule are grouped in one subset, otherwise into another. This is the case of the classification and regression trees (CART), which are a form of binary recursive partitioning based on univariate rule induction (Breiman *et al.*, 1984). A simple classification tree is shown in Figure 4.5 as an example; it is made of just three nodes (t_1, t_2, and t_3) and splits the objects into three classes (class 1, class 2, and class 3).

On the other hand, multivariate rule induction finds a partition of the data that is based on a linear combination of all the variables instead of just one variable and is useful if there are collinearities between the variables. Each partitioning searches for the vector that best separates the data into pure subsets and the separation rules correspond to hyperplanes that increasingly isolate the class subspaces in the data space.

In realistic situations, both for univariate and multivariate approaches, the number of nodes in the tree can be very large; the solution to this can be a sort of optimization and simplification of the tree (pruning) by reducing the number of rules (nodes) when a less than optimal purity is reached. However, CART analysis has several advantages over other classification methods: it is scale-independent and non-parametric; it gives intuitive classification models that consist of a graph where

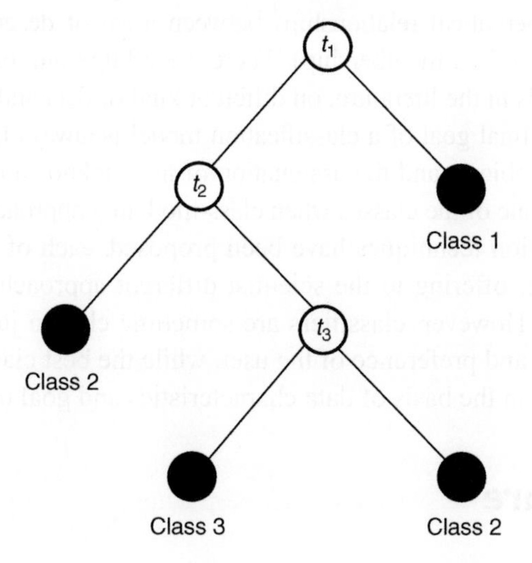

Figure 4.5 Example of classification tree made of three nodes (t_1, t_2, and t_3) for a generic data set comprising three classes (class 1, class 2, and class 3).

each node is represented, associated to the classification rule of the node; moreover, CART identifies splitting variables with an exhaustive search of all the possibilities. Consequently, the most discriminating variables can be easily recognized and a sort of variable selection is applied, since in order to assign new objects, only the splitting variables considered in the classification rules will be used.

New classifiers

Among new classification approaches, we can cite extended canonical variate analysis (ECVA), which has been recently proposed as a modification of the standard canonical variate analysis method (Nørgaard *et al.*, 2007). The modified CVA method forces the discriminative information into the first canonical variates and the weight vectors found in the ECVA method hold the same properties as weight vectors of the standard CVA method, but the combination of the suggested method with, for example, LDA as a classifier gives an efficient operational tool for classification and discrimination of collinear data.

Classification and influence matrix analysis (CAIMAN) is a new classifier based on leverage-scaled functions (Todeschini *et al.*, 2007). The leverage of each object is a measure of the object distance from the model space of each class; consequently, exploiting the leverage properties, CAIMAN models each class by means of the class leverage matrix and calculates the leverage of objects with respect to each class space. Moreover, in order to face non-linear boundaries between classes, the CAIMAN approach has been developed for defining a new mathematical concept called hyper-leverage, which basically extract information from the space defined by the leverages themselves.

Conclusions

Multivariate classification is one of the basic methodologies in chemometrics and consists in finding mathematical relationships between a set of descriptive variables and a qualitative variable (class membership). There are a huge number of applications of classification methods in the literature, on different kind of data and with different aims, even if basically the final goal of a classification model is always the separation of two (or more) classes of objects and the assignation of new unknown objects in one of the defined classes (or none of the classes when class-modeling approaches are applied).

Several classification techniques have been proposed, each of them with different properties and skills, offering to the scientist different approaches for solving classification problems. However, classifiers are sometime chosen just on the basis of a personal knowledge and preference of the user, while the best classification approach should be preferred on the basis of data characteristics and goal of analysis.

Nomenclature

n number of samples (objects)

p number of variables

G	number of classes
$\mathbf{X}$	data matrix ($n \times p$)
x_{ij}	element of $\mathbf{X}$, representing the value of the j-th variable for the i-th object
$\mathbf{c}$	class vector ($n \times 1$), constituted by G different labels or integers, each representing a class
$\mathbf{C}$	unfolded class matrix ($n \times G$)
c_{ig}	element of $\mathbf{C}$, representing the membership of the i-th object to the g-th class
$\bar{\mathbf{x}}_\mathbf{g}$	centroid of the g-th class
$\mathbf{S_g}$	covariance matrix of the g-th class
$\mathbf{S_p}$	pooled covariance matrix
n_g	number of objects belonging to the g-th class
n_{gk}	number of objects belonging to class g and assigned to class k
n_M	number of objects belonging to the largest class
P_g	prior class probability

Indices on n, p, G run as follows:
$i = 1, \ldots, n$
$j = 1, \ldots, p$
g or $k = 1, \ldots, G$

References

Barker M, Rayens WS (2003) Partial least squares for discrimination. *Journal of Chemometrics*, **17**, 166–173.

Breiman LJ, Friedman JH, Olsen R, Stone C (1984) *Classification and Regression Trees.* Belmont, CA: Wadsworth International Group, Inc.

Burges CJC (1998) A Tutorial on Support Vector Machines for Pattern Recognition. *Data Mining and Knowledge Discovery*, **2**, 121–167.

Cover TM, Hart PE (1967) Nearest neighbor pattern classification. *IEEE Transactions on Information Theory*, **13**, 21–27.

Derde MP, Massart DL (1986) UNEQ: a disjoint modelling technique for pattern recognition based on normal distribution. *Analytica Chimica Acta*, **184**, 33–51.

Fisher RA (1936) The use of multiple measurements in taxonomic problems. *Annals of Eugenics*, **7**, 179–188.

Frank IE, Todeschini R (1994) *The Data Analysis Handbook.* Amsterdam: Elsevier.

Friedman JH (1989) Regularized discriminant analysis. *Journal of the American Statistical Association*, **84**, 165–175.

Hand DJ (1997) *Construction and Assessment of Classification Rules.* Chichester: Wiley.

Hecht-Nielsen R (1987) Counter-propagation networks. *Applied Optics*, **26**, 4979–4984.

Jennrich RJ (1977) Stepwise discriminant analysis. In: *Statistical Methods for Digital Computers* (Enslein K, Ralston A, Wilf HF, eds). NewYork: Wiley & Sons.

Jolliffe IT (1986) *Principal Component Analysis.* New York: Springer-Verlag.

Kohonen T (1988) *Self-Organization and Associative Memory.* Berlin: Springer Verlag.

Kowalski BR, Bender CF (1972) The K-nearest neighbor classification rule (pattern recognition) applied to nuclear magnetic resonance spectral interpretation. *Analytical Chemistry*, **44**, 1405–1411.

Mardia KV, Kent JT, Bibby JM (1979) *Multivariate Analysis*. London: Academic Press.

Massart DL, Kaufman L (1983) *The Interpretation of Analytical Chemical Data by the Use of Cluster Analysis*. New York: Wiley.

McLachlan G (1992) *Discriminant Analysis and Statistical Pattern Recognition*. New York: Wiley.

Nørgaard L, Bro R, Westad F, Engelsen SB (2007) A modification of canonical variates analysis to handle highly collinear multivariate data. *Journal of Chemometrics*, **20**, 425–435.

Todeschini R (1989) K-nearest neighbour method: the influence of data transformations and metrics. *Chemometrics and Intelligent Laboratory Systems*, **6**, 213–220.

Todeschini R, Ballabio D, Consonni V, Mauri A, Pavan M (2007) CAIMAN (classification and influence matrix analysis): a new approach to the classification based on leverage-scaled functions. *Chemometrics and Intelligent Laboratory Systems*, **87**, 3–17.

Vapnik V (1999) *The Nature of Statistical Learning Theory*. New York: Springer-Verlag.

Wold H (1966) Estimation of principal components and related models by iterative least squares. In: *Multivariate Analysis* (Krishnaiah PR, ed.). New York: Academic Press.

Wold S (1976) Pattern recognition by means of disjoint principal components models. *Pattern Recognition*, **8**, 127–139.

Zupan J (1994) Introduction to artificial neural network (ANN) methods: What they are and how to use them. *Acta Chimica Slovenica*, **41**, 327–352.

Zupan J, Gasteiger J (1993) *Neural Networks for Chemists: An Introduction*. Weinheim: VCH-Verlag.

Zupan J, Novic M, Ruisánchez I (1997) Kohonen and counterpropagation artificial neural networks in analytical chemistry. *Chemometrics and Intelligent Laboratory Systems*, **38**, 1–23.

5 Calibration Transfer Methods

Frans van den Berg and Åsmund Rinnan

Introduction

The subject of calibration transfer methods in chemometric model building really falls under economics. Constructing a high-quality inverse multivariate calibration model such as partial least squares (PLS) in the presence of unknown interfering signals requires tens, hundreds or sometimes up to a thousand samples plus reference values collected over a long period of time. This is a big investment. Calibration transfer focuses on preserving this investment by keeping the model valid over time for the same instrument (model maintenance) or by sharing the cost where a model developed on one system (the primary or master (M) instrument) is applied to one or more other systems of a similar nature (the secondary or slave (S) instrument).

The relevance of this subject is emphasized by the fact that in 2007 over 70 references with primary focus on calibration transfer appeared in the literature, and a relatively large number of these indicate connections with industry. Most of these papers are found in the analytical chemistry/chemometric and spectroscopy literature focusing on computational methodologies; two comprehensive reviews are available by de Noord (1994) and by Feudale *et al.* (2002). Moreover, just like chemometric data analysis in general, the potential of calibration transfer is being recognized by the outside world (Park *et al.*, 1999; Duponchel *et al.*, 1999; Fontaine *et al.*, 2004; Bergman *et al.*, 2006; Alamar *et al.*, 2007).

Economics also play a role in the reverse direction: since we are trying to minimize our expenses on the collection of sample and reference analysis, the model

Infrared Spectroscopy for Food Quality Analysis and Control
ISBN: 978-0-12-374136-3

will be suboptimal compared to full recalibration (e.g. on a secondary instrument) and, moreover, an independent evaluation via a test set or uncertainty estimation by resampling is typically not feasible due to the small number of samples involved. This makes an understanding of the mathematical operations involved in calibration transfer and the effects on spectroscopic data crucial for proper and safe use.

The transfer issue

A number of scenarios could result in a multivariate calibration model becoming invalid. This would occur, for instance, if the original instrument is replaced by a new one. The responses from two instruments for the same sample measured under the same conditions will be different and multivariate calibration models will thus not necessarily be valid for this new instrument. It must be stated that big improvements have been achieved by instrument manufacturers on hardware harmonization in recent years. Diode lasers for wavelength alignment, internal (reflection) standards for intensity corrections, and charged-coupled device (CCD) similarity matching in the factory by characteristics comparison are just a few measures on offer to improve instrument-to-instrument compatibility. Nevertheless, the instability of one and the same unit over time is another problem which can seriously affect the performance of a model. Small continuous changes (instrumental drift) and sudden changes (response shifts caused by repairs or replacements, for example) in the instrument still cause signals to change, leading to increased prediction errors without proper model maintenance.

The most straightforward solution, of course, would be to re-calibrate for the all new measurement conditions or to expand the original model for the new situation. Unfortunately, this is also the most expensive solution and sometimes technically impossible. Standardization and calibration transfer methods have been developed aimed at eliminating the need for a full recalibration and to preserve the information collected in an existing model.

Even if instrumental hardware is matched well, sampling for in-process measurements could still render the calibration model invalid. For example, interfacing to a process stream via different routes such as multiplexers and/or fiber optics will not always be in the hands of the instrument manufacturer. Bend angles of the fibers and optical components for different sampling points will differ, all influencing the detector responses in a unique way. This is precisely the direction where new developments in process monitoring and control on near-infrared (NIR) are heading: various measurement points for similar streams in the factory based on multiplexers or a family of relatively cheap CCD-based systems that depend on one global calibration (Bouveresse *et al.*, 1998). The cost-saving aspect of calibration transfer is thus still very much active.

Data set and transfer set selection

To illustrate the model/calibration transfer concepts and methods, in this chapter we will use a well-established data set as an example. This data set, generated by Wülfert *et al.* (1998), consists of ternary mixtures with known concentrations expressed

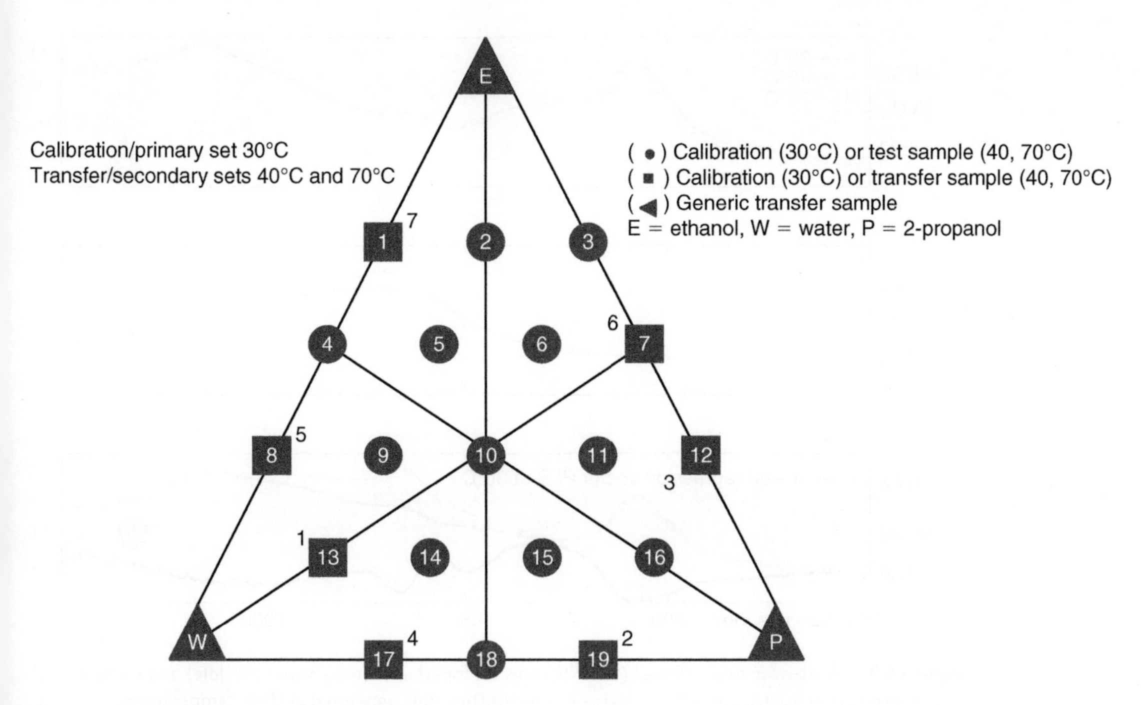

Figure 5.1 Experimental design of ternary mixtures.

as molar fractions of ethanol (E), water (W) and 2-propanol (P), measuring the short-wave near-infrared spectra (SW-NIR; Figure 5.1). The complete data set, available from the internet for non-commercial use, consists of 22 samples measured at five temperatures. In this chapter we will only use three temperatures, assuming that the recordings at 30°C represent our standard calibration conditions (primary instrument), the measurements at 40°C represent modest deviations in the signal (model maintenance), and those at 70°C represent considerable instrumental differences (secondary instrument). The corner points (pure components) will not be used in modeling. This leaves 19 samples for model building at 30°C. In this chapter we will focus on the prediction of ethanol concentration only.

The spectral range used in this chapter goes from 850 to 1050 nm; some example spectra of the three temperature/instrumental conditions are presented in Figure 5.2. As can be seen from Figure 5.2, the temperature has a profound effect on the signals. It is also observed that water is overall the strongest influence in this wavelength region with characteristics distinct from those of ethanol; the spectral characteristics of pure 2-propanol (not shown) are comparable to those of ethanol (Wülfert *et al.*, 1998). In contrast to the original work, the only spectral correction applied in this chapter is a baseline/offset removal based on the average value for the first 10 data points (850–860 nm) subtracted from each spectrum individually.

One could expect this ternary system to be of relative low complexity due to the closure relation for the molar fractions E + W + P = 1. However, due to the chemical interactions (mainly hydrogen bonding) the chemical rank has been shown to be higher than three (or two after mean centering; Wülfert *et al.*, 2000a, 2000b).

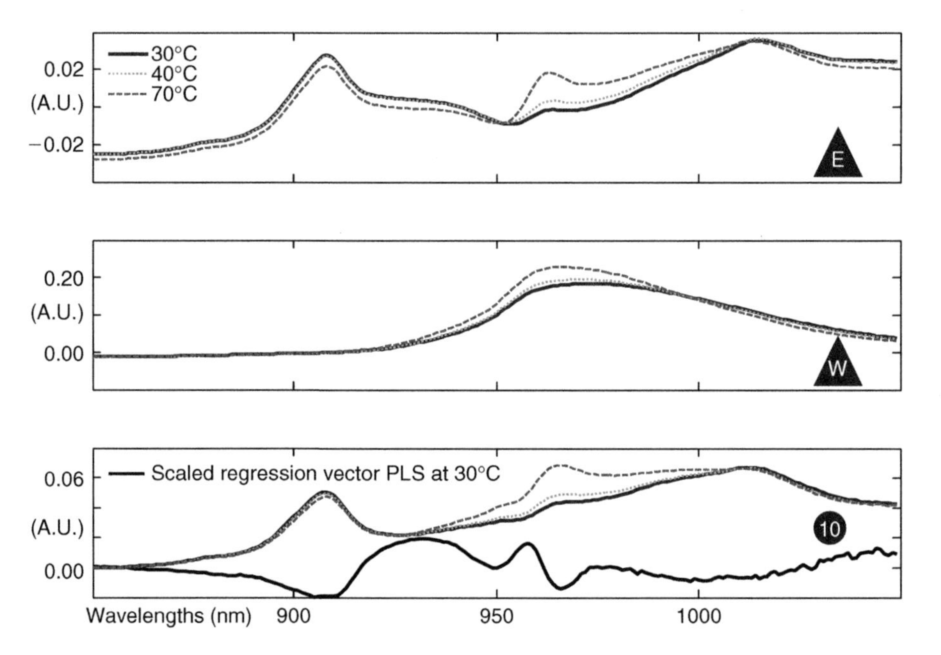

Figure 5.2 Short-wave near-infrared (SW-NIR) spectra for ethanol (top), water (middle) and mixture 10 plus scaled partial least squares (PLS) regression vector (bottom), measured at three temperatures.

It should be noted that our objective is not to find the ultimate model performance at each condition/temperature. We use this data set as a general illustration of model/calibration transfer concepts and methods.

An important step in calibration transfer can be, depending on the methodology employed, selection of samples that are measured in both conditions (e.g. primary and secondary instruments). From an operational point of view generic standards in the form of such as polystyrene standard materials, certified sample materials, easily reproducible mixtures, etc. are preferred. However, it is essential that these generic standards are compatible with regular samples from an information point of view, both in desired (e.g. the concentration to be predicted) as well as undesired properties (e.g. scatter properties), which is not always a simple requirement. In this chapter we will try the corner point in the ternary mixture design as generic standards (Figure 5.1).

A more practical approach is to select a number of regular samples to compute transfer functions, and the obvious choice for this would be a subset selected from the calibration set on the primary instrument. Subset selection algorithms are also of interest outside calibration transfer (Allesø *et al.*, 2007). The most often used algorithms in the literature are the leverage-based methods, the Kennard–Stone selection (Kennard and Stone, 1969) algorithm and principal properties selection (Carlson *et al.*, 2001). The Kennard–Stone algorithm adds samples that are farthest away from the previously selected set of points. It starts by selecting two samples that have the largest squared distance (e.g. the Euclidean norm) between them. In the next step an additional sample is selected based on maximizing the summed distances towards all the previously selected samples. This procedure is continued until the desired number of objects is found. Hence, the Kennard–Stone algorithm is expected to initially select design points from the outer regions/periphery of the sample space, successively going to the inner regions, and may therefore be regarded as a peeling method, spanning the relevant chemical space in the data set.

Principal properties selection is based on principal component analysis (PCA) and deflation of the data matrix $\mathbf{X}$ by orthogonal projection. The first step is to perform PCA on the calibration set. The row-vector with the sample/spectrum $\mathbf{x}_k$ $(l \times n)$ being closest to the loading vector of the first principal component—in this chapter expressed as the highest squared correlation coefficient between spectrum and loading vector—is selected as the first sample in the subset. In the next step all samples $\mathbf{x}_i$ are deflated by a projection on this selected sample:

$$\hat{\mathbf{x}}_i = \mathbf{x}_i - \frac{\mathbf{x}_k \mathbf{x}_i^T}{\mathbf{x}_k \mathbf{x}_k^T} \mathbf{x}_k \tag{5.1}$$

These two steps, i.e. PCA and deflation, are repeated until the desired number of samples has been selected. Hence, the characteristics of each selected sample are given by the properties of the successive first loading vectors in question, also denoted as the principal properties of the sample set. From the definition of PCA this loading basis is selected to describe the direction of maximum variance, which is approximated by the one sample best describing this variance. In the second step this variance direction is removed from the data via the deflation step. It should be noted that the sample selected will be empty/all zero after deflation ($\mathbf{x}_k = \mathbf{x}_i$) by equation (5.1), and will thus not play a role in future selections. Overall, it is expected that the uniqueness of the selected samples will again span the relevant chemical space in $\mathbf{X}$.

In this chapter we use the principal properties algorithm to determine a transfer subset of seven samples based on the 30°C SW-NIR spectral data set, picking samples 13, 19, 12, 17, 8, 7 and 1, in that order (Figure 5.1). To simplify our study we will only work with this particular transfer set. However, the user should be aware that the number of samples in the transfer set and different selection criteria can have a profound effect on the results (Siano and Goicoechea, 2007). Prediction of the ethanol contents in the remaining 12 samples at 40°C and 70°C will be used for performance evaluation. For the curious reader: the Kennard–Stone algorithm would pick samples {7, 8}, 14, 17, 19, 4 and 1.

Calibration transfer case study

In this section we will present and discuss a number of potential model transfer methods; the performance of the different trials is collected in Table 5.1. Figure 5.3 shows the model performance for the 30°C data set. If we apply no correction the model maintenance set at 40°C performs reasonably, while the model clearly underperforms for the secondary instrument at 70°C, which is to be expected due to the spectral differences in Figure 5.1.

Table 5.1 Comparison of the performance of different calibration transfer methods

Calibration transfer method	Ethanol		
	$RMSP_{CV}$ 30°C	$RMSP_P$ 40°C	$RMSP_P$ 70°C
(A) CS 30°C, PLS, $F = 4$, $R^2 = 0.99$	**0.018**		
No transfer		0.042	0.151
Offset-and-slope correction based on TS		0.040	0.205
(B) CS 30°C + TS 40°C PLS, $F = 5$, $R^2 = 0.99$	0.018	0.034	
(B) CS 30°C + TS 70°C PLS, $F = 3$, $R^2 = 0.95$	0.050		0.066
(C) CS 30°C 2nd derivative, PLS, $F = 4$, $R^2 = 0.99$	0.017		
No transfer		0.117	0.536
Offset-and-slope correction based on TS		0.009	0.020
(D = A) CS 30°C, PLS, $F = 4$, $R^2 = 0.99$	0.018		
DS based on TS		0.013	0.022
DS based on GTS		0.189	0.194
PDS based on TS (window 40°C = 5, 70°C = 51)		0.018	0.033
PDS based on GTS (window 40°C = 1, 70°C = 47)		0.036	0.098

$RMSP_{CV}$, root mean squared error, leave-one-out cross-validation; $RMSP_P$, root mean squared error, prediction; CS, calibration set; R^2, squared correlation reference-predicted value; F, factors in partial least squares; TS, seven-sample transfer set; PLS, partial least squares; GTS, three-sample generic transfer set.

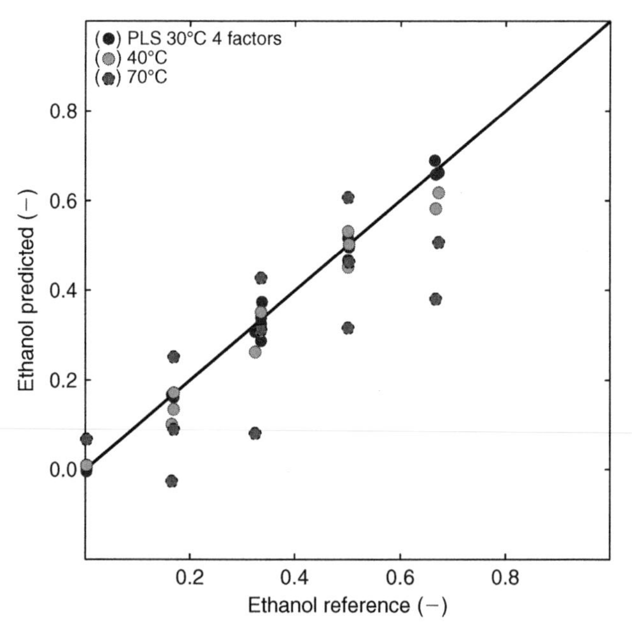

Figure 5.3 Predicted versus reference value for ethanol concentration with no transfer.

An often applied correction principle in model maintenance is to compute a simple univariate offset-and-slope correction $y_p = b_0 + b_1 \cdot y_{ref}$ between the predicted values and the determined reference values of control samples. New samples can be transferred by the simple operation $y_T = (y_p - b_0) \cdot b_1^{-1}$. It should be noticed that this correction can be based on any set of samples, e.g. quality control samples that are analyzed anyhow, which can be a considerable operational and financial advantage, though care must be taken to have enough span in this ad-hoc transfer set to get a good estimate for the correction function. As seen from Table 5.1(A) no gain is found in an offset-and-slope correction as expected from the results in Figure 5.3.

Another obvious tactic for instrument standardization is to include the transfer set in the calibration set in an attempt to expand the model to the new measurement conditions. This strategy requires that a limited number of samples are representative, a goal supported by the principal properties selection method (but notice that again it is not a requirement to have a transfer set based on calibration samples). From Table 5.1(B) it can be seen that this method works well for model maintenance but not so well for the primary–secondary instrument situation.

Data pre-processing by removing variation in the signal that is not constant or by carrying predictive information (Chapter 2) is (or should be) an integral part of calibration transfer since the difference between primary and secondary instruments is by definition not useful (Zachariassen *et al.*, 2005). A simple trick to improve transferability of data is local centering, where the spectra are mean centered towards the average spectral response on the system itself rather than on the average of the calibration set, thus removing the differences in the centers of gravity from multivariate signals stemming from different systems (Bergman *et al.*, 2006). Here we will test a 21-point symmetric window-wise second-order polynomial filter computing the smoothed second-order derivative to remove differences in baseline along the signals (so-called Savitzky–Golay method; Table 5.1(C); Figure 5.4). The predictive performance is initially worse but, as is obvious from Figure 5.4, the offset-and-slope correction is very efficient for the processed data for both transfers in our relatively simple data set.

To give the reader an impression of what is happening, Figure 5.5 shows the first two sample score values from a PCA on the master set, second derivative data. Projected on the loadings from this PCA model are the test sets for both secondary instruments and the generic sample sets. As can be seen from Figure 5.5, the ternary mixture triangle is easily recognized, despite some twists. It shows how the 40°C samples are close to but not overlapping with the calibration set (a typical model maintenance situation) while the 70°C set is far away and rotated but still in the same conformation. It also shows that direct transfer is not an easy task and that the generic standards used in this case study are probably not suitable for determining the correct transfer applicable to the real samples.

By far the greatest interest within the field of chemometrics for the potential of calibration transfer methods for NIR spectrometry was raised by a series of studies conducted by Wang and co-authors (Wang *et al.*, 1991; Wang and Kowalski, 1993a, 1993b). They introduced the idea of direct standardization (DS)

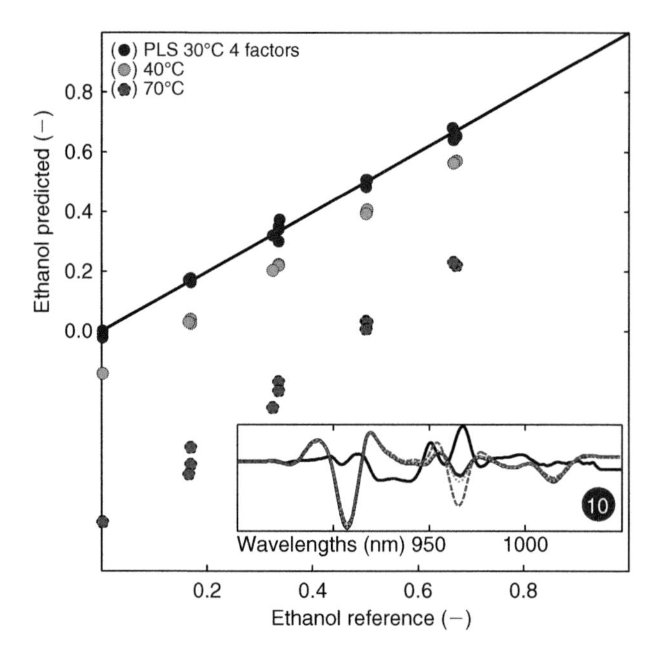

Figure 5.4 Predicted versus reference value for ethanol concentration—second-order derivative spectrum. Insert: second-order derivative for mixture 10 plus scaled partial least squares (PLS) regression vector, measured at three temperatures.

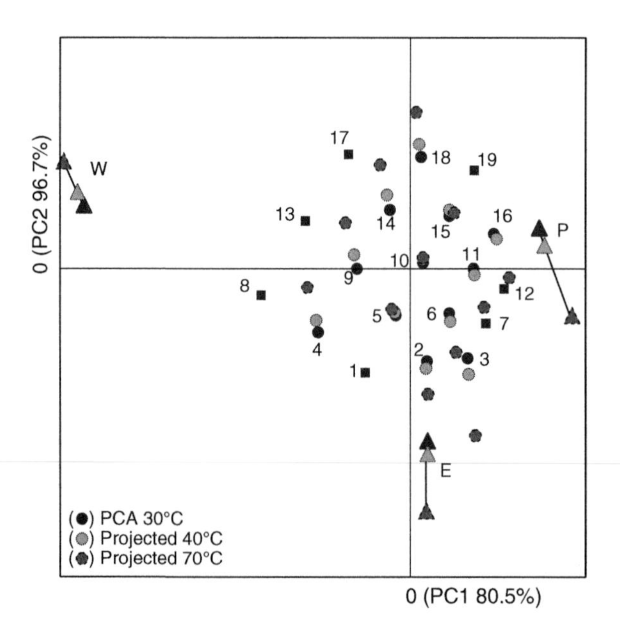

Figure 5.5 Principal component analysis (PCA) score plot calibration data set at 30°C, and estimated score values for transfer data sets at 40°C, 70°C, and generic samples.

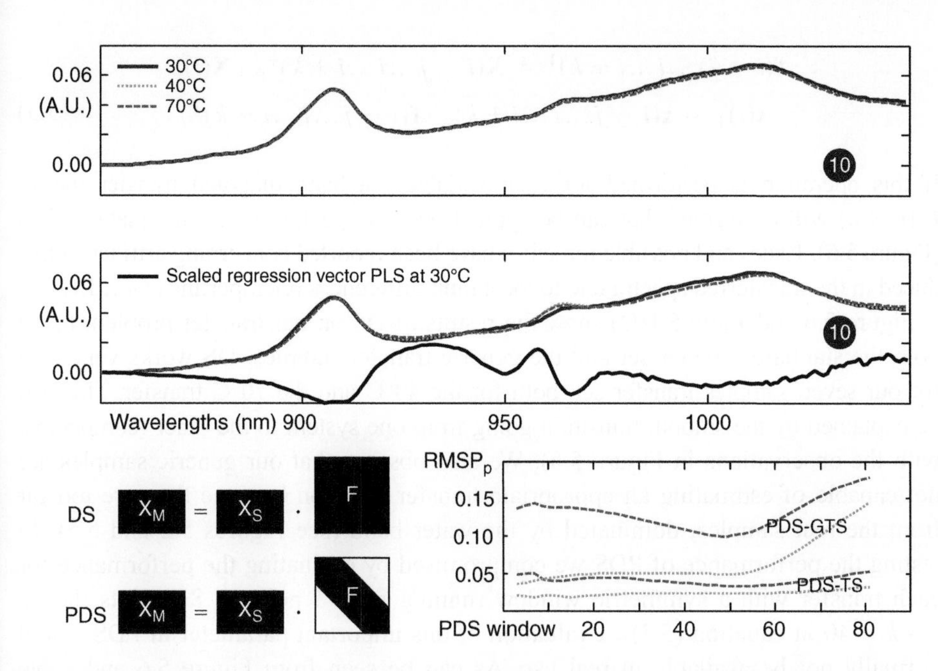

Figure 5.6 Transferred spectra mixture 10 with direct standardization with transfer data set (top) and generic data set (middle); piecewise direct standardization prediction errors for different window sizes are shown (bottom).

and piecewise direct standardization (PDS). In DS a transfer matrix **F** ($n \times n$) is computed between the same samples measured on two systems (Figure 5.6):

$$
\begin{aligned}
\mathrm{X_P} &= \mathrm{X_S} \cdot \mathrm{F} \\
\mathrm{F} &= X_{\mathrm{S}}^{+} \cdot \mathrm{X_P} \\
\mathrm{x_T} &= \mathrm{x_S} \cdot \mathrm{F}
\end{aligned}
\tag{5.2}
$$

where $\mathbf{X}_\mathrm{P}$ ($m \times n$) is a transfer set measured on the primary instrument and $\mathbf{X}_\mathrm{S}$ ($m \times n$) is the same set measured on the secondary instrument. A new measurement on the secondary instrument, row-vector $\mathbf{x}_\mathrm{S}$ ($l \times n$), can be transformed to resemble a measurement on the primary instrument and thus be employed in the original calibration model. To determine **F** the Moore–Penrose pseudo-inverse is applied (alternatively one could use PCR, PLS or any other stabilized regression method). This step can easily lead to numerical instabilities translating into poor results, especially if the number of transfer samples m is much smaller than the number of variables in the spectrum n and/or if generic samples are used. This observation led to the alternative model PDS where the transfer for each variable i in the spectrum of the primary instrument is estimated from a window $i - j \ldots i \ldots i + k$ on the secondary instrument by a much smaller (and thus more stable) inversion step:

$$
\mathbf{X}(i)_\mathrm{P} = (i - j \ldots i \ldots i + k)_\mathrm{S} \cdot \mathbf{f}(i - j \ldots i \ldots i + k)^\mathrm{T}
$$

$$\mathbf{f}(i - j)\ldots i\ldots i + k)^{\mathrm{T}} = \mathbf{X}(i - j\ldots i\ldots i + k)^{+}{}_{\mathrm{P}} \cdot \mathbf{X}(i)_{\mathrm{P}}$$
$$\mathbf{x}(i)_{\mathrm{T}} = \mathbf{x}(i - j\ldots i\ldots i + k)_{\mathrm{S}} \cdot \mathbf{f}(i - j\ldots i\ldots i + k)^{\mathrm{T}} \tag{5.3}$$

If this operation is performed for each variable i a band diagonal transfer matrix $\mathbf{F}$ $(n \times n)$ will be formed that can be applied on new spectra just as in equation (5.2) (Figure 5.6). Issues and possible remedies have been reported concerning artifacts introduced in the transferred spectra due to local rank differences (Gemperline *et al.*, 1996).

Figure 5.6 and Table 5.1(D) show the results of DS on our transfer problem using both the standard transfer set and the generic transfer samples. DS works very well for our seven samples transfer set, both for the 40°C and the 70°C transfer. This can be explained by the smooth transition going from one system to the other (comparable with the observations in Figure 5.5). We also observe that our generic samples are not capable of estimating an appropriate transfer function because they are too far from the real samples, dominated by the water band (see Figures 5.2 and 5.5). In testing the performance of PDS we compromised by evaluating the performance for each transfer with a symmetric window running from 1 point to 81 points (hence $j = k = 40$ in equation (5.3)). Evaluation of this important parameter in PDS would normally not be available in real use. As can be seen from Figure 5.6 and Table 5.1(D), where the best solution is included, none of the PDS models performed as well as DS; apparently the ethanol-specific information is not found in one smaller region of the spectrum and local differences in the spectral responses can be modeled from one global correction in our data set.

However, the data in our case study is just one example to illustrate some methods reported in literature. The performance of these methods might be completely different for other spectroscopic data problems. For example, PDS, one of the most widely used transfer methods, is often employed as a reference for other novel techniques (Feudale *et al.*, 2002) in contrast with the results presented here. However, from Figure 5.2 we can observe that the information in SW-NIR is distributed over the entire range and the noise level is low in these measurements (despite the limited signal-to-noise ratio). This makes our problem different from the conventional NIR (1100–2500 nm range) where the chemical rank can be quite different for different spectral regions and PDS is thus probably more successful. This is again different from fundamental IR spectroscopy with distinct spectral features. Therefore the first step in any successful calibration transfer is to gain knowledge and insight on the data and the idea behind the methods (as in all data analysis questions).

Other calibration transfer methods from the literature

Several alternative methods for calibration transfer have been presented in the literature, some of which will be presented briefly in this section. As stated before, spectral pre-processing can be of vital importance for achieving model maintenance and calibration transfer. Swierenga *et al.* (1999) took a different pre-processing approach by selecting variables in their Raman signal, using a simulated annealing algorithm that are least influenced by instrumental differences while maintaining a high predictive ability. However this idea—the so-called robust multivariate calibration models in a subspace of the original full spectral domain (Swierenga *et al.*, 1998)—requires samples measured on the secondary instrument, but not necessarily the samples as recorded on the primary instrument.

A simple univariate approach to standardization was developed and patented by Shenk and Westerhaus (1989). It involves a single correction factor at each wavelength channel to adjust for intensity differences in the same spectra measured on two systems. In order to account for small wavelength shifts, a further local correction procedure was developed that performs a wavelength index conversion followed by a spectral intensity transformation at each wavelength. An alternative method by Nørgaard (1995), called single wavelength standardization, corrects for intensity differences that vary across the wavelength axis by a simple linear regression correction of the responses of both instruments at each channel.

An alternative solution to transfer was formulated by Blank *et al.* (1996) based on an idea positioned between signal pre-processing and a transfer algorithm. They used a finite impulse response (FIR) filter to match a spectrum on the secondary instrument with the primary system (be aware that the concept of FIR here is slightly different from classical causal filtering theory in using a symmetric window surrounding the target point). These two spectra do not have to be from the same sample, which leads to the idea of standard-less transfer. The authors suggest filtering out the differences between spectral signals, leaving only the predictive information.

Duponchel *et al.* (1999) use a window-wise approach similar to PDS, but they employ a single-hidden-layer feed-forward neural network to establish the local regression model. They achieve good results using just five samples in the transfer set. Another promising approach combining DS and pre-processing (or rather domain transformation) was introduced by Tan and Brown (2001), who use discrete wavelet-base decomposition on a transfer set on both instruments to split the original signal of length 2^N into approximation spectra (also called the signal) and detail spectra (the noise) both of length 2^{N-1}. The next step is to compute DS transfers for the approximation and the detail blocks. These transfers can then be used on new samples from the secondary instrument. The authors selected an inverse wavelet transformation on the standardized approximation and detail to get back to an original spectrum of length 2^N which can be employed in the master calibration model.

Orthogonal signal correction (OSC) is a data analytical method developed to remove variation by projection from a signal that has no predictive information for (is orthogonal to) a reference value. The spectral differences between the primary and secondary instruments are precisely that the OSC has been successfully used in the calibration transfer issue (Sjöblom *et al.*, 1998), however the OSC does need reference measurements on the secondary instrument, but not necessarily samples that were also measured on the primary instrument. A related method was used by Andrew and Fearn (2004), in which a so-called repeatability file containing the average of a number of samples measured on different instruments is used (note that this can also be generic standards, but they have to be the same samples measured on each machine). A PCA on this repeatability file will thus show the differences between instruments and the data can be corrected for these disturbing spectral directions by projection.

Conclusions

Construction of a good multivariate calibration model requires a considerable amount of effort, time, and money. This fact makes calibration transfer methods a permanent subject of high relevance. Several strategies and methodologies are presented in this chapter for solving the standardization problem to a satisfactory degree. However, important aspects such as selection of transfer standards and the compatibility of the spectroscopic technique with the transfer algorithms do require input and understanding of the user for a successful calibration transfer implementation.

Nomenclature

b_0	offset in univariate correction
b_1	slope in univariate correction
$\mathbf{f}$	one row in transfer matrix
$\mathbf{F}$	transfer matrix in DS and PDS
$\mathbf{x}_i$	row-vector with sample/spectrum i
$\mathbf{X}$	matrix with samples/spectra as rows
y	reference value/concentration

Superscripts

T	vector/matrix transpose
+	Moore–Penrose pseudo-inverse of a matrix

References

Alamar MC, Bobelyn E, Lammertyn J, Nicolaï BM, Moltó E (2007) Calibration transfer between NIR diode array and FT-NIR spectrophotometers for measuring the soluble solids contents of apple. *Postharvest Biology and Technology*, **45**, 38–45.

Allesø M, Rantanen J, Aaltonen J, Cornett C, van den Berg F (2007) Solvent subset selection for polymorph screening. *Chemometrics and Intelligent Laboratory Systems*, in press.

Andrew A, Fearn T (2004) Transfer by orthogonal projection: making near-infrared calibrations robust to between-instrument variation. *Chemometrics and Intelligent Laboratory Systems*, **72**, 51–56.

Bergman EL, Brage H, Josefson M, Svensson O, Sparén A (2006) Transfer of NIR calibrations for pharmaceutical formulations between different instruments. *Journal of Pharmaceutical and Biomedical Analysis*, **41**, 89–98.

Blank ThB, Sum ST, Brown SD, Monfre SL (1996) Transfer of near-infrared multivariate calibrations without standards. *Analytical Chemistry*, **68**, 2987–2995.

Bouveresse E, Casolino C, de la Pezuela C (1998) Application of standardisation methods to correct the spectral differences induced by a fibre optic probe used for the near-infrared analysis of pharmaceutical tablets. *Journal of Pharmaceutical and Biomedical Analysis*, **18**, 35–42.

Carlson R, Carlson J, Grennberg A (2001) A novel approach for screening discrete variations in organic synthesis. *Journal of Chemometrics*, **15**, 455–474.

Duponchel L, Ruckebusch C, Huvenne JP, Legrand P (1999) Standardisation of near-IR spectrometers using artificial neural networks. *Journal of Molecular Structure*, **480/481**, 551–556.

Feudale RN, Woody NA, Tan H, Myles AJ, Brown SD, Ferre J (2002) Transfer of multivariate calibration models: a review. *Chemometrics and Intelligent Laboratory Systems*, **64**, 181–192.

Fontaine J, Hörr J, Schrimer B (2004) Amino acid contents in raw materials can be precisely analyzed in a global network of near-infrared spectrometers: collaborative trials prove the positive effects of instrument standardization and repeatability files. *Journal of Agricultural and Food Chemistry*, **52**, 701–708.

Gemperline PJ, Cho JH, Aldridge PK, Sekulic SS (1996) Appearance of discontinuities in spectra transformed by the piecewise direct instrument standardization procedure. *Analytical Chemistry*, **68**, 2913–2915.

Kennard RW, Stone LA (1969) Computer aided design of experiments. *Technometrics*, **11**, 137–148.

Noord OE de (1994) Tutorial—Multivariate calibration standardization. *Chemometrics and Intelligent Laboratory Systems*, **25**, 85–97.

Nørgaard L (1995) Direct standardisation in multi wavelength fluorescence spectroscopy. *Chemometrics and Intelligent Laboratory Systems*, **29**, 283–293.

Park RS, Agnew RE, Gordon FJ, Barnes RJ (1999) The development and transfer of undried grass silage calibrations between near infrared reflectance spectroscopy instruments. *Animal Feed Science and Technology*, **78**, 325–340.

Shenk JS, Westerhaus MO (1989) US Patent 4866644, September 12.

Siano GG, Goicoechea HC (2007) Representative subset selection and standardization techniques. A comparative study using NIR and a simulated fermentative process UV data. *Chemometrics and Intelligent Laboratory Systems*, **88**, 204–212.

Sjöblom J, Svensson O, Josefson M, Kullberg H, Wold S (1998) An evaluation of orthogonal signal correction applied to calibration transfer of near infrared spectra. *Chemometrics and Intelligent Laboratory Systems*, **44**, 229–244.

Swierenga H, de Groot PJ, de Weijer AP, Derksen MWJ, Buydens LMC (1998) Improvement of PLS model transferability by robust wavelength selection. *Chemometrics and Intelligent Laboratory Systems*, **41**, 237–248.

Swierenga H, de Weijer AP, Buydens LMC (1999) Robust calibration model for on-line and off-line prediction of Poly(Ethylene Terephthalate) yarn shrinkage by Raman spectroscopy. *Journal of Chemometrics*, **13**, 237–249.

Tan HW, Brown SD (2001) Wavelet hybrid direct standardization of near-infrared multivariate calibrations. *Journal of Chemometrics*, **15**, 647–663.

Wang Y, Kowalski BR (1993a) Standardization of second-order instruments. *Analytical Chemistry*, **65**, 1160–1174.

Wang Y, Kowalski BR (1993b) Temperature-compensating calibration transfer for near-infrared filter instruments. *Analytical Chemistry*, **65**, 1301–1303.

Wang Y, Veltkamp DJ, Kowalski BR (1991) Multivariate instrument standardization. *Analytical Chemistry*, **63**, 2750–2756.

Wülfert F, Kok WTh, Smilde AK (1998) Influence of temperature on vibrational spectra and consequences for the predictive ability of multivariate models. *Analytical Chemistry*, **70**, 1761–1767.

Wülfert F, Kok WTh, de Noord OE, Smilde AK (2000a) Linear techniques to correct for temperature-induced spectral variation in multivariate calibration. *Chemometrics and Intelligent Laboratory Systems*, **51**, 189–200.

Wülfert F, Kok WTh, de Noord OE, Smilde AK (2000b) Correction of temperature-induced spectral variation by continuous piecewise direct standardization. *Analytical Chemistry*, **72**, 1639–1644.

Zachariassen CB, Larsen J, van den Berg F, Balling Engelsen S (2005) Use of NIR spectroscopy and chemometrics for on-line process monitoring of ammonia in low methoxylated amidated pectin production. *Chemometrics and Intelligent Laboratory Systems*, **76**, 149–161.

Infrared (IR) Spectroscopy—Near-Infrared Spectroscopy and Mid-Infrared Spectroscopy

6

Mengshi Lin, Barbara A Rasco, Anna G Cavinato, and Murad Al-Holy

Introduction

Infrared spectroscopy (IR) is conventionally divided into three wavelength regions: the near-infrared (NIR: 750–2500 nm or 13 333–4000 cm^{-1}), mid-infrared (MIR: 2500–25 000 nm or 4000–400 cm^{-1}), and far-infrared (25–1000 μm or 400–10 cm^{-1}). The distinction between these three regions can vary depending upon the type of

Infrared Spectroscopy for Food Quality Analysis and Control
ISBN: 978-0-12-374136-3

instrumentation used to acquire IR spectral information. Other distinctions are based upon radiation properties; for example, it is widely agreed that NIR covers electromagnetic radiation with the range between 750 and 2500 nm. However, some users define NIR range as 650–2500 nm, which overlaps with part of the red light region of the visible spectrum (400–750 nm). In principle, the NIR region starts from where the human eye generates no visual response. With wavelengths of 650 nm and higher, the response of the human eye is so low that this tail end of the visible spectrum is often included as part of the NIR region (Raghavachari, 2001).

In some publications, spectroscopy within the range of 600–2500 nm is referred to as "far-visible spectroscopy" and sometimes as "overtone vibrational spectroscopy." Also, because of the significant number of applications within the wavelength range from 600 to 1100 nm, a new spectral region, the short-wavelength near-infrared region (SW-NIR) is separately designated. The American Society for Testing Materials defines short-wavelength near-infrared region (SW-NIR: 700–1100 nm) to include part of the visible light region (700–750 nm). This chapter will primarily focus on instrumentation for near-infrared spectroscopy and mid-infrared spectroscopy, including historical developments in IR instrumentation and how IR instruments were designed for applications in food quality analysis and control.

History of infrared instrumentation

Early interest in IR technology and instrumentation can be traced to the nineteenth century, when William Herschel first observed IR radiation on a photographic plate. Herschel's experiments were aimed at filtering the heat from a telescope but along with these developments he demonstrated that light radiation exists beyond the visible spectrum (Ciurczak, 2001).

During the first half of the twentieth century, interest in IR-based analytical methods developed and was mainly focused in the MIR region (Rabkin, 1987). The first custom-made IR instrument appeared in industrial laboratories during the 1930s. During World War II, the development of photoelectric detectors, primarily lead sulfide (PbS) detectors usable in the IR region, made it possible to record IR spectra. This advance revived scientific interest in IR and renewed research into IR spectroscopy.

The first commercial IR instrument was the Beckman IR-1 marketed in 1942 for analyses in the rubber and petroleum fields. The Beckman IR-1 was both an infrared and visible spectrophotometer but it was not widely sold because of governmental restrictions. The IR-1 used spherical mirrors and had long focal lengths and optical paths to affect dispersion. In 1944, Perkin-Elmer Inc. introduced the Model 12 infrared spectrometer that used aspherical shaped mirrors, making this a more compact instrument with higher resolution due to smaller focal lengths and shorter beam paths. The Model 12 was a single-beam instrument designed for point-by-point measurement. However, highly trained personnel were required for routine operation (Rabkin, 1987).

The first double-beam IR instrument, the Perkin-Elmer 21, was sold in 1947 and was the first widely adopted IR spectrophotometer (McClure, 2001). A double-beam optical system provided flexibility, making it a versatile and popular analytical instrument in almost every branch of analytical chemistry including applications in food and agriculture (Beckman *et al.*, 1977). In the mid-1950s, Wilbur Kaye of Beckman Instruments published two important papers demonstrating the underlying theoretical principles of IR spectroscopy and instrumental analysis (Barton, 2002).

Later in the 1950s, NIR became recognized for its potential in quantitative analysis (Osborne and Fearn, 1986). In its very first applications, NIR was simply employed as an accessory to other optical devices such as UV/Vis and MIR spectrometers. NIR instruments became differentiated from MIR when PerkinElmer and Cary manufactured commercial visible and NIR instruments separately (Barton, 2002).

A major breakthrough for NIR quantitative analysis of foods originated from the efforts of Karl Norris at the US Department of Agriculture (USDA) in the 1950s to develop a moisture meter for cereal products. Norris investigated the optical properties of dense, light-scattering materials. However, difficulties were encountered from interference in the moisture measurements as a result of absorptions due to other constituents such as protein, starch, and oil. The problem was finally solved when computer-aided multivariate statistical regression methods were developed to correlate NIR spectral features with reference values. This improvement not only made the water analysis possible, but also permitted the determination of other components such as protein, starch, and crude lipid simultaneously in grains and food products with relatively low moisture content (Osborne and Fearn, 1986). In 1975, McClure and Shenk expanded these applications to other agricultural products, including tobacco and forages.

In the 1970s and 1980s, there was an explosion of agricultural applications for NIR developed through government and university research programs, with the first official NIR method being adopted by the Canadian Grain Commission for protein testing in wheat. Another major development occurred when Dr Wetzel at Kansas State University developed a field test instrument for wheat protein testing and quality evaluation of flour that could be used at grain elevators and other remote locations. The American Association of Cereal Chemists (AACC) accepted NIR methods in 1982 for grain analysis (Osborne and Fearn, 1986). The research of the following individuals was particularly important during this period: McClure (tobacco and forage), Norris (cereals, meats, and forage), Williams (cereals and pulses), Tkachuk (grain), Shenk (forage), Wetzel (cereals), Birth (corn, fruits, and vegetables), Murry (forage), Barton (forage), and Osborne (cereals) (Osborne and Fearn, 1986; McClure, 1994).

In the 1980s, simple, stand-alone NIR instruments for chemical analysis appeared in the market propelled by the introduction of light fiber optics. Improved monochromators and detectors finally made NIR a more practical technique for the applications in agricultural and food science. For example, the forage industry fully adopted the NIR methods in the early 1980s (McClure, 1994).

Prior to the 1980s, conventional MIR spectrometers were dispersive instruments that measured the amount of absorbed energy based upon the fact that the frequency of IR light varies when passing through a monochromator. Most of these early MIR instruments were double-beam dispersive grating systems similar to the systems used in UV/Vis spectroscopy. Starting in the late 1960s, an important technological advance occurred with Fourier transform infrared spectrometers (FTIR) becoming popular. Some of the advantages of FTIR include speed of analysis, a high signal-to-noise (S/N) ratio, sensitivity, cost, and precision (Smith, 1996). Most modern MIR instruments are FTIR using interferometers instead of monochromators (Coates, 1997). As a result of this, the term "MIR spectroscopy" and "FTIR spectroscopy" are used interchangeably. MIR refers specifically to an electromagnetic spectral range of 4000–400 cm^{-1} which is the typical range of FTIR instruments. Although not popular anymore, dispersive MIR instruments still remain on the market.

Currently, various techniques are used to separate polychromatic NIR light into monochromatic light for both qualitative and quantitative analysis of food samples, including light-emitting diode (LED), diffraction grating, interferometer, diode-array or acousto-optic tunable filter (AOTF), permitting expansion of IR analyses into new arenas. For example, in the mid-1990s, PerkinElmer introduced the first Fourier transform near-infrared (FT-NIR) instrument, which produces better testing results than those obtained with traditional dispersive instruments.

With recent technological advances in both hardware and software design, much attention has been focused on developing more compact, portable, and robust NIR and MIR spectrometer systems with advanced and sophisticated software to support much faster spectral data processing and analysis. Improvements in microprocessors made it possible in recent years to analyze spectral data in seconds, a task that would have taken several hours in the 1990s. To date, NIR and MIR have become so popular that they have found practical applications in virtually all branches of agricultural and food industries.

Optical systems in infrared instruments

A typical NIR instrument consists of a radiation source, a wavelength selection device such as a monochromator, a sample holder, a photoelectric detector that measures the intensity of detected light and converts it to electrical signals, and a computer system that acquires and processes spectral data. Traditional dispersive grating MIR spectrometers had similar optical systems. An optical diagram of a traditional dispersive grating MIR apparatus is shown in Figure 6.1. The IR light emitted

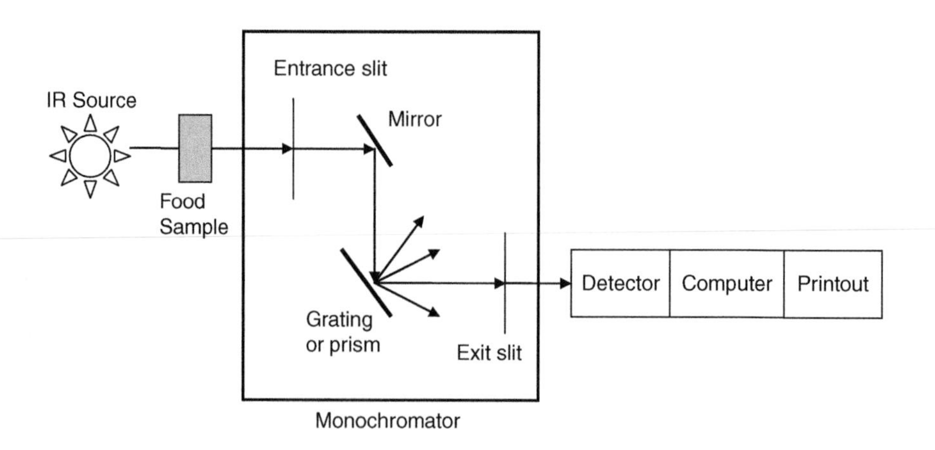

Figure 6.1 Dispersion of a single infrared beam by a diffracting grating. The angle of diffraction of lights depends on both the wavelength and the order of diffraction. Adapted from White (1990).

from an IR source passes through a sample, then through a monochromator that contains optical devices such as a prism or grating. The light is guided through the prism or grating and separated into a spectrum of different wavelengths. An exit slit is employed to allow light at specific wavelengths to be segmented into readily measurable components and pass through to reach the detector. In situations where a grating or prism is used, these can be rotated to allow the light at different wavelengths to pass through the exit slit (Smith, 1996).

Radiation source

The requirements of an ideal radiation source for an IR instrument include a wide wavelength range preferably covering the whole NIR or MIR measurement range, a strong and stable intensity of light from the source at different wavelengths over a short and long period of time, and a power supply which provides consistent stable energy to the source.

Two major types of radiation sources are used in analytical spectroscopy, either a continuum source or a line source (Table 6.1). Continuum sources emit light with a relatively stable and continuous intensity over a wide range of wavelengths and are most common for molecular absorption and fluorescence spectrometric instrumentation. Line sources emit at only a few discrete wavelengths of light and the intensity of the emitted light varies at different wavelengths (Robinson *et al.*, 2005).

Continuum sources, such as tungsten halogen lamps and common incandescent radiation sources are widely employed in NIR spectrometers due to their low cost and high intensity. For example, many NIR instruments use quartz halogen lights as NIR radiation sources. A typical quartz halogen lamp consists of a tungsten wire filament and iodine vapor that is enclosed in a quartz envelope or bulb.

Usually a tungsten lamp with a quartz bulb is satisfactory for NIR measurements in food analysis. In a standard tungsten filament lamp, the tungsten evaporates from the filament and then deposits itself on the wall of the lamp. Tungsten halogen lamps provide more stable performance and longer life than standard tungsten filament lamps (McClure, 2001). The halogen gas provides some cleaning action by removing the evaporated tungsten on the wall of the lamp and redepositing it back on the filament,

Table 6.1 Radiation sources used in analytical spectroscopy

Type	Source	Region
Continuum sources	Tungsten filament lamp	Visible light and NIR
	The deuterium tungsten light	UV, NIR
	High pressure mercury or xenon arc lamps	UV and visible
	Heated solid ceramics or heated wires	MIR
Line sources	Hollow cathode lamps	UV and visible
	Electrodeless discharge lamps	UV and visible
	Light-emitting diode (LED)	UV and visible, IR
	Sodium or mercury vapor lamps	UV and visible
	Lasers	UV and visible, IR, Raman

From Robinson *et al.* (2005).

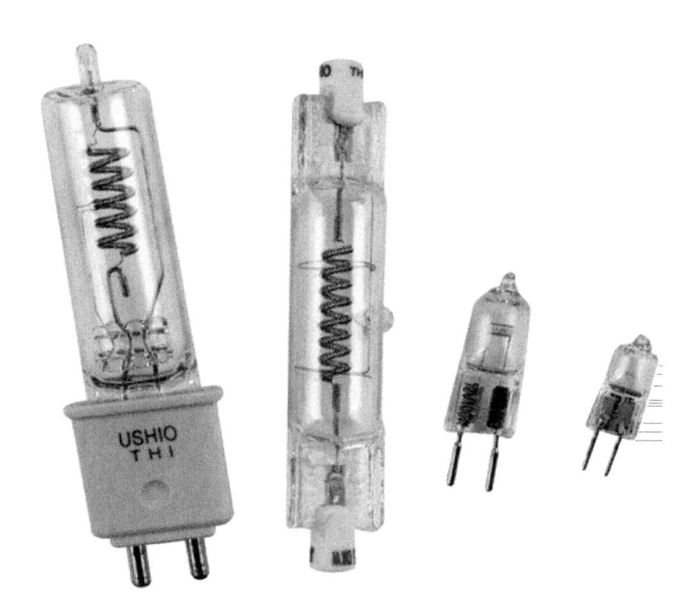

Figure 6.2 Four quartz tungsten halogen (QTH) lamps for NIR instruments. QTH lamps are good visible/NIR sources due to their smooth spectral curve and stable output. These lamps use a doped tungsten filament inside a quartz envelope and are filled with a rare gas and a small amount of halogen. Permission to use this photo has been granted by Newport Corporation. (www.newport.com).

making the intensity of the source much higher and the output of the lamp more stable. The typical light range of these sources is between 400 and 5000 nm (Robinson *et al.*, 2005). Figure 6.2 shows four commercial quartz tungsten halogen (QTH) lamps. However, tungsten halogen lamps have some major disadvantages, such as the large amount of heat generated, relatively short lifetime, and "drift" problems (e.g. the spectral content of the output drifts over time) (Cen and He, 2007).

Line sources or such as lasers, laser diodes, and LEDs are commonly used in instruments dedicated to specific applications where full spectrum acquisition is not necessary and normally used for moisture and crude lipid measurements in grain and oilseed analysis. Using a line source makes it possible to develop portable and relatively inexpensive analytical devices (McClure *et al.*, 2002). Lasers are common sources that provide coherent line radiation for NIR and MIR instruments. A major advantage of laser sources is that they emit monochromatic radiation with high intensity. Two types of IR laser sources are widely used in NIR spectroscopy: a gas phase laser and a solid-state laser. The tunable carbon dioxide laser is a gas laser while the diode laser is a solid-state source. The helium neon (HeNe) lasers (Figure 6.3), for instance, emit highly monochromatic radiation and are widely used in red, infrared, and far-infrared regions. HeNe lasers have an emission that is determined by neon atoms by means of a resonant transfer of excitation of helium. LED is a semiconductor diode that emits discrete, incoherent, and narrow spectral bands. LEDs and lasers differ in spot size. Different spot sizes of lasers can be achieved since lasers are inherently collimated; while LED light diverges as it leaves the source, making it difficult to focus all energy into a spot (Bozkurt and Onaral, 2004).

Figure 6.3 A helium neon (HeNe) laser. Photo courtesy of Meredith Instruments (Glendale, AZ, USA).

Two common types of radiation sources used in many MIR instruments are electrically heated rigid ceramic rods and coiled wires. Commercial MIR ceramic sources come with a variety of sizes and shapes, including different types of cylindrical rod or tubes (e.g. the Nernst glower and the Globar). The Nernst glower has been replaced by the Globar which is made of silicon carbide (SiC) that operates at about 1100°C. The shape of electrically heated coiled wires is similar to that of an incandescent light bulb filament. These coiled wires are heated electrically in air to a temperature of approximately 1100°C, which may result in a "burn out" problem as they age. To provide more uniform light output over time, in some cases, coiled wires are wrapped around a ceramic rod for support (Robinson *et al.*, 2005).

Wavelength selection devices

The mechanism of light separation plays an important role in IR instrumental design. Various types of NIR and MIR instruments are accordingly classified as filter-based instruments, monochromator-based instruments, Fourier transform instruments, and other instruments (Wetzel, 2001).

Filters

There are two major types of filters used in NIR to select wavelengths of the light: optical interference filters and electronically tunable filters. Interference filters consist of multiple thin layers of dielectric material with different refractive properties. They are simple and cheap choices for NIR light selection (Wetzel, 2001). Electronically tunable filters are a group of devices whose spectral transmission can be electronically controlled by adjusting voltage, acoustic signal, and other parameters (Gata, 2000). Two prevailing electronically tunable filters are liquid crystal tunable filter (LCTF) and acousto-optical tunable filter (AOTF). A typical LCTF consists of a stack of polarizers and tunable retardation liquid crystal plates; while an AOTF utilizes radio frequency (RF) acoustic waves to separate a single wavelength of light from a broadband radiation source through a crystal. The selected wavelength of light is dependent upon the frequency of the RF applied to the crystal. As a consequence, the wavelength of the filtered light can be selected by adjusting the frequency (Gata, 2000; Workman and Burns, 2001).

Monochromator

A monochromator is an optical device used to disperse light with wide range of wavelengths into monochromatic light at different wavelengths. The word "monochromator" is derived from the Greek roots of "mono-" which means single, "chroma" which means "color," and the Latin suffix "-ator" which means "an agent". In optics, dispersion is a phenomenon that results in the separation of light into spectral components of different wavelengths, depending upon the speed of the light at each wavelength. In some cases, dispersion is referred to as chromatic dispersion because of its wavelength-dependent nature.

Prism

The most commonly used prisms are silicate glass for visible and NIR instruments and prisms made of NaCl or potassium bromide (KBr) for MIR instruments. Historically, the prisms were the most widely used dispersion elements in a monochromator. A glass prism can disperse IR light into a spectrum of light with different wavelengths (Scotter, 1997). However, prisms have a major drawback in that their useful wavelength range is limited. Therefore, in modern NIR and MIR spectrometers, prisms have been replaced by diffraction gratings or by interferometers that are used in Fourier transform systems because of their capacity for linear dispersion or their higher performance in terms of sensitivity and signal-to-noise ratio.

Diffraction grating

A diffraction grating is an optical element that disperses or separates light radiation into its constituent wavelengths of lights. Each wavelength of dispersed light occupies a specific position in space (Figure 6.4). The polychromatic IR light that hits the

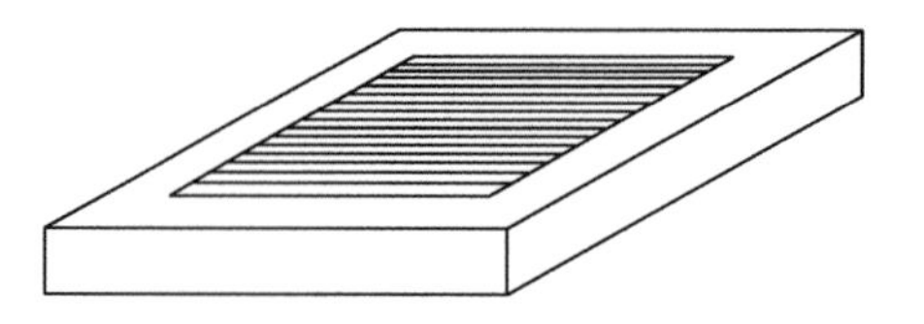

Figure 6.4 A typical diffraction grating used in near-infrared instruments.

grating is dispersed and each constituent light of different wavelength is reflected from the grating at a slightly different angle.

A typical diffraction grating is a piece of substrate made of glass, metallic, or ceramic material the surface of which has been cut into closely spaced and replicated parallel grooves and coated with a reflecting material such as aluminum. Diffraction gratings can be used to spatially disperse light of different wavelengths. The use of diffraction gratings can be traced back to the late eighteenth century. However, early diffraction gratings had the major limitation that they were relatively imprecise, yielding ghostlike spectra instead of distinctive spectral readings. In the later nineteenth century, Henry A Rowland, the first physics professor at Johns Hopkins University, and one of the finest physicists of his day, became famous in part for building devices that could accurately engrave gratings, making the production of high-quality gratings possible. His contribution led to significant advances in analytical spectroscopy (Hendricks, 2000). Most modern diffraction gratings are molded from masters, producing replicate gratings, or manufactured with new technologies such as holographic engraving or semiconductor lithography.

The grooves on the diffraction grating are identical in size and the whole grating is relatively compact (Figure 6.4). A typical grating for the IR region may have 50–200 grooves per mm. The quality and uniformity of spacing of the grooves are critical to the performance of the grating. The dispersion efficiency of a grating largely depends on the distance between adjacent grooves and the angle of the grooves. The spacing is fixed and cannot be altered without changing the grating. Diffraction gratings perform better and are more precise and efficient than prisms because gratings provide a linear dispersion of wavelengths. However, diffraction gratings have a major limitation due to multiple order of diffraction, where light of different wavelengths may leave the grating at the same angle, travelling along the same light path without being separated by the grating. This can be minimized by using cut-off filters that eliminate higher order diffraction wavelengths (White, 1990).

IR incident beam that hits the diffraction grating is dispersed on the diffraction grating (Figure 6.5). Each wavelength of dispersed light bounces away from the grating at a specific angle. A small selected range of light with specific wavelengths at the correct angle will be allowed to pass the exit slit and strike the IR detector.

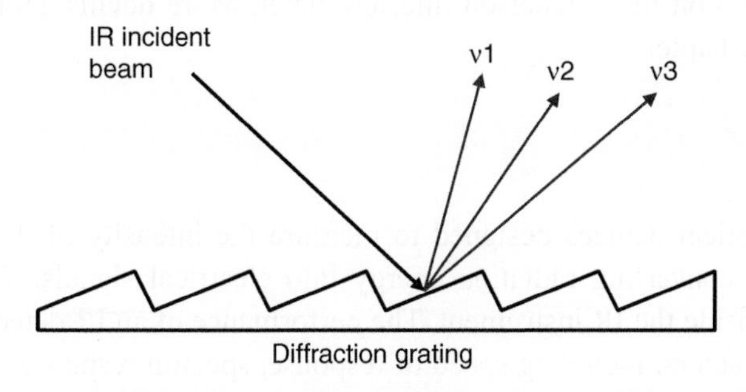

Figure 6.5 Dispersion of a single infrared beam by a diffraction grating. The angle of diffraction of lights depends on both the wavelength and the order of diffraction. Adapted from White (1990).

Monochromator

A monochromator is typically composed of a dispersion element, an entrance slit and an exit slit, and lenses or mirrors (Figure 6.1). Prisms and gratings are widely used as dispersion elements to disperse or spread IR light radiation. Lenses and mirrors are used to focus the IR light. In a dispersive instrument, the IR light is emitted from the source, enters the monochromator through the entrance slit, and penetrates into the dispersion element such as a prism or grating. There, the light is dispersed into different wavelengths. The exit slit allows a very narrow wavelength of dispersed light to pass through and hit the detector. One major advantage of the monochromators is that they can be easily used to select a pure color of light over a wide wavelength range (Osborne and Fearn, 1986).

A dispersive spectrometer is based upon a monochromator system. Some NIR instruments use stand-alone grating systems while some choose double-beam dispersive systems like UV/Vis systems. In some ways, NIR instrumentation is quite similar to that of UV/Vis systems. Many modern commercial UV/Vis instruments extend their ability of spectral acquisition into the NIR range, with the same detector being used to collect spectral information from the entire UV/Vis and NIR range. Likewise, some MIR instruments also extend their detection range to the NIR range.

A major disadvantage of dispersive MIR instruments (Figure 6.1) is that the exit slit restricts the range of possible wavenumbers that can be used, thus significantly reducing the amount of IR light that reaches the detector. Besides, only a small amount of light can be measured at a time and each wavenumber is measured for only a small fraction of the total measurement time (Smith, 1996). However, all these problems mentioned above have been well solved by FTIR-based systems that are based on a Michelson interferometer with a movable mirror to acquire an interferogram.

Interferometer

The interferometer was invented by Albert Abraham Michelson in the famous Michelson–Morley experiment (Smith, 1996). FTIR and FT-NIR spectrometers operate by applying Fourier transform to interferograms obtained from a Michelson interferometer with a movable mirror. Most modern MIR instruments are FTIR instruments based on the Michelson interferometer. More details about FT systems are provided in Chapter 7.

Detectors

Detectors are optical devices designed to measure the intensity of the IR light that strikes them by converting radiation energy into electrical signals. The detector is usually placed inside the IR instrument. The performance of an IR detector is dependent upon many factors, including speed of response, spectral response, limit of detection, and temperature of operation (McClure, 2001). The selection of detectors used in different IR instruments depends on the range of wavelengths to be measured.

There are two major types of IR detectors classified according to the principles of operation: thermal detectors and photon-sensitive detectors (photodiodes). Thermocouples, bolometers, thermistors, Golay cells, and pyroelectric devices such as those based on deuterated triglycine sulfate (DTGS) are examples of thermal detectors; while silicon photodiode, indium gallium arsenide (InGaAs), lead selenide (PbSe), mercury cadmium telluride (MCT), and indium antimonide (InSb) are photon-sensitive semiconducting detectors (Robinson *et al.*, 2005).

Bolometers and microbolometers are very sensitive electrical thermometers that operate based upon changes in electrical resistance, making them very suitable for IR radiation detection. The modern microbolometers have dimensions of a few micrometers and respond quickly. They are widely used for far-IR detection. Thermocouples and thermistors are also based upon measurement of a thermoelectric effect while Golay detectors are based on the detection of thermal expansion of the material. Semiconductor detectors are very sensitive and can reach a steady electrical signal within a very short time, thus permitting the rapid scan of a series of wavelengths, which is important for gathering IR spectral information (Robinson *et al.*, 2005).

Photodetectors are more commonly used in NIR systems. A silicon photodiode array detector works most efficiently in the visible and SW-NIR range of 16 700–9000 cm^{-1} (Table 6.2). Other commonly used detectors in NIR instruments include InGaAs, DTGS, and PbSe devices. The response of InGaAs begins around 12 000 cm^{-1} and ends around 6000 cm^{-1} depending upon the specific configuration of the instrument.

For MIR instruments, pyroelectric detectors such as DTGS (Figure 6.6) detectors are the most widely used. For example, a DTGS detector with a cesium iodide window (DTGS/CsI) covers the wavenumber range from 6400 to 200 cm^{-1}, which includes part of the NIR range, the entire MIR region, and part of the far-IR range. The MCT detectors (Figure 6.7) can dramatically improve the sensitivity, resulting in the recovery of high-resolution spectral information. This significantly reduces the sampling time when IR throughput is low and sample measurements must be made at high speed (www.newport.com).

Table 6.2 Commonly used detectors in infrared instruments

NIR region	Responsivity range (cm^{-1})	MIR region	Responsivity range (cm^{-1})
Silicon	16 700–9000	DTGS/KBr	12 000–350
InGaAs	12 000–6000	DTGS/CsI	6400–200
PbSe	11 000–2000	MCT	11 700–400
InSb	115 000–1850	Photoacoustic	10 000–400
MCT	117 000–400		
DTGS/KBr	12 000–350		

NIR, near-infrared; MIR, mid-infrared; DTGS, deuterated triglycine sulfate; InGaAs, indium gallium arsenide; PbSe, lead selenide; InSb, indium antimonide; MCT, mercury cadmium telluride.
From Thermo Scientific, Madison, WI, USA; Newport Corporation, Irvine, CA, USA; Robinson *et al.* (2005).

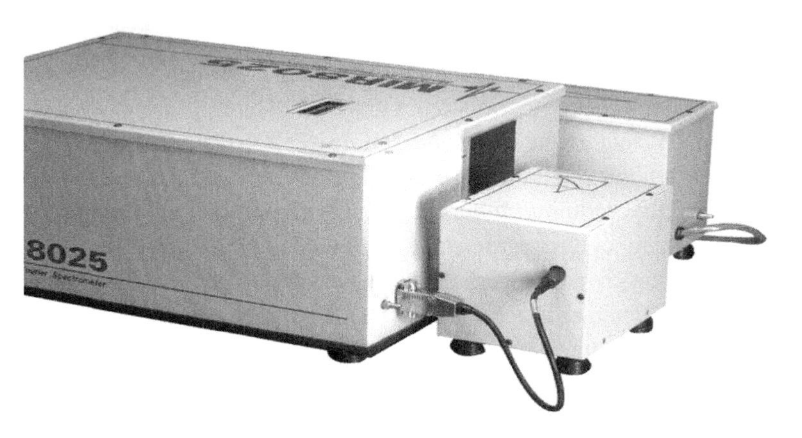

Figure 6.6 A deuterated triglycine sulfate (DTGS) detector for a mid-infrared spectrometer that exhibits large and spontaneous electrical polarization signals when the incident infrared beam affects its polarization.Permission to use this photo has been granted by Newport Corporation. All rights reserved (www.newport.com).

Figure 6.7 A liquid nitrogen cooled mercury cadmium telluride (MCT) detector. It has a broad spectral response close to that of deuterated triglycine sulfate (DTGS) but acquires spectral data 8 times faster than DTGS.Permission to use this photo has been granted by Newport Corporation. All rights reserved (www.newport.com).

Single-beam and double-beam optics

IR optics consists of a specific arrangement of IR instrumental components such as single-beam or double-beam optics. A single-beam arrangement is used for all spectroscopic emission systems, while double-beam optics are common for spectroscopic absorption systems.

Figure 6.8 shows a diagram of a single-beam transmittance apparatus that is well suited for the reliable quantitative analysis of food samples. The PerkinElmer IR instrument model 12 was such a single-beam instrument. Single-beam IR instruments

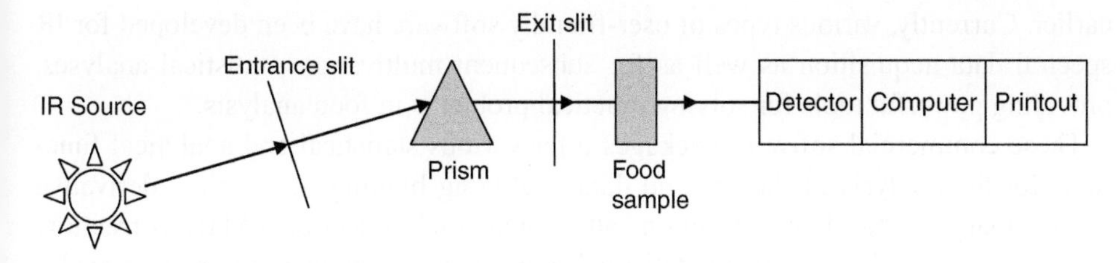

Figure 6.8 A single-beam transmittance apparatus.

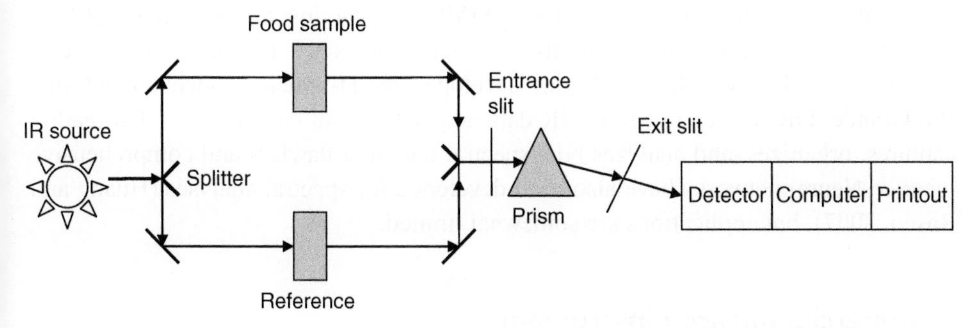

Figure 6.9 A double-beam transmittance apparatus.

were quite popular until 1949 when double-beam instruments became available. A major problem associated with early single-beam systems was the fluctuation of the intensity of the source radiation, which led to an analytical error called "drift." In addition, it required tedious manual replotting to obtain a standardized spectrum. These problems were solved by the use of double-beam systems (Figure 6.9). In double-beam optics, the light intensity of the IR beam is measured before and after passing through a food sample by splitting the source radiation with the aid of a beam splitter (a half mirror). The light is usually split into two beams of equal energy, with one passing through the reference side called "the reference beam," and the other passing through the sample called "the sample beam." After passing through the sample, the sample beam merges with the reference beam and together they pass through a monochromator then to a detector. The ratio of the reference beam to the sample beam can be measured, and this ratio is not influenced by the fluctuation of radiation source intensity and other factors such as the drift. By constantly comparing the relative intensity of the sample and reference beams, the double-beam systems offer more stable and rapid measurements than single-beam systems (Workman and Burns, 2001; Robinson *et al.*, 2005).

Software

IR spectral data acquired with computerized IR systems need to be processed and analyzed using software packages. In the early days of IR spectroscopy, IR instruments were not well supported with computerized programs for data analysis such as those being used to today. Improvements in microprocessors made it possible to analyze sets of spectral data in seconds that had taken much longer time a few years earlier. Currently, various types of user-friendly software have been developed for IR spectral data acquisition as well as for subsequent multivariate statistical analyses, providing powerful tools for solving practical problems in food analysis.

These commercial software packages offer various statistical and analytical functions for the analysis of the spectral data, including binning, smoothing, derivative transformations, baseline correction, attenuated total reflection (ATR) correction, normalization, and multivariate statistical tools such as principal component analysis (PCA), principal component regression (PCR), partial least squares (PLS), soft independent modeling of class analogy (SIMCA), evolving factor analysis (EFA), two-dimensional (2D) correlation, etc. (Martens and Naes, 1989). For example, a commercial software called DeLight, developed by D-Squared Development Inc. (La Grande, OR, USA), integrates IR data acquisition and data analysis that readily captures, organizes, and analyzes NIR spectral data in a flawless and comprehensive manner. Neural networks have also been developed for spectral analyses (Huang and Rasco, 2007), but applications are somewhat limited.

Commercial infrared instruments

Since the first commercial IR instrument, the Beckman IR-1, was launched in 1942, many IR instruments have been manufactured and sold. With recent developments and advances in analytical chemistry and IR technologies, NIR and MIR instruments have been widely employed to meet the needs of the pharmaceutical, agricultural, food and beverage, chemical, and other industries.

Table 6.3 shows a list of the names and models of some major NIR instruments and their manufacturers, including Foss NIRSystems, Thermo Fisher Scientific, PerkinElmer, Hitachi High Technologies, Analytical Spectral Devices, and other companies. Most of these manufacturers exhibit their new IR instruments at international analytical conferences such as the Pittsburgh Conference (Pittcon), which focuses on analytical chemistry and applied spectroscopy and is held annually. Like other industrial sectors, analytical chemistry instrumentation has experienced several mergers over the past two decades. For instance, in 2007 Thermo Electron and Fisher Scientific combined to create Thermo Fisher Scientific, becoming a leading provider of instruments, equipment, reagents, software and services for research and analysis.

NIR instruments can be classified into two major categories: laboratory analyzers for research purposes and process analyzers for use in the production lines or operation in harsh manufacturing conditions. Laboratory NIR analyzers generally come with broad scanning range and adjustable scanning rate for testing a variety of samples; while process analyzers are specifically optimized for defined applications.

Laboratory analyzers

NIR laboratory analyzers are widely used in research and development (R&D), quality control (QC), process control in food processing plants, and academic laboratories because they are simple, quick, reagentless, flexible, and require little or no sample preparation.

Table 6.3 Some major manufacturers and models of near-infrared instruments

Company	Products
Analytical Spectral Devices Inc (ASD)	QualitySpec® Pro, OSB, and iP; LabSpec® 5000/5100 series
Axsun Technologies	IntegraSpec™ NIR Spectrometers
Brimrose Corporation	Luminar AOTF-NIR Spectrometers
Büchi Corporation	NIRFlex N-500; SpectraAlyzer
B&W Tek Inc.	BWS Series Pro UV-Vis-NIR Analyzer; i-Spec Series; portable NIR Spectrometer
Decagon Devices Inc.	AquaLab DA7200 NIR Analyzer
D-Squared Development	DPA20; Grain analyzer
Foss NIRSystems Inc	Infratec™ 1241 Grain Analyzer; NIRSystem model 6500; etc.
Hitachi High Technologies	U-4100 UV-Vis-NIR Spectrophotometer
Kett Corporation	KJT Series Near-infrared Analyzer
Jasco	The New V600 range of UV-Visible-NIR Spectrophotometers
LT Industries	LT-NIR
Malvern Instruments Ltd	The SyNIRgi Near Infrared Chemical Imaging (NIR-CI) systems
m.u.t. GmbH	TRISTAN spectrometers
NIR Technology Australia	Benchtop NIR Analyzer; NIT-38 Analyzers; the Series 1000 Alcohol Analyzer
Ocean Optics Inc.	NIR 256, 512 Spectrometers
PerkinElmer Inc.	Lambda 25/35/45 UV/Vis Systems; Lambda 950, 850 and 650 UV/Vis/NIR
Perten Instruments	Diode Array 7200 Full-spectrum NIR instrument
Thermo Fisher Scientific Inc.	Antaris Near-Infrared Analyzers
Unity Scientific Inc.	SpectraStar™ Series Near Infrared Analyzers
Varian Inc.	Cary range of UV-Vis-NIR Spectrophotometers
Zeltex Inc.	ZX-440, 800 Near-Infrared Analyzer

Figure 6.10 shows a commercial laboratory NIR composition meter that is designed for desktop use. It can be used to measure one or multiple food components depending on the number of measurement filters used in the instrument. The measurement of moisture content in a food product is conducted by focusing NIR light on the sample and determining how much light the sample absorbs. The properties of reflected light are dependent upon the characteristics of food composition in the sample. The moisture content can therefore be determined by measuring the NIR light attenuation caused by the sample. For example, a food sample with a moisture content of 10% will absorb and reflect NIR light differently from a sample with a moisture content of 15%. The NIR composition meter collects and measures this reflected light and converts it to moisture content values using specialized software.

Process analyzer

NIR and MIR process analyzers have been widely used in food processing and on-line quality control settings due to their heavy-duty and rugged design, high

Figure 6.10 A commercial lab NIR composition meter—KJT-270 (Kett Inc., Villa Park, CA, USA).

performance and flexibility. These IR process analyzers are specifically designed for routine analyses in production lines, factory floors, loading docks, warehouses, and manufacturing settings. Instead of being tested in the lab, samples in the production status are analyzed conveniently on-line and in real-time.

For example, recently, an on-line NIR testing system (Figure 6.11) has been developed by Analytical Spectral Devices (ASD) together with the USDA to predict meat tenderness in real-time at the processing plant. This system allows meat processors to analyze the tenderness of beef carcasses non-destructively and at the grading stand. In addition, the US Meat Animal Research Center has developed a method with this technology to predict the shear force of longissimus muscle slice at 14-day postmortem in cattle (*Bos taurus*). This new method helps beef processors to predict beef tenderness more accurately on nearly 100% of the carcasses on-line, thus significantly improving the quality of beef products (www.asdi.com).

The new Thermo Nicolet Antaris Near-infrared analyzer (Figure 6.12) manufactured by Thermo Fisher Scientific Inc. uses a unique method-development-sampling (MDS) system for solids, liquids, powders, pastes, and tablets analysis. Its hand-held diffuse reflectance fiber optic probe can take sample measurements directly in the production settings or can penetrate through packaging materials. This product line represents an industry-driven transfer of spectroscopy from laboratory to industry with fit-for-purpose NIR analyzers.

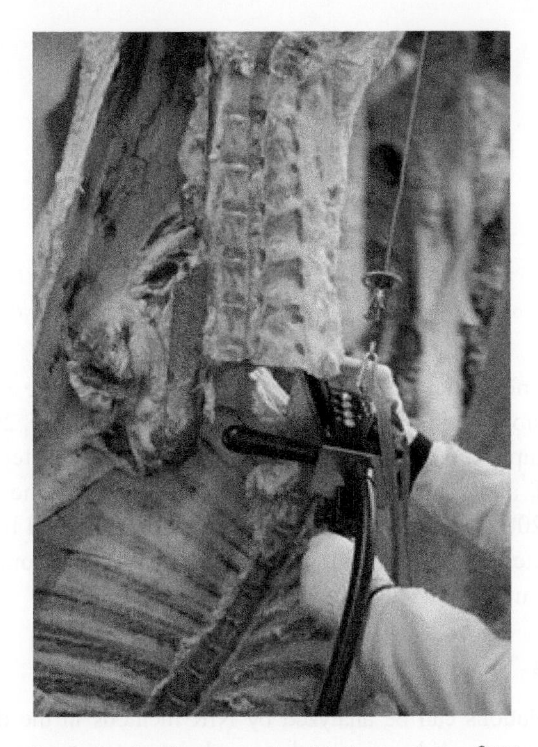

Figure 6.11 On-line testing for beef tenderness quality with the QualitySpec® BT system by ASD Inc. Photo courtesy of ASD Inc. (www.asdi.com).

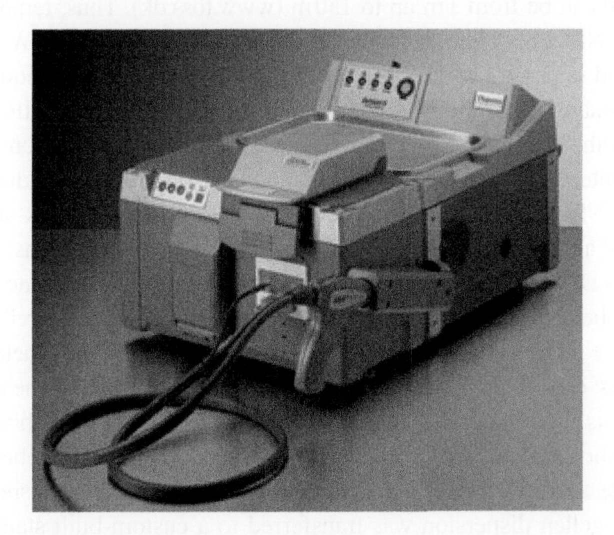

Figure 6.12 Thermo Nicolet Antaris FT-NIR analyzers (Thermo Fisher Scientific, Inc. Waltham, MA, USA).

Sampling techniques of IR methods

Because NIR measurements are usually conducted on bulk, raw and untreated food samples, how to sample the food products and present them to IR instruments for

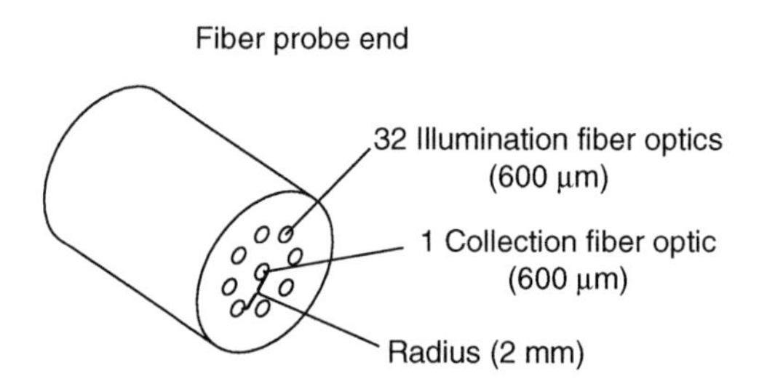

Figure 6.13 Schematic of a fiber optic probe used to acquire near-infrared spectra.

measurement becomes crucial. Two basic sampling methods are used for IR spectroscopy: transmission mode and diffuse reflectance mode (Rolfe, 2000). When light is incident on a sample it may be reflected, absorbed, or transmitted. The sum of the reflected, absorbed, and transmitted energy must be equal to the initial energy of the light (Tuchin, 2000). Thus, the amount of radiation absorbed by the sample can be measured by detecting the amount of energy transmitted through the sample or reflected from the sample.

Liquid samples

Liquids, gels, or solutions can be analyzed by NIR methods in the diffuse reflectance mode or in transmission mode by pouring the sample into a transparent cuvette. NIR light from the instrument is usually transferred to the sample by a fiber optic emitter. The length of the fiber optic can be from 1 m up to 150 m (www.foss.dk). Thus, remote sampling is possible with an NIR fiber optic probe connected to an NIR instrument. With a fiber optic probe, many food samples can be conveniently measured by NIR technique with little or no sample preparation. Because quartz optical fibers are transparent to NIR radiation, they are bundled together to acquire spectral information over a large area on food surfaces. The low OH-content quartz fibers are single filaments 100–600 μm in diameter and are widely used for NIR applications (Robinson *et al.*, 2005). Because NIR signals travel at the speed of light along the fiber optic probe, measurements are almost instantaneous.

For instance, a DPA-20 spectrophotometer (D-Squared Development Inc.) coupled with a fiber optic probe can be used to acquire NIR spectra in the diffuse reflectance mode. The probe (Figure 6.13) contains 32 illumination fibers (600 mm in diameter) arranged in concentric circles; each fiber is 2 mm away from a central pickup fiber. The diameter of the fiber optic probe is about 1 cm. An internal tungsten bulb illuminates the fibers.

Figure 6.14 shows a schematic representation of a fiber optic probe connected to a DPA-20 in the test cell for NIR spectral acquisition of a gellan dispersion (Huang *et al.*, 2003). A gellan dispersion was transferred to a custom-built stainless steel test cell with a water jacket surrounding the sample holder of the cell. The fiber optic probe was immersed into the gel dispersion from the top of the test cell to record NIR spectra.

Solid samples

Solid samples are analyzed by NIR methods with reflectance measurements. The collected radiation may be either produced by specular reflectance or diffuse reflectance.

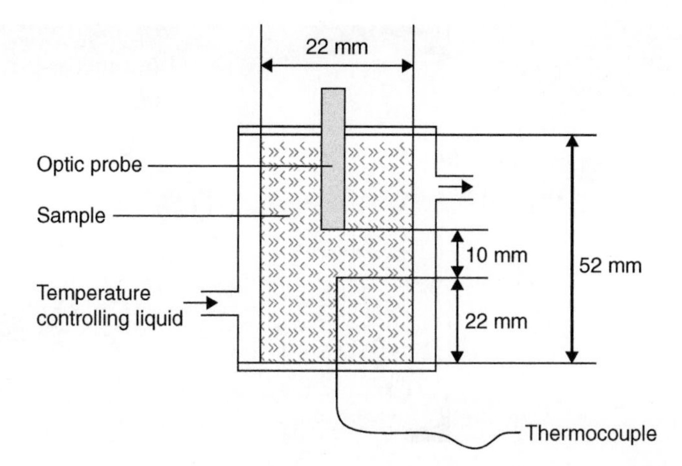

Figure 6.14 Schematic representation of positions of a fiber optic probe and a thermocouple probe in the test cell for near-infrared spectral acquisition of a tested hydrocolloid (Huang *et al.*, 2003).

Specular reflectance occurs at the surface of a sample and the radiation does not penetrate the sample. The specular reflectance in spectrophotometry is to be avoided since it provides no absorption information. Diffuse reflectance radiation penetrates into the sample and interacts with the sample before being reflected back to the surface. When radiation encounters particles in the sample are much larger than the wavelength, the radiation will propagate in all directions—a phenomenon called "scattering." Though scattering is generally a source of error in optical experiments, it may help enhance the weak absorption bands, thus providing useful diffuse reflectance spectra of biological samples since the actual path length is much larger than the sample thickness (Osborne and Fearn, 1986).

In diffuse reflectance mode, solid samples are usually placed in a sample cup for NIR measurement. Alternatively food samples are measured by placing the fiber optic probe directly in contact with the sample. The light penetration depth in a specific sample is a function of the geometry of the optical probe and the scattering characteristics of the sample (Birth and Hecht, 2001).

For example, Figure 6.15 shows a powerful probe (D-Squared Development, Inc.) utilized for SW-NIR spectral acquisition on coho salmon tissue. SW-NIR spectra were recorded with a probe of a customized spectrophotometer. The probe features four tungsten bulbs placed concentrically around a central fiber optic bundle. Compared with the fiber optic probe shown in Figure 6.13, this customized probe is much larger and more powerful. For this experiment all spectral measurements were recorded in diffuse reflectance mode. During the measurement, the probe was independently positioned on top of the skin, scales, and muscle tissues of the coho samples to acquire diffuse reflected signals.

Figure 6.16 illustrates a study where a diffuse reflectance NIR method was used to measure the quality indices of apples. An intact apple was measured by a Vis/NIR system consisting of a quartz halogen light source, a fruit holder, and a non-scanning polychromatic/diode array spectrometer. The apple was placed on the holder and irradiated by the light from below. Light entered the fruit, diffused in the apple flesh

Figure 6.15 A powerful probe (D-Squared Development Inc.) utilized for short-wavelength near-infrared (SW-NIR) spectral acquisition on coho salmon muscle (*Oncorhynchus kisutch*) through skin and scales.

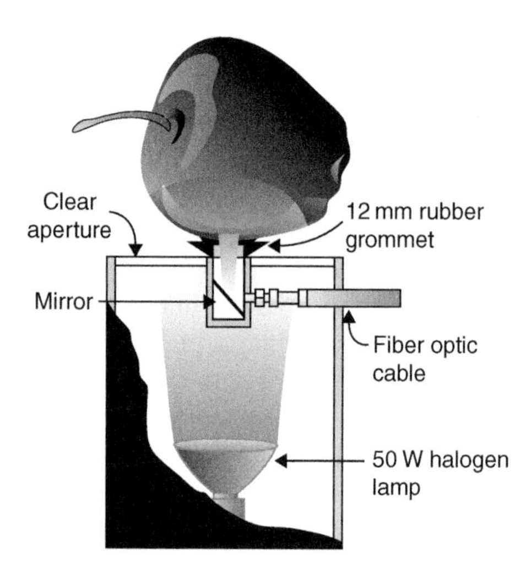

Figure 6.16 The visible/near-infrared (Vis/NIR) measurement system (McGlone *et al.*, 2002).

and then exited from the fruit. The light source and detector were on the same side, avoiding the measurement of surface reflections and so allowing subsurface penetration of the light into the apple flesh (McGlone *et al.*, 2002).

Gas samples

Traditionally, NIR technology was not widely used for gas analysis due to the low density of gas samples and the relatively weak overtone bands. However, recent developments of highly sensitive NIR instruments with tunable diode lasers could help to solve this problem. This new technique significantly increases the source brightness and thus improves the spectral resolution. For example, gas samples are filled within gas cells for NIR measurements using long fiber optic sample cells (www.goaxiom.com). Fiber optic coupled low volume gas cells can provide extremely low volume and rapid sample exchange. The design was specifically used for NIR analysis of gas samples.

Infrared band and spectral interpretation

IR spectroscopy is a technique based upon the overtones and vibrations of the atoms of a molecule when passing IR radiation through a tested sample. The energy at which any peak in an absorption spectrum appears corresponds to the frequency of a vibration of a part of the sample molecule (Stuart, 1997). In the IR region, various fundamental molecular vibrations, including those from C–H, O–H, N–H, C=O, and other functional groups can be detected (Weyer, 1985; Murray and Williams, 2001). For example, when a food is irradiated with NIR light, it absorbs the light with frequencies matching characteristic vibrations of particular functional groups, whereas the light of other frequencies will be transmitted or reflected (Foley *et al.,* 1998). Therefore, the biochemical components of a food tissue determine the amount and frequency of absorbed light and the quantity of reflected or transmitted light can be used to infer the chemical composition of that food tissue (André and Lawler, 2003).

Representative SW-NIR spectra collected on hot smoked king salmon portions are shown in Figure 6.17a. The prominent water band at about 985 nm arises from the $2\upsilon_1 + \upsilon_3$ combination transition where υ_1 is the symmetric O–H stretch, υ_3 is the antisymmetric O–H stretch, and υ_2 is the O–H bending mode. The second derivative spectra for hot smoked king salmon portions are shown in Figure 6.17b. The peak at about 985 nm and weaker absorption bands arising from $2\upsilon_1 + \upsilon_2 + \upsilon_3$ and $3\upsilon_1 + \upsilon_3$ observed near 840 nm and 750 nm can be assigned to the presence of water in the sample tissue (Lin *et al.*, 2003).

In the MIR region (4000–400 cm^{-1}), there are many absorption bands arising from various biochemical functional groups such as those present in water, lipids, proteins, polysaccharides, and nucleic acids. The prominent absorption peaks around 3400 cm^{-1} are primarily from water. The absorption peaks around 2960, 2929, and 1740 cm^{-1} are from fatty acids (Zeroual *et al.*, 1994; Kansiz *et al.*, 1999). Peaks around 1650 and 1550 cm^{-1} are from amide I and II vibrations of protein or peptides. The peaks between 1200 and 900 cm^{-1} are believed to correspond to stretching vibrations of the phosphate and the vibrations of polysaccharide moieties (Schmitt and Flemming, 1998; Lin *et al.*, 2005).

Both NIR and MIR spectroscopy can be implemented as non-destructive and non-invasive measurements, require minimal or no sample preparation, and only need a few seconds for spectral collection. A major advantage of NIR over MIR is that NIR light can penetrate much farther into a sample than MIR radiation. Hampton *et al.* (2002–2003) showed that the maximum penetration depth of the DPA20 probe into fish tissue is 13 mm. Thus, NIR allows the use of long path length in spectral acquisition of samples in various packaging materials such as glass, plastic materials, films, and others that are transparent to NIR light. Furthermore, the use of NIR spectroscopy with fiber

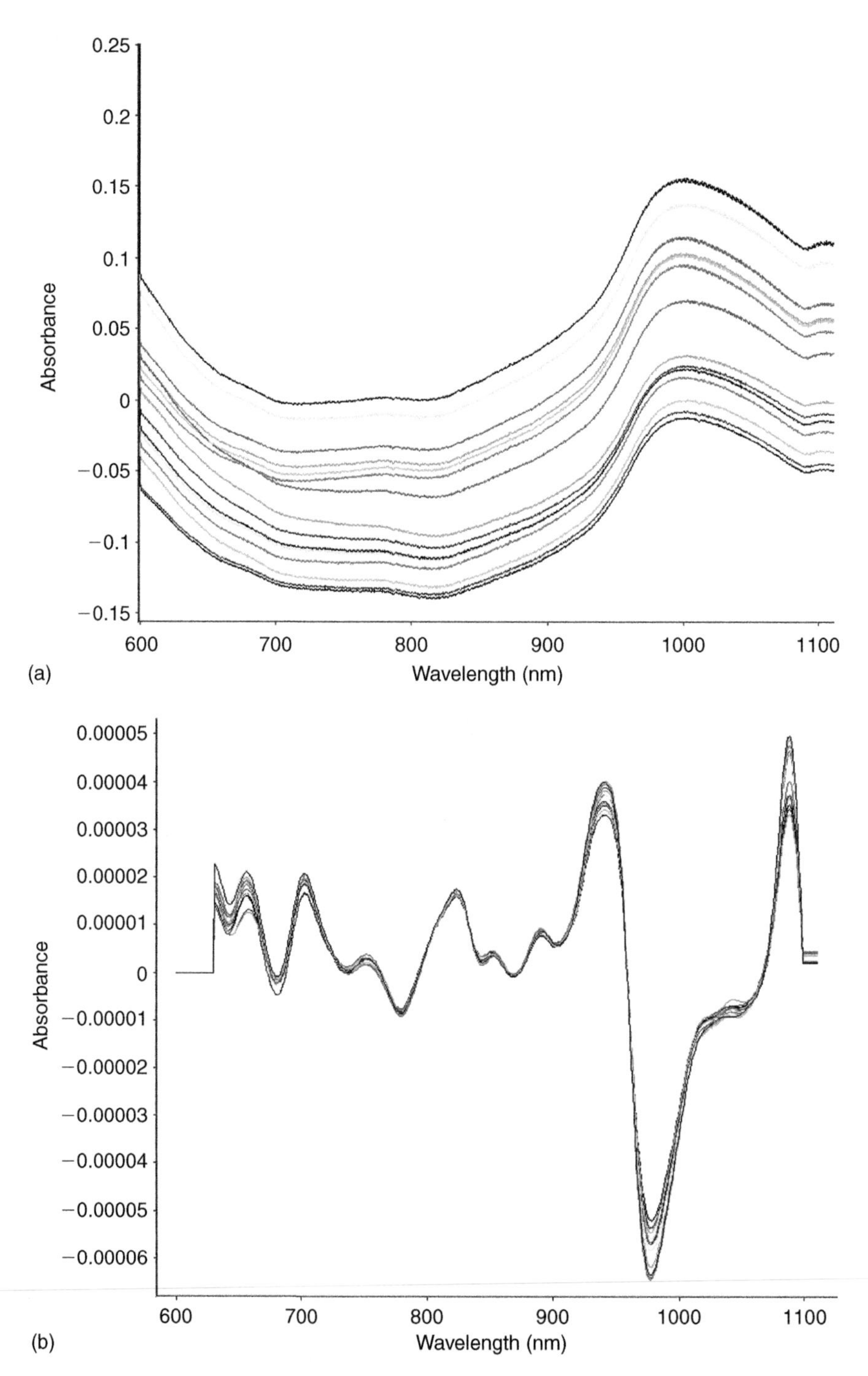

Figure 6.17 Original short-wavelength near-infrared (SW-NIR) reflectance spectra (a) and second derivative transformation (b) for hot smoked king salmon (*Oncorhynchus tshawytscha*) (Lin *et al.*, 2003).

optic probes permits users to collect full spectra of an intact food sample. For example, NIR light can penetrate through fish skin and scale into muscle tissue, permitting analysis of intact whole fish and fish fillet. In contrast, although MIR provides better specificity, the radiation light of MIR has a very short penetration depth (usually a few micrometers) and cannot penetrate through the glass, plastics, and other materials.

Although NIR spectroscopy has longer penetration depth, it does not come without limitations. Most NIR instruments provide limited selectivity and cannot be used for accurate measurement for food components with content of less than 1%. Besides, NIR methods require data calibration using reference values collected by traditional chemical methods and each food component needs a separate calibration. In addition, chemometric analysis with complicated mathematical data processing techniques is often regarded as a "black box approach" and is confusing to users who want to gain in-depth knowledge about NIR analytical techniques.

Conclusions

With technological advances in hardware and software design over the recent decades, NIR and MIR spectroscopy have been established as rapid, accurate, non-invasive, non-destructive, and environmentally friendly techniques and are increasingly used in food quality analysis and control. NIR and MIR technologies are among the most practicable and important analytical techniques to be implemented in the agricultural and food industries. Development of simple, rapid and non-invasive IR methods for safety and quality detection in foods will greatly assist food processors to produce safe and high-quality food products.

Dedication

We would like to dedicate this chapter to the memory of our dear friend, colleague and long time collaborator, Dr David Mayes in recognition of his many important contributions to the field. Dr Mayes developed the instrumentation, chemometric algorithms and analytical software used by many of us in the USA for IR spectroscopic analysis.

References

André J, Lawler IR (2003) Near infrared spectroscopy as a rapid and inexpensive means of dietary analysis for a marine herbivore, dugong (*Dugong dugon*). *Marine Ecology Progress Series*, **257**, 259–266.

Barton FE (2002) Theory and principles of near infrared spectroscopy. *Spectroscopy Europe*, **14**, 12–18.

Beckman AO, Gallaway WS, Kaye W, Ulrich WF (1977) History of spectrophotometry at Beckman instruments, Inc. *Analytical Chemistry*, **49**, 280A–300A.

Birth GS, Hecht HG (2001) The physics of near-infrared reflectance. In: *Near-infrared Technology in the Agricultural and Food Industries,* 2nd edn (Williams PC, Norris KH, eds). St. Paul, MN: AACC, pp. 2–8.

Bozkurt A, Onaral B (2004) Safety assessment of near infrared light emitting diodes for diffuse optical measurements. *BioMedical Engineering OnLine*, **3**, http://www.biomedical-engineering-online.com/content/3/1/9.

Cen H, He Y (2007) Theory and application of near infrared reflectance spectroscopy in determination of food quality. *Trends in Food Science & Technology*, **18**, 72–83.

Ciurczak EW (2001) Principles of NIR spectroscopy. In: *Handbook of Near-Infrared Analysis*, 2nd edn (Burns DA, Ciurczak EW, eds). Boca Raton, FL: CRC Press, pp. 8–9.

Coates J (1997) Vibrational spectroscopy: Instrumentation for infrared and Raman spectroscopy. In: *Analytical Instrumentation Handbook*, 2nd edn (Ewing GW, ed.). New York: Marcel Dekker, Inc.

Foley WJ, McIlwee AP, Lawler IR, Aragones L, Woolnough AP, Berding N (1998) Ecological applications of near infrared spectroscopy—a tool for rapid, cost effective prediction of the composition of plant and animal tissues and aspects of animal performance. *Oecologia*, **116**, 293–305.

Gata N (2000) Imaging spectroscopy using tunable filters: a review. *SPIE*, **4056**, 50–64.

Hampton KA, Wutzke JL, Cavinato AG, Mayes DM, Lin M, Rasco BA (2002–2003) Characterization of optical probe light penetration depth for noninvasive analysis. *Eastern Oregon Science Journal*, **XVIII**, 14–18.

Hendricks M (2000) Spectral illuminations. *Johns Hopkins Magazine*. Available from: http://www.jhu.edu/~jhumag/0400web/02.html. Accessed October 24, 2007.

Huang Y, Rasco BA (2007) Application of neural networks in food science. *Critical Reviews in Food Science and Nutrition*, **47**, 113–126.

Huang Y, Tang J, Swanson BG, Cavinato AG, Lin M, Rasco BA (2003) Near infrared spectroscopy: a new tool for studying physical and chemical properties of polysaccharide gels. *Carbohydrate Polymers*, **53**, 281–288.

Kansiz M, Heraud P, Wood B, Burden F, Beardall J, McNaughton D (1999) Fourier transform infrared microspectroscopy and chemometrics as a tool for the discrimination of cyanobacterial strains. *Phytochemistry*, **52**, 407–417.

Lin M, Cavinato AG, Huang Y, Rasco BA (2003) Predicting sodium chloride content in commercial king (*Oncorhynchus tshawytscha*) and chum (*O. keta*) hot smoked salmon fillet portions by short-wavelength near-infrared (SW-NIR) spectroscopy. *Food Research International*, **36**, 761–766.

Lin M, Al-Holy M, Chang S-S, Huang Y, Cavinato AG, Kang D-H, Rasco BA (2005) Detection of *Alicyclobacillus* isolates in Apple juice by Fourier transform infrared spectroscopy. *International Journal of Food Microbiology*, **105**, 369–376.

Martens H, Naes T (1989) *Multivariate Calibration.* New York: John Wiley & Sons.

McClure WF (1994) Near-infrared spectroscopy. In: *Spectroscopic Techniques for Food Analysis* (Wilson RH, ed.). New York: VCH Publishers, Inc., pp. 13–52.

McClure WF (2001) Near-infrared instrumentation. In: *Near-infrared Technology in the Agricultural and Food Industries*, 2nd edn (Williams PC, Norris KH, eds). St. Paul, MN: AACC, pp. 96–98.

McClure WF, Moody D, Stanfield DL, Kinoshita O (2002) Hand-held NIR spectrometry. Part II: An economical no-moving parts spectrometer for measuring chlorophyll and moisture. *Applied Spectroscopy*, **56**, 720.

McGlone VA, Jordan RB, Martinsen PJ (2002) Vis/NIR estimation at harvest of pre- and post-storage quality indices for 'Royal Gala' apple. *Postharvest Biology and Technology*, **25**, 135–144.

Murray I, Williams PC (2001) Chemical principals of near-infrared technology. In: *Near-infrared Technology in the Agricultural and Food Industries*, 2nd edn (Williams PC, Norris KH, eds). St. Paul, MN: AACC, pp. 22–23.

Osborne BG, Fearn T (1986) *Near Infrared Spectroscopy in Food Analysis.* New York: John Wiley & Sons Inc.

Rabkin YM (1987) Technological innovation in science: the adoption of infrared spectroscopy by chemists. *Isis*, **78**, 31–54.

Raghavachari R (2001) *Near-infrared Applications in Biotechnology.* New York: Marcel Dekker, Inc.

Robinson JW, Frame EMS, Frame GM II (2005) *Undergraduate Instrumental Analysis*, 6th edn. New York: Marcel Dekker.

Rolfe P (2000) In vivo near-infrared spectroscopy. *Annual Review of Biomedical Engineering*, **2**, 715–754.

Schmitt J, Flemming H-C (1998) FTIR-spectroscopy in microbial and material analysis. *International Biodeterioration and Biodegradation*, **41**, 1–11.

Scotter CNG (1997) Non-destructive spectroscopic techniques for the measurement of food quality. *Trends in Food Science & Technology*, **8**, 285–292.

Smith BC (1996) *Fundamental of Fourier Transform Infrared Spectroscopy.* Boca Raton, FL: CRC Press, Inc.

Stuart B (1997) *Biological Applications of Infrared Spectroscopy.* Chichester: John Wiley & Sons, Ltd.

Tuchin V (2000) *Tissue Optics: Light Scattering Methods and Instruments for Medical Diagnosis.* Bellingham, WA: SPIE Press.

Wetzel DL (2001) Contemporary Near-Infrared Instrumentation. In: *Near-infrared Technology in the Agricultural and Food Industries*, 2nd edn (Williams PC, Norris KH, eds). St. Paul, MN: AACC, pp. 129–144.

Weyer LG (1985) Near-infrared spectroscopy of organic substances. *Applied Spectroscopy Reviews*, **21**, 1–43.

White R (1990) *Chromatography/Fourier Transform Infrared Spectroscopy and Its Applications.* New York: Marcel Dekker, Inc.

Workman JJ, Burns DA (2001) Commercial NIR instrumentation. In: *Handbook of Near-Infrared Analysis*, 2nd edn (Burns DA, Ciurczak EW, eds). Boca Raton, FL: CRC Press, pp. 53–56.

Zeroual W, Choisy C, Doglia SM, Bobichon JH, Angiboust JR, Manfait M (1994) Monitoring of bacterial growth and structural analysis as probed by FT-IR spectroscopy. *Biochimica et Biophysica Acta*, **112**, 171–178.

Fourier Transform Infrared (FTIR) Spectroscopy

7

Anand Subramanian and Luis Rodriguez-Saona

Infrared Spectroscopy for Food Quality Analysis and Control
ISBN: 978-0-12-374136-3

Introduction

The concept of Fourier transform infrared (FTIR) spectroscopy has been known about for more than a century. It began with the invention of the interferometer by Michelson in the 1880s (Michelson, 1891, 1892). Soon after, Lord Rayleigh proposed that interference pattern produced by the interferometer could be converted into a spectrum using Fourier transformation (Rayleigh, 1892). Despite these early inventions, it took more than 60 years for FTIR spectroscopy to gain recognition as a potent analytical tool. It attracted widespread interest and attention only after the years following World War II when several independent lines of development converged, resulting in the emergence of FTIR spectroscopy as a useful tool.

Advances in FTIR instrumentation, mathematical transformations, sampling techniques and computer technology ensued in the following years. The invention of fast Fourier transform (FFT), an improvement over the discrete Fourier transform, by Cooley and Tukey (1965) improved the performance of these early FTIR spectrometers. The first commercial FTIR spectrometers were available in the late 1960s. Dispersive instruments that had been widely used since the late 1940s were then slowly supplanted by FTIR spectroscopy.

Today, due to rapid commercial development and extensive research, FTIR spectroscopy is considered to be one of the most powerful techniques for chemical analysis. Because of its simplicity, sensitivity, versatility, and speed of analysis, its applications in biological analysis, including food, are growing at a rapid pace. The growing number of research papers on applications of FTIR spectroscopy in food is a testimony to this. FTIR spectroscopy has attracted tremendous attention in the last couple of decades in spite of tough competition from other spectroscopic techniques and mass spectroscopy.

FTIR spectroscopy is no longer restricted to chemists and spectroscopists. Many specialists and non-specialists from diverse disciplines are starting to adopt this technique. This chapter explains the concepts of FTIR spectroscopy, with emphasis on the history, principles and theories, instrumentation, and applications. Fourier transform near-infrared (FT-NIR) and Fourier transform mid-infrared (FT-MIR) spectroscopy are also explained with relevant examples. The primary aim of this chapter is to introduce the fundamentals of FTIR spectroscopy. While several chapters later in the book detail the applications of FTIR spectroscopy in animal and plant food products, in this chapter examples related to food safety are provided to support the discussions.

A brief history of Fourier transform infrared spectroscopy

Several publications and articles are available on the history of FTIR spectroscopy. Most recently, Ferraro published a very detailed account (Ferraro, 1999). In addition, several books and publications have discussed the history of FTIR spectroscopy in great detail (Bell, 1972; Griffiths and de Haseth, 1986; Johnston, 1991; Christy *et al.*, 2001). All the publications agree that the root of IR spectroscopy is the discovery IR radiation by Sir William Herschel (1800). Herschel then developed prism-based techniques for measuring IR spectra. The first mid-infrared (MIR) spectrometer was constructed as early as 1833 by Melloni following his observation that NaCl is IR-transparent (Christy *et al.*, 2001). FTIR spectroscopy was born with the invention of the interferometer by Michelson (Michelson, 1891, 1892). He used the interferometer to measure accurately the wavelength of light, which earned him the Nobel Prize in Physics in 1907. Michelson also used the interferometer to collect several interferograms. Since then it has taken more than half a century for the technique to establish itself as an analytical tool.

Soon after the construction of the first working interferometer by Michelson, Rayleigh (1892), who also worked on the development of interferometers at the same time as Michelson, recognized that by computing the Fourier transform (FT) of the interference pattern it may be possible to obtain a spectrum. He was the first to relate an interference pattern to a spectrum through Fourier transformation. However, calculation of FT was a complex task for Michelson. To accomplish it, he constructed a harmonic analyzer in the late 1890s. This was an analog device that could perform Fourier transform, as described elsewhere (Michelson, 1898; Johnston, 1991). Despite these developments, FTIR spectroscopy saw a decline in the decades prior to World War II, mainly due to the lack of computing and instrumentation technology. Many other competing techniques, including dispersive techniques were being investigated and developed.

Fortuitously, several streams of advancements converged after the War and combined to promote FTIR spectroscopy. By 1949, Claude Shannon of Bell Laboratories USA had developed information theory (Shannon, 1948). This detailed the requirements of data sampling, which is very important to the data analysis methods of FTIR spectroscopy. Developments in instrumentation science occurred with the discovery of the multiplex advantage or Fellgett advantage of interferometers by Peter Fellgett in 1951 (Fellgett, 1958) and the throughput advantage or Jacquinot advantage by Pierre Jacquinot in 1954 (Jacquinot, 1954). These concepts revived the field of FTIR spectroscopy. One of the major breakthroughs in the field of FT computation occurred in 1965 with the invention of FFT or Cooley–Tukey algorithm by James Cooley and John Tukey (Cooley and Tukey, 1965). This algorithm significantly increased the resolution while reducing the analysis time. FTIR instrumentation also saw notable improvements in the 1960s with the introduction of helium-neon (He-Ne) lasers, better IR detectors, and analog-to-digital converters.

The numerous developments that occurred in the decade following World War II spurred the commercialization of FTIR spectroscopy. Several researchers and companies worked on commercial applications for FTIR from the mid-1960s. Peter Griffiths, Raul Curbelo, Lawrence Mertz, and many others were key contributors to this field (Mertz, 1965a; Griffiths *et al.*, 1972). The introduction of the first commercial FTIR spectrometer (FTS 14) by Biorad Company of Cambridge, MA, USA in 1969 attracted widespread attention and generated considerable interest in the use of IR spectroscopy for analytical applications. In the following years many other companies, including Nicolet Instruments (Madison, WI, USA), PerkinElmer Corp. (Norwalk, CT, USA), Bruker (Billerica, MA, USA), Mattson Instruments (Now Thermo Electron; Madison, WI, USA), and Midac (Irvine, CA, USA) began manufacturing and marketing FTIRs.

The development of rapid scanning interferometers by Mertz in the 1960s (Mertz, 1967a) significantly improved the speed of FTIRs. Various instrumental and computational improvements, including digital signal processing, FTIR software, diagnostic features, etc., occurred in the following decades that improved the resolution, signal-to-noise ratio, sensitivity and speed of detection. The introduction of the IR microscope made microsampling and analysis possible. Developments of step scanning interferometers and dynamic alignment have now enabled kinetic studies, two-dimensional spectroscopy and photoacoustic depth profiling. Miniaturization and the lower cost of production has extended its applications to industry and quality control labs. As a standalone technique, FTIR spectroscopy has established itself as a primary analytical tool and offers great potential for integration with other analytical techniques, providing new research opportunities as well as novel applications.

The interferometer

The interferometer was first constructed by Albert Abraham Michelson (Michelson, 1891), the first American to win the Nobel Prize. What started as a preparation for a class demonstration in 1878 at the American Naval Academy at Annapolis, resulted in the invention of the interferometer a couple of years later at the University of Berlin. Michelson designed the interferometer not to perform IR spectroscopy but to investigate the existence of "luminiferous aether," a medium that was thought to be essential for the propagation of light waves. In a famous experiment, known as the "Michelson–Morley" experiment, it was shown that there is no evidence for the existence of luminiferous aether, a result that proved many previous conclusions wrong and raised several questions. Michelson also used the interferometer for many other purposes, including measuring the diameters of stars, developing standards for length measurement, etc. Later in the nineteenth century Michelson and Lord Rayleigh (Rayleigh, 1892) recognized the potential of the interferometer to provide interference patterns of samples, which could then be converted to spectra. In the following decades, the use of interferometers to obtain spectra was commandeered into the development of FTIR spectroscopy. It is right to say that the invention of interferometers revolutionized the field of IR spectroscopy. Although several other designs were developed and commercialized, most modern interferometer designs are still based on the original interferometer constructed by Michelson.

Construction and working principle

An interferometer is an optical device that allows the controlled generation of interference patterns or interferograms. The construction and working of the interferometer

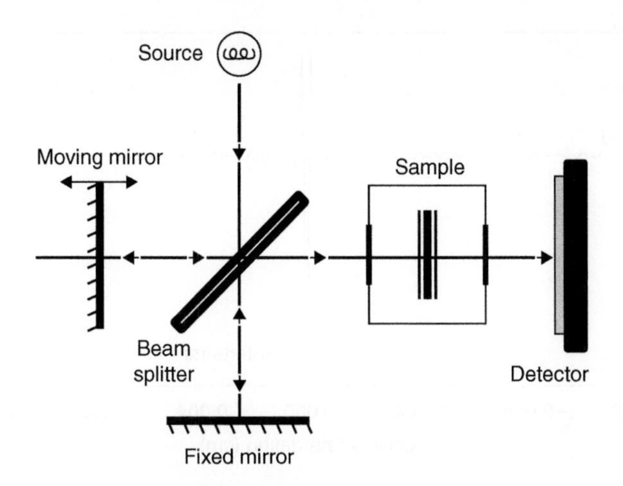

Figure 7.1 Schematic diagram of a Michelson interferometer.

have been explained in detail by several authors (Griffiths and de Haseth, 1986; Johnston, 1991; Smith, 1996). Briefly, it consists of a source, beam splitter, a fixed mirror, and a moving mirror as shown in Figure 7.1. The source emits light in the IR region when electricity is passed through it. The beam splitter, as the name suggests, serves to split the incident IR light into two. The mirrors are aligned so as to reflect the light waves in a direction that would allow recombination of the waves at the beam splitter. The movable mirror is capable of moving along the axis, away from and towards the beam splitter. One half of the light passes through the beam splitter and is reflected by a stationary mirror back to the beam splitter. The other half of the light is reflected on to the moving mirror, which in turn reflects the light back to the beam splitter. The two reflected beams from the mirrors recombine at the beam splitter.

The difference in distance travelled by the two light beams, created due to the movement of the mirror, is called the optical path difference (OPD) or optical retardation. The recombined beam passes through the sample and is finally detected by the detector.

When the two mirrors are at the same distance (zero path difference; ZPD) the reflected beams are in phase and hence interfere constructively. The intensity of the beam is the highest during constructive interference. This occurs when the OPD between the two mirrors is an integer (n) multiple of the wavelength (λ). Conversely, when the two light beams from the mirrors are out of phase with each other, they interfere destructively, leading to a beam of low intensity. Complete destructive interference occurs when the path difference is a ($n + 1/2$) multiple of the wavelength. At other path differences both constructive and destructive interferences take place and the resulting light intensity varies in the form of a cosine wave.

The plot of the intensity of light (in volts) over the OPD is known as an interferogram. In short, an interferogram is obtained by adding large number of sinusoidal waves of intensity at different wavelengths. A typical interferogram obtained

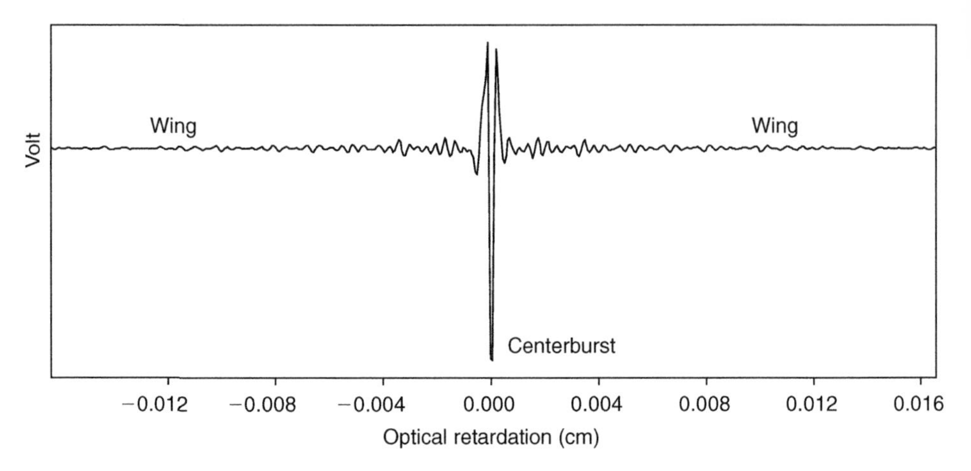

Figure 7.2 Typical interferogram of a modern FTIR spectrometer. The total intensity of the source is shown by the centerburst, which does not contain any signal from sample. The wings of the interferogram contain signal from the sample.

using an FTIR spectrometer is shown in Figure 7.2. An interferogram represents one forward motion of the mirror until the point of ZPD and backward motion to the initial position. The point of highest intensity, called the centerburst, occurs at ZPD, where all waves constructively interfere. The centerburst provides information about the total amount of energy from the source. However, it does not contain any signal from the sample. The regions on either side of the centerburst are called the wings of the interferogram, where both constructive and destructive interference take place at varying levels. Ideally the wings, which carry the signal from the sample, should be identical on either side of the centerburst.

Advantages and disadvantages of interferometry

The advantages of an interferometer stem from two major concepts: (1) the multiplex advantage or the Fellgett advantage and (2) the throughput advantage or the Jacquinot advantage, and these in turn result in several other advantages that significantly improve the efficiency of a spectrometer.

Fellgett advantage

One of the most important practical advantages of FTIR spectroscopy is the multiplex or Fellgett advantage identified by Peter Fellgett during his doctoral studies at Cambridge University, UK, between 1948 and 1951 (Johnston, 1991). Fellgett's attempts to measure the IR spectra of stars was limited by the poor sensitivity of the available detectors, which were capable of recording only a narrow band of wavelengths at a time. The long detection times directly influenced the signal-to-noise ratio (SNR). This problem prompted Fellgett to explore ways to detect multiple wavelengths simultaneously with the aim of reducing the detection time and hence the random noise. After investigating several different avenues, Fellgett heard about interferometers and quickly recognized that they could be a solution to his problem. He then went on to show that, unlike a conventional dispersive instrument in which the spectrum is recorded one wavelength at a time, an interferometer can measure the whole interferogram at the same time. He also showed that the SNR, which is a measure of signal quality, at a particular wavenumber is proportional to the square root of the time spent observing that wavenumber.

Fellgett's proposal to use an interferometer to simultaneously measure the whole wavelength range later gave rise to the necessity of Fourier transformation to convert the interferogram to an IR spectrum. Detailed literature on multiplexing were published by Fellgett in the 1950s (Fellgett, 1958).

Jacquinot advantage

The second advantage of interferometry that contributes to the high SNR of FTIR spectroscopy is the throughput advantage or Jacquinot advantage discovered by Pierre Jacquinot in the late 1940s and published in 1954 (Jacquinot, 1954). Jacquinot wanted to find a cheaper means to perform high-resolution spectroscopy. He tried several types of interferometers to achieve higher resolution and throughput. He proposed that by producing only two interfering beams in the interferometer, a coded version of the spectrum could be produced, which on Fourier transformation would provide the actual spectrum. Thus, Jacquinot discovered the principle of interferometric spectroscopy.

The throughput advantage of interferometers is their ability to pass all the IR radiation through the sample and detect them at once. Unlike the dispersive instruments there are no slits to limit the wavenumber range or the intensity of IR radiation that strikes the detector. The circular aperture wheel in the interferometer does not significantly change the amount of IR radiation passing through it. Thus Jacquinot identified that quantitatively an interferometer can transmit a higher intensity IR radiation at high spectral resolution than a monochromator.

Other advantages

Another practical advantage of interferometry, called the registration advantage, was identified by Janine Connes. This explains the ability of the modern interferometers to accurately and precisely determine the sampling position using a He-Ne laser as a reference (Connes and Connes, 1966). Several other advantages of the interferometer stem from this, including fast scan time, high resolution, high wavenumber accuracy, large scan range, and high sensitivity.

Disadvantages of interferometry

Despite its great potential and several advantages, interferometry does have a few limitations, although they are not considered major problems in recent times. Interferometric analysis requires the use of high-performance computers capable of performing complex calculations. Advancements in computer technology have helped in overcoming this limitation. The second limitation of interferometry is that certain interferograms are complex and cannot be interpreted visually. With the advent of new mathematical spectral treatment procedures and multivariate statistical procedures, the interferogram can be transformed to a more interpretable form. Furthermore, many types of software are available today that can perform these operations in a few seconds.

Dispersive and multiple beam instruments

The field of spectroscopy began with dispersive prism type instruments called spectroscopes. These instruments consisted of a source, a slit to limit the radiation, a collimator to direct the beam, a prism to disperse the light to its component wavelengths, and a condenser to condense the beam onto the detector plate. Multiple prisms can be used to increase the dispersion. Prism type spectrometers were the instruments of choice for visible spectroscopy in the late nineteenth century. However they suffered from several disadvantages: the resolution of measurement varied across the spectrum and the slit assembly reduced the energy of the beam significantly.

Diffraction gratings were another category of dispersive instruments. They consisted of a transparent plate etched with densely packed parallel lines. The number of lines can range from a few dozens to several thousand lines per millimeter. Unlike prisms, which caused dispersion of light by refraction, diffraction gratings dispersed light by the process of diffraction from the fine lines on the plate. The diverging light waves from the lines interfere in the same way as in an interferometer. A diffraction grating produces a spectrum with a smooth change in wavelength but the intensity of the spectra is low due to loss of energy during diffraction. Although the blazed diffraction grating introduced in the early twentieth century increased the intensity of the spectrum, the shortage of good diffraction gratings and narrow spectral range of gratings limited their development.

Both prism spectrometers and diffraction grating spectrometers were commonly used by analytical chemists during the 1920s and 1930s.

Several types of interferometers evolved at the beginning of the twentieth century due to the growing interest in spectroscopy. Michelson himself developed another type of interferometer (Michelson, 1902), which he called the "echelon spectroscope," to observe the structure of spectral lines directly and without any mathematical analysis. Interferometers with multiple beams were also becoming popular due to their ability to produce high-intensity and high-resolution spectra. The Fabry–Pérot etalon and Lummer–Gehrcke interferometer were two of the very famous multiple beam interferometers investigated and have been discussed in detail by Johnston (1991). These instruments produced a direct spectrum from an incoming beam of light that was relatively straightforward and easy to interpret. However, they were limited by their narrow spectral range, the performance of the mirrors, and the precision of interferometer construction.

The Michelson interferometer, a two-beam interferometer, was complicated. However, it was very good in its resolving power and wavenumber-sorting ability over a broad range. A prism spectroscope produces an infinite number of beams. As the number of beams reduces the energy throughput increases. With at least two beams required to produce an interference pattern, the Michelson interferometer has the highest efficiency or throughput. It provides a more complicated spectrum than the other three types but it has the simplest sine or cosine Fourier transform.

Slow and rapid scan interferometers

The first generation of commercial FTIRs produced in the late 1960s used a slow scanning interferometer. The scan speeds were as low as 4 μm/s (Griffiths and de Haseth, 1986). The scan mirror was moved at increments of OPD using a driving mechanism, stopped to allow mirror stabilization, and then scanned. These types of interferometers were well suited for weak or small signals. However, they had a very slow sampling rate, which not only slowed down the analysis but also added uncertainties in the spectra due to variations in the environment such as temperature change, electrical fluctuations, etc. This spurred the development of rapid scan interferometers. Pioneering work in rapid scanning interferometry was done by Lawrence Mertz at Block Engineering (Cambridge, MA, USA). Unlike slow scan interferometers, in rapid scan interferometers the mirror velocity is high enough to enable each wavenumber to be modulated in the audio-frequency range. It does not have any auxiliary modulators or phase-sensitive amplifiers. The reciprocal drive mechanism used in rapid scan interferometers provides reproducible interferograms. The mirror velocity is about 0.158 cm/s (Griffiths and de Haseth, 1986). All the modern rapid-scanning interferometers are developments over the first interferometer designed by Mertz (1965b). Rapid scan systems are suited for low- or medium-resolution MIR spectroscopy where the energy is high and slow scan interferometers work well in the far-infrared range (Griffiths and de Haseth, 1986; Johnston, 1991).

Mathematical processing of interferograms

Because the Michelson interferometer yields a very complex interferogram, mathematical transformations are often needed to improve the characteristics of the spectra and make them more interpretable. There are numerous transformations that can be performed on the interferograms and the details and the mathematics have been presented by many authors (Bell, 1972; Griffiths and de Haseth, 1986; Johnston, 1991). The two most common mathematical operations, apodization and phase correction, are briefly discussed below.

Apodization

In order to perform FT properly and obtain a complete spectrum, an optical path length of infinity should be used and infinite data points should be collected. Since this is impossible, the interferogram is generally truncated and Fourier transformed over a finite limit. This causes oscillations to appear around the sharp spectral features. These oscillations on either side of the sharp band are called side lobes or feet. The process of removing the side lobes or the feet around the spectral band is called apodization ("a podi" in Greek means "no feet"). The interferogram is multiplied

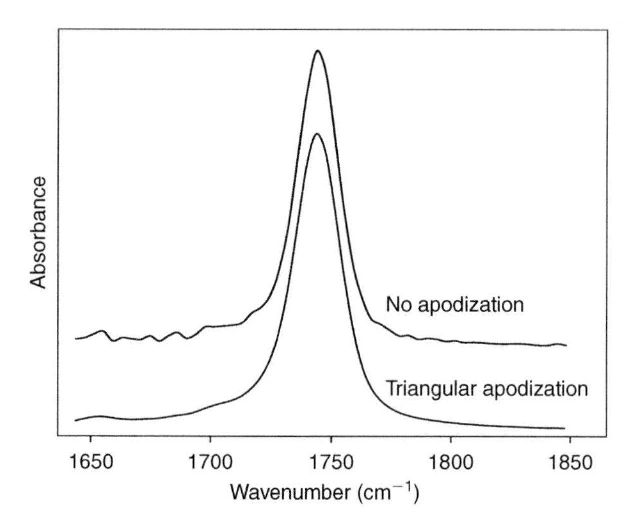

Figure 7.3 An illustration of the effect of triangular apodization on the spectra of corn oil in the carbonyl region.

by an apodization function, which prevents the appearance of side lobes after Fourier transformation. A comparison of MIR spectra of corn oil without apodization (also called "Boxcar" apodization) and with triangular apodization in the carbonyl region (1760–1740 cm^{-1}) is shown in Figure 7.3. Many other apodization methods exist, such as Norton–Beer, Happ–Genzel, etc. and the use of a specific apodization function is interferogram- and application-dependent. The drawback of apodization is that it worsens the spectral resolution. Hence, apodization is actually a compromise made to compensate for incomplete data.

Phase correction

Ideally an interferometer provides an interferogram that is symmetric on either side of the centerburst. Thus by measuring one side of the interferogram, the complete spectrum can be constructed. This process reduces the time significantly. However, interferograms are almost always never symmetric due to imperfections and disturbances in the interferometer, especially in the beam splitter. The process of removing the asymmetry in the interferogram is called phase correction. It is generally done by applying a mathematical function to the single-sided interferogram. Algorithms developed by Mertz (1967b) and Forman (1966) are the two most popular phase correction procedures. Phase correction allows measurement of single-sided interferograms without any uncertainties due to asymmetry.

Fourier analysis

Interferogram obtained using an interferometer needs to be converted to a spectrum, which is a representation of the intensity over wavelength or frequency, to draw inferences about the sample. Fourier transformation (FT) is a mathematical procedure that is applied to the interferograms to obtain the spectrum. Essentially, FT breaks down the interferogram provided by the interferometer into sine waves for each wavelength in the light. These sine waves are arranged over the wavelength to produce the conventional spectrum. In the early years of IR spectroscopy Michelson's harmonic analyzer was the most convenient method to compute Fourier transformation. FTIR spectroscopy suffered due to the lack of proper instrumentation, computational problems, and long calculation times in transforming the interferogram to spectra. While making spectroscopic measurements took only a few minutes, decoding the interferogram required hours or days. Even with the post-War developments in computer technology, performing FT was still limited by the slow speed of the computers as well as the complexity of the discrete FT procedure itself. Interference curves have been converted to spectrum by FT since the 1890s but the commercialization of FTIRs took root only after the discovery of fast Fourier transforms, which simplified the calculations significantly.

Fast Fourier transforms

Until the mid-twentieth century discrete FT was applied to interferograms to obtain the spectrum. Discrete FT is a complex mathematical procedure. In short, Fourier transforming the interferogram involves integrating the interferogram between the limits of zero and maximum path difference. It converts a signal that was recorded as intensity with respect to time (interferogram) into a plot showing intensity with respect to frequency (spectrum). This technique by itself was redundant and the long calculation times severely hampered the analysis. The discovery of FFT by Cooley and Tukey (1965) allowed a breakthrough in the application of interferometry to spectroscopy. An interferogram with n number of points can be represented in an FT as an n-vector. This vector has to be multiplied using an $n \times n$ matrix, in which each row represents the sinusoidal function for each wavenumber, in order to obtain the transformed spectrum. This approach requires a total of n^2 complex multiplication and addition operations. FFT takes advantage of the fact that the $n \times n$ matrix is ordered and cyclical and hence can be factored. Furthermore, if n is an integer power of 2, then the possibility of performing FT on a computer adds significant advantages. Additional information on FT, FFT and their application in spectroscopy can be found in many books (Bell, 1972; Duffieux, 1983; Brigham, 1988; Bracewell, 2000).

FFT reduces the analysis time significantly. For example, to Fourier transform an interferogram with 4096 data points, a total of 16777216 (i.e. 4096^2) multiplications and additions have to be performed. FFT reduces this to 14796 (i.e. $4096 \times \log 4096$) multiplications and additions. Apart from increasing the speed this also reduces amount of data to be handled. Thus, FFT simplified the complex calculations and enabled spectra with a million points to be obtained by the early 1970s. Real-time FFT has also enabled system adjustment and optimization before the actual experiment.

The intensity of calculations and time requirements aside, both discrete FT and FFT provide exactly the same results. FFT is limited by the requirement for interferograms with numbers of points in the power of 2 due to the binary mode of calculations in computers. Hence interferograms with 8, 16, 32, ... 2^n points are best suited for performing FFT. Interferograms with different numbers of data points can also be Fourier transformed after the process of "zero-filling," which involves extending the interferogram to the nearest 2^n points by filling zeros.

Applications of FFT

Even after the publication of the FFT algorithm by Cooley and Tukey (1965), it was still unknown in the field of spectroscopy. It was Forman (1966) who called the algorithm to the attention of Fourier transform spectroscopists in an article entitled "Fourier transform technique and its applications to Fourier spectroscopy." Since then FFT has revolutionized not only spectroscopy but various other fields. Workers in the fields of instrumentation chromatography, microscopy, spectroscopy, X-ray diffraction, and electrochronography all use FFT. In spectroscopy FFT extended the applications of FTIR spectroscopy to near-infrared (NIR) regions and visible light regions. Prior to its development, due to the difficulties in calculations at higher wavenumbers, high resolving power spectroscopy was limited to the far-infrared region (Bell, 1972).

Today FFT is applied in virtually every field in some form. Brigham (1988) describes FFT as being ubiquitous because of the diversity of disciplines in which it is used, including mechanics, acoustics, signal processing, numerical methods, and electromagnetics.

Fourier transform infrared spectroscopy

The fundamentals of FTIR spectroscopy have been presented earlier in this chapter. This section explains the construction of FTIR hardware, the collection of spectra, the fundamentals of FT-NIR and FT-MIR and their applications.

FTIR hardware

FTIR hardware and instrumentation have changed significantly over the years. Today several types and configurations of instruments are available, many of which have specific applications. It is essential to understand the basic construction of an FTIR. A schematic diagram of a FTIR spectrometer is shown in Figure 7.4. The major components are the IR source, beam splitter, detector and reference laser. The setup includes reflecting mirrors at various points to direct the path of IR light. The light from the source passes through the aperture wheel and hits a mirror that directs the light onto the beam splitter. The recombined light from the interferometer is then directed by mirrors into the sample compartment and is finally detected by the detector.

Many different types of materials and components are available for FTIR construction. This section briefly describes most of the important components of an FTIR spectrometer, namely the interferometer, source, beam splitter, detector, and laser, along with their advantages and disadvantages.

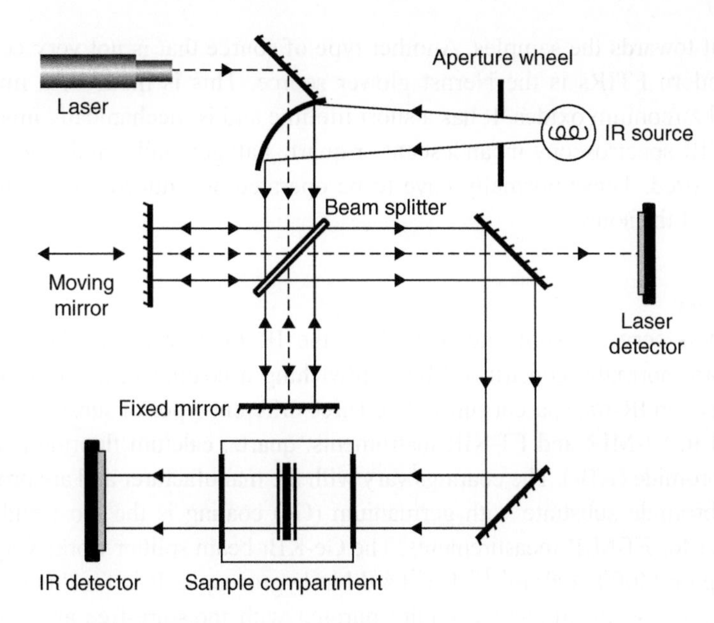

Figure 7.4 Optical layout of a typical FTIR spectrometer.

Interferometers

The components of an interferometer and their functions were described in detail earlier in this chapter. One of the most important components is the moving mirror. It is essential to control its position precisely in order to obtain an accurate measurement of spectra. Two types of mechanisms are commonly used to move the mirror: (1) air bearings and (2) mechanical bearings. In air bearings the movement of the mirror is completely pneumatic. The mirror shaft floats in a stream of air, which moves the mirror. Air bearings are frictionless but expensive and require a constant source of clean and dry air. They are also prone to disturbance due to vibrations. The mechanical bearing system commonly uses a ball-bearing. This is relatively inexpensive but is susceptible to wear. It is important to note that despite the disadvantages mentioned above, modern FTIRs are very sturdy and can work very well for years in any type of configuration.

Sources

The function of the IR source in the spectrometer is to emit IR radiation. Most IR sources work based on the generation of heat due to resistance of the source to conduction of current. The resistance heats up the source (to above 800°C) causing it to emit IR radiation. Because of the high operating temperatures a cooling system is needed. Based on the cooling system FT-MIR sources are of two types: water-cooled or air-cooled. Water-cooled sources are called globar sources and are made of silicon carbide. These provide high throughput but require a constant flow of water for cooling and are expensive. The latest FTIRs use sources made of ceramic or nichrome wire. These are air-cooled and are relatively inexpensive. Normally, sources are mounted in front of a concave mirror in order to capture escaping light and direct it towards the samples. Another type of source that is not very commonly used in modern FTIRs is the Nernst glower source. This is made of a mixture of yttrium and zirconium oxides. It has a short lifetime and is mechanically unstable.

For FT-NIR spectroscopy incandescent or quartz halogen bulbs in the power range 5–50 W are used. These normally have to be operated at reduced voltage to extend the lifetime of the source.

Beam splitters

Beam splitters serve to split and recombine the IR light waves in the interferometer. They are normally constructed by sandwiching a coating of a semi-transparent material between IR-transparent substrates. Three different types of substrates are commonly used in FT-MIR and FT-NIR instruments: quartz, calcium fluoride (CaF_2) and potassium bromide (KBr). The coatings vary with the manufacturer and are proprietary. Potassium bromide substrate with germanium (Ge) coating is the most widely used beam splitter for FT-MIR measurements. The Ge-KBr beam splitter works very well in the MIR region (4000–400 cm^{-1}). It is hard but hygroscopic. It is for this reason that many of the FTIR spectrometers require purging with moisture-free air or nitrogen. Cesium iodide (CsI), another substrate for FT-MIR beam splitters, has a wider range than KBr (4000–200 cm^{-1}). However, it is soft and hygroscopic and hence its use is not encouraged. FT-NIR instruments commonly use quartz or CaF_2 substrates.

Detectors

The function of the detector is to transduce the light intensity received by it to electrical signal. Two most commonly used detectors in both FT-NIR and FT-MIR instruments are the deuterated triglycine sulfate (DTGS) detectors and the mercury cadmium telluride (MCT) detectors. In a DTGS detector a change in the intensity of IR radiation striking the detector will cause a proportional change in temperature. Change in temperature in turn will cause change in the dielectric constant of DTGS and hence its capacitance. This change in capacitance is measured as the detector response in voltage. DTGS detectors are very simple and inexpensive but have relatively slow response and low sensitivity. The MCT detector is a semiconductor and the electrons present in it absorb IR light and move from valence band to conduction band. These electrons in the conduction band generate an electrical current proportional to the IR intensity. MCT detectors are more sensitive and faster than DTGS detectors and hence provide a spectrum with higher SNR.

Disadvantages of MCT detectors are that most of them have a narrow bandwidth based on their composition and they saturate very easily. They are very sensitive to temperature and require cooling with liquid nitrogen. Improper cooling will result in a noisy signal. MCT detectors cost more than DTGS detectors.

Several other types of detectors are available for FT-NIR: PbSe (lead selenide), PbS (lead sulfide), InSb (indium antimonide), and InGaAs (indium gallium arsenide). InGaAs detectors are very fast and have a very high sensitivity. PbSe and PbS fall between DTGS and InGaAs or MCT detectors. A table summarizing the speed of response and sensitivity of these detectors has been published by McCarthy and Kemeny (2001). The electrical signal produced by the detector is converted to voltage, amplified, processed and converted from analog to digital using analog-to-digital converters. The digitized signals are then Fourier transformed.

Laser

Modern FTIR instruments are equipped with a red He-Ne laser. It gives off light at exactly 15 798.637 cm^{-1}. The laser serves two purposes. First, since its wavenumber is known precisely, it acts as an internal wavenumber standard based on which other wavenumbers are measured. The wavenumber reproducibility of most FTIR spectrometers is $\pm 0.01\, cm^{-1}$ or better. Second, it is used to determine the position of moving mirror and hence the OPD. The detector response is measured at every zero-crossing of the laser signal. It is also used for checking and aligning the optics in the interferometer.

Collection of spectra

Most FTIR instruments are single beam. The background and the spectrum are collected at different times. The spectrum of the sample obtained by Fourier transforming the interferogram is called the single-beam spectrum and represents the signal from the sample as well as from the instrument and environment. The single-beam spectrum is ratioed against the background spectrum obtained without the sample, to obtain the actual spectrum of the sample. A schematic diagram of the sequence of steps involved in obtaining a spectrum of a sample is shown in Figure 7.5. The SNR is directly proportional to the square root of the number of scans. Hence, typically multiple scans are added together. During co-addition noise, which is random positive and negative signals, cancels out while the signal intensity remains the same. Smith (1996) suggests that co-adding around 100 scans or more is in general enough to obtain a spectrum with good SNR. This depends, however, on the instrument. For example, an IR microscope, which is much more sensitive, may require co-adding more scans to obtain a good SNR.

Advantages and limitations

Modern FTIR instruments are very simple to use, rapid, very sensitive, and have a high throughput. They provide well-defined and consistent spectra with better wavelength accuracy than dispersive instruments or previous versions of FTIRs. Two primary reasons for these advantages are the Fellgett advantage and Jacquinot advantage, which were discussed in detail earlier in this chapter. Despite these advantages the application of FTIR in analysis is limited by its shortcomings. Some of its disadvantages are as follows:

- FTIR cannot detect atoms and monatomic ions, elements, and inert gases such helium and argon.
- FTIR cannot detect diatomic molecules such as N_2 and O_2. However, in certain cases this can be seen as an advantage as it eliminates the need for vacuum for analysis.

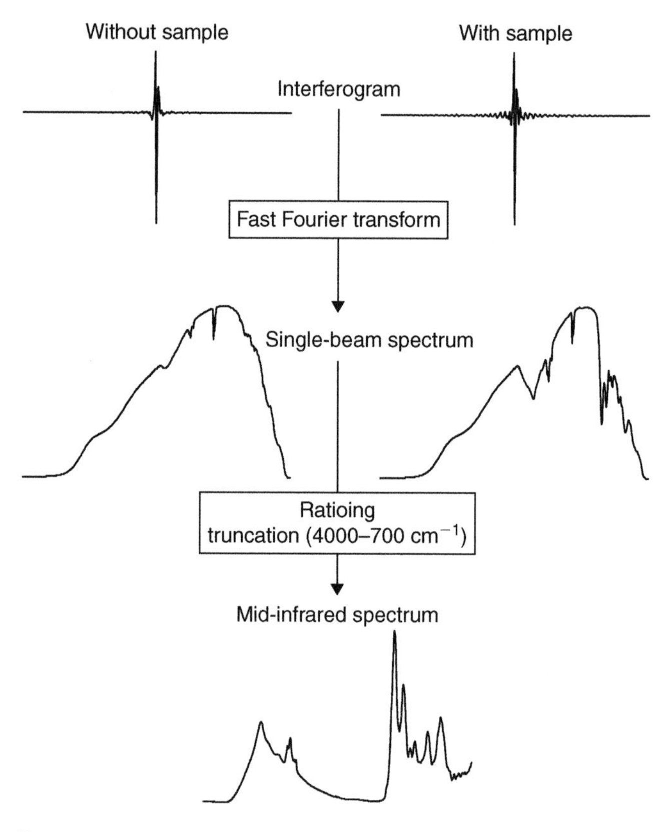

Figure 7.5 An illustration of how a mid-infrared spectrum is obtained from the interferogram.

- Biological samples including food are complex mixtures and hence their FTIR spectra are complicated with overlapping peaks and signal masking.
- Most biological samples contain water, which has a strong absorption band that can mask certain important signals. Often sample preparation procedures are required to reduce the effect of water.
- Since most FTIRs are single-beam instruments, a change in the environment (carbon dioxide and water vapor) can occur during the experiment, causing uncertainties in the spectra.

Fourier transform near-infrared spectroscopy

Near-infrared spectroscopy started almost at the same time as the field of spectroscopy. In fact Sir William Herschel (Herschel, 1800) discovered the heating effect of IR light in the NIR region. Over the past several decades development of improved NIR instrumentation and optics, diffuse reflection technique, and advanced chemometric methods has resulted in the evolution of NIR spectroscopy as a routine laboratory technique. The history of NIR spectroscopy has been presented by several authors in the *Handbook of Near-Infrared Analysis* (Burns and Ciurczak,

2001). While FT-MIR was being widely investigated, the analytical applications of FT-NIR were virtually unexplored until Karl Norris used FT-NIR and chemometrics to characterize agricultural products (Ben-Gera and Norris, 1968). Today FT-NIR spectroscopy is widely employed in various industries including chemical, pharmaceutical, food and beverage industries, for rapid analysis and quality control.

This section provides a brief discussion on the fundamentals and applications of FT-NIR spectroscopy with relevant examples in food safety.

Fundamentals of FT-NIR spectroscopy

NIR spectroscopy involves studying the absorption of compounds in the NIR range (10 000–4000 cm^{-1}) of the electromagnetic spectrum. An FT-NIR spectrometer is very similar to an FT-MIR spectrometer and the minor differences in the hardware were discussed in the previous section of this chapter. A typical NIR spectrum consists of overtone and combination bands of fundamental vibrations. A band can be produced at frequencies 2 to 3 times the fundamental frequency (overtone). The majority of the overtone peaks in a NIR spectrum are due to O–H, C–H, S–H, and N–H stretching modes. Two or more vibrations can combine through addition and subtraction of energies to give a single band (combination).

An NIR spectrum is complex and is marked by broad overlapping peaks and large baseline variations, which makes interpretation difficult. However, mathematical processing such as derivatization and deconvolution can be applied to improve spectral characteristics. A typical FT-NIR absorbance spectrum of *Bacillus cereus* and its second derivative are shown in Figure 7.6. Derivatization of raw spectra removes baseline shifts, improves the peak resolution, and reduces variability between replicates. Overtone bands of C–H groups of fatty acids appear in the regions 8600–8150 (first overtone) and 5950–5600 cm^{-1} (second overtone). The regions 7400–7000 and 4350–4033 cm^{-1} can be attributed to combination bands of C–H, typically from

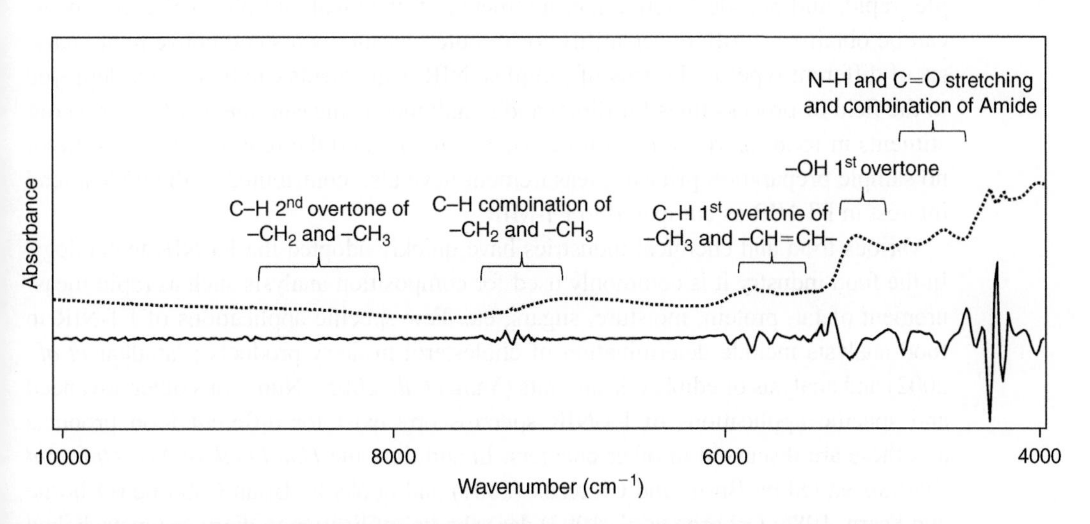

Figure 7.6 Fourier transform near-infrared (FT-NIR) diffuse reflectance absorbance raw (dotted line) and 2nd derivative (solid line) spectra of *Bacillus cereus*.

fatty acids and carbohydrates. Bands of O–H groups have been identified in the spectral regions 5200–5100 (first overtone) and 5190 cm^{-1} (O–H stretch). Stretching and combination vibrations of N–H and C=O of amide A/I and amide B/II of proteins absorb between 5000 and 4500 cm^{-1}. Several other overtone and combination bands have been identified in the FT-NIR spectrum. A comprehensive list of vibration modes and band assignments of different functional groups related to agricultural products can be found in the *Handbook of Near-Infrared Analysis* (Shenk *et al.*, 2001).

NIR spectroscopy is not as sensitive as MIR spectroscopy because NIR radiation penetrates the sample more than MIR radiation. NIR bands are approximately 10–100 times less intense than MIR bands. This can be very useful in direct analysis of highly absorbing bulk and porous samples with little or no sample preparation. The NIR region contains weak and broad overtone and combination bands that make it difficult to identify and associate IR frequencies with specific chemical group. Furthermore, very robust calibration techniques are often needed for accurate NIR analysis. Hence, analyzing NIR data often involves applying chemometrics and multivariate statistical techniques such as principal component analysis (PCA), including soft independent modeling of class analogy (SIMCA), and partial least squares (PLS) to draw interference about the composition of the sample.

Modern statistical software packages available for spectral analysis make it very simple to analyze complex spectra to draw meaningful information. The speed of IR spectroscopy, lower cost and simple procedure of FT-NIR spectroscopy and the user-friendly analysis software packages available have made FT-NIR a preferred method for the rapid analysis of food composition and quality.

Applications of FT-NIR

FT-NIR spectroscopy is an appealing technology for the food industry because simple, rapid, and non-destructive measurements of chemical and physical components can be obtained. It offers versatility for remote measurements and convenient analysis of different types and forms of samples. NIR instruments can be readily deployed to the field or process lines for direct and simultaneous measurements of several constituents in food matrices. Its non-destructive nature and the requirement for little or no sample preparation prior to measurement have also contributed to the widespread interest in FT-NIR as compared to FT-MIR.

Various food and chemical industries have quickly adopted the FT-NIR technology. In the food industry it is commonly used for composition analysis such as rapid measurement of fat, protein, moisture, sugars, etc. Few specific applications of FT-NIR in food analysis include determination of cholesterol in dairy products (Paradkar *et al.*, 2002) and analysis of edible oils and fats (Yang *et al.*, 2005). Numerous other advanced and specific applications of FT-NIR spectroscopy exist for different food products and these are discussed in other chapters. In addition, the *Handbook of Near-Infrared Analysis* edited by Burns and Ciurczak (2001) and books by Brian Osborne (Osborne and Fearn, 1986; Osborne *et al.*, 1993) describe its application in diary products, baked products, beverages and several other non-food materials.

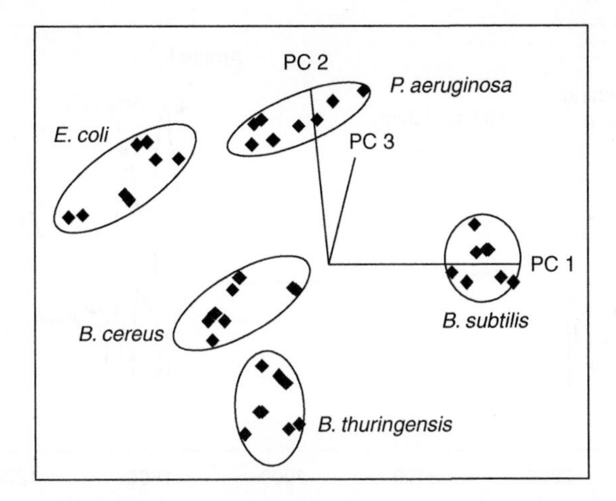

Figure 7.7 Classification of bacteria using Fourier transform near-infrared (FT-NIR) combined with soft independent modeling of class analogy multivariate analysis.

Another application of FT-NIR spectroscopy that has attracted the attention of a few researchers is the characterization of microorganisms. The spectra of bacteria show highly specific fingerprint patterns based on structural and biochemical components, which can enable discrimination up to strain level. Research on the capability of FT-NIR to characterize and discriminate bacteria and other biomolecules has been limited compared with that using FT-MIR, probably because of the difficulties in the interpretation of FT-NIR spectra.

One common and very important application from the food safety point of view is the rapid differentiation of bacterial species. FT-NIR spectra represent the ratio of various chemical groups present in the sample which enables subtle differences to be observed. Multivariate classification methods such as SIMCA have enabled clustering of samples based on biochemical differences while reducing random noise. SIMCA creates a three-dimensional (3D) model of the samples over the first three principal components that explain the most amount of differences in the samples. An example of the application of diffuse reflectance FT-NIR spectroscopy to discriminate various bacterial species using SIMCA is shown in Figure 7.7. All the five samples formed distinct clusters that were well separated from other clusters in 3D space. With suitable sample preparation it is possible to push the limits further.

Rodriguez-Saona and co-workers have demonstrated that by treating the samples with ethanol and filtering the samples to eliminate food matrix effects bacteria can be differentiated to strain level (Rodriguez-Saona *et al.*, 2001, 2004). This technique could be used to develop bacterial IR fingerprint libraries that would enable rapid identification of bacterial samples. It has great potential as a rapid tool to monitor the safety of food supply. One major limitation of FT-NIR in bacterial analysis is that it requires relatively large biomass to provide good spectra of the sample. This makes analysis of single colonies or cells difficult. Additional information on fundamental theory, instrumentation, and application of NIR spectroscopy is available in many books (Osborne and Fearn, 1986; Burns and Ciurczak, 2001; Chalmers and Griffiths, 2002) and articles by NIR publications (www.nirpublications.com).

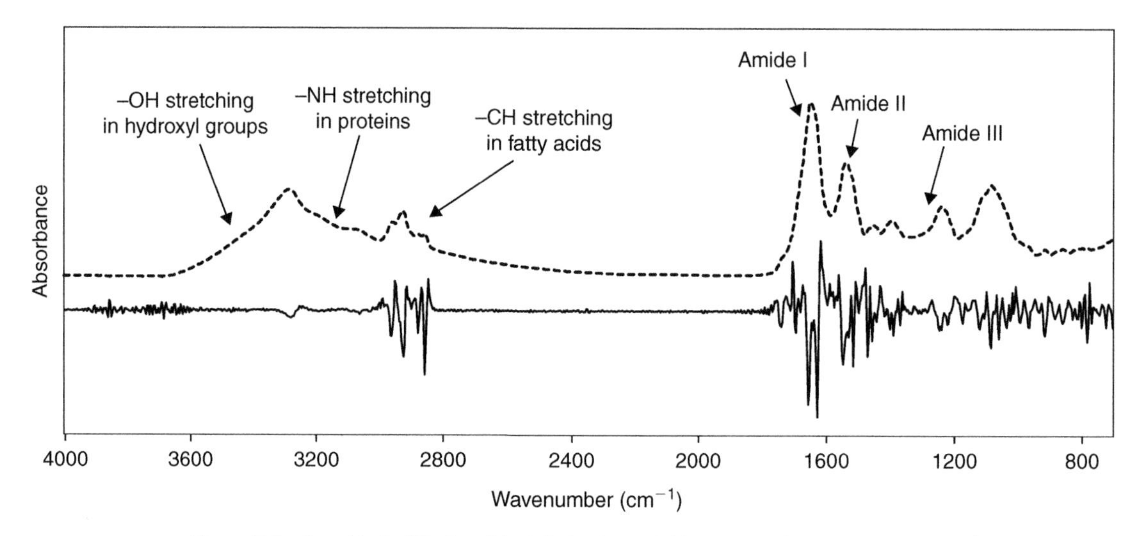

Figure 7.8 Raw (dashed line) and 2nd derivative (solid line) mid-infrared (4000–700 cm^{-1}) spectra of *Salmonella* Enteritidis prepared in distilled water and measured on a three-bounce zinc selenide attenuated total reflectance crystal.

Fourier transform mid-infrared spectroscopy

Since the beginning of spectroscopy, MIR spectroscopy has attracted tremendous interest due to its ability to provide information-rich spectra that enable structural characterization of molecules. Advances in FTIR instrumentation combined with the development of powerful multivariate data analysis makes this technology ideal for large-volume, rapid screening and characterization of minor food components at ppb levels. The detailed history of the discoveries and developments in FTIR spectroscopy presented earlier in this chapter mainly applied to MIR spectroscopy. This section briefly discusses the fundamental principles that form the basis of FT-MIR spectroscopy and applications specifically in food safety.

Fundamentals of FT-MIR spectroscopy

FT-MIR spectroscopy monitors the fundamental vibrational and rotational stretching of molecules, which produces a chemical profile of the sample. The MIR (4000–400 cm^{-1}) is a very robust and reproducible region of the electromagnetic spectrum in which very small differences in composition of samples can be reliably measured. Molecules absorb MIR energy and exhibit stretching, bending, twisting, rocking, and scissoring motions at one or more locations in the spectra, depending on several factors including bond configuration, location, etc. It is rich in information that helps in analyzing the composition and determining the structure of chemical molecules. As an example, a typical FT-MIR spectrum of *Salmonella* Enteritidis and its second derivative are shown in Figure 7.8. FT-MIR spectra reflect the total biochemical composition of the sample, with bands due to major cellular constituents such as water, lipids, polysaccharides, acids, etc. The region from 4000 to 3100 cm^{-1} consists of absorbance from O–H and N–H stretching vibrations of hydroxyl groups and amide A of proteins, respectively. Protein bands also appear in the regions 1700–1550 cm^{-1} (amide I and amide II) and 1310–1250 cm^{-1} (amide III). The C–H stretching vibrations of $-CH_3$ and $> CH_2$ functional groups appear between 3100 and 2800 cm^{-1}. The spectral range 1250–800 cm^{-1} consists of signals from phosphodiesters and carbohydrates. The region from 1200 to 600 cm^{-1} is called the "fingerprint region" as it contains signals that are distinct between each sample and are highly conserved within each sample. An FT-MIR spectrum is easier to interpret than an FT-NIR spectrum. However, chemometric analysis may still be required to further its applications and draw meaningful information.

Applications of FT-MIR

MIR spectroscopy is the FTIR spectroscopic method of choice in applications dealing with structural characterization. Applications for FT-MIR spectroscopy in food analysis are diverse but FT-MIR is relatively recent to food analysis and is not as commonly used as FT-NIR. In the last couple of decades the number of researches on application of FT-MIR for food analysis has grown significantly. Some of the general areas of application include studying the interactions of food components, the quantification of nutrients and several other specific compounds in foods, structural characterization of food molecules, determination of the quality of raw materials and additives, and detecting the adulteration or authenticity of foods.

Joseph Irudayaraj's group has done extensive research on application of FT-MIR spectroscopy for the analysis of many food products including honey (Sivakesava and Irudayaraj, 2002; Tewari and Irudayaraj, 2004), maple syrup (Paradkar *et al.*, 2002; Paradkar *et al.*, 2003), edible oils and fats (Yang and Irudayaraj, 2001; Yang *et al.*, 2005). Applications of FT-MIR in different types of foods are presented in several chapters later in this book.

The ability of FT-MIR spectroscopy to detect, identify, and characterize bacteria was established by Naumann (1984). Bacteria have shown highly specific MIR spectral patterns that may be unique for individual strains. The IR spectra are the result of bands of fundamental vibrational transitions associated mainly with functional groups. FTIR allows for the chemically based discrimination of intact microbial cells and produces complex biochemical fingerprints that are distinct and reproducible for different bacteria. The complex FTIR spectra reflect the total biochemical composition of the microorganism, with bands due to major cellular constituents such as lipids, proteins, nucleic acids, polysaccharides, and phosphate-carrying compounds. Mariey *et al.* (2001) and more recently Burgula *et al.* (2007) have reviewed the application of FT-MIR spectroscopy for bacterial detection. Detailed tables of major publications concerning the use of FTIR and various sampling techniques for the discrimination, classification, and identification of microorganisms were generated by both authors.

The potential of FT-MIR spectroscopy combined with multivariate analysis to predict viable spore concentrations in samples treated by pressure-assisted thermal processing (PATP) and thermal processing (TP) based on differences in biochemical

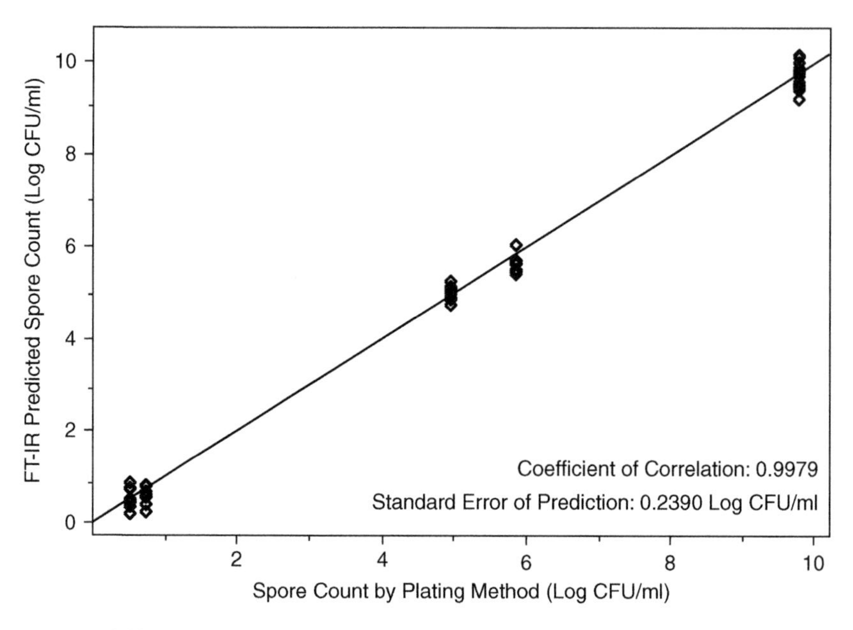

f0090

Figure 7.9 Partial least squares regression plot for correlation between Fourier transform mid-infrared (FT-MIR) spectrum and standard plate count of *Bacillus amyloliquefaciens* Fad 82 spores during heat inactivation.

composition has been demonstrated (Subramanian *et al.*, 2006). Partial least squares regression (PLSR) models developed based on standard plate count data and FT-MIR spectra had correlation of coefficients >0.99 and standard errors of cross-validation ranging between $10^{0.2}$ and $10^{0.5}$ CFU/mL. A PLSR model for correlation between FT-MIR spectrum and standard plate count of *Bacillus amyloliquefaciens* Fad 82 spores during heat inactivation is shown in Figure 7.9. Another recent publication by Subramanian *et al.* (2007) reported the possibility of discriminating bacterial spores to the strain level using FT-MIR spectroscopy and SIMCA procedure. FT-MIR spectra and multivariate analyses were used to monitor biochemical changes in bacterial spores, especially in dipicolinic acid content during inactivation by PATP and TP. A few other researchers have attempted to extend this technique to monitor bacterial spores and their composition (Thompson *et al.*, 2003; Perkins *et al.*, 2004, 2005). In addition, a book by Mantsch and Chapman (1996) and numerous publications by Naumann and his group provide extensive information on bacterial identification and characterization using FT-MIR spectroscopy (Naumann *et al.*, 1991, 1996; Naumann, 2000, 2001).

Comparison of FT-NIR and FT-MIR

FT-NIR and FT-MIR hardware is essentially the same. The main difference in the analysis is due to the difference in wavenumbers or the energy of the beam, which causes the molecules to respond differently. Unlike an MIR spectrum which consists of clear and strong signals in many cases, an NIR spectrum is complex with weak and overlapping signals comprising of overtones and combination absorptions. Typical FT-NIR (Figure 7.6) and FT-MIR (Figure 7.8) spectra were shown earlier and some of the fundamental absorption bands were highlighted. NIR spectra have a relatively weak absorption due to the water overtones enabling analysis of high-moisture foods. NIR is also less influenced by atmospheric carbon dioxide, allowing NIR instruments to be operated without any purging systems to create a moisture- and carbon dioxide-free environment inside the spectrometer. NIR bands are 10–100 times less intense than their corresponding MIR bands. This provides a built-in dilution series that has enabled direct analysis of samples that are highly absorbing and strongly light scattering without dilution or extensive sample preparation. A few studies have compared FT-NIR and FT-MIR for specific applications; including discrimination of edible oils and fats (Yang *et al.*, 2005), measurement of essentials oils (Schulz *et al.*, 2003), detection of adulteration in maple syrup (Paradkar *et al.*, 2002), and adulteration of virgin olive oil (Yang and Irudayaraj, 2001). Sivakesava *et al.* (2004) compared the use of MIR and NIR for the classification of *Bacillus*, *Lactobacillus*, *Saccharomyces*, *Micrococcus*, and *Escherichia* and reported that NIR methods were easier to use, while MIR gave better discrimination between the tested samples. The choice between FT-MIR and FT-NIR is primarily dependent on the type of sample and analysis.

Sampling techniques

Many different sampling methods exist. Each technique has its own advantages and limitations and hence its use is application-specific. This section provides a brief description of the various sampling techniques available for use in both FT-NIR and FT-MIR. A few examples of applications in food safety for some of the sampling techniques are provided in Table 7.1. Further information on the sampling techniques is available in several basic FTIR text books (Johnston, 1991; Smith, 1996; Stuart, 2004). Many FTIR accessory manufacturers also provide technical notes on their website on various sampling techniques.

Transmittance

Transmittance is probably the simplest of all sampling techniques. It involves passing the IR radiation directly through the sample and detecting on the other side (Figure 7.10a). It provides spectra with high SNR and is relatively cost effective. It is suited for analyzing solid, liquid, and gaseous samples. However, transmission technique is limited by the sample thickness. Samples within the range 1 to 20 μg are suited for transmission analysis (Smith, 1996). Transmission also involves careful and time-consuming sample preparation. Generally, an IR-transparent material such as KBr is used as a substrate for the sample. Solid samples are mixed with KBr or mixed with a mulling agent such as Nujol (mineral oil). In the latter case mull (sample + oil) is ground and sandwiched between two KBr pallets for analysis. Another method of applying the sample is by casting films on a KBr window. Samples prepared in a

Table 7.1 Some recent examples of applications of different FTIR sampling techniques for analysis of microorganisms

Sampling technique	Application	Reference
Attenuated total reflectance	Differentiation of *Salmonella enterica* serovars	Baldauf *et al.*, 2007
	Identification and quantification of food-borne pathogens	Gupta *et al.*, 2006
	Identification of food-borne yeasts	Kümmerle *et al.*, 1998
	Quantitative detection of microbial spoilage in beef	Ellis *et al.*, 2004
	Investigation of interactions between antimicrobial agents and bacterial biofilms	Suci *et al.*, 1998
Diffuse reflectance	Discrimination of *Bacillus* species	Winder *et al.*, 2004
	Detection of the dipicolinic acid biomarker in *Bacillus* spores	Goodacre *et al.*, 2000
	Detection and identification of bacteria in juice matrix	Rodriguez-Saona *et al.*, 2004
Transmittance	Differentiation of *Salmonella enterica* serovars	Baldauf *et al.*, 2006
	Identification of sporulated and vegetative bacteria	Foster *et al.*, 2004
	Identification of lactic acid bacteria	Dziuba *et al.*, 2006
Photoacoustic spectroscopy	Identification of bacteria	Foster *et al.*, 2003
	Differentiation and detection of microorganisms	Irudayaraj *et al.*, 2002
Microspectroscopy	Characterization of microorganisms	Yu and Irudayaraj, 2005
	Characterization and identification of microorganisms	Ngo-Thi *et al.*, 2003
	Species level analysis of food-borne microbial communities	Wenning *et al.*, 2006
	Identification of yeasts	Wenning *et al.*, 2002

solvent are applied on the KBr window and allowed to dry prior to analysis. Films can also be cast by applied heat and pressure. For liquid samples, special types of sealed liquid cells are available, which are typically made of KBr and create a very thin layer of liquid through which IR light is passed. Gas cells, usually 10 cm long, are available for analysis of gaseous samples. FTIRs with gas cells are now commonly used to monitor air quality and pollution.

Reflectance

Reflectance is the reverse of transmittance in which IR light reflected back from the sample is measured. Unlike transmittance, reflectance involves easier and faster sample preparation. It is non-destructive and is not influenced by the sample thickness. Frequently, special and expensive accessories are required for reflectance, which limit its application. The SNR is typically lower than transmittance. Furthermore, the depth of penetration into the sample is not exactly known and the surface of the sample influences the spectra more than the interior. These disadvantages are normally shadowed by the simplicity and speed of this technique. Based on the type of reflectance from the sample, reflectance technique also includes specular reflectance, diffused reflectance, and attenuated total reflectance.

Attenuated total reflectance

Attenuated total reflectance (ATR) is one of the most commonly used sampling techniques in recent times. When an IR beam travels from a medium of high refractive index (e.g. zinc selenide crystal) to a medium of low refractive index (sample), some amount of the light is reflected back into the low refractive index medium. At a particular angle of incidence, almost all of the light waves are reflected back. This phenomenon is called total internal reflection. In this condition, some amount of the light energy escapes the crystal and extends a small distance (0.1–5 μm) beyond the surface in the form of waves. This invisible wave is called evanescent wave. The intensity of the reflected light reduces at this point. This phenomenon is called attenuated total reflectance. When the sample is applied on the crystal some amount of the IR radiation penetrating beyond the crystal is absorbed by the sample. This absorbance is translated into the IR spectrum of the sample. A clean, empty crystal is normally used for collection of background spectrum. An illustration of a multi-bounce ATR accessory with sample is shown in Figure 7.10b. An example FT-MIR spectrum of *Salmonella* Enteritidis obtained using a zinc selenide (ZnSe) ATR crystal was shown earlier (Figure 7.8). ATR is very commonly used for food analysis. Some examples of its food safety applications are listed in Table 7.1.

Several different ATR configurations are available with different crystal materials. Commonly used materials for crystals include zinc selenide, germanium, silicon, diamond, and KRS-5 (thallium iodide or thallium bromide). The properties of each material and its applications are available from most FTIR accessory manufacturers. A list of FTIR accessory companies and their web addresses is given in Table 7.2. ATR enables analysis of solid and liquid samples. Unlike transmittance it is not influenced by sample thickness. It is very convenient and simple to use and hence is widely applied in food analysis. Since ATR spectra are from the surface of the sample, this technique is limited by homogeneity and thickness of the sample. With suitable sample preparation, ATR can provide spectra with well-defined features.

Diffuse reflectance

Diffuse reflectance occurs when the IR beam is reflected back from the sample surface in random direction. It involves both absorption and scattering. The scattering is due to the rough sample surface, which is the category that most foods belong to. Diffuse reflectance is well suited for solid and powder materials. A diffuse reflectance accessory employs two flat mirrors to direct the light and one concave focusing mirror exactly above the sample cup to concentrate the IR beam on to the sample. The sample is mixed with KBr and placed in a sample cup. A spectrum of pure KBr

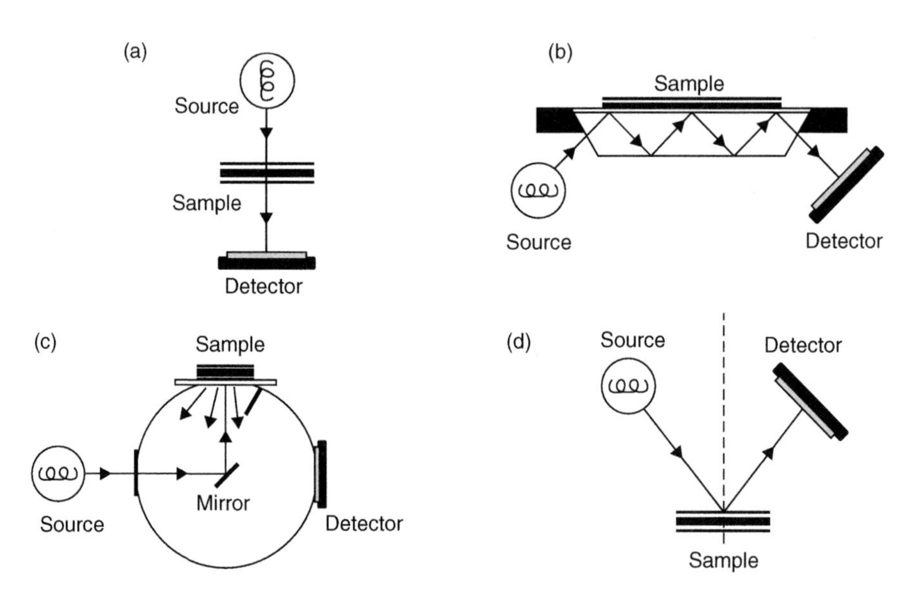

Figure 7.10 Schematic diagram of various FTIR sampling techniques: (a) transmission, (b) attenuated total reflectance, (c) diffuse reflectance in an integrating sphere, and (d) specular reflectance.

powder serves as the background. The packing density and particle size influence the intensity of the output beam and hence the spectral intensity. Diffuse reflectance offers a simple and quick method for analysis of bulk and coarse samples.

The integrating sphere, a diffuse reflectance accessory, improves diffuse reflectance analysis further. It consists of a highly reflective spherical enclosure that includes a mirror to direct the IR beam onto the sample. The sample is normally separated from the sphere by a thin IR-transparent plate or it could be open, allowing the IR beam to interact directly with the sample. The reflected light bounces many times before reaching the detector. The sphere enables spatial integration of the light reflected from the sample, hence the name integrating sphere. A baffle is placed between the sample and the detector to avoid the first reflection of the sample from directly reaching the detector. A schematic diagram of an integrating sphere is shown in Figure 7.10c and a detailed description is provided by Hanssen and Snail (2001).

Integrating spheres are available for both FT-MIR and FT-NIR applications from many FTIR accessories manufacturers. It is an expensive accessory but offers advantages such as the combined measurement of diffuse and specular reflectance and the analysis of inhomogeneous samples as is frequently used for analysis of solid, bulky, rough, or powdery food material.

Specular reflectance

Specular reflectance is a type of reflectance technique that occurs when the angle of incidence of the IR radiation incident on the sample is equal to the angle at which it is reflected back (Figure 7.10d). Reflectance on a mirror surface is a typical example of specular reflectance. A classic specular reflectance accessory consists of mirrors that direct the light onto the sample. The IR radiation reflected from the sample surface

Table 7.2 Some FTIR and FTIR accessory manufacturing companies

Manufacturer	Website
FTIR manufacturers	
Applied Instrument Technologies	www.orbital-ait.com/
Aspectrics	www.aspectrics.com
Block Engineering	www.blockeng.com/
Bruker Optics	www.bruker.com
Hamilton Sundstrand	www.hs-ait.com/
Jasco Inc.	www.jascoinc.com/
Midac Corporation	www.midac.com/
PerkinElmer Inc.	www.perkinelmer.com/
Shimadzu	www.shimadzu.com/
Thermo Scientific	www.thermo.com/
Varian Inc.	www.varianinc.com/
Accessory manufacturers	
Axiom Analytical	www.goaxiom.com
CIC Photonics	www.cicp.com
Harrick Scientific	www.harricksci.com/
MTEC Photoacoustics	www.mtecpas.com/
Newport Corporation	www.newport.com
Pike Technologies	www.piketech.com/
Remspec Corporation	www.remspec.com/
Resultec Analytic Equipment	www.resultec.de/
Smiths Detection	www.smithsdetection.com/
Specac	www.specac.com/
S.T. Japan	www.stjapan.de/

is directed towards the detector by another mirror. A reflective surface such as gold is normally used to obtain the background spectrum. Sample preparation for specular reflectance is very simple. However, this technique requires a sample with a smooth surface. Some samples have a coating on the surface. In such cases the light beam passes through the coating and is reflected off the surface, involving both absorbance and reflectance. This phenomenon is called double transmission.

Specular reflectance is mainly used in analyzing polymers and its applications in food analysis are limited.

Photoacoustic spectroscopy

The photoacoustic effect was observed by Alexander Graham Bell almost at the same time as the invention of the Michelson interferometer (Bell, 1880). However, its application in FTIR evolved only in the last two decades. In a photoacoustic accessory the sample is placed in a cup and covered with an IR-transparent window. The accessory is then filled with helium or air. The IR beam is then directed on to the sample, which heats up the sample. Helium absorbs the heat from the sample and expands, causing increase in pressure and creating currents. The movement of helium within the sample cell causes sounds that are transduced into electrical signal using a microphone. The microphone signal is plotted against the path difference to obtain the interferogram. The spectrum of carbon black is used as a background. The interferogram is Fourier-transformed to get the spectra.

Photoacoustic spectroscopy (PAS) can be applied to solids, liquids, and gases. It is a non-destructive technique which requires little or no sample preparation. PAS is affected by the size of the sample. The sensitivity reduces with increase in sample size. Several other factors such as the instrument optics, sample properties (thickness, preparation, transparency, etc.), and composition of the food sample influence PAS analysis. Applications of PAS in food are very scarce. A couple of researchers have attempted to detect and identify bacteria using PAS. These applications are listed in Table 7.1.

FTIR microspectroscopy

The introduction of the IR microscope by Digilab and Spectra-Tech in the early 1980s pushed the capabilities of FTIR further by enabling visual and IR analysis of micron level samples (Messerschmidt and Harthcock, 1988). Today, the IR microscope has gained wide acceptance as a very efficient microanalytical tool for identifying and characterizing chemical and biological samples.

The construction of an IR microscope in principle is the same as most optical microscopes. IR radiation from the spectrometer is directed onto the sample through a series of mirrors and lenses. The light emerging from the sample is channeled into a detector. The detector in most modern microscopes is an MCT detector, as described earlier in this chapter. The IR microscope is a very expensive accessory. However, it has diversified the possible applications of FTIR spectroscopy by enabling microanalysis, and increasing the sensitivity and speed of detection. Modern microscopes allow both transmission and reflectance studies. Applications of FTIR microspectroscopy in food are still evolving and so are its applications in food safety or microbial characterization. A few examples of its applications in microbial detection and characterization are listed in Table 7.1.

New sampling techniques have also been developed for bacterial analysis. Ngo-Thi *et al.* (2000) used a novel stamping technique to prepare the samples for microspectroscopy. The stamping tool enabled transfer for a few cells from the colonies on agar plate onto an IR-transparent plate for microscopic analysis. FTIR microspectroscopy with its unique advantages has great potential as analytical tool and has opened up a new area of research. More information on the growth, developments, fundamentals, and applications of FTIR microspectroscopy can be found elsewhere (Katon, 1996; Smith, 1996).

Factors affecting FTIR spectroscopy

Obtaining the highest possible SNR and a good enough resolution for a specific analysis are important factors while using FTIR spectroscopy for analysis. Several factors contribute to acquiring a good FTIR spectrum. The SNR is the ratio of the height of a band in the spectrum to the height of the noise at some point on the baseline of the spectrum. As said before, the SNR is directly proportional to the square root of the number of scans. Hence, by co-adding several scans it is possible to achieve high SNR. SNR is also directly proportional to the square of the time spent for measuring a data point. Modern rapid scanning spectrometers are extremely quick thereby increasing the SNR. Spectral resolution is a property of the instrument and it is its ability to distinguish closely positioned features in the spectra. The greater the OPD, the greater the spectral resolution. During our research on food analysis using FTIR we have found 8 cm^{-1} or 4 cm^{-1} to be good resolutions to use. The SNR and resolution themselves are directly related. High-resolution spectra will have more noise due to the divergence of light in the optics, electronic noise, and use of apertures. Another factor to be considered is that co-adding scans and obtaining high-resolution spectra improve the quality of the spectra at the cost of time. Hence it is essential to collect a few trial spectra before actual experiment to optimize the parameters.

Commercial FTIR instruments and accessories

Since the introduction of the first commercial FTIR spectrometer by Biorad in 1969 (Christy *et al.*, 2001), several companies started manufacturing FTIRs. A field that was mainly composed of physicists diversified following commercialization of FTIRs. Several analytical chemists adopted FTIR spectroscopy by the 1970s. Soon FTIR spectroscopy became a very popular and convenient laboratory tool. Today many companies manufacture fast, robust, sensitive, and user-friendly FTIR spectrometers that work for a relatively long with minimal maintenance. A few of the FTIR manufacturing companies and their web addresses are listed in Table 7.2.

With time, FTIR spectroscopy started to be employed in numerous disciplines of science and engineering. As the samples diversified the demand for special FTIR accessories grew. In the last couple of decades several companies have begun manufacturing accessories that are designed to perform specific types of analysis. The names and websites of the major companies manufacturing accessories for wide brands of FTIR spectrometers are provided in Table 7.2.

Conclusions

FTIR and sampling instrumentation have been constantly evolving over the years. Special heated crystals (FatIR™ from Harrick Scientific) are now available for analysis of fats and oils. ATR accessories have been further improved by incorporating multi-bounce crystals in which the light bounces on the sample many times, thereby increasing absorbance. Remote sampling and analysis is now possible with the advent of fiber optic probes. The FTIR microscope has developed significantly in the last two decades. It is now widely recognized as a multifunctional accessory that allows sampling by transmission, reflectance, ATR, and grazing angle. FTIR software interface has also improved significantly. Almost all data acquisition softwares allow the manipulation of the optics such as auto-alignment and self-diagnostics. They also allow spectral processing such as subtractions, differentiation, deconvolution, etc. Many manufacturers are developing FTIR spectrometers dedicated for specific applications such as moisture measurement. Such initiatives are on the rise due to the increasing demand for rapid and online food quality monitoring.

Technology integration is another area that is receiving widespread attention in recent times. FTIR spectroscopy has been successfully coupled with other analytical techniques, such as gas chromatography, liquid chromatography, and thermogravimetric analysis. Such integrations have extended the analytical capabilities of FTIR. Infrared imaging or spectrochemical imaging is another evolving technology. It uses an IR microscope to construct a complete image of the sample in the MCT detector. At each array element a complete IR spectrum of the sample is acquired. Thus each point in the image can be used to extract a complete spectrum. With its speed of data collection, improved spatial resolution and SNR, FTIR imaging potentially has a very wide range of applications and can offer significant advantages to the analyst. These new technologies are information rich investigative techniques. They find application in many branches of the physical science but are still foreign to analysis of food samples. Tremendous growth and diversification of FTIR technology has increased the versatility of the technique. With this great potential and still much to be explored in food analysis and control, the future of FTIR spectroscopy in the food sciences is very promising.

References

Baldauf NA, Rodriguez-Romo LA, Yousef AE, Rodriguez-Saona LE (2006) Differentiation of selected Salmonella enterica serovars by Fourier transform mid-infrared spectroscopy. *Applied Spectroscopy*, **60**, 592–598.

Baldauf NA, Rodriguez-Romo LA, Männig A, Yousef AE, Rodriguez-Saona LE (2007) Effect of selective growth media on the differentiation of Salmonella enteric serovars by Fourier-transform mid-infrared spectroscopy. *Journal of Microbiological Methods*, **68**, 106–114.

Bell AG (1880) On the production and reproduction of sound by light: the photophone. *Proceedings of the American Association for Advancement of Science*, **29**, 115–136.

Bell RJ (1972) *Introductory Fourier Transform Spectroscopy.* New York: Academic Press Inc.

Ben-Gera I, Norris KH (1968) Direct spectrophotometric determination of fat and moisture in meat products. *Journal of Food Science*, **33**, 64–67.

Bracewell RN (2000) *The Fourier Transform and its Applications.* New York: McGraw-Hill.

Brigham EO (1988) *The Fast Fourier Transform and its Applications.* New Jersey: Prentice Hall.

Burgula Y, Khali D, Kim S, Krishnan SS, Cousin MA, Gore JP, Reuhs BL, Mauer LJ (2007) Review of mid-infrared Fourier transform infrared spectroscopy application for bacterial detection. *Journal of Rapid Methods and Automation in Microbiology*, **15**, 146–175.

Burns DA, Ciurczak EW (2001) *Hand Book of Near-infrared Analysis.* New York: Taylor & Francis.

Chalmers JM, Griffiths PR (2002) *Handbook of Vibrational Spectroscopy.* New York: John Wiley & Sons.

Christy AA, Ozaki Y, Gregoriou VG (2001) *Modern Fourier Transform Infrared Spectroscopy.* New York: Elsevier.

Connes J, Connes P (1966) Near-infrared planetary spectra by Fourier spectroscopy. I. instruments and results. *Journal of the Optical Society of America*, **56**, 896–910.

Cooley JW, Tukey JW (1965) An algorithm for the machine calculation of complex Fourier series. *Mathematics of Computation*, **19**, 297–301.

Duffieux PM (1983) *The Fourier Transform and its Applications to Optics.* New York: John Wiley & Sons.

Dziuba B, Babuchowski A, Nalecz D, Niklewicz M (2006) Identification of lactic acid bacteria using FTIR spectroscopy and cluster analysis. *International Dairy Journal*, **17**, 183–189.

Ellis DI, Broadhurst D, Goodacre R (2004) Rapid and quantitative detection of the microbial spoilage of beef by Fourier transform infrared spectroscopy and machine learning. *Analytica Chimica Acta*, **514**, 193–201.

Fellgett P (1958) A contribution to the theory of the multiplex spectrometer. *Journal of Physics and Radium*, **19**, 187–191.

Ferraro JR (1999) History of Fourier transform-infrared spectroscopy. *Spectroscopy,* **14**, 28–40.

Forman ML (1966) Fast Fourier-transform technique and its application to Fourier spectroscopy. *Journal of the Optical Society of America*, **56**, 978–983.

Foster NS, Thompson SE, Valentine NB, Amonette JE, Johnson TJ (2003) Identification of bacteria using statistical analysis of Fourier transform mid-infrared transmission and photoacoustic spectroscopy data. *Abstracts, 58th Northwest regional meeting of the American Chemical Society,* Bozeman, MT, USA, p. 44.

Foster NS, Thompson SE, Valentine NB, Amonette JE, Johnson TJ (2004) Identification of sporulated and vegetative bacteria using statistical analysis of Fourier transform mid-infrared transmission data. *Applied Spectroscopy*, **58**, 203–211.

Goodacre R, Shann B, Gilbert RJ, Timmins EM, McGovern AC, Alsberg BK, Kell DB, Logan NA (2000) Detection of dipicolinic acid marker in *Bacillus* spores using Curie-point pyrolysis mass spectrometry and Fourier transform infrared spectroscopy. *Analytical Chemistry*, **72**, 119–127.

Griffiths PR, de Haseth JA (1986) *Fourier Transform Infrared Spectrometry.* New York: John Wiley & Sons.

Griffiths PR, Foskett CT, Curbelo R, Dunn ST (1972) Rapid-scan infrared Fourier transform spectroscopy. *Applied Spectroscopy Reviews*, **6**, 31–78.

Gupta MJ, Irudayaraj JM, Schmilovitch Z, Mizrach A (2006) Identification and quantification of foodborne pathogens in different food matrices using FTIR spectroscopy and artificial neural networks. *Transactions of the ASABE*, **49**, 1249–1255.

Hanssen LM, Snail KA (2001) Integrating spheres for mid- and near-infrared and near-infrared reflection spectroscopy. In: *Handbook of Vibrational Spectroscopy* (Chalmers JM, Griffiths PR, eds). New York: Wiley & Sons, pp. 1175–1191.

Herschel W (1800) Observations tending to investigate the nature of the Sun, in order to find the causes or symptoms of its variable emission of light and heat; with remarks on the use that may possibly be drawn from solar observations. *Philosophical Transactions of the Royal Society of London*, **1**, 49–52.

Irudayaraj J, Yang H, Sakhamuri S (2002) Differentiation and detection of microorganisms using Fourier transform infrared photoacoustic spectroscopy. *Journal of Molecular Structure*, **606**, 181–188.

Jacquinot P (1954) The luminosity of spectrometers with prisms, gratings, or Fabry Perot etalons. *Journal of the Optical Society of America*, **44**, 761–765.

Johnston SF (1991) *Fourier Transform Infrared: A Constantly Evolving Technology.* New York: Ellis Horwood.

Katon JE (1996) Infrared microspectroscopy: a review of fundamentals and applications. *Micron*, **27**, 303–314.

Kümmerle M, Scherer S, Seiler H (1998) Rapid and reliable identification of food-borne yeasts by Fourier-transform infrared spectroscopy. *Applied and Environmental Microbiology*, **64**, 2207–2214.

Mantsch HH, Chapman D (1996) *Infrared Spectroscopy of Biomolecules.* New York: Wiley-Liss.

Mariey L, Signolle JP, Travert AJ (2001) Discrimination, classification, identification of microorganisms using FTIR spectroscopy and chemometrics. *Vibrational Spectroscopy*, **26**, 151–159.

McCarthy WJ, Kemeny GJ (2001) Fourier transform spectrometers in the near-infrared. In: *Handbook of Near-Infrared Analysis* (Burns DA, Ciurczak EW, eds). New York: Taylor & Francis, pp. 71–90.

Mertz L (1965a) Astronomical infrared spectrometer. *Astronomical Journal*, **70**, 548.

Mertz L (1965b) *Transformations in Optics.* New York: John Wiley & Sons.

Mertz L (1967a) Rapid scanning Fourier transform spectrometry. *Journal of Physics*, **28** (Suppl 3), 88.

Mertz L (1967b) Auxiliary computation for Fourier spectrometry. *Infrared Physics,* **7**, 17–23.

Messerschmidt RG, Harthcock M (1988) *Infrared Microspectroscopy: Theory and Applications.* New York: Marcel Dekker.

Michelson AA (1891) On the application of interference-methods to spectroscopic measurements, -I. *Philosophical Magazine*, **31**, 338–346.

Michelson AA (1892) On the application of interference-methods to spectroscopic measurements, -II. *Philosophical Magazine*, **34**, 280–299.

Michelson AA (1898) A new harmonic analyser. *Philosophical Magazine*, **45**, 85–91.

Michelson AA (1902) *Light Waves and their Uses.* Chicago: The University of Chicago Press.

Naumann D (1984) Some ultrastructural information on intact, living bacterial cells and related cell-wall fragments as given by FTIR. *Infrared Physics*, **24**, 233–238.

Naumann D (2000) FTIR spectroscopy in microbiology. In: *Encyclopedia of Analytical Chemistry* (Meyers RA, ed.). Chichester: John Wiley & Sons, pp. 101–131.

Naumann D (2001) FT-infrared and FT-Raman spectroscopy in biomedical research. In: *Infrared and Raman Spectroscopy of Biological Material* (Gremlich HU, Yan B, eds). New York: Marcel Dekker, pp. 323–377.

Naumann D, Helm D, Labischinski H, Giesbrescht P (1991) The characterization of microorganisms by Fourier-transform infrared spectroscopy (FT-IR). In: *Modern Techniques for Rapid Microbiological Analysis* (Nelson WH, ed.). New York: Wiley-VCH, pp. 43–96.

Naumann D, Schultz CP, Helm D (1996) What can infrared spectroscopy tell us about the structure and composition of intact bacterial cells. In: *Infrared Spectroscopy of Biomolecules* (Mantsch HH, Chapman D, eds). New York: Wiley-Liss, pp. 279–310.

Ngo-Thi NA, Kirschner C, Naumann D (2000) FT-IR microspectrometry: a new tool for characterizing microorganisms. *Proceedings of the SPIE*, **3918**, 36–44.

Ngo-Thi NA, Kirschner C, Naumann D (2003) Characterization and identification of microorganisms by FT-IR microspectrometry. *Journal of Molecular Structure*, **661/662**, 371–380.

Osborne BG, Fearn T (1986) *Near Infrared Spectroscopy in Food Analysis.* New York: Longman Scientific & Technical.

Osborne GB, Fearn T, Hindle PH (1993) *Practical NIR Spectroscopy with Applications in Food and Beverage Analysis.* New York: Longman Scientific & Technical.

Paradkar MM, Sakhamuri S, Irudayaraj J (2002) Comparison of FTIR, FT-Raman, and NIR spectroscopy in a Maple syrup adulteration study. *Journal of Food Science*, **67**, 2009–2015.

Paradkar MM, Sivakesava S, Irudayaraj J (2003) Discrimination and classification of adulterants in maple syrup with the use of infrared spectroscopic techniques. *Journal of the Science of Food and Agriculture*, **83**, 714–721.

Perkins DL, Lovell CR, Bronk BV, Setlow B, Setlow P, Myrick ML (2004) Effects of autoclaving on bacterial endospores studied by Fourier transform infrared microspectroscopy. *Applied Spectroscopy*, **58**, 749–753.

Perkins KL, Lovell CR, Bronk BV, Setlow B, Setlow P (2005) Fourier transform infrared reflectance microspectroscopy study of *Bacillus subtilis* engineered without dipicolinic acid: the contribution of calcium dipicolinate to the mid-infrared absorbance of *Bacillus subtilis* endospores. *Applied Spectroscopy*, **59**, 893–896.

Rayleigh L (1892) On the interference bands of approximately homogeneous light: In a letter to Prof. A. Michelson. *Philosophical Magazine*, **34**, 407–411.

Rodriguez-Saona LE, Khambaty FM, Fry FS, Calvey EM (2001) Rapid detection and identification of bacterial strains by Fourier transform near-infrared spectroscopy. *Journal of Agricultural and Food Chemistry*, **49**, 574–579.

Rodriguez-Saona LE, Khambaty F, Fry F, Dubois J, Calvey EM (2004) Detection and identification of bacteria in a juice matrix with Fourier transform-near infrared spectroscopy and multivariate analysis. *Journal of Food Protection*, **67**, 2555–2559.

Schulz H, Schrader B, Quilitzsch R, Pfeffer S, Krüger H (2003) Rapid classification of Basil chemotypes by various vibrational spectroscopy methods. *Journal of Agricultural and Food Chemistry*, **51**, 2475–2481.

Shannon C (1948) The mathematical theory of communication. *Bell Systems Technical Journal*, **27**, 623–656.

Shenk JS, Workman JJ, Westerhaus MO (2001) Application of NIR spectroscopy to agricultural products. In: *Handbook of Near-Infrared Analysis* (Burns DA, Ciurczak EW, eds). New York: Taylor & Francis, pp. 419–474.

Sivakesava S, Irudayaraj J (2002) Classification of simple and complex sugar adulterants in honey by mid-infrared spectroscopy. *International Journal of Food Science and Technology*, **37**, 351–360.

Sivakesava S, Irudayaraj J, DebRoy C (2004) Differentiation of microorganisms by FTIR-ATR and NIR spectroscopy. *Transactions of the ASAE*, **73**, 951–957.

Smith BC (1996) *Fundamentals of Fourier Transform Infrared Spectroscopy.* New York: CRC Press.

Stuart B (2004) *Infrared Spectroscopy: Fundamentals and Applications.* New York: Wiley.

Subramanian AS, Ahn J, Balasubramaniam VM, Rodriguez-Saona L (2006) Determination of spore inactivation during thermal and pressure-assisted thermal processing using FT-IR spectroscopy. *Journal of Agricultural and Food Chemistry*, **54**, 10300–10306.

Subramanian AS, Ahn J, Balasubramaniam VM, Rodriguez-Saona L (2007) Monitoring biochemical changes in bacterial spore during thermal and pressure-assisted thermal processing using FT-IR spectroscopy. *Journal of Agricultural and Food Chemistry*, **55**, 9311–9917.

Suci PA, Vrany JD, Mittelman MW (1998) Investigation of interactions between antimicrobial agents and bacterial biofilms using attenuated total reflection Fourier transform infrared spectroscopy. *Biomaterials*, **19**, 327–339.

Tewari J, Irudayaraj J (2004) Quantification of saccharides in multiple floral honeys using Fourier transform infrared microattenuated total reflectance spectroscopy. *Journal of Agricultural and Food Chemistry*, **52**, 3237–3243.

Thompson SE, Foster NS, Johnson TJ, Valentine NB, Amonette JE (2003) Identification of bacterial spores using statistical analysis of Fourier transform infrared photoacoustic spectroscopy data. *Applied Spectroscopy*, **57**, 893–899.

Wenning M, Seiler H, Siegfried S (2002) Fourier-transform infrared microspectroscopy, a novel and rapid tool for identification of yeasts. *Applied Environmental Microbiology*, **68**, 4717–4721.

Wenning M, Theilmann V, Scherer S (2006) Rapid analysis of two food-borne microbial communities at the species level by Fourier-transform infrared microspectroscopy. *Environmental Microbiology*, **8**, 848–857.

Winder CL, Goodacre R (2004) Comparison of diffuse-reflectance absorbance and attenuated total reflectance FT-IR for the discrimination of bacteria. *Analyst,* **129**, 1118–1122.

Yang H, Irudayaraj J (2001) Comparison of near-infrared, Fourier transform-infrared and Fourier transform-Raman methods for determining olive pomace oil adulteration in extra virgin olive oil. *Journal of the American Oil Chemists' Society*, **78**, 889–895.

Yang H, Irudayaraj J, Paradkar MM (2005) Discriminant analysis of edible oils and fats by FTIR, FT-NIR and FT-Raman spectroscopy. *Food Chemistry*, **93**, 25–32.

Yu C, Irudayaraj J (2005) Spectroscopic characterization of microorganisms by Fourier transform infrared microspectroscopy. *Biopolymers*, **77**, 368–377.

PART II

Applications

8

Meat and Meat Products

Rahul Reddy Gangidi and Andrew Proctor

Introduction

Meat is a premium, high-demand food throughout the world. Meats have high-quality proteins, which contain essential amino acids, and are a very good source of vitamin B, dietary iron, and zinc. The meat-processing industry is one of the largest agricultural and food-processing industries in the world and infrared (IR) spectroscopy can play an important role in maintaining consistent product quality. For example, determination of fat content is an important quality factor that can be measured quickly by Fourier transform infrared spectroscopy (FTIR). A typical fat measurement by a chemical extraction takes 3–4 h, and is laborious and off-line, involving toxic and

Infrared Spectroscopy for Food Quality Analysis and Control
ISBN: 978-0-12-374136-3

hazardous chemicals. With the use of rapid spectroscopic methods, the same results could be obtained in 2–5 minutes. In addition, other characteristics of the sample can be determined simultaneously. Although, IR spectroscopic methods are just one of many spectroscopic methods, IR-based methods are well accepted due to their versatility in predicting multiple attributes accurately and precisely in processing plants and laboratories.

Infrared spectroscopy involves absorption of a certain wavelength of IR radiation by the specific chemical functional groups, based on their dipole moment. The advantages of using IR spectroscopy for meat and meat product quality and control are its simplicity (does not require sample preparation) and speed, it is non-destructive to the sample, and it is non-invasive. It can also be used as an online, at-line, and inline process analyzer with additional fiber-optic capability, the cost of analysis is low per sample, it is non-ionizing, non-chemical reaction inducing, non-heating, and safe (low intensity of energy), and it does not require hazardous gas cylinders or toxic chemicals (i.e. it is green technology). Most IR instruments are bench top with some models being portable.

IR reflectance spectroscopy has the limitation that it is a surface analysis technique, with little depth penetration into the sample, it is affected by high moisture content, which absorbs IR radiation and by the surface free water, due to specular reflectance from reflective surfaces and high absorption. Although near-infrared (NIR) transmittance has better penetration, it is also affected by the higher depth, high moisture content, and less energy transmission due to reflectance from partially transparent meat matrices.

Near-infrared (0.7–2.5 μm; 12 900–4000 cm^{-1}) spectroscopy is often coupled with visible spectroscopy (0.38–0.7 μm) to measure the components and quality attributes of meat and meat products. It is further classified into NIR reflectance spectroscopy and NIR transmission spectroscopy. NIR can be non-dispersive (filter-based instrumentation), dispersive and use Fourier transform-based instrumentation. Mid-infrared (MIR) (2.5–50 μm; 4000–200 cm^{-1}) spectroscopy, on the other hand, is generally used with Fourier transformation, a complex mathematical technique converting time domain data into frequency domain, and reflectance-based techniques, diffuse reflectance and attenuated total reflectance; and very rarely with dispersive and non-dispersive instrumentation. Most recent advanced instrumentation involves combining mid- and near-infrared spectroscopy (FTIR and FT-NIR). Fourier transform and dispersive, NIR and MIR spectroscopy often require the application of chemometrics (multivariate statistics) to extract the spectral information for qualitative and quantitative analysis. NIR and MIR spectroscopy is often complemented with visible spectroscopy, nuclear magnetic resonance spectroscopy, Raman spectroscopy and gas chromatography for specific applications.

Far-infrared (2.5–50 μm; 200–10 cm^{-1}) radiation is used for the processing and cooking of meat and food products and has very limited practical spectroscopic application.

Infrared spectroscopy is utilized for proximate analysis of meat and meat products, mainly moisture, fat, and protein, and in some cases minerals. It has been used to determine meat quality, pH, fatty acid profile, appearance and color, muscle characteristics, such as water-holding capacity, intramuscular fat, tenderness, and microbial spoilage. In addition, it has been used to detect the adulteration of low-cost meat in premium meat, such as adulteration of beef with kangaroo and pork or skeletal muscle with organ meat, and to identify fresh meats from frozen-then thawed meats, adulteration with non-meat protein (e.g. from vegetable and dairy origin) and fecal contamination. It has also been used to detect spinal cord, which is a central nervous system tissue and is prohibited in meat due to concerns about the transmission of bovine spongiform encephalopathy (BSE). Proximate and quality attributes of meats from other species have also been analyzed with IR spectroscopy.

It is beyond the scope of this chapter to include all the meat-related applications of IR. However, an effort has been made to include the common applications and solutions to the problems. Most of the IR meat applications are tabulated for quick reference. Thermo Fisher Scientific Inc. is not responsible for the contents in this chapter. NIR reflectance and NIR transmission spectroscopy instruments were from NIR Systems Inc. (Silver Springs, Maryland, USA), unless mentioned otherwise.

Beef and beef products

Proximate composition

Norris's group demonstrated the applicability of NIR transmission spectroscopy to the measurement of the fat, protein, water content in emulsified meats (Ben-Gera and Norris, 1968). This work was further advanced by Kruggel *et al.* (1981) in emulsified and ground meats; and by Lanza (1983) in ground and homogenized meats with NIR reflectance spectroscopy. Moisture was predicted using multiple linear regression with 1732, 1700, 1990, and 1218 nm with an $r = 0.985$ and standard error of prediction (SEP) of 0.48%. Similarly protein (1376, 1772, 1524, 1804 nm) was predicted with an $r = 0.899$ and SEP of 0.53%, and fat content (1720, 1308, 2388, 1890 nm) was predicted with an $r = 0.998$ and SEP of 0.23% (Lanza, 1983). Since the moisture and fat content were predicted with a low error compared with the protein content, it would be advantageous to measure the protein from moisture and fat.

Packaged beef

Much of the meat in grocery stores is packaged with polyethylene and related materials to prevent oxygen from interacting with the meat, to hold meat odors within the package, and to provide a clear view of the meat. Isaksson *et al.* (1992) used NIR (1100–2500 nm) and NIR transmission spectroscopy (850–1100 nm) to measure the protein, fat, and water content in packaged homogenized beef. The laminate of polyamide (PA)/ethylenevinyl alcohol (EVOH)/PA/polyethylene (PE) was applied to the beef that was 85 μm thick. The PE layer was in contact with the beef whereas the PA layer was in contact with the atmosphere. The thickness of the beef samples without laminates was ~14 mm. A lean beef with 6% fat and a fatty beef with 14% fat were used in the study. Also products with 33% lean beef in fatty beef and

67% lean beef in fatty beef were prepared for a total of 100 samples. The spectra were collected at room temperature of 22°C. Principal component regression (PCR) analysis was used as the statistical technique. Fat was measured with a very high degree of accuracy and was least affected by the laminate despite having the common $-CH_2$ chemical groups. The laminate packaging marginally affected the water and protein measurements, probably due to the refractive index and the presence of amides in the laminate, but the accuracy of the measurement was good.

NIR reflectance and NIR transmission spectroscopy can be used to measure moisture, protein and fat content at room temperature. Laminate $-CH_2$ bands were found at 1200, 1700–1750, and 2300–2400 nm and were least affected by the presence of the meat. Based on this study, NIR reflectance spectroscopy could be used for other laminate packaged meat and meat products with thickness greater than 14 mm.

Online NIR spectroscopy for beef analysis

Lean meat is blended with tallow or fatty beef to obtain beef with various fat contents. An online NIR instrument at the blender would provide an accurate measurement of fat content. Tøgersen *et al.* (1999) adapted the work of Isaksson *et al.* (1996) for the measurement of water and fat content, and calculated protein content from the total, fat, and moisture values by using five filters (1441, 1510, 1655, 1728, 1810 nm) with an NIR online spectrometer. The 1441 and 1510 nm filters were used for moisture measurement, the 1728 nm filter for fat measurement and the low-absorbing 1655 and 1810 nm were used as references. The beef samples with fat (7–26%), moisture (58–75%), and protein (15–21%) were analyzed by NIR at the beef grinder outlet. Fat and moisture content values were determined with multiple linear regression. The prediction error for fat was 0.82–1.49%, water was 0.94–1.33%, and protein was 0.35–0.70%. Figure 8.1 shows a typical filter based online NIR gauge. The specific filtered NIR wavelength from sensor (A) is directed onto the sample (B) and the reflected light from the sample is detected by the NIR-sensitive lead sulfide detector in the sensor. The source (a quartz-halogen lamp), the wheel containing wavelength-specific filters, the optics to direct the light onto the sample and gather and then direct the light reflected from the sample to the detector, and the electronic circuits to process the detector signal are all in the sensor (Carlson, 1978). The electronically processed response from the sensor can be digitally displayed in a remote box (C)100 or on an optional computer (D).

Semi-frozen beef

A great deal of beef is frozen post mortem to prevent microbial growth and chilled to age or tenderize the beef. Continuing earlier work with fresh beef analysis at room temperature, Tøgersen *et al.* (2003) utilized an NIR spectrometer for the measurement of the fat, moisture, and protein in semi-frozen meat. Frozen meat at −7°C was placed in a refrigerator/freezer combination to obtain beef at temperatures of −5, −2, 0, 2, and 10°C. The beef samples had fat ranges of 7.66–22.91%, moisture ranges of 59.36–71.48%, and protein ranges of 17.04–20.76%. A scanning NIR spectrometer (1100–2500 nm) was used to select the wavelengths appropriate

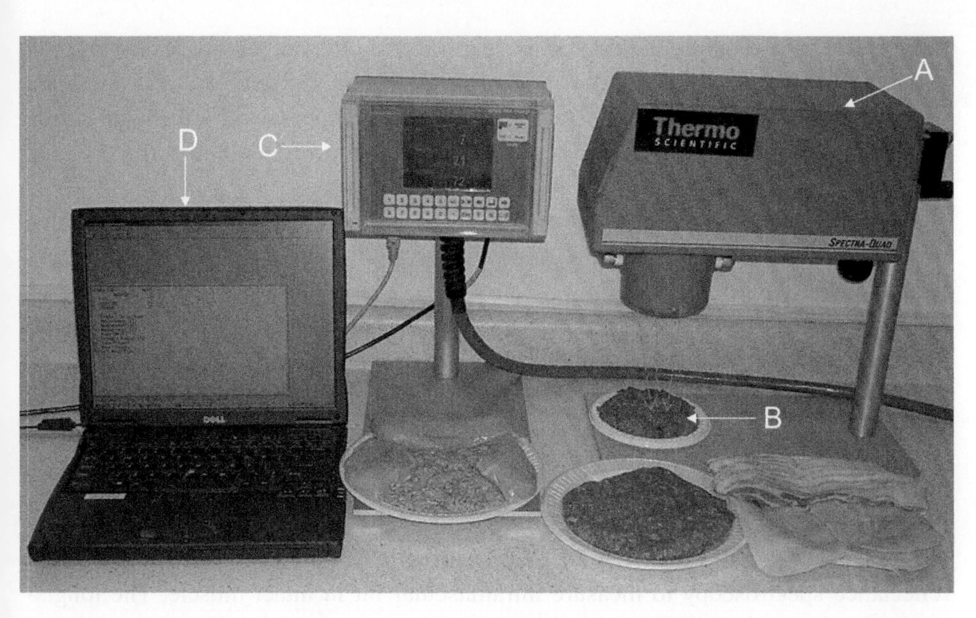

Figure 8.1 Thermo Scientific Spectra-Quad™ (A), a rugged at-line and online filter wheel-based industrial near-infrared gauge, analyzing meat sample (B), with remote display (C), and optional remote computer display (D). (Photo printed with permission from Thermo Fisher Scientific Inc., Minneapolis, MN, USA.)

to the measurement of fat and moisture, and in addition that would be least influenced by the crystallization of the water. The O–H sensitive ranges (1400–1600 and 1900–2050 nm) were excluded from the filter selection range. The filter wavelengths of 1630, 1728, 1810, 2100 and 2180 nm were selected. Using partial least squares regression 2 (PLS2) on the filter data, an error of 0.48–1.11% for fat, 0.43–0.97% for moisture, and 0.41–0.47% for protein were observed. The errors are similar to those for the room temperature measurements of fat, moisture, and protein.

Compensation for temperature fluctuation and water phase shifts

Temperature variation is known to affect spectral data either in terms of spectral shifts or in spectral intensity. It is therefore very important to develop models or methods that are not influenced by temperature fluctuation. Segtnan *et al.* (2005) used beef samples ($n = 100$) at −1, +1, and +3°C, and NIR reflectance spectroscopy (1100–2350 nm) to correct for temperature variation. Homogenized meat samples were placed in the polythene bag and then in the NIR sample holder cell. Four additional samples were selected with the constituent variation in the entire range; and the spectra of each of the four samples were collected at −1, +1, and +3°C. The difference spectra of −1°C is calculated by subtracting the spectrum at −1°C from the spectrum of the middle temperature, i.e. +1°C. Similarly, the difference spectra is calculated for +3°C; and the difference spectra are computed for each of the four additional samples. The difference spectra when incorporated with a complex simulation into the 100 spectra gave similar results to that of the model utilizing all the 100 sample spectra at all three temperatures. This modeling was found to be better than random addition of noise to the data to compensate for extraneous variables, such as temperature, not related to the response variables, such as moisture, fat, and protein content. The procedure not only compensated for the temperature fluctuation but also for the change of state from frozen, semi-frozen, and unfrozen in the measurement of protein, fat, and water content.

A robust method would be to develop a model with as many extraneous variations for the same response variable. This approach minimizes the error in an unstable and unpredictable environment. However it requires a large data set and also requires additional factors or latent vectors in the multivariate statistical model development. This work could be applied to other meats and meat products.

Quality attributes

Intramuscular fat or marbling

Beef grade classifications are based on appearance, which is mostly influenced by the intramuscular fat. In a novel NIR application, Rødbotten *et al.* (2000) used NIR reflectance spectroscopy to measure intramuscular fat in intact muscle. The longissimus dorsi muscles of 127 carcasses with a fat content of 1–14% were analyzed by NIR reflectance spectroscopy between 1100 and 2500 nm at 4 nm resolution. A slice of 4–5 cm diameter and 1.5 cm thickness was placed in the sample holder, and a 1 cm^2 area was exposed to the radiation. Partial least squares (PLS) models were developed with leave-one-out cross-validation and the data prior to the PLS modeling, multiplicative scatter correction was performed. Multiplicative scatter correction corrects for multiplicative and additive scatter related affects. The correlation for pre-rigor and post-rigor samples ranged from 0.76 to 0.84 and root mean square error of prediction (RMSEP) was 1.2%. The results were significant as the data was collected on heterogeneous intact muscle, and in earlier studies the fat content of homogenous ground beef was predicted with an accuracy greater than 0.98 (R^2). The absorbance wavelengths of 1152–1248, 1376–1460, 1676–1776, and 2248–2440 nm were prominent in the measurement of fat content.

Sensory meat tenderness determination

Tenderness is the most important beef sensory quality. Naes and Hildrum (1997) used NIR reflectance spectroscopy to classify the meat samples based on tenderness (very tender, intermediate, or tough), which were determined by the sensory panel. Ten minutes post mortem, carcasses were given low-voltage electrical stimulation (80 V, 14 Hz, 32 s) and other carcasses were not given electrical stimulation. Longissimus dorsi muscle was chilled at 4°C for 26 h to prevent cold shortening and later NIR spectra were collected on 4.5-cm-diameter slices with an InfraAnalyzer 500 (Norderstedt, Germany) NIR reflectance spectrometer between 1100 and 2500 nm. The samples were stored at −40°C after NIR analysis and later heated at 70°C for 50 min or 75 min and were presented to a 12-member sensory panel at 20°C. Prior to sensory analysis, some samples were aged at 4°C for 7 and 14 days. NIR spectra were collected and sensory panel responses were recorded on these samples as well. Approximately 67, 47, and 55% of the samples were correctly classified by NIR as tough, intermediate, and tender. With only two tough and tender degrees, 100% of the samples were classified correctly, however the large proportion of intermediate samples were classified as tender or tough.

This approach is a rapid, simple, low-cost approach to predict the tenderness. However, the method would need to be improved to enhance the accuracy of the PCR model. Furthermore when tenderness was predicted a correlation of 0.65 and an error of 1.2% were observed for a scale of 1–9, with 1 being most tough and 9 being the most tender. The application of FTIR and its derivatives may improve the results in some cases.

Instrumental beef tenderness determination

Park *et al.* (1998) used NIR reflectance spectroscopy to determine the Warner–Bratzler shear force (WBSF) of longissimus thoracis steaks with 79% correct classification based on <6 kg (tender) and >6 kg (tough). A 2.5-cm-thick longissimus thoracis steak was used for the WBSF measurement and another similar steak was used for NIR analysis. The steaks were thawed for 24 h at 4°C, broiled to initial temperature of 40°C on one side, turned and broiled to a final temperature of 70°C. Steaks were cooled and stored at 4°C and six cores of 1.27 cm diameter parallel to the longitudinal orientation of the muscles were prepared. A core was sheared perpendicular to the muscle fiber orientation with Instron Universal Testing Machine (Instron Corp., Canton, MA, USA). For NIR spectral data collection, initially a 38-mm-diameter core was obtained and from that 8-mm-diameter circular slices were obtained. The NIR spectra were obtained between 1100 and 2498 nm with 2 nm resolution. PLS regression analysis was performed on the calibration, cross-validation with leave-one-out approach was used, and then the model was tested with external validation samples. Multiple linear regression analysis was also performed to identify important spectral peaks related to the WBSF.

The WBSF values had a range of 2–11.7 kg, with a mean of 5.5 and a standard deviation of 2.2 kg. Tough steaks absorbed radiation between 1100 and 1350 nm. A PLS calibration model with six PLS factors predicted WBSF with an R^2 of 0.67 and an SEP of 1.2 kg, and validation R^2 of 0.63 and SEP of 1.3 kg. Eighty-three per cent of the samples were correctly predicted as tender and 75% of the samples were correctly predicted as tough with an overall accuracy of 79% for tough and tender samples, assuming the steaks with shear force <6 kg were classed as tender and >6 kg as tough. Multiple linear regression identified 1854, 1688, 1592, and 2140 nm as the prominent wavelengths for the prediction of WBSF and R^2 of the model was 0.67 with 89% of the samples were correctly classified. Similar results were also found by Byrne *et al.* (1998), Liu *et al.* (2003a), Rødbotten *et al.* (2000); however Leroy *et al.* (2003) and Tornberg *et al.* (2000) did not find a significant relationship between NIR reflectance spectroscopy and WBSF.

Cooking end-point temperature

The processing or cooking end-point temperature is critical for the safe consumption of the cooked beef. Higher cooking temperatures and longer cooking times lower the palatability of the beef; while lower temperatures and shorter times increase the risk of food poisoning. Therefore an optimum time–temperature combination is required for the production of safe and acceptable beef. Thyholt and Isaksson (1998) used offline chemical extraction based on dry extract spectroscopy by infrared reflection (DESIR) in the visible/near-infrared region (400–2500 nm) with multivariate statistical analysis in the determination of the end-point temperature (EPT). Longissimus lumborum muscle 1.5 h post mortem was stored for 26 h at 15°C and then frozen at −40°C. The muscle was thawed at 4°C for 24 h and visible fat was removed, ground with 5 mm plate diameter and a sample of 100 g at 8 mm thickness was packed in the polythene bag. Samples were placed in a water bath and heated at 1°C/min to temperatures between 65.6 and 75.6°C. The sample was kept at the final temperature (EPT) for 4 min. The cooked meat with the juices was centrifuged for 30 min at 6750 ***g*** at 10°C and later filtered and vacuum dried at 20°C. The dried material was analyzed with NIR reflectance spectroscopy. The EPT was predicted with correlation of 0.965, and an RMSEP error of 0.74°C, with 8 PLS factors and in the Vis-NIR region of (400–2500 nm). The bands at 2270, 2304, and 2500 nm were the most influential and these represent the protein. There were also less significant bands at 940, 2242, 2304, and 2490 nm, which represent amine, amide, and proteins. Because of the heat treatment, the water-holding capacity of the meat was reduced due to the denaturation of the sarcoplasmic and myofibrillar proteins.

Detection of adulteration and contamination

Spinal cord contamination

Proctor's group (Gangidi *et al.*, 2003) used attenuated total reflectance (ATR)-FTIR (4000–700 cm^{-1}) to detect spinal cord contamination in ground beef (Figure 8.2). Spinal cord and other central nervous system tissues are prohibited in beef due to concerns about the transmission of bovine spongiform encephalopathy (USDA/FSIS, 2004). The phosphates (P–O–C) at 1050 cm^{-1} and amide (N–H) stretches at 3400–3600 cm^{-1} of sphingomyelin, a major amino lipid present in spinal cord, contributed to the detection (Figure 8.3). Spinal cord at levels of 20–100 ppm were added to the beef and was mixed thoroughly. Ground beef without the spinal cord was used as control. An ATR-FTIR spectrum at 8 cm^{-1} resolution was collected with Impact 410 FTIR (Thermo Fisher Scientific, Madison, WI, USA) (Figure 8.2). The developed model had an R-value ranging from 0.87 to 0.94 and an error of 16–23 ppm. Ninety per cent of spinal cord-containing samples were correctly identified and none of the samples not having spinal cord were identified as having spinal cord, when predicted spinal cord at levels less than 23 ppm were considered as not having spinal cord (Gangidi; 2005).

A similar study was conducted with the Nexus 670 Infrared spectrometer (Thermo Fisher Scientific, Madison, WI, USA) in the NIR region (5400–10 000 cm^{-1}). Second derivatives of the spectral data were used and the correlation between predicted and added spinal cord content was 0.90–0.94 and the detection limit was between 19 and 21 ppm (Gangidi *et al.*, 2005). When predicted samples with less than 21 ppm were considered as not having spinal cord, 87% of the samples were correctly classified as having the spinal cord content, and 33% of the samples were misclassified as having spinal cord when they did not (Gangidi, 2005).

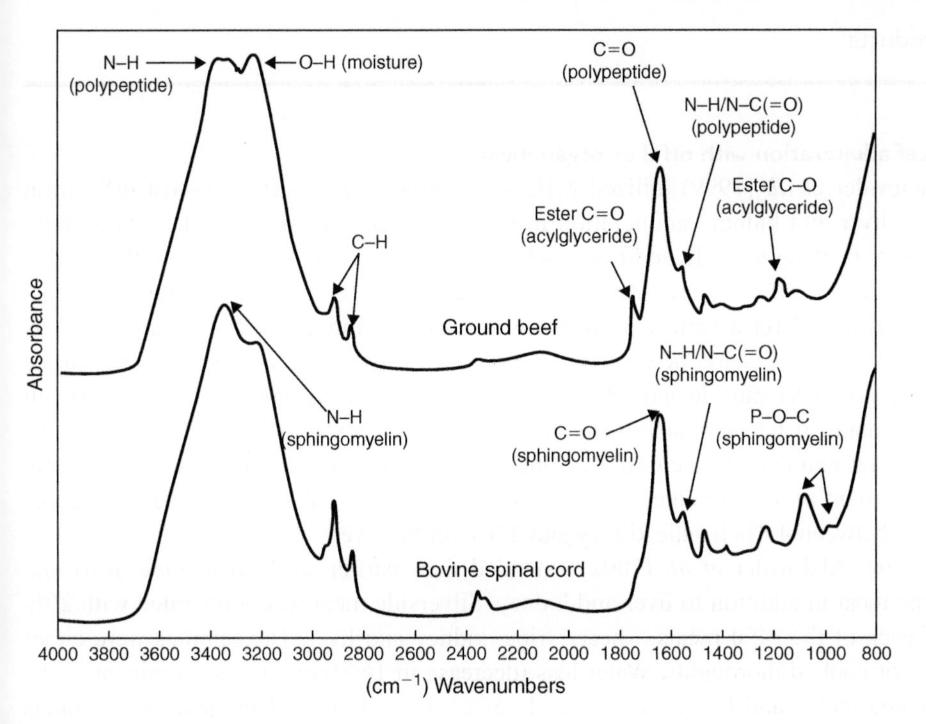

Figure 8.2 Ground beef and bovine spinal cord attenuated total reflectance Fourier transform infrared (ATR-FTIR) spectra (800–4000 cm^{-1}) obtained at 8 cm^{-1} resolution (Gangidi *et al.*, 2003). Copyright permission obtained from Wiley-Blackwell Publishing Ltd.

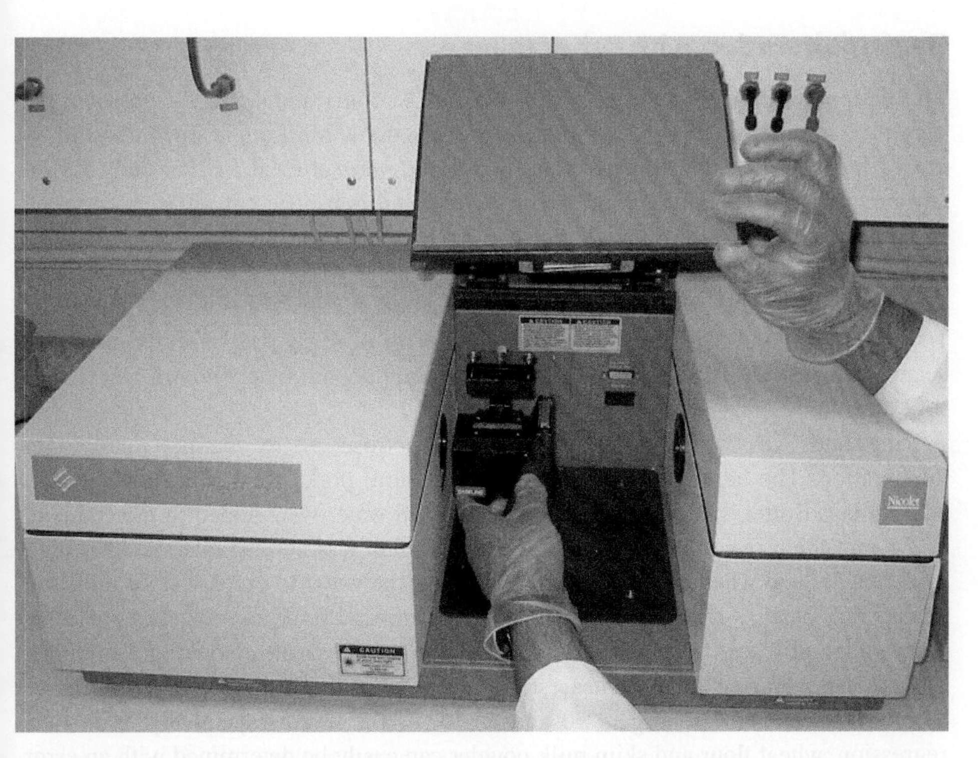

Figure 8.3 Beef spectrum in the mid-infrared region (4000–750 cm^{-1}) is collected with Impact 410 FTIR. (Courtesy of Thermo Fisher Scientific, Inc.; Jonietz, 2003.) Permission obtained from Copyright Clearence Center/Technology Review.

Beef adulteration with offal or organ meat

Al-Jowder *et al.* (1999) utilized MIR spectroscopy to identify low-cost offal meat (e.g. liver and kidney) adulteration in beef. Silverside, brisket, and beef neck were used as beef samples. In addition, each beef muscle sample was mixed with 10–90% liver and 10–90% kidney meat. The work reported here showed that MIR spectroscopy is useful for a variety of different analyses of minced beef, ox kidney, and ox liver. Upon application of principal component analysis (PCA) and canonical variate analysis (CVA), muscle and offal tissues can be readily distinguished. It was difficult to distinguish between silverside and brisket, but neck muscle, possibly due to its lower fat content, can be easily distinguished. PLS regression was used to quantify the amount of added kidney and liver separately in two calibrations. The SEP values were between 4.8% for the kidney and 4.0% for the liver.

Later, Al-Jowder *et al.* (2002) extended their earlier work to include heart and tripe meat in addition to liver and kidney. Silverside meat was adulterated with 20% of each of the offal meats – heart, tripe, kidney and liver. The samples were either raw or cooked thoroughly. Water loss (decrease in 1650 cm^{-1}) was prominent in the cooked meats and hence higher levels of cooking decreased the statistical models ability to identify the offal meat. PLS with linear discriminate analysis can correctly distinguish offal-adulterated meat from unadulterated meat in raw and cooked meats 96% of the time.

Beef adulteration with other meat

McElhinney *et al.* (1999b) used NIR (1100–2498 nm) and MIR spectroscopy to identify and quantify beef adulterated with low-cost lamb. Longissimus dorsi muscle of lamb and beef semimembranous muscle were utilized in the study. In an earlier study (Al-Jowder *et al.*, 1999), it was found that different muscles from the same animal species were indistinguishable. Homogeneous 100% beef, 5, 10, 20% lamb in beef and 100% lamb were prepared. The Vis-NIR spectra were collected at 400–2500 nm and MIR spectra were collected in the 800–2000 cm^{-1} region. Using the NIR and MIR raw spectral data an error of 4.1% was observed and with second derivative NIR and MIR data an eight component model showed an error of 0.91% lamb.

Ding and Xu (2000) utilized Vis-NIR spectroscopy to identify hamburger adulteration. The adulterants were low-cost mutton, pork, skim milk powder, and wheat flour. Butter, salt, white pepper powder, and water were added to minced beef and 5-mm-thick raw hamburger was prepared, which was cooked later. A 30% skim milk powder and wheat flour paste were added to the water to prepare skim milk and wheat flour paste. Additions of 5, 15, and 25% minced pork, mutton, and pastes of skim milk and wheat flour were added to the hamburger prior to cooking. Canonical discriminate analysis and K nearest numbers identified adulterated hamburgers from unadulterated hamburgers 90 and 92.7% of the time, respectively. With PLS regression, wheat flour and skim milk powder can easily be determined with an error of 0.5–1.7% when compared to the mutton or pork, which were determined with an error of 2.9–4.6%.

Distinguishing fresh from frozen-then-thawed beef

Frozen-then-thawed meat has poorer texture and flavor than fresh meat and it is difficult to identify fresh from thawed meat when it is homogenized or minced. Downey and Beauchêne (1997) attempted to identify fresh meat from frozen-then-thawed meat. Longissimus dorsi muscle was frozen to −18°C and thawed by placing the sample for 8 h at room temperature and later at +4°C for 18 h. The unfrozen samples were stored at 4°C. The samples were analyzed in the Vis-NIR spectral region (650–1100 nm) and with a fiber-optic probe. The samples were scanned close to 4°C. Overall 61–64% of the samples were correctly classified. However, when spectral data with multiplicative scatter correction was used along with PCA factorial discriminate analysis, none of the frozen samples were classified as fresh and only 19% of the fresh samples were misclassified as frozen.

Microbial spoilage

In a novel study, Ellis *et al.* (2004) used horizontal attenuated total reflectance (HATR)-FTIR to determine microbial spoilage in homogenized ground beef. The beef FTIR microbial detection study was based on a similar study on homogenized chicken meat (Ellis *et al.*, 2002). Beef obtained from a grocery store was blended with a homogenizer "as is" (i.e. collagen and fatty material were not removed). The beef was uniformly pressed to a thickness of ~5 mm and exposed to open air at room temperature (22°C) on a Petri dish for 24 h. HATR-FTIR spectra of the sample were collected every hour for 24 h (i.e. 0–24 h) with an IFS28 spectrometer (Bruker Ltd, Coventry, UK) with a deuterated triglycine sulfate detector at 16 cm^{-1} resolution and 256 scans were co-added/averaged. The zinc selenide crystal had 10 internal reflections and with sample penetrative depth of approximately 1 μm. Six replicated spectra were collected. The bacterial counts were determined every hour for 24 h. The spectral preprocessing procedure, scaling to maximum absorbance, was conducted with the MATLAB software (The MathWorks Inc., Natick, MA, USA). A variable selection procedure of a genetic algorithm coupled to multiple linear regression was performed.

The initial total viable count was 2×10^5 CFU cm^{-2} (5.31 log_{10}TVC), and the microbial count after 24 h was 2×10^7 CFU cm^{-2} (7.31 log_{10}TVC). FTIR/PLS data predicted with an error of 0.4–0.6 log_{10}TVC. The genetic algorithm identified the 1413 and 1405 cm^{-1} peaks as influencing microbial detection; these peaks are indicative of amide–CN due to protein degradation by microorganisms. Initial and final pH of the beef was 5.43 and the pH is known to influence microbial growth. The study determined microbial counts based on the extent of microbial spoilage of the substrate (i.e. biochemical changes in the meat). However, studies addressing detection of microbial population independent of spoilage would be useful. This could be done by adding microorganism to fresh meat and then immediately investigating the feasibility of obtaining a microbial count by FTIR. This would minimize the dependency on microbial spoilage to determine microbial count, and would be useful in addressing bioterrorism issues.

FTIR in transmission mode and with classification statistics was found to be a rapid and inexpensive technique for identifying and classifying *Staphylococcus* species and *Pseudomonas* species grown in culture media (Naumann *et al.*, 1991). In addition to total microbial counts, identifying the specific species and classes of microbes directly on the meat would be an excellent spectroscopic technique for protecting consumers from harmful microorganisms and meat processors from expensive meat recalls.

Table 8.1 lists some infrared spectroscopic applications suitable for beef and beef products.

Table 8.1 Near- and mid-infrared spectroscopy to determine beef and beef product quality

Instrumental method	Quality attribute	Reference
FTIR/Machine learning	Microbial spoilage; Total viable counts	Ellis *et al.*, 2004
MIR and chemometrics	Identification offal meat—heart, kidney, liver in beef muscle	Al-Jowder *et al.*, 1999
FTIR	Quantitation of lamb in beef minced meat	McElhinney *et al.*, 1999a
FTIR	Spinal cord detection in ground meat	Gangidi *et al.*, 2003
FTIR	Monitoring tenderization process	Iizuka and Aishima, 1999
IR reflection/Dry extracts	Non-bovine meat in beef patties	Thyholt and Isaksson, 1998
FTIR/Microscopy/ preprocessing	Separation and characterization of physical scatter and chemical constituent information with extended multiplicative scatter correction	Kohler *et al.*, 2005
FTIR/ATR	Monitor enzyme-based proteolysis and tenderization in beef	Iizuka and Aishima, 2000
NIR	Differentiation of frozen and unfrozen beef	Thyholt and Isaksson, 1997
Vis/NIR	Kangaroo meat identification in beef	Ding and Xu, 1999
NIR reflectance/Dry extract	Identification pork, chicken and mutton in minced beef	Thyholt *et al.*, 1997
NIR	Beef hamburger adulteration	Ding and Xu, 2000
NIIRS	Tenderness of ground meat	Prieto *et al.*, 2007
NIR transmittance	Free fatty acid composition	Sierra *et al.*, 2007
Vis/NIR	Beef muscle characterization	Xia *et al.*, 2007
NIR	Oxen chemical composition	Prieto *et al.*, 2006
Vis/NIR	Beef muscle quality attributes—pH 24, L0, L60 color values; pH 3, sarcomere length, cooking loss, Warner Bratzler shear force	Andrés *et al.*, 2008
Vis/NIR online	Beef longissimus tenderness	Shackelford *et al.*, 2005
NIR reflectance	Fatty acid composition	Realini *et al.*, 2004
NIR reflectance and transmittance	Technological and organoleptic properties	Leroy *et al.*, 2003
NIR reflectance/Vis	Color, texture and sensory characteristics of steaks	Liu *et al.*, 2003a
NIR reflectance	Muscle type identification, dry matter, crude protein, ash	Alomar *et al.*, 2003
NIR reflectance	Chemical composition of semi-frozen ground beef	Tøgersen *et al.*, 2003
NIR reflectance	Quality attributes—sensory and instrumental tenderness, flavor and organoleptic acceptability	Byrne *et al.*, 1998
NIR/Vis	Discrimination between fresh and frozen-then-thawed beef	Downey and Beauchêne, 1997
NIR	Fat, protein and moisture measurement in meat patties	Oh and Großklaus, 1995
NIR	Sensory characteristics	Hildrum *et al.*, 1994
Vis/NIR	Quantization of lamb in beef minced meat	McElhinney *et al.*, 1999b
NIR	Moisture, protein, fat and calories in ground beef	Lanza, 1983
NIR	Physical and chemical characteristics of beef cuts	Mitsumoto *et al.*, 1991
FT-NIR	Spinal cord detection in ground meat	Gangidi *et al.*, 2005
NIR	Pre-rigor conditions	Tornberg *et al.*, 2000

Table 8.1 Continued

Instrumental method	Quality attribute	Reference
NIR	pH	Tornberg *et al.*, 2000
NIR reflectance	Beef quality attributes	Rødbotten *et al.*, 2000
NIR	Intramuscular fat content	Prevolnik *et al.*, 2005
NIR/Online	Proximate composition in ground beef	Westad *et al.*, 2004
NIR/Inline	Proximal composition of ground beef	Hildrum *et al.*, 2004
Vis/NIR	Pasture- or corn silage-fed cattle by meat analysis	Cozzolino, 2002
NIR reflectance	Moisture, crude protein, intramuscular fat	Cozzolino and Murray, 2002
NIR/Diode array	Tenderness classification	Rødbotten *et al.*, 2001
NIR reflectance	Tenderness and other quality attributes	Venel *et al.*, 2001
NIR/NIT	Fat, protein moisture in polythene-wrapped beef	Isaksson *et al.*, 1992
NIR reflectance	Low cost temperature and moisture state (frozen, semi-frozen) in homogenized beef	Segtnan *et al.*, 2005
NIT	Identifying undesirable heterogeneity in homogeneous meat along with measurement of moisture, fat and protein content	Davies *et al.*, 1998
NIR	Maximum temperature of previous heat treatment in beef	Isaksson *et al.*, 1989
NIR/NIT	Maximum temperature of previous heat treatment in beef in wet and freeze-dried meat	Ellekjaer and Isaksson, 1992
NIR	Hardness, tenderness, juiciness in fresh and frozen-then-thawed beef	Hildrum *et al.*, 1995
NIR reflectance	Fatty acid content in beef neck lean	Windham and Morrison, 1998

NIR, near-infrared; NIT, near-infrared transmittance; FTIR, Fourier transform infrared spectroscopy; ATR, attenuated total reflectance.

Pork and pork products

Proximate composition

The proximate analysis of pork produces results similar to those obtained by IR-based methods used on beef as described earlier (Lanza, 1983). Ortiz-Somovilla *et al.* (2007) utilized NIR reflectance spectroscopy to successfully measure fat, protein and moisture contents in pork sausages. Lean pork meat from Iberian and Landrace, which is a standard breed, were frozen and later thawed. The standard pork at levels of 0, 25, 50, 75, and 100% was added to the Iberian pork, and later sausage ingredients—fine salt (400 g), dextrose (18 g), thyme (36 g), nitrifier (36 g), wine (0.35 L), garlic (90 g), ground black pepper (45 g), polyphosphate (18 g), and ascorbic acid (9 g), were added to minced pork (18 kg). The mixed sausage is left to stay or ripen at 4°C for 24 h. One hundred samples (300 g each) were obtained. The samples were homogenized, resulting in homogenized samples made from minced samples prior to the collection of the NIR spectral data.

The samples were analyzed with Perten 7000 NIR/Vis spectrometer (Perten Instrument, Huddinge, Sweden) that can collect data between 400 and 1700 nm. Eighty of the 100 samples were used in the calibration and the remaining samples were utilized as external validation data set. Modified PLS with cross-validation in groups was performed. The fat content in samples varied from 8 to 31.7%, with mean of 20.27% and a standard deviation of 7.31%; protein content varied from 12.7 to 20.5%, with a mean of 16.7% and a standard deviation of 2%; and moisture range of 50.2 to 68.4% with a mean at 58.9% and a standard deviation of 5.3%. The PLS model can predict fat, protein and moisture with an R^2 of approximately 0.98, 0.90, and 0.97 and SEP of 1.2, 0.85, and 0.9%, respectively. Residual predictive deviation was greater than 3 in most cases, suggesting very good predictability of the PLS model. Residual predictive deviation is the relationship between the standard deviation of reference samples and standard error of cross-validation. In a related study, Gaitán-Jurado *et al.* (2007) observed similar results for the intact sliced and homogenized pork dry-cured sausages.

Quality attributes

Fatty acid composition

In pork, saturated fatty acids are preferred to the unsaturated fatty acids, as they are more oxidatively stable. Furthermore, unsaturated fatty acids may cause undesirable softness in meat. Ripoche and Guillard (2001) used FTIR and FT-NIR to measure pork fatty acid content. Meat from pigs fed with variable amounts of sunflower oil or no sunflower oil was studied for variations in pork fat content in the samples. Back fat and breast fat were analyzed. A 2-mm-thick back fat sample and a breast fat sample were frozen and placed in polythene bags and later thawed at 2°C. Diffuse reflectance NIR spectra were directly collected on the samples in polythene bags. Lipids were extracted with 2:1 chloroform/methanol solvent. Infrared spectra were collected with BOMEM MB100 spectrophotometer. Diffuse reflectance and transmission NIR spectra were collected between 11000 and 4000 cm^{-1} (900–2500 nm) with a resolution of 8 cm^{-1}. Diffuse reflectance NIR spectra were directly collected on the samples in polythene bags.

Lipids were extracted with 2:1 chloroform/methanol solvent and transmission spectra collected by placing in the 0.5 cm thermostatic cell. FTIR spectra (6000–900 cm^{-1}) were collected at 4 cm^{-1} resolution with zinc selenide ATR crystal. Fatty acid composition of the samples was measured by gas chromatography with a flame ionization detector. PLS regression analysis was performed between the spectral data and the fatty acids concentration. A FTIR band between 2825 and 3967 cm^{-1} (characteristic of CH_2, CH_3 stretches, and CH stretches, as in *cis* HC=CH) and 711–1853 cm^{-1} stretches (characteristic of C=O stretch and CH stretch as in *trans* HC=CH) were used to predict fatty acid composition. Transmission NIR bands between 1362 and 1480 nm (7342–6756 cm^{-1}) (characteristic of CH_3 and CH_2 stretches), 1687–1855 nm (5927–5390 cm^{-1}) (characteristic of CH stretch), and 2115–2172 nm (4728–4604 cm^{-1}) (characteristic of HC = CH bands) showed prominent peaks or bands related to the fat. CH_3, CH_2, CH, and –HC=CH– bands are commonly found, but may not be specific to fat. The saturated fatty acid (SFA) content ranged from 29.9 to 45.9%, with a mean of 38.6%; monounsaturated fatty acid (MUFA) content varied from 38.9 to 58.4%, with a mean of 49.9%; polyunsaturated fatty acid content (PUFA) varied from 7.9 to 20.4% with a mean of 11.6; C16:0 ranged from 20.4 to 27.6, with a mean of 23.8%; C18:0 ranged from 8.3 to 17.8%, with a mean of 13.3%; C18:1 ranged from 37.2 to 55.2%, with a mean of 46.9%; and C18:2 ranged from 7.1 to 19.3%, with a mean of 10.6%.

The ATR-FTIR prediction of calibration and external validation samples was approximately 0.90 (R^2) and the error was between 0.5 and 0.7%. Individual saturated fatty acids, such as C16:0 and C18:0 prediction were less accurate. A similar trend was also observed with transmission NIR, except for higher calibration errors of 1.14% and 1.17% with MUFA and C18:1, respectively. Diffuse reflectance NIR predicted SFA, PUFA, and oleic acid (C18:1) and linoleic acid (C18:2), with an accuracy of 0.67, 0.62, 0.72, and 0.79 and an error of 3.1, 3.6, 4.3, and 2.7%, respectively. This could be probably due to interference from heterogeneous compounds present in the slice of the back fat. Nine PLS factors were utilized for SFA, MUFA, and PUFA and individual fatty acids required 11–15 PLS factors. However, when NIR reflectance predicted values were utilized to classify back fat into good, medium, or bad categories, PUFA and C18:2 values could be used to identify good and medium samples from bad samples with 100% accuracy. Similar results were found with intramuscular fat of pork loins (González-Martín *et al.*, 2005) and subcutaneous fat of Iberian breeds (González-Martín *et al.*, 2003) with fiber-optic NIR reflectance spectroscopy.

Meat pH

The pH is known to affect the meat quality, with higher pH resulting in dark, firm, dry (DFD) meat, whereas lower pH results in pale, soft, exudative (PSE) meat due to denaturation of the proteins. Andersen *et al.* (1999) used NIR reflectance spectroscopy to measure pH in pork. Forty-nine longissimus dorsi and semimembranous muscles were analyzed for pH. Unhomogenized longissimus dorsi pH ranged from 6.67 to 5.37, with a mean of 5.69 and standard deviation of 0.25 units. Samples were utilized "as is" or ground twice through plate with 2-mm holes. The pH was measured the day after slaughter. The unhomogenized semimembranous muscle pH was 6.97–5.47, with a mean of 5.77 and standard deviation of 0.27. Homogenized muscles had approximately the same range, mean, and standard deviation as that of unhomogenized muscles. The NIR spectra between 1000 and 2630 nm were collected with an MB series 160 FT-NIR instrument with a diffuse IR sample compartment made by the BOMEM (Quebec City, Canada) with 4 cm^{-1} resolution and surface measure area of approximately 13 cm^2. A single 5177 cm^{-1} peak predicted pH with an R-value of 0.5. The homogeneous longissimus dorsi gave a slightly better result of $r = 0.55$.

The pH of the pork could be determined for the following reasons: (1) Longer wavelengths are more sensitive to pH mainly because of the lower availability of surface free water at high pH compared with at low pH. In addition, at lower energies or shorter wavelengths, the water reflects less light or absorbs more light than at low pH. A negative correlation between absorbance and pH is observed for high pH at longer wavelengths. (2) In the high energy or shorter wavelength region, less light is absorbed by the water in meat. This is advantageous as higher surface free water is available at lower pH as meat denatures at low pH. Furthermore, denatured muscle at low pH is known to inhibit light penetration and thus enhance the amount of reflected light. In contrast, longer wavelengths are reflected less, or not at all, from samples with high surface free moisture.

A multiplicative scatter correction was performed on the data, prior to the PLS regression analysis and the model was cross-validated using the leave-one-sample-out methodology. For unhomogenized longissimus dorsi, homogenized longissimus dorsi, unhomogenized semimembranous and homogenized semimembranous muscle samples, an *R*-value of 0.73, 0.77, 0.79, and 0.73 and RMSEP of 0.10, 0.10, 0.08, 0.09, respectively, were observed. Three PLS factors were utilized for the unhomogenized longissimus dorsi and for other samples a single PLS factor was utilized. The predicted RMSEP values for the muscle tissue are comparable to that of the reference values, i.e. 0.1 units, irrespective of unimpressive *r*-values, which could be due to the narrow range of pH values of the muscles. The NIR reflectance spectroscopy method precisely predicted pH.

Water-holding capacity

Pork with a high water-holding capacity (WHC) has better appearance and higher yield than pork with a low water-holding capacity.

Fourier transform infrared spectroscopy

Pedersen *et al.* (2003) utilized FTIR to study the water-holding capacity of pork. A wide range of pork water-holding capacities, initial pH and post-mortem pH values were obtained by giving pigs 0.3 mg adrenaline/kg live weight, 16 h prior to slaughter, or exercising on a treadmill for 14–20 min prior to slaughter. Control animals were also used (i.e. without any treatment). Adrenaline is known to affect glycogenolysis and thus influences the ultimate pH. At 45 min post mortem, the carcasses were chilled at 4°C. FTIR spectra and drip loss were recorded on longissimus dorsi muscle. Drip loss was measured by placing a 2.5-cm-thick slice of muscle taken 24 h post mortem and placed in a water-permeable holder or net in a plastic bag for 48 h at +4°C. Percentage loss of weight is drip loss. A zinc selenide ATR system was used to collect the spectral data using 4000–750 cm^{-1}. PCA, PLSR, and interval-PLSR (IPLSR) were utilized to develop regression models to determine the drip loss. IPLSR is similar to PLSR and is performed only on sections of FTIR spectra compared to the entire FTIR spectrum in PLSR.

Myofibrillar protein showed a peak at 3300 cm^{-1} due to O–H group adsorbed to myofibrillar protein and N–H groups of polypeptides. Specific peaks of myofibrillar protein due to amide I carbonyl can be observed at 1650 cm^{-1} and NH vibrations due to amide II can be observed at 1540 cm^{-1}. The bands between 1160 and 1080 cm^{-1} are due to glycogen. A plot of PC1 vs. PC4 with 900–1800 cm^{-1} spectral region identified adrenaline-injected meats from control and exercised meats. These results suggest that FTIR can provide early identification of undesirable low-quality PSE meat.

The spectral regions of 1396–1317 cm^{-1} and 1072–993 cm^{-1} predicted WHC with an *r*-value of 0.89 and an error of 0.85, and showed a WHC range of 0.7–8%. In addition, the spectral region of 1800–900 cm^{-1} predicted WHC with an *r*-value of 0.89

and an RMSEP error of 0.86%. The IR region at 1360 cm^{-1} (due to de-protonated carboxylic group and thus related to pH) and the spectral region at 1020 cm^{-1} (due to CO stretching of glycogen, which could be comparatively high 45 min post mortem) are probably the bands best related to predicting WHC. However, the entire MIR spectral region provided less accurate and precise results. Therefore, WHC can be measured by FTIR with pH- and glycogen-specific spectral regions.

Near-infrared spectroscopy

Forrest *et al.* (2000) used NIR to measure the drip loss in the pork and WHC by a combination of NIR data in the 900–1800 nm region and PLS. They predicted WHC with an *r*-value of 0.84 and RMSEP error of 1.8%. The wavelengths 1792, 1704, 1356, 1597, 1473, 1425, 1693, 1536, and 951 nm were prominent in WHC measurement. Geesink *et al.* (2003) used NIR reflectance spectroscopy to classify pork samples as less than 5% or higher than 7% WHC with 100% correct classification.

The Rendement Napole (RN) gene

Pork derived from pigs carrying the RN gene has lower pH and low protein content, resulting in lower meat yields due to the minimal water-holding capacity (Josell *et al.*, 2000). Pork from RN gene carrier pigs can be identified by Vis-NIR spectroscopy coupled with PLS and neural networks. The pork from RN gene carriers and non-carriers has been analyzed using NIR. Spectral differences at 1430–1462 nm and 1880–2000 nm were seen in pork from RN gene carriers and non-carriers, when spectra were collected 30 min post mortem. Similarly, halothane gene-containing carcasses could also be identified by NIR.

Table 8.2 lists some infrared spectroscopic methods applied to pork and pork products.

Chicken and chicken products

Proximate composition

Proximate analysis of chicken and chicken products is similar to that described for beef and pork meat and meat products (Kruggel *et al.*, 1981; Lanza, 1983; Isaksson *et al.*, 1992).

Quality attributes

Chilled and frozen storage effects

Appropriate time–temperature combination storage is necessary for safe and qualitative meat. Liu *et al.* (2004a) continued their earlier work (Liu *et al.*, 2000) with chilled or frozen stored samples, and used two-dimensional (2D) Vis-NIR spectroscopy to measure the quality of chilled or frozen meats. In 2D correlation, both *x* and *y* axes are wavelengths and usually describe the same spectrum; in some cases different but related spectra are used. The spectra chosen would summarize the variations due to an attribute or attributes, for example intra- and inter-spectral variations due to temperature changes.

Table 8.2 Infrared spectroscopy techniques. Near and mid-infrared spectroscopy to determine pork and pork product quality

Instrumental method	Quality attribute	Reference
FTIR	C22:5, C22:6 marine fatty acids in pork fat	Flåtten *et al.*, 2005
FTIR	Monitoring lipid oxidation in fat	Guillen and Cabo, 2004
FTIR	Fatty acid composition of pork fat	Ripoche and Guillard, 2001
FTIR	Intramuscular fat variability	Geers *et al.*, 1995
NIR reflectance	Moisture, fat and protein content in intact and homogenized sausages	Gaitán-Jurado, 2007
IR	Characterization of nitrosyl heme pigments, produced in heat-processed cured ham	Burge and Smith, 1992
NIR/Vis	pH, intramuscular fat, drip loss, CIE L, a, b values	Savenije *et al.*, 2006
NIR/Vis and fiber optics	Texture and color of dry cured ham	Garcia-Rey *et al.*, 2005
NIR reflectance	Monitor pork quality changes early postmortem, drip loss, muscle metabolism. Intramuscular fat measurement	Hoving-Bolink *et al.*, 2005
NIR/Fiber optic	Intramuscular fatty acid in Iberian pork	González-Martín, 2005
NIRS/Fiber optic	Fatty acids in subcutaneous fat of Iberian swine	González-Martín *et al.*, 2003
NIR reflectance	Classification of meats based on water-holding capacity	Geesink *et al.*, 2003
Vis/NIR reflectance	RN-phenotype identification in pig carcasses	Josell *et al.*, 2000
NIR	Moisture, protein, fat and calories	Lanza, 1983
NIR/AOTF/Online	Moisture	Kestens *et al.*, 2007
NIR reflectance	Proximate analysis of pork sausages	Ortiz-Somovilla *et al.*, 2007
NIR	pH after 24 h post mortem, L, a, b, drip loss after 24 and 48 h in minced and intact muscles	Čandek-Potokar *et al.*, 2006
NIR	Intramuscular fat content	Prevolnik *et al.*, 2005
NIR/InfraAnalyzer 260	Moisture and fat	Czarnik-Matusewicz and Korniewicz, 1998
NIR	Sodium chloride content in sausages	Ellekjaer *et al.*, 1993
Vis/NIR	Chemical composition and genotype identification	McDevitt *et al.*, 2005
Vis/NIR	Water-holding capacity and warmed over flavor	Brøndum *et al.*, 2000a, 2000b.
NIT	Identifying undesirable heterogeneity in homogeneous meat along with measurement of moisture, fat and protein content	Davies *et al.*, 1998

NIR, near-infrared; NIT, near-infrared transmittance; FTIR, Fourier transform infrared spectroscopy; ATR, attenuated total reflectance; AOTF, acoustic-optical tunable filter.

Five hundred and twenty-five poultry carcasses were put in three treatment groups (Liu *et al.*, 2004b). In the first treatment the carcasses were stored at 4, 0, −3, −12, and −25°C for 2 days, in the second treatment the carcasses were stored for 7 days, and in the third treatment, which is similar to second treatment, carcasses were stored at −18°C for an additional 7 days. The right breasts were used for NIR spectral analysis and left breasts for sensory and cooking analysis. The spectra for each treatment–temperature combination were used to give 15 averaged spectra, which were analyzed by 2D correlation. The samples aged at 4°C had greater tenderness, probably because the protein denaturation within the myofibril enzymes had increased N–H/O–H interactions as shown by increased spectral intensities at 1400 to 1600 nm, or because of the loss of moisture from melting of ice during the freeze–thaw cycle. Asynchronous 2D spectra changes with cooking, chilling, and freezing resulted in degradation of the C–H peak at 1200 and 1330 nm from the destruction of heme pigments at higher temperature. Increased absorption at 1465 and 1960 nm due to O–H/N–H vibrations from protein and water interactions showed protein denaturation. These results suggest that the temperature profile of the chicken products, such as maximum cooking temperature or frozen or chilled product can be determined with NIR reflectance spectroscopy and 2D correlation spectroscopy, in addition to the wavelength selection for these particular attributes.

Tender and tough meat has been identified with NIR reflectance spectroscopy and 2D correlation spectroscopy. Tender meat, exhibiting less than 8 kg shear pressure, was prominent at 1120, 1275, 1450, 2000, and 2230 nm. Tougher meat, with greater than 8 kg shear pressure, was identified by bands at 1440, 1860, and 2300 nm.

Synchronous two-dimensional correlation spectroscopy

Two-dimensional (2D) correlation spectroscopy is also used for quality inspection and consists of synchronous and asynchronous spectra. A synchronous 2D correlation spectra characterizes similar variations in both the spectral data (i.e. variations being consistently negative or positive in both cases). Autopeaks appear at the diagonal position, which represents dynamic variations in the spectral data. The synchronous off-peaks or cross-peaks appear at off-diagonal positions. These peaks occur when spectral intensities at two different wavelengths (or resonant wavelengths) have similar trends. The positive cross-peaks appear when the intensities at the two wavelengths have similar trends (i.e. either decreasing or increasing). The negative cross-peaks occur when the spectral intensities at the two wavelengths are dissimilar (i.e. when the spectral intensity at one wavelength is increasing while the spectral intensity at the other wavelength is decreasing).

Asynchronous two-dimensional correlation spectroscopy

An asynchronous 2D correlation spectrum comprises exclusively off-diagonal cross-peaks that are entirely a result of effects related to perturbation or in this case time-dependent systematic variations in spectral intensities at different wavelengths. These intensities are not the same as spectral intensities that are resonance peaks observed as diagonal peaks in synchronous 2D correlation spectrum. They only show up when the spectral intensities are dissimilar, and a positive peak indicates that the peak in wavelength 1 occurs after the event occurred at wavelength 2. A positive asynchronous cross-peak occurs when the peak at wavelength 2 appears after the event occurred at wavelength 1.

Advantages and limitations of two-dimensional correlation spectroscopy

One of the advantages of 2D correlation is that the interaction between different spectral regions can be effortlessly observed which is not possible with typical intensity vs. wavelength spectra. Even in 3D spectra of time, intensity, and wavelength, spectral variations with time for a particular wavelength can only be clearly observed and analyzing changes within a spectrum would be difficult. However 2D correlation spectroscopy requires a sophisticated chemometrics package and results in two plots—synchronous and asynchronous—which are difficult to analyze due to the complex contours. Nevertheless, 2D spectral analysis is a required and important method in understanding biochemical changes and in variable selection for statistical modeling (Liu *et al.*, 2000).

Thermal processing

Proper heat treatment is vital for the meat processors to provide safe and nutritious products. Liu *et al.* (2000) used Vis-NIR spectroscopy and 2D correlation spectroscopy to understand the spectral changes due to thermal cooking treatments. Fresh thin slices (1 cm thick and 3.8 cm diameter) were obtained from a healthy, wholesome chicken carcass. The uncooked slice was cut to fit into a quartz window-clad cylindrical cup. The slices were stored at 0°C in a polythene bag and then cooked at a constant air temperature of 150°C for 3, 6, 9, 12, 15, and 18 min. The sample was immediately removed from the oven once the set time was reached and the surface temperature measured with a J-type thermocouple. It was then stored in a polythene bag and cooled to room temperature for subsequent analysis. The spectra were collected with NIR reflectance spectroscopy from 400 to 2500 nm with 2 nm resolution on a rotating sample holder. With increasing cooking time a higher surface temperature was observed. The spectra were subtracted from an average spectrum as data in the 2D correlation analysis in both x and y dimensions. The spectral analysis was performed with the KG2D correlation program (School of Science, Kwansei-Gakuin University, Nishinomiya, Japan) installed in Grams/32 software (Thermo Fisher Scientific, Waltham, MA, USA). A 2D correlation spectroscopy threshold of 30% of the maximum point in the contour map was selected. Too high a threshold resulted in non-selection of the information filled finer details; and too low a threshold results in minor features resulting from noise and baseline distortion to be included in the contour map.

Cooking resulted in a decrease in spectral intensities due to first and second overtones of C–H stretching and combination of C–H stretching at 1655, 1195, and 1360 nm, respectively. Spectral variations at these wavelengths are different in water, suggesting that the C–H variations are from heme groups in deoxymyoglobin and oxymyoglobin. The early changes in these peaks coincide with the denaturation of the meat heme myoglobin protein pigments.

Processing parameters—T_{max}, C and F

The T_{max}, F, and C values are indicators of cooking and processing quality of the processed meats. T_{max}, is the maximum internal meat cooking temperature; The 'F' is the equivalent process temperature at 121°C and is defined as the time-based integral function of the temperature of the sample and Z_f. The 'C' value is the cooking value of equivalent process temperature at 100°C. It is a time-based integral function of temperature and Z_c. A higher C value indicated greater cooking temperature and/or a longer cooking time. Z_f and Z_c are the required rise in temperature for ten-fold decrease in microbial content and increase in reaction rate, respectively.

Chen and Marks (1997) used NIR reflectance spectroscopy to predict C (minutes), F (minutes) and T_{max} (°C) values related to the cooking profile of chicken patties. Z_f values ranged from 7 to 13°C. A temperature of 10°C was selected for predictive modeling. The values range from 24 to 30°C and 26°C was selected for predictions. Ninety-eight ground, formed, and frozen, thick, 6.2 cm diameter chicken breast patties were placed in a polythene bag and thawed overnight at 3°C and stored at 25°C for 120 min. The samples were cooked in a convection oven at air temperatures of 135, 149, 163, 177, 191, 204 and 218°C at end-point temperatures of 50, 55, 60, 65, 70, 75, and 80°C for a total of 49 air–product temperature combinations. The end-point temperature was measured with a T-type thermocouple which was placed half-way between the top and bottom of each sample in a radial direction. After the set temperature was reached, samples were immediately removed, weighed, and resealed in polythene bags and then cooled with running water. The sample was cut with a diameter of 3.2 cm and placed in a NIR reflectance spectroscopy sample holder and data collected between 400 and 2500 nm with 2 nm resolution.

The C and F values were found to be non-linear; hence log(C) and log(F) values were used to produce linearity in data to develop predictive statistical modeling. The C-values (minutes) ranged from 0.02 to 1.69 with a mean of 0.28 and a standard deviation of 0.32. The computed log(C) values then ranged from −1.62 to 0.23 with a mean of −0.79 and standard deviation of 0.48. Similarly F-values (minutes) ranged from 9.42×10^{-8} to 1.77×10^{-4}, with a mean and standard deviation of 7.24×10^{-5}. The corresponding computed log(F) values ranged from −7.03 to −2.75, with a mean of −5.23 and stand deviation of 1.10. T_{max} (°C) values ranged from 50.4 to 91.4, with a mean of 67.4 and standard deviation of 10.9. The temperature variation between top and bottom of the patty was less than 6°C.

A modified PLS in the Vis-NIR region was used for multivariate prediction of the T_{max}, C, and F values. A modified PLS standardizes the residual after each factor is calculated. R^2 values were 0.97, 0.97, 0.97, and standard error of calibration (SEC) were 0.08, 0.19 and 1.95 for log(C), log(F) and T_{max}, respectively. Preprocessing steps such as mulitiplicative scatter correction and standard normal variate gave similar or slightly more accurate results. Similarly, performing first and second derivatives on the raw data gave similar results or slightly less accurate results. Cooking value, processing time, and maximum internal temperature of the meat can be predicted accurately and precisely with NIR reflectance spectroscopy.

Another 33 validation samples not included in the calibration samples were predicted with modified PLS models and it was found that SEP values were slightly higher at 1.38, 1.32, and 1.30 times the SEC values for log(C), log(F), and T_{max}. These results suggest that the end-point temperature can be determined non-destructively and non-invasively with Vis-NIR spectroscopy and hence it has great potential in online process analysis.

Texture measurement

Chen and Marks (1998) continued their earlier work by using NIR reflectance spectroscopy to measure cooking loss and Kramer shear properties—yield loss, yield deformation, and yield energy, of chicken patties, which were measured after various degrees of cooking as described earlier (Chen and Marks, 1997). Cooking loss was simply measured by subtracting the cooked weight from pre-cooked weight and is reported as percentage weight loss.

A modified Vis-NIR PLS was utilized on the first derivative to predict cooking loss, yield force, deformation, and energy. A Kramer shear press, with multiple blades and a cell, was attached to a compression testing machine (Instron model 1011, with Series IX control software; Instron Corp., Canton, MA, USA), and tests were conducted with a cross-head speed of 100 mm/s. A sample was placed inside and sheared. Yield force and deformation were computed. Yield energy was computed by integrating the yield force vs. deformation. The yield point was determined as the point of maximum force (Lyon and Lyon, 1996). The T_{max} and $\log(C)$ values were similar to those observed in the earlier study (Chen and Marks, 1997). The cooking loss values ranged from 4.1 to 18.6, with a mean of 10.8 and standard deviation of 3.66. The yield force (newtons) ranged from 60 to 430, with a mean of 263 and a standard deviation of 104. The yield deformation (mm) values ranged from 3.95 to 6.30, with a mean of 4.88 and standard deviation of 0.51; and yield energy values (Nm) ranged from 0.13 to 0.96 with a mean of 0.55 and a standard deviation of 0.21. Cooking loss, yield force, yield deformation, and yield energy prediction showed R^2 of 0.91, 0.97, 0.53, and 0.90 and SEP of 1.28, 30.44, 0.32, and 0.11, respectively. Except the yield deformation, the other parameters could be predicted with a high accuracy. In addition, cooking loss was highly correlated ($r = 0.91$) to $\log(C)$ and yield force and energy were highly correlated (0.83–0.86) to both T_{max} and $\log(C)$. Physical changes due to mass or water loss and denaturation of proteins in the patties during thermal processing (50–80°C) were observed as good predictors of cooking loss and texture.

Liver analysis of unhealthy carcasses

An abnormal liver is a good indicator of an unhealthy chicken. Dey *et al.* (2003) utilized NIR reflectance spectroscopy to identify normal and abnormal or septicemic chicken carcasses. Abnormal chicken carcasses are dark red or bluish in color, dehydrated and stunted, and very high speeds of poultry processing may not allow a thorough visual inspection. However NIR reflectance spectroscopy can identify abnormal chicken carcasses with 94% correct predictability. In this study (Dey *et al.*, 2003), 100 each of normal and abnormal livers were procured, and NIR reflectance spectroscopy spectral data were collected between 400 and 2500 nm. Another 50 livers each of normal and abnormal carcasses were selected for external validation. PCA coupled to neural network classification analysis was used to classify the chicken carcasses. Three out of 50 carcasses were misclassified as normal when abnormal chicken carcasses were identified by visual observation of the livers. However with histopathological examination of the livers, 100% of the abnormal livers were identified correctly. NIR reflectance spectroscopy with PCA and neural networks could be utilized as a rapid technique to identify abnormal from normal chicken carcasses based on the liver spectral data.

Microbial spoilage

Ellis *et al.* (2002) utilized FTIR and machine learning, a combination of variable selection and classification multivariate statistical methods, to detect microbial count in homogenized chicken breast meat. The microbial content was determined accurately and precisely with FTIR. In an investigation similar to the beef study discussed earlier (Ellis *et al.*, 2004), homogenized chicken breast meat was exposed to air at room temperature for 24h and samples analyzed every hour for the 24h. The total viable count of microbials was conducted with the Lab M Blood agar base method, in which 50μL of the sample is incubated at 25°C for 48h.

Amide I at 1640 cm^{-1}, amide II at 1550 cm^{-1}, and amines at 1240 and 1088 cm^{-1} due to proteolysis were the most significant peaks in the determination of the TVC counts. A PLS model based on the spectral data predicted with a low error of 0.15 to 0.27 log_{10}TVC units. The microbial TVC values ranged from 7×10^6 CFU g^{-1} (6.85 log_{10}TVC) to 2×10^9 CFU g^{-1} (9.31 log_{10}TVC) and the pH values were 5.87 and 6.67, respectively, initially and after 24h.

Table 8.3 lists some infrared spectroscopic applications for chicken meat and chicken products.

Miscellaneous applications

Bone adulteration of meat

Adulteration of meat with bone can affect fat, protein, and moisture measurements but can be corrected. Crane and Duganzich (1986) emulsified meat samples with alkaline reagents. Fat, protein, moisture, and carbohydrate content were measured with an NIR spectrometer (Super Scan type 10600, Foss, Hillerod, Denmark). Carbohydrate content was most affected by the presence of bone in meat and this was confirmed with additional studies with added tricalcium phosphate ($Ca_3(PO_4)_2$), which is similar to hydroxylapatite ($Ca_5(PO_4)_3(OH)$), a major bone constituent. Fat, protein, and water data had to be corrected by a multiplying factor of 0.96, 1.05, and 0.97, respectively, when 3.6% of the tricalcium phosphate, which is similar to the phosphate amount present in the intact bone-in-meat carcass, was added to boneless meat. This study could be further extended to detect and measure the bone content in meat and corrected for the measurement in fat, protein, and water content online, i.e. without requiring chemical extraction prior to the NIR measurements.

Dimensions of the NIR fiber-optic reflectance probe

Shackelford *et al.* (2004) found that the size of the probe affects repeatability of the data when 35-mm and 3-mm-diameter probes were utilized for lamb longissimus muscle analysis. Spectral data were collected with Model A108310 LabSpec Pro portable spectrophotometer (Analytical Spectral Devices, Inc.; Boulder, CO, USA) in the range of 450–2500nm range. The 35-mm probe had a greater precision (>0.88) between 660 and 1326nm than the 3-mm probe (<0.55). This could be due

Table 8.3 Near- and mid-infrared spectroscopic utilization on chicken and chicken products

Instrumental method	Quality attribute	Reference
FTIR/Machine learning	Microbial spoilage; total viable counts	Ellis *et al.*, 2002
FTIR/Microscope	Myowater and protein secondary structure changes	Wu *et al.*, 2007
Visible/NIR	Moisture, fat and protein in breast and thigh muscle	Chen and Massie, 1993
IR	Lard content in meat and meat products	Kamal *et al.*, 1988
Vis/NIR	Color change in cooked chicken muscles	Swatland, 1983
Vis/NIR	Identification of minced and intact broiler meat from non-broiler meat	Ding *et al.*, 1999
Vis/NIR 2D correlation spectroscopy	Monitor thawing frozen chickens	Liu and Chen, 2001
NIR reflectance	Fat content in broiler chickens based on freeze-dried breast muscle	Abeni and Bergoglio, 2001
Vis/NIR	Previous thermal treatments on patties	Chen and Marks, 1997
NIR	Chill storage effects on breast meat	Lyon *et al.*, 2001
NIR reflectance	Tenderness of pectoralis major or breast muscle	Meullenet *et al.*, 2004
Vis/NIR	Monitoring chicken carcass in storage	Chen *et al.*, 1996
Vis/NIR reflectance	Chicken carcass inspection system	Chen *et al.*, 2000
Vis/NIR 2D spectroscopy	Monitor thermal treatments of chicken	Liu *et al.*, 2000
Vis/NIR 2D spectroscopy	Monitor frozen and chilled storage of chicken	Liu *et al.*, 2004a
Vis/NIR	pH, color, shear force, tough and tender classification of cooked and raw muscles	Liu *et al.*, 2004b
NIR	Authentication	Fumière *et al.*, 2000
Vis/NIR reflectance	Moisture, fat and protein in breast and thigh muscle	Cozzolino *et al.*, 1996
Vis/NIR 2D spectroscopy	Monitor meat quality in cold storage	Liu and Chen, 2000
NIR/Chemometrics	Classification of meat based on tenderness	Naes and Hildrum, 1997
NIR/NIT	Protein, fat, and water in plastic-wrapped homogenized meat	Isaksson *et al.*, 1992
Vis/NIR	Fecal contamination on chicken skins	Liu *et al.*, 2003b
NIR	Moisture, fat, protein, identification of carcasses fed *n*-3 fatty acids enriched diet over normal diet	Berzaghi *et al.*, 2005
Vis/NIR	Cooking loss, instrumental tenderness yield force, yield energy of cooked chicken patties	Chen and Marks, 1998

NIR, near-infrared; NIT, near-infrared transmittance; FTIR, Fourier transform infrared spectroscopy.

to the larger surface area analyzed with the 35-mm diameter probe. The repeatability was low, with both probes in the 1326–2500 nm spectral region but the 35-mm probe provided better precision in the short-wave NIR region than the 3-mm probe.

FTIR-photoacoustic spectroscopy

Fat, protein, and moisture content in beef and pork were analyzed with a novel FTIR-photoacoustic spectroscopy (Hong and Irudayaraj, 2001). In this spectroscopic technique, a sample is exposed to IR radiation and the sound waves generated by the heating due to molecular IR absorption are measured as well as subsequent temperature fluctuations and pressure oscillations in a helium-sealed chamber. The sound waves are recorded with a sensitive microphone and then converted to an electrical signal. In addition, this technique can be used to analyze depth profiles between 7 and 64 μm, which is a subsurface analysis only. FTIR-photoacoustic spectroscopy was a better technique than the ATR method in high moisture containing meat samples. The technique may not have practical online applications, due to its complicated instrumentation.

Table 8.4 lists some IR spectroscopic applications for miscellaneous products.

Table 8.4 Near- and mid-infrared spectroscopy to determine miscellaneous meat quality

Instrumental method	Quality attribute	Reference
FTIR	Phospholipid content	Villé *et al.*, 1995
FTIR	Species identification in raw homogenized meats	Downey *et al.*, 2000
NIR	Lamb sensory characteristics	Andrés *et al.*, 2007
FTIR	Meat water-holding capacity	Pedersen *et al.*, 2003
MIR spectroscopy/Transmission	Fat and protein in prepared milk like meat emulsion	Mills *et al.*, 1984
NIR reflectance	Chemical composition of freeze-dried ostrich meat	Viljoen *et al.*, 2005
NIR reflectance	NIR reflectance probe dimensions	Shackelford *et al.*, 2005
NIR/Online	Protein, fat, water	Tøgersen *et al.*, 1999
NIR	Sensory quality of meat sausages	Ellekjaer *et al.*, 1994
NIR	Collagen solubility and concentration measurement	Young *et al.*, 1996
NIR reflectance	Fatty acid content in rabbit meat	Pla *et al.*, 2007
NIR reflectance	Discrimination between conventional and organic rabbit production systems	Pla *et al.*, 2007
NIR	Tenderness, post-rigor and water-holding status	McGlone *et al.*, 2005
NIR	Intramuscular fat content	Prevolnik *et al.*, 2005
Vis/NIR	Species identification in raw meat	Arnalds *et al.*, 2004
NIR reflectance	Species identification in raw homogenized meats	McElhinney *et al.*, 1999a
NIR	Sample heterogeneity	Martínez *et al.*, 1998
NIR reflectance/Dry extract	Meat speciation	Thyholt *et al.*, 1997
NIR	Fat quantization	Afseth *et al.*, 2005
NIR	Heme and non-heme iron content measurement in raw meat	Hong and Yasumoto, 1996
NIR	Multivariate scatter correction, inverse kubelka-munk transformation in high prediction of fat content	Geladi *et al.*, 1985
NIR	Total lipid content, oleic and palmitic acids, dry matter or moisture in goose fatty livers	Molette *et al.*, 2001
Vis/NIR	Meat tenderness	Aignel *et al.*, 2003
NIR/NIT	Meat composition and grading	Marno, 2007

NIR, near-infrared; MIR, mid-infrared; NIT, near-infrared transmittance; FTIR, Fourier transform infrared spectroscopy.

Conclusions

Meat proximate analysis and quality can be readily measured with IR spectroscopy. However the accuracy and precision of quality attributes, such as tenderness and pH, could be improved. Proximate composition and quality attributes of unconventional meats such as kangaroo, camel, bison, and game fowls could be measured with minimal modifications to the existing techniques. There are opportunities and a need to develop rapid online IR techniques for microbial spoilage analysis, total microbial count measurement, and the identification of specific microbial species.

References

Abeni F, Bergoglio G (2001) Characterization of different strains of broiler chicken by carcass measurements, chemical and physical parameters and NIRS on breast muscle. *Meat Science*, **57**, 133–137.

Afseth NK, Segtnan VH, Marquardt BJ, Wold JP (2005) Raman and near-infrared spectroscopy for quantification of fat composition in a complex food model system. *Applied Spectroscopy*, **59**, 1324–1332.

Aignel D, Faure P, Laumonier P (2003) Method and device for determining meat tenderness. *US Patent*, **6**,563,580.

Al-Jowder O, Kemsley EK, Wilson RH (1997) Mid-infrared spectroscopy and authenticity problems in selected meats: a feasibility study. *Food Chemistry*, **59**, 195–201.

Al-Jowder O, Defernez M, Kemsley EK, Wilson RH (1999) Mid-infrared spectroscopy and chemometrics for the authentication of meat products. *Journal of Agricultural and Food Chemistry*, **47**, 3210–3218.

Al-Jowder O, Kemsley EK, Wilson RH (2002) Detection of adulteration in cooked meat products by mid-infrared spectroscopy. *Journal of Agricultural and Food Chemistry*, **50**, 1325–1329.

Alomar D, Gallo C, Castañeda M, Fuchslocher R (2003) Chemical and discriminant analysis of bovine meat by near infrared reflectance spectroscopy (NIRS). *Meat Science*, **63**, 441–450.

Andersen JR, Borggaard C, Rasmussen AJ, Houmøller LP (1999) Optical measurements of pH in meat. *Meat Science*, **53**, 135–141.

Andrés S, Murray I, Navajas EA, Fisher AV, Lambe NR, Bünger L (2007) Prediction of sensory characteristics of lamb meat samples by near infrared reflectance spectroscopy. *Meat Science*, **76**, 509–516.

Andrés S, Silva A, Soares-Pereira AL, Martins C, Bruno-Soares AM, Murray I (2008) The use of visible and near infrared reflectance spectroscopy to predict beef M longissimus thoracis et lumborum quality attributes. *Meat Science*, **78**, 217–224.

Arnalds T, McElhinney J, Fearn T, Downey G (2004) A hierarchical discriminant analysis for species identification in raw meat by visible and near infrared spectroscopy. *Journal of Near Infrared Spectroscopy*, **12**, 183–188.

Ben-Gera I, Norris KH (1968) Direct spectrophotometric determination of fat and moisture in meat products. *Journal of Food Science*, **33**, 64–67.

Berzaghi P, Zotte AD, Jansson LM, Andrighetto I (2005) Near-infrared reflectance spectroscopy as a method to predict chemical composition of breast meat and discriminate between different n-3 feeding sources. *Poultry Science*, **84**, 128–136.

Brøndum J, Byrne DV, Bak LS, Bertelsen G, Engelsen SB (2000a) Warmed-over flavour in porcine meat a combined spectroscopic, sensory and chemometric study. *Meat Science*, **54**, 83–95.

Brøndum J, Munck L, Henckel P, Karlsson A, Tornberg E, Engelsen SB (2000b) Prediction of water-holding capacity and composition of porcine meat by comparative spectroscopy. *Meat Science*, **55**, 177–185.

Burge DL, Smith JS (1992) Characterization of model nitrosylheme pigments with visible, infrared and 15N Fourier transform nuclear magnetic resonance spectroscopy. *Journal of Muscle Foods*, **3**, 123–131.

Byrne CE, Downey G, Troy DJ, Buckley DJ (1998) Non-destructive prediction of selected quality attributes of beef by near-infrared reflectance spectroscopy between 750 and 1098 nm. *Meat Science*, **49**, 399–409.

Čandek-Potokar M, Prevolnik M, Škrlep M (2006) Ability of near infrared spectroscopy to predict pork technological traits. *Journal of Near Infrared Spectroscopy*, **14**, 269–277.

Carlson RE (1978) Moisture analyzing method and apparatus. US Patent 4,097,743.

Chen H, Marks BP (1997) Evaluating previous thermal treatment of chicken patties by visible/near-infrared spectroscopy. *Journal of Food Science*, **62**, 753–780.

Chen H, Marks BP (1998) Visible/near-infrared spectroscopy for physical characteristics of cooked chicken patties. *Journal of Food Science*, **63**, 279–282.

Chen YR, Massie DR (1993) Visible/near infrared reflectance spectroscopy for the detrmination of moisture, fat, protein in chicken breast and thigh muscle. *Transactions of American Society of Agricultural Engineers*, **36**, 863–869.

Chen YR, Huffman RW, Park B (1996) Changes in the visible/near-infrared spectra of chicken carcasses in storage. *Journal of Food Process Engineering*, **19**, 121–134.

Chen YR, Hruschka WR, Early H (2000) A chicken carcass inspection system using visible/near-infrared reflectance: In-plant trials. *Journal of Food Process Engineering*, **23**, 89–99.

Cozzolino D (2002) Visible and near infrared spectroscopy of beef longissimus dorsi muscle as a means of dicriminating between pasture and corn silage feeding regimes. *Journal of Near Infrared Spectroscopy*, **10**, 187–193.

Cozzolino D, Murray I (2002) Effect of sample presentation and animal muscle species on the analysis of meat by near infrared reflectance spectroscopy. *Journal of Near Infrared Spectroscopy*, **10**, 37–44.

Cozzolino D, Murray I, Paterson R, Scaife JR (1996) Visible and near infrared reflectance spectroscopy for the determination of moisture, fat and protein in chicken breast and thigh muscle. *Journal of Near Infrared Spectroscopy*, **4**, 213–223.

Crane B, Duganzich DM (1986) The effect of bone in meat samples upon analytical results for fat, protein and water by foss Super-Scan type 10600. *Meat Science*, **18**, 181–190.

Czarnik-Matusewicz HW, Korniewicz A (1998) The use of the InfraAlyzer 260 Whole Grain for water and fat determination in pork meat. *Journal of Near Infrared Spectroscopy*, **6**, 83–86.

Davies AMC, Cowe IA, Withey RP, Eddison CG (1998) Commodity testing and sub-sample homogeneity system for the Meatspec analyzer. *Journal of Near Infrared Spectroscopy*, **6**, 69–75.

Dey BP, Chen YR, Hsieh C, Chan DE (2003) Detection of septicemia in chicken livers by spectroscopy. *Poultry Science*, **82**, 199–206.

Ding HB, Xu RJ (1999) Differentiation of beef and kangaroo meat by visible/near infrared spectroscopy. *Journal of Food Science*, **64**, 814–817.

Ding HB, Xu RJ (2000) Near-infrared spectroscopic technique for detection of beef hamburger adulteration. *Journal of Agricultural and Food Chemistry*, **48**, 2193–2198.

Ding HB, Xu RJ, Chan DKO (1999) Identification of broiler chicken meat using a visible/near infrared spectroscopic technique. *Journal of the Science for Food and Agriculture*, **79**, 1382–1388.

Downey G, Beauchêne D (1997) Discrimination between fresh and frozen-then-thawed beef m. longissimus dorsi by combined visible-near infrared reflectance spectroscopy: A feasibility study. *Meat Science*, **45**, 353–363.

Downey G, McElhinney J, Fearn T (2000) Species identification in selected raw homogenized meats by reflectance spectroscopy in the mid-infrared, near-infrared, and visible ranges. *Applied Spectroscopy*, **54**, 894–899.

Ellekjaer MR, Isaksson T (1992) Assessment of maximum cooking temperatures in previously heat treated beef. Part 1: Near infrared spectroscopy. *Journal of the Science of Food and Agriculture*, **59**, 335–343.

Ellekjær MR, Hildrum KI, Næs T, Isaksson T (1993) Determination of the sodium chloride content of sausages by near infrared spectroscopy. *Journal of Near Infrared Spectroscopy*, **1**, 65–75.

Ellekjaer MR, Isaksson T, Solheim R (1994) Assessment of sensory quality of meat sausages using near infrared spectroscopy. *Journal of Food Science*, **59**, 456–464.

Ellis DI, Broadhurst D, Kell DB, Rowland JJ, Goodacre R (2002) Rapid and quantitative detection of the microbial spoilage of meat by Fourier transform infrared spectroscopy and machine learning. *Applied and Environmental Microbiology*, **68**, 2822–2828.

Ellis DI, Broadhurst D, Goodacre R (2004) Rapid and quantitative detection of the microbial spoilage of beef by Fourier transform infrared spectroscopy and machine learning. *Analytica Chimica Acta*, **514**, 193–201.

Flåtten A, Bryhni EA, Kohler A, Egelandsdal B, Isaksson T (2005) Determination of C22:5 and C22:6 marine fatty acids in pork fat with Fourier transform mid-infrared spectroscopy. *Meat Science*, **69**, 433–440.

Forrest JC, Morgan MT, Borggaard C, Rasmussen AJ, Jespersen BL, Andersen JR (2000) Development of technology for the early post mortem prediction of water holding capacity and drip loss in fresh pork. *Meat Science*, **55**, 115–122.

Fumière O, Sinnaeve G, Dardenne P (2000) Attempted authentication of cut pieces of chicken meat from certified production using near infrared spectroscopy. *Journal of Near Infrared Spectroscopy*, **8**, 27–34.

Gaitán-Jurado AJ, Ortiz-Somovilla V, España-España F, Pérez-Aparicio J, Pedro-Sanz EJD (2007) Quantitative analysis of pork dry-cured sausages to quality control by NIR spectroscopy. *Meat Science*, **78**, 391–399.

Gangidi (2005) Novel measurement and production of commercially important bovine lipids. PhD dissertation, University of Arkansas.

Gangidi RR, Proctor A, Pohlman FW (2003) Rapid determination of spinal cord content in ground beef by attenuated total reflectance Fourier transform infrared spectroscopy. *Journal of Food Science*, **68**, 124–127.

Gangidi RR, Proctor A, Pohlman FW, Meullenet JF (2005) Rapid determination of spinal cord content in ground beef by near-infrared spectroscopy. *Journal of Food Science*, **70**, c397–c400.

García-Rey RM, García-Olmo J, Pedro ED, Quiles-Zafra R, Castro MDLD (2005) Prediction of texture and colour of dry-cured ham by visible and near infrared spectroscopy using a fiber optic probe. *Meat Science*, **70**, 357–363.

Geers R, Decanniere C, Villé H, Hecke PV, Bosschaerts L (1995) Variability within intramuscular fat content of pigs as measured by gravimetry, FTIR and NMR spectroscopy. *Meat Science*, **40**, 373–378.

Geesink GH, Schreutelkamp FH, Frankhuizen R, Vedder HW, Faber NM, Kranen RW, Gerritzen MA (2003) Prediction of pork quality attributes from near infrared reflectance spectra. *Meat Science*, **65**, 661–668.

Geladi P, MacDougall D, Martens H (1985) Linearization and scatter-correction for near-infrared reflectance spectra of meat. *Applied Spectroscopy*, **39**, 491–500.

González-Martín I, González-Pérez C, Hernández-Méndez J, Alvarez-García N (2003) Determination of fatty acids in the subcutaneous fat of Iberian breed swine by near infrared spectroscopy (NIRS) with a fibre-optic probe. *Meat Science*, **65**, 713–719.

González-Martín I, González-Pérez C, Alvarez-García N, González-Cabrera JM (2005) On-line determination of fatty acid composition in intramuscular fat of Iberian pork loin by NIRs with a remote reflectance fibre optic probe. *Meat Science*, **69**, 243–248.

Guillén MD, Cabo N (2004) Study of the effects of smoke flavourings on the oxidative stability of the lipids of pork adipose by means of Fourier transform infrared spectroscopy. *Meat Science*, **66**, 647–657.

Hildrum KI, Nilsen BN, Mielnik M, Næs T (1994) Prediction of sensory characteristics of beef by near-infrared spectroscopy. *Meat Science*, **38**, 67–80.

Hildrum KI, Isaksson T, Næs T, Nilsen BN, Rødbotten M, Lea P (1995) Near infrared reflectance spectroscopy in the prediction of sensory properties of beef. *Journal of Near Infrared Spectroscopy*, **3**, 81–87.

Hildrum KI, Nilsen BN, Westad F, Wahlgren NM (2004) In-line analysis of ground beef using a diode array near infrared instrument on a conveyor belt. *Journal of Near Infrared Spectroscopy*, **12**, 367–376.

Hong JH, Yasumoto K (1996) Near-infrared spectroscopic analysis of heme and nonheme iron in raw meats. *Journal of Food Composition and Analysis*, **9**, 127–134.

Hong Y, Irudayaraj J (2001) Characterization of beef and pork using Fourier-transform infrared photoacoustic spectroscopy. *Lebensmittel-Wissenchaft Und-Technologie*, **34**, 402–409.

Hoving-Bolink AH, Vedder HW, Merks JWM, Klein WJKD, Reimert HGM, Frankhuizen R, Broek WHAMVD, Lambooij EE (2005) Perspective of NIRS measurements early post mortem for prediction of pork quality. *Meat Science*, **69**, 417–423.

Iizuka K, Aishima T (1999) Tenderization of beef with pineapple juice monitored by Fourier transform infrared spectroscopy and chemometric analysis. *Journal of Food Science*, **64**, 973–977.

Iizuka K, Aishima T (2000) Comparing beef digestion properties of four proteolytic enzymes using infrared spectrometry and chemometric analysis. *Journal of the Science of Food and Agriculture*, **80**, 1413–1420.

Isaksson T, Ellekjær MHR, Hildrum KI (1989) Determination of the previous maximum temperature of heat-treated minced meat by near infrared reflectance spectroscopy. *Journal of the Science of Food and Agriculture*, **49**, 385–387.

Isaksson T, Miller CE, Næs T (1992) Nondestructive NIR and NIT Determination of protein, fat, and water in plastic-wrapped, homogenized meat. *Applied Spectroscopy*, **46**, 1685–1694.

Isaksson T, Nilsen BN, Tøgersen G, Hammond RP, Hildrum KI (1996) On-line, proximate analysis of ground beef directly at a meat grinder outlet. *Meat Science*, **43**(Suppl 3–4), 245–253.

Jonietz E (2003) Meat Monitor. *Technology Review,* July/August, 19.

Josell A, Martinsson L, Borggaard C, Andersen JR, Tornberg E (2000) Determination of RN phenotype in pigs at slaughter-line using visual and near-infrared spectroscopy. *Meat Science*, **55**, 273–278.

Kamal M, Youssef E, Omar MB, Skulberg A, Rashwan M (1988) Detection and evaluation of lard in certain locally processed and imported meat products. *Food Chemistry*, **30**, 167–180.

Kestens V, Charoud-Got J, Bau A, Bernreuther A, Emteborg H (2007) Online measurement of water content in candidate reference materials by acoustic-optical tunable filter near-infrared spectrometry (AOTF-NIR) using pork meat calibrants controlled by Karl Fischer titration. *Food Chemistry*, **106**, 1359–1365.

Kohler A, Kirschner C, Oust A, Martens H (2005) Extended multiplicative signal correction as a tool for separation and characterization of physical and chemical information in Fourier transform infrared microscopy images of cryo-sections of beef loin. *Applied Spectroscopy*, **59**, 707–716.

Kruggel WG, Field RA, Riley ML, Radloff HD, Horton KM (1981) Near infrared reflectance determination of fat, protein and moisture in fresh meat. *Journal of Association of Official Analytical Chemists*, **64**, 692–696.

Lanza E (1983) Determination of moisture, protein, fat, and calories in raw pork and beef by near infrared spectroscopy. *Journal of Food Science*, **48**, 471–474.

Leroy B, Lambotte S, Dotreppe O, Lecocq H, Istasse L, Clinquart A (2003) Prediction of technological and organoleptic properties of beef longissimus thoracis from near-infrared reflectance and transmission spectra. *Meat Science*, **66**, 45–54.

Liu Y, Chen YR (2000) Two-dimensional correlation spectroscopy study of visible and near-infrared spectral variations of chicken meats in cold storage. *Applied Spectroscopy*, **54**, 1458–1470.

Liu Y, Chen YR (2001) Two-dimensional visible/near-infrared correlation spectroscopy study of thawing behavior of frozen chicken meats without exposure to air. *Meat Science*, **57**, 299–310.

Liu Y, Chen YR, Ozaki Y (2000) Two-dimensional visible/near-infrared correlation spectroscopy study of thermal treatment of chicken meats. *Journal of Agricultural and Food Chemistry*, **8**, 901–908.

Liu Y, Lyon BG, Windham WR, Realini CE, Dean T, Pringle S, Duckett S (2003a) Prediction of color, texture, and sensory characteristics of beef steaks by visible and near infrared reflectance spectroscopy. A feasibility study. *Meat Science*, **65**, 1107–1115.

Liu Y, Windham WR, Lawrence KC, Park B (2003b) Simple algorithms for the classification of visible/near-infrared and hyperspectral imaging spectra of chicken skins, feces, and fecal contaminated skins. *Applied Spectroscopy*, **57**, 1609–1612.

Liu Y, Barton FE, Lyon BG, Windham WR, Lyon CE (2004a) Two-dimensional correlation analysis of visible/near-infrared spectral intensity variations of chicken breasts with various chilled and frozen storages. *Journal of Agricultural and Food Chemistry*, **52**, 505–510.

Liu Y, Lyon BG, Windham WR, Lyon CE, Savage EM (2004b) Prediction of physical, color, and sensory characteristics of broiler breasts by visible/near infrared reflectance spectroscopy. *Poultry Science*, **83**, 1467–1474.

Lyon BG, Lyon CE (1996) Texture evaluations of cooked, diced broiler breast samples by sensory and mechanical methods. *Poultry Science*, **75**, 812–819.

Lyon BG, Windham WR, Lyon CE, Barton FE (2001) Sensory characteristics and near infrared spectroscopy of broiler breast meat from various chill-storage regimes. *Journal of Food Quality*, **24**, 435–452.

Marno H (2007) Online recording of wavelength absorption spectra in meat. World Patent W02007000166.

Martínez ML, Garrido-Varo A, Pedro ED, Sánchez L (1998) Effect of sample heterogeneity on near infrared meat analysis: the use of the RMS statistic. *Journal of Near Infrared Spectroscopy*, **6**, 313–320.

McDevitt RM, Gavin AJ, Andrés S, Murray I (2005) The ability of visible and near infrared reflectance spectroscopy to predict the chemical composition of ground chicken carcasses and to discriminate between carcasses from different genotypes. *Journal of Near Infrared Spectroscopy*, **13**, 109–117.

McElhinney J, Downey G, Fearn T (1999a) Chemometric processing of visible and near infrared reflectance spectra for species identification in selected raw homogenised meats. *Journal of Near Infrared Spectroscopy*, **7**, 145–154.

McElhinney J, Downey G, O'Donnell C (1999b) Quantitation of lamb content in mixtures with raw minced beef using visible, near and mid-infrared spectroscopy. *Journal of Food Science*, **64**, 587–591.

McGlone VA, Devine CE, Wells RW (2005) Detection of tenderness, post-rigor age and water status changes in sheep meat using near infrared spectroscopy. *Journal of Near Infrared Spectroscopy*, **13**, 277–285.

Meullenet JF, Jonville E, Grezes D, Owens CM (2004) Prediction of the texture of cooked poultry pectoralis major muscles by near-infrared reflectance analysis of raw meat. *Journal of Texture Studies*, **35**, 573–585.

Mills BL, van-de-Voort FR, Kakuda Y (1984) The quantitative analysis of fat and protein in meat by transmission infrared analysis. *Meat Science*, **11**, 253–262.

Mitsumoto M, Maeda S, Mitsuhashi T, Ozawa S (1991) Near-infrared spectroscopy determination of physical and chemical characteristics in beef cuts. *Journal of Food Science*, **56**, 1493–1496.

Molette C, Berzaghi P, Zotte AD, Remignon H, Babile R (2001) The use of near-infrared reflectance spectroscopy in the prediction of the chemical composition of goose fatty liver. *Poultry Science*, **80**, 1625–1629.

Naes T, Hildrum KI (1997) Comparison of multivariate calibration and discriminant analysis in evaluating NIR spectroscopy for determination of meat tenderness. *Applied Spectroscopy*, **51**, 350–357.

Naumann D, Helm D, Labischinski H (1991) Microbiological characterization by FTIR spectroscopy. *Nature*, **351**, 81–82.

Oh EK, Großklaus D (1995) Measurement of the components in meat patties by near infrared reflectance spectroscopy. *Meat Science*, **41**, 157–162.

Ortiz-Somovilla V, España-España F, Gaitán-Jurado AJ, Pérez-Aparicio J, De-Pedro-Sanz EJ (2007) Proximate analysis of homogenized and minced mass of pork sausages by NIRS. *Food Chemistry*, **101**, 1031–1040.

Park B, Chen YR, Hruschka WR, Shackelford SD, Koohmaraie M (1998) Near-infrared reflectance analysis for predicting beef longissimus tenderness. *Journal of Animal Science,* **7**, 2115–2120.

Pedersen DK, Morel S, Andersen HJ, Engelsen SB (2003) Early prediction of water-holding capacity in meat by multivariate vibrational spectroscopy. *Meat Science*, **65**, 581–592.

Pla M, Hernández P, Ariño B, Ramírez JA, Díaz I (2007) Prediction of fatty acid content in rabbit meat and discrimination between conventional and organic production systems by NIRS methodology. *Food Chemistry*, **100**, 165–170.

Prevolnik M, Čandek-Potokar M, Škorjanc D, Velikonja-Bolta S, Škrlep M, Žnidaršiča T, Babnika D (2005) Predicting intramuscular fat content in pork and beef by near infrared spectroscopy. *Journal of Near Infrared Spectroscopy*, **13**, 77–86.

Prieto N, Andrés S, Giráldez FJ, Mantecón AR, Lavín P (2006) Potential use of near infrared reflectance spectroscopy (NIRS) for the estimation of chemical composition of oxen meat samples. *Meat Science*, **74**, 487–496.

Prieto N, Andrés S, Giráldez FJ, Mantecón AR, Lavín P (2007) Discrimination of adult steers (oxen) and young cattle ground meat samples by near infrared reflectance spectroscopy (NIRS). *Meat Science*, **79**, 198–201.

Realini CE, Duckett SK, Windham WR (2004) Effect of vitamin C addition to ground beef from grass-fed or grain-fed sources on color and lipid stability, and prediction of fatty acid composition by near-infrared reflectance analysis. *Meat Science*, **68**, 35–43.

Ripoche A, Guillard AS (2001) Determination of fatty acid composition of pork fat by Fourier transform infrared spectroscopy. *Meat Science*, **58**, 299–304.

Rødbotten R, Nilsen BN, Hildrum KI (2000) Prediction of beef quality attributes from early post mortem near infrared reflectance spectra. *Food Chemistry*, **69**, 427–436.

Rødbotten R, Mevik BH, Hildrum KI (2001) Prediction and classification of tenderness in beef from non-invasive diode array detected NIR spectra. *Journal of Near Infrared Spectroscopy*, **9**, 199–210.

Savenije B, Geesink GH, Palen GJPVD, Hemke G (2006) Prediction of pork quality using visible/near-infrared reflectance spectroscopy. *Meat Science*, **73**, 181–184.

Segtnan VH, Bjørn-Helge M, Isaksson T, Naes T (2005) Low-cost approaches to robust temperature compensation in near-infrared calibration and prediction situations. *Applied Spectroscopy*, **59**, 816–825.

Shackelford SD, Wheeler TL, Koohmaraie M (2004) Development of optimal protocol for visible and near-infrared reflectance spectroscopic evaluation of meat quality. *Meat Science*, **68**, 371–381.

Shackelford SD, Wheeler TL, Koohmaraie M (2005) On-line classification of US Select beef carcasses for longissimus tenderness using visible and near-infrared reflectance spectroscopy. *Meat Science*, **69**, 409–415.

Sierra V, Aldai N, Castro P, Osoro K, Coto-Montes A, Oliván M (2007) Prediction of the fatty acid composition of beef by near infrared transmittance spectroscopy. *Meat Science*, **78**, 248–255.

Swatland HJ (1983) Fiber optic spectrophotometry of color changes in cooked chicken muscles. *Poultry Science*, **62**, 957–959.

Thyholt K, Isaksson T (1997) Differentiation of frozen and unfrozen beef using near infrared spectroscopy. *Journal of the Science for Food and Agriculture*, **73**, 525–532.

Thyholt K, Isaksson T (1998) Detecting non-bovine meat in beef patties by dry extract spectroscopy by infrared reflection—a preliminary model study. *Journal of Near Infrared Spectroscopy*, **6**, 361–362.

Thyholt K, Indahl UG, Hildrum KI, Ellekjær MR (1997) Meat speciation by near infrared reflectance spectroscopy on dry extract. *Journal of Near Infrared Spectroscopy*, **5**, 195–208.

Thyholt K, Enersen G, Isaksson T (1998) Determination of endpoint temperatures in previously heat treated beef using reflectance spectroscopy. *Meat Science*, **48**, 49–63.

Tøgersen G, Isaksson T, Nilsen BN, Bakker EA, Hildrum KI (1999) On-line NIR analysis of fat, water and protein in industrial scale ground meat batches. *Meat Science*, **51**, 97–102.

Tøgersen G, Arnesen JF, Nilsen BN, Hildrum KI (2003) On-line prediction of chemical composition of semi-frozen ground beef by non-invasive NIR spectroscopy. *Meat Science*, **63**, 515–523.

Tornberg E, Wahlgren M, Brøndum J, Engelsen SB (2000) Pre-rigor conditions in beef under varying temperature and pH-falls studied with rigometer, NMR and NIR. *Food Chemistry*, **69**, 407–418.

USDA/FSIS (US Dept of Agriculture/Food Safety and Inspection Service) (2004) Prohibition of the use of specified risk materials for human food and requirements for the disposition of non-ambulatory disabled cattle. *Federal Register*, **69**, 1862–1873.

Venel C, Mullen AM, Downey G, Troy DJ (2001) Prediction of tenderness and other quality attributes of beef by near infrared reflectance spectroscopy between 750 and 1100 nm; further studies. *Journal of Near Infrared Spectroscopy*, **9**, 185–198.

Viljoen M, Hoffman LC, Brand TS (2005) Prediction of the chemical composition of freeze dried ostrich meat with near infrared reflectance spectroscopy. *Meat Science*, **69**, 255–261.

Villé H, Maes G, Schrijver RD, Spincemaille G, Rombouts G, Geers R (1995) Determination of phospholipid content of intramuscular fat by Fourier Transform Infrared spectroscopy. *Meat Science*, **41**, 283–291.

Westad F, Nilsen BN, Wahlgren NM, Hildruma KI (2004) Short communication: Removal of conveyor belt near infrared signals in in-line monitoring of proximal ground beef composition. *Journal of Near Infrared Spectroscopy*, **12**, 377–380.

Windham WR, Morrison WH (1998) Prediction of fatty acid content in beef neck lean by near infrared reflectance analysis. *Journal of Near Infrared Spectroscopy*, **6**, 229–234.

Wu Z, Bertram HC, Bocker U, Ofstad R, Kohler A (2007) Myowater dynamics and protein secondary structural changes as affected by heating rate in three pork qualities: a combined FT-IR microspectroscopic and 1 H NMR relaxometry study. *Journal of Agricultural and Food Chemistry*, **55**, 3990–3997.

Xia JJ, Berg EP, Lee JW, Yao G (2007) Characterizing beef muscles with optical scattering and absorption coefficients in VIS-NIR region. *Meat Science*, **75**, 78–83.

Young OA, Barker GJ, Frost DA (1996) Determination of collagen solubility and concentration in meat by near infrared spectroscopy. *Journal of Muscle Foods*, **7**, 377–387.

Suggested reading

Aberle ED, Forrest JC, Gerrard DE, Mills EW (2001) *Meat Science.* Dubuque, Iowa: Kendall/Hunt Publishing.

Lawrie RA, Ledward DA (2006) *Lawrie's Meat Science*, 7th edn. Cambridge, England: Woodhead Publishing.

9

Fish and Related Products

Musleh Uddin and Emiko Okazaki

Introduction

Infrared (IR) spectroscopy is the subset of spectroscopy that deals with the infrared region of the electromagnetic (EM) spectrum. IR spectroscopy covers a range of techniques, the most common being a form of absorption spectroscopy. As with all spectroscopic techniques, it can be used to identify compounds or investigate the composition of samples. The IR portion of the EM spectrum is divided into three regions; the near-, mid-, and far-infrared, named after their relation to the visible spectrum. The far-infrared, lying adjacent to the microwave region, has a low energy and is used for rotational spectroscopy. The mid-infrared is used to study fundamental vibrations and the associated rotational–vibrational structure. The higher energy near-infrared can excite overtone or harmonic vibrations. The names and classifications of these subregions are merely conventions, being neither strict divisions nor based on exact molecular or electromagnetic properties. In the seafood industry, near-infrared (NIR) spectroscopy is the widely used region for quantitative and qualitative analysis of fish and related products.

NIR spectroscopy has become established during the last decade as one of the most important tools of modern industrial analysis, especially for online and inline analysis. The reasons for this are that it yields essentially real-time results, is reagent-less and non-destructive, and can yield information about model compliance.

Infrared Spectroscopy for Food Quality Analysis and Control
ISBN: 978-0-12-374136-3

NIR light is defined as the wavelength region from 750 to 2500 nm, lying between the visible and IR regions. The use of NIR spectroscopy has allowed the development of a non-destructive, rapid, and sensitive method for the analysis of organic materials with little or no sample preparation. The method is based on the simple fact that organic molecules absorb light.

The absorption bands are the result of overtones or combinations of overtones originating in the fundamental MIR region of the spectrum. Overtones are especially common due to hydrogenic stretching vibrations or combinations involving stretching and bending modes of C–H, O–H, N–H or C=O groups (Osborne *et al.*, 1993; Williams and Norris, 2001). As a result, NIR spectra are complex, and any peak of interest is typically overlapped by one or more interfering peaks. In order to handle this complexity, multivariate data analysis is used to calibrate the spectra and to rapidly identify and test quality or quantify constituent levels in samples.

In the ultraviolet (UV) and the visible (Vis) spectral regions electronic transitions are induced; in the IR spectral region molecules can be excited by light absorption of distinct wavelengths to rotate and vibrate (normal modes) as well as to form combination and higher harmonic (overtone) vibrations. In the NIR region combination vibrations of different basic oscillations or higher harmonics with double, triple, etc. the frequency of fundamental oscillations are observed. The overtone spectra obtained are typical of a molecule or a characteristic molecular group. Therefore, by suitable selection of the wavelength of the irradiating light traces of gases in mixtures can be selectively detected.

Another rapid method, called Fourier transform infrared (FTIR) spectroscopy, allows rapid analysis like NIR spectroscopy, but the wavelengths are longer. Fundamental chemical bonding can be more specifically determined in NIR spectroscopic analysis. In some instruments NIR spectroscopy and FTIR can be perfomed together using the same instrument. Also the development of several chemometric software programs have made it easy to analyze correctly complex data. The Unscrambler, Vision, NSAS (Near-Infrared Spectral Analysis Software), and MatLab are the major chemometric software packages for data analysis of NIR spectra.

Recently, NIR spectroscopy has become a well-accepted method for the analysis of the chemical constituents of food (Osborne *et al.*, 1993; Kays *et al.*, 2005). NIR spectroscopy has also gained a foothold as a quantitative method in food analysis, although comparatively little is known concerning its applicability to seafood. In this chapter we provide a summary of studies on seafood quality evaluation using NIR spectroscopy. Future prospects for this method are also discussed.

Quantitative analysis

Chemical composition of fish

Applications of NIR spectroscopy in the seafood industry are focused mainly on quantitative analysis rather than qualitative aspects. Determinations of the chemical composition of fish or fishery products have been reported by various researchers, the evaluation of fat content being one of the prime targets. Chemical composition and freshness are key parameters of fish quality as well as raw materials for feed for aquaculture. However, the fat content is the most important subject in this regard. Quality assurance programs require methods for the simple and rapid analysis of raw materials and final products. Therefore, evaluation of the fat content by non-destructive analysis is required for quick estimation and avoidance of damage in commercial fish. Most commercially important fish have already been examined by NIR spectroscopy and the results are promising, therefore NIR spectroscopy has become a common technique in the seafood industry. Atlantic salmon, for example, is one of the most commercially important fish studied extensively by NIR spectroscopy for chemical composition analysis (Isaksson *et al.*, 1995).

NIR transmittance spectroscopy was used to determine the average fat content in farmed Atlantic salmon fillets with skin and scales by Downey (1996) and Wold *et al.* (1996), who used a wide range of fish where the fat content was 5.7–17.6% and weight range 1.0–5.4 kg. A partial least squares (PLS) regression resulted in a multivariate prediction correlation of 0.97 and a root mean square error of cross-validation (RMSECV) of 0.75%. Results showed that NIR transmittance is suited to the determination of the fat content non-destructively in whole salmon fillets with skin and scales. Wold and Isaksson (1997) evaluated the average crude fat and moisture content in the muscle of whole Atlantic salmon (*Salmo salar*) rather then fillets. Using the same PLS regression technique for the 49 whole salmon resulted in an RMSECV of 1.12% fat ($R = 0.87$) and 0.98% moisture ($R = 0.86$). Several studies have revealed the analytical accuracy of the chemical composition for Atlantic salmon using reflectance, interactance, or transflectance NIR spectroscopy (Sollid and Solberg, 1992; Huang *et al.*, 2002). Determination of fat in live Atlantic salmon using non-invasive NIR spectroscopy is one of the biggest successes in this field (Solberg *et al.*, 2003).

In contrast, several Asian scientists have evaluated NIR spectroscopy as a rapid technique to determine the chemical composition of various fish species (especially the fat content) other than salmon. Shimamoto *et al.* (2003a, 2003b, 2004) reported a series of studies of fat content determination in fresh and frozen skipjack tuna, glazed and thawed bigeye tuna, and mackerel using both standard and portable NIR spectrophotometers. Accurate data analysis was obtained for all types of fish examined using both NIR instruments. However, almost no attempts have been made to determine the fat content of small pelagic fishes which form a large part of fishery products and also are an important part of the by-catch or underutilized fish (Peng *et al.*, 2004; Christos *et al.*, 2005). It should be noted that sardine, a pelagic fish, is an important fish species as a feed material in aquaculture as well as being an important human food source. In modern tuna fish culture the use of sardine as a food is common practise. Various types of canned food, sausages, fishcakes, and other fish meat gel products are also prepared using sardine species.

In our study (Uddin *et al.*, 2007), a surface interactance fiber-optic accessory with a 9-mm-diameter probe was designed to analyze the chemical constituents of small pelagic fish species over a wide range of sizes and weights. One hundred and sixty fresh sardine (*Sardinops melanostictus*) samples were used in this study and the fat content of the samples was found to range between 2.64 and 25.52%. In this study (Uddin *et al.*, 2007), NIR spectra of the intact sardine samples were collected in transmittance mode from 400 to 1100 nm with a scanning monochromator

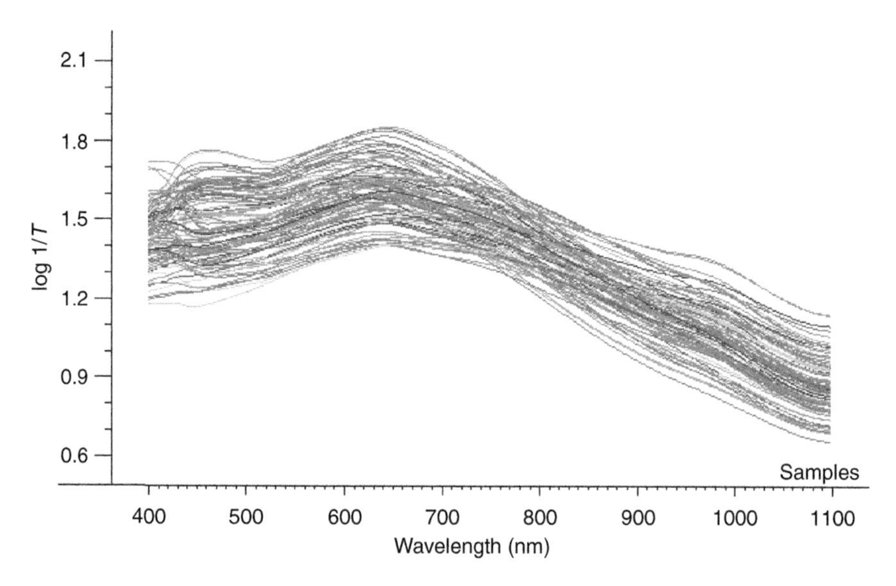

Figure 9.1 Original spectra of intact sardine samples.

NIRSystems 6500 at 2-nm spectral increments equipped with a surface interactance fiber-optic accessory. All of the subsequent operations were carried out at 5°C. Predictive equations for fat were developed by PLS regression with leave-one-out cross-validation to avoid overfitting of the model.

Two different methods of spectral treatment were considered: multiplicative scatter correction (MSC) (Geladi *et al.*, 1985) combined with smoothing and Savitzky–Golay (Savitzky and Golay, 1964) second derivative with second-order polynomials or smoothing with the derivative only. Figure 9.1 shows the original sardine spectra for all samples, while Figure 9.2 shows the corresponding second derivative spectra. The original spectra are rather featureless, possess significant baseline shifts, and the absorbance values show a decreasing trend as a function of increasing wavelength. Some of the spectra have a less steep run-off, being almost parallel with the wavelength axis. There is only one weak peak centered in the NIR region around 964 nm, which is the absorption band of water (Williams, 1996; Williams and Norris, 2001; Uddin *et al.*, 2005b, 2006a, 2007; Okazaki and Uddin, 2006a). To resolve these spectra and cancel out the baseline shifts second derivatives were applied. By this treatment, the baseline shift was eliminated; however, spectra are still spaced far apart. The reason for this is a phenomenon called scattering, which alters the effective pathlength significantly, thereby increasing the sample absorbance. In general, the spectra show a high level of noise despite the smoothing applied, but there are two peaks that became clearly visible, the first one at 926 nm and the second one at 964 nm, which are absorption bands for fat and water, respectively (Williams and Norris, 2001).

To reduce scattering, MSC treatment was applied to the spectra as can be seen in Figure 9.3. Scattering was removed to some extent and the water peak, now appearing at 968 nm, has become more intensive, whereas the 926 nm peak for fat became smaller.

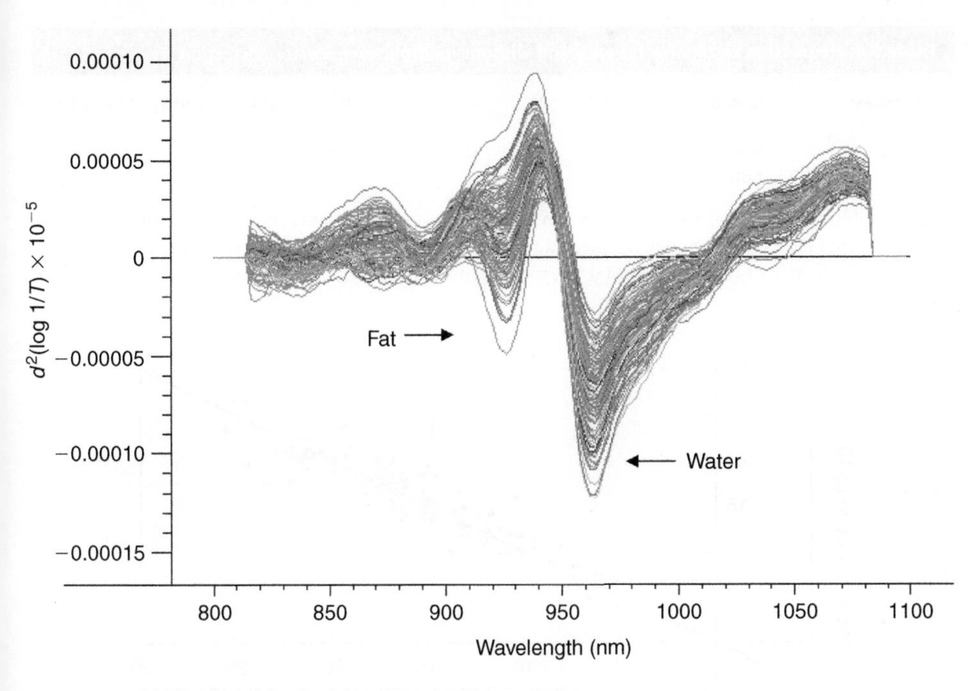

Figure 9.2 Second derivative spectra of intact sardine samples.

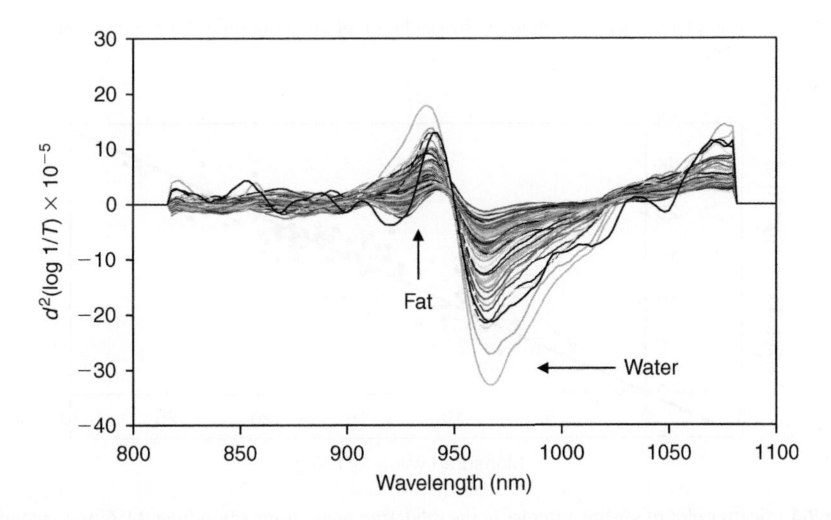

Figure 9.3 Multiplicative scatter correction-treated second derivative spectra of intact sardine samples.

Regression results for fat with and without MSC treatment are shown in Table 9.1 and the corresponding scatter plots are displayed in Figures 9.4 and 9.5. Both models show relatively good performances with regression coefficients higher than 0.9 and errors less than 1% on a fresh weight basis. Therefore, a valid case of non-destructive fat determination of intact sardines has been demonstrated with fiber optics that makes non-invasive on-site measurements possible. This rapid technique could allow fat content measurement of small pelagic fish, enabling several applications.

Table 9.1 Regression results for fat with and without multiplicative scatter correction treatment

Treatment	Spectral range (nm)	N	F	R	R^2	SEC (g/100 g)	RMSECV
sm + d^2	800–1000	160	4	0.92	0.85	0.74	0.82
MSC + sm + d^2	800–1100	160	5	0.92	0.85	0.76	0.96

MSC, multiplicative scatter correction; sm, smoothing; d^2, second derivative; N, sample number; F, number of factors in the model; R, correlation coefficient; R^2, coefficient of determination; SEC, standard error of calibration; RMSECV, root mean square error of cross-validation.

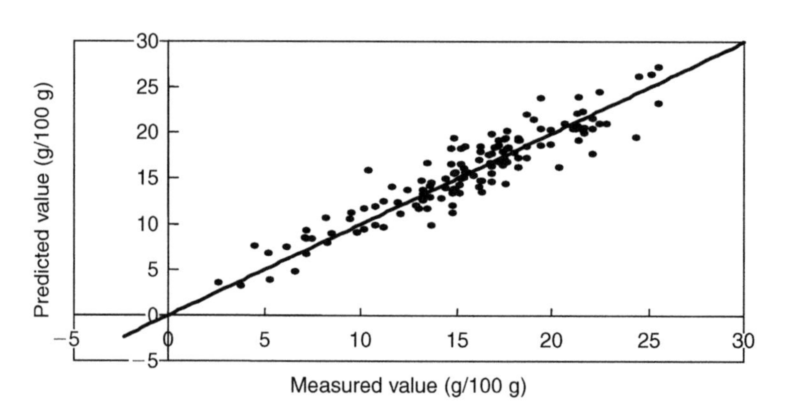

Figure 9.4 Scatter plot of sardine samples in the validation phase using smoothing and second derivative as spectral transformation.

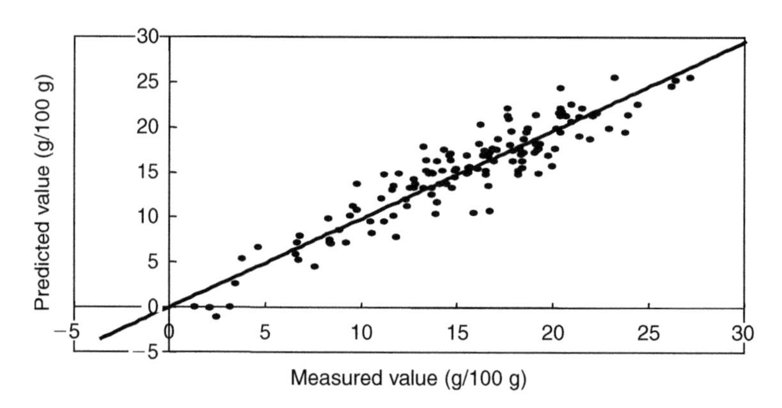

Figure 9.5 Scatter plot of sardine samples in the validation phase using smoothing, MSC and second derivative as spectral transformation.

A rapid method is also required for determination of the iodine value (IV) and saponification value (SV) of fish oils in the food industry and recently Endo *et al.* (2005) minimized this requirement by using NIR spectroscopy. In this technique, a PLS regression calibration model is developed based on a spectral range due to the C–H bond. The model is validated by comparing the IV and SV of a series of fish oils predicted by the PLS model with the values obtained by titration methods (Endo *et al.*, 2005) of the Japan Oil Chemists' Society. Predicted IV and SV were completely

Table 9.2 Statistical characteristics for water and crude protein reference data of the surimi samples

Constituent	*N*	Max	Min	Mean	SD	Units
Crude protein	52	17.57	13.71	15.87	0.85	g/100 g
Water	52	77.19	73.17	74.82	0.66	g/100 g

N, number of samples; SD, standard deviation.

consistent with chemically determined IV and SV. Moreover, the NIR technique showed higher accuracy and reproducibility than the titration method. According to their conclusions, the NIR technique is suitable for IV and SV determination of fish oils as well as vegetable oils and can be carried out within 2 min.

Surimi and minced fish

"Surimi" is an intermediate fish product, used primarily for the preparation of the traditional gel food called "Kamaboko" and more recently used for the production of seafood analogs (fabricated food). Surimi is gaining more prominence worldwide, because of its high protein quality, low fat content, and convenience for consumers. Surimi gelation is associated with temperature-induced structural changes of the protein; however, water which comprises about 73–80% of surimi plays a vital role in gel formation (Luo *et al.*, 2001). Recently Uddin *et al.* (2006b) employed NIR spectroscopy (400–1100 nm) directly on surimi using a surface interactance fiber-optic accessory to determine the water and protein contents. This was a model experiment where imitated surimi samples were used. Different types (percentage of water and protein content) of surimi were prepared by the addition of appropriate amount of water. The reason why NIR spectroscopy is well suited for assessing the presence of water or protein is due to the specificity of the O–H and N–H or C–H bondings. Predictive equations were developed using PLS regression with excellent predictions for protein and water. Regression coefficients were higher than 0.98, errors were small and RPD (ratio of the standard deviation in the reference data for the validation set to the RMSECV) value for protein was well over 8 and that for water was 7.6, which can therefore be used for analytical purposes. In this study imitated thawed surimi samples were used for the water and protein analysis, however, surimi blocks were stored in a frozen state and there is a need to determine those constituents rapidly in commercial frozen surimi. From this point of view, we (Okazaki and Uddin, 2006b) performed a complete study on frozen and thawed blocks of surimi. In this study 52 blocks of commercial SA (finest grade surimi) grade walleye pollack (*Theragra chalcogramma*) frozen surimi from different lots were collected from Maruha Co., Tokyo, Japan, and upon arrival at the laboratory were stored at −20°C until analysis. Water and crude protein were determined in triplicate for each sample using accepted reference methods. Chemical values were averaged, resulting in 52 values for the surimi samples (Table 9.2).

Compared with previous reports (Uddin *et al.*, 2006b), where imitated surimi samples were used, the main difference is that the standard deviation of crude protein and water in this sample set is much smaller. The reason for this is because of the manufacturing process, where both values are controlled to be within specified limits. Samples were measured according to the procedure described by Uddin *et al.* (2006b). Before scanning each sample, a ceramic tile was measured as a reference. Care was taken to exclude stray light and to remove excess water from the surface of the thawed samples. Two spectra were recorded for each sample and were averaged for subsequent analysis. Thus 52 spectra were available for calculations. There were two sample sets for spectral measurements; one for frozen and one for thawed samples. Spectra of the frozen samples were recorded with samples frozen, whereas thawed samples were measured only after the appropriate thawing time. Spectra were stored as optical density units ($\log 1/R$), where R represents the percent reflected radiation. Operation of the spectrometer and collection of spectra were performed using the "VISION" software package (NIRSystems, MD, USA). Predictive equations for water and crude protein were developed by PLS regression with leave-one-out cross-validation to avoid overfitting of the model. Application of representative and separate validation sample sets was not possible due to the relatively small number of samples and large spectral variation. Two different methods of spectral treatments were considered: MSC (Geladi *et al.*, 1985) combined with Savitzky–Golay (Savitzky and Golay, 1964) first or second derivative with second-order polynomials or with the derivative only. For frozen samples the MSC treatment and a derivative (second order) window of 10 left and 10 right points (20–20 nm) from the center point of the derivative window produced the clearest results. In the case of thawed samples, only the size of the derivative window was different, which was set at 15 left and 15 right side points (30–30 nm) from the center point of the derivative window.

As a first step, the 400 nm to 698 nm region was cut off as spectra in this wavelength region displayed noisy features, which reduced calibration accuracy. This smaller region was then further optimized for frozen thawed samples and water and crude protein calibrations made separately to retain the most relevant information for these constituents. Calibration statistics included the standard error of calibration (SEC), correlation coefficient (R), coefficient of determination (R^2) and RMSECV. Optimum calibrations were selected by minimizing RMSECV.

The absorbance values of $\log 1/R$ spectra in Figure 9.6a spread from 0.17 OD to 0.37 OD value at 700 nm. This difference is maintained along the entire spectral range up to 1098 nm. Some spectra deviate from being parallel with the majority, especially below 850 nm. Two regions of interest can be distinguished, the first one being around 912 nm and the second one around 1026 nm. The first of these bands is associated with protein absorptions and can be assigned to C–H stretching third overtone. The second band, however, is a bit more ambiguous to assign, and will be discussed in more detail later. After MSC treatment and second derivative transformation, the set of spectra shown in Figure 9.7a was obtained, which possesses more features. They are, in order of increasing wavelength as follows: 760 nm absorption of water due to O–H stretching third overtone (Osborne *et al.*, 1993), around 798 nm a weak absorption band which will be assigned later (see below), and around 904 nm again protein absorption becomes apparent, with somewhat shifted wavelength owing

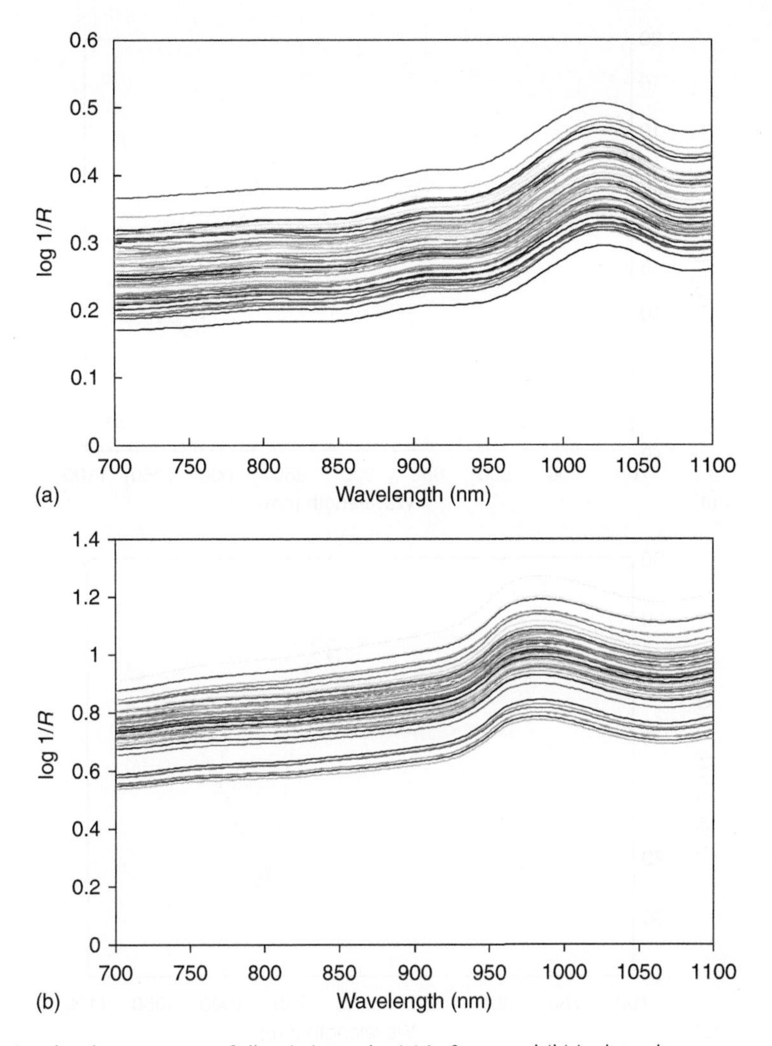

Figure 9.6 Absorbance spectra of all surimi samples (a) in frozen and (b) in thawed states.

to properties of the derivative treatment. The band at 1026 nm features strongly just like in the original spectra. Also, there are four spectra, which show a different pattern around the 904 nm and 1024 nm bands. This difference is attributable to the MSC treatment, which changed the pattern of the above four spectra, resulting in higher values around the absorption bands stated above, which in turn became more negative after the second derivative transformation.

To investigate the origin of the 1024 nm band, spectra of freeze-dried surimi and surimi with 40% and 75% water content were recorded. If the second derivative absorption spectra of freeze-dried surimi (about 8% water content) and those containing 40% and 75% water (Figure 9.8) can be examined, respectively, some information can then be revealed about the origin of the 1024 nm band. All spectra were measured in the frozen state, and several important areas were distinguished. The first one, in increasing order of wavelengths, is a weak band near 800 nm. This band has an increasing intensity as the water content in the samples increases. This is

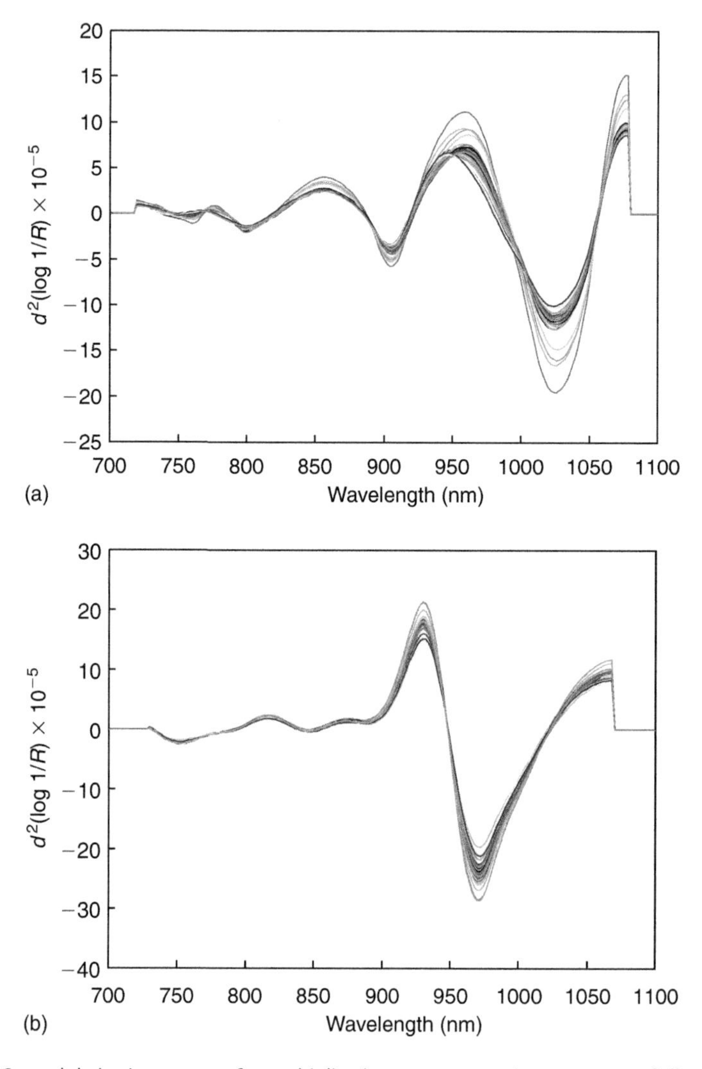

Figure 9.7 Second derivative spectra after multiplicative scatter correction treatment of all surimi samples (a) in frozen and (b) in thawed states.

followed by the 906–912 nm band intervals. This band has already been assigned to protein previously, however, it should be mentioned that in freeze-dried surimi, which has more than a 6% carbohydrate content, the contribution to this band becomes more pronounced, since CH vibrations are abundant in carbohydrates. As the water content increases, the intensity of this band decreases and the center wavelength is shifted to smaller wavelengths. The most intensive band is seen around 1024 nm. In the case of freeze-dried surimi, there is a small, flat peak, whose intensity becomes bigger with the higher water content. Therefore, this band is due to absorption by frozen water.

In thawed surimi spectra (Figure 9.6b) some major differences are displayed compared with frozen ones. First, absorption values are much higher and the spread of spectra at the beginning and at the end of the spectral region is considerably greater,

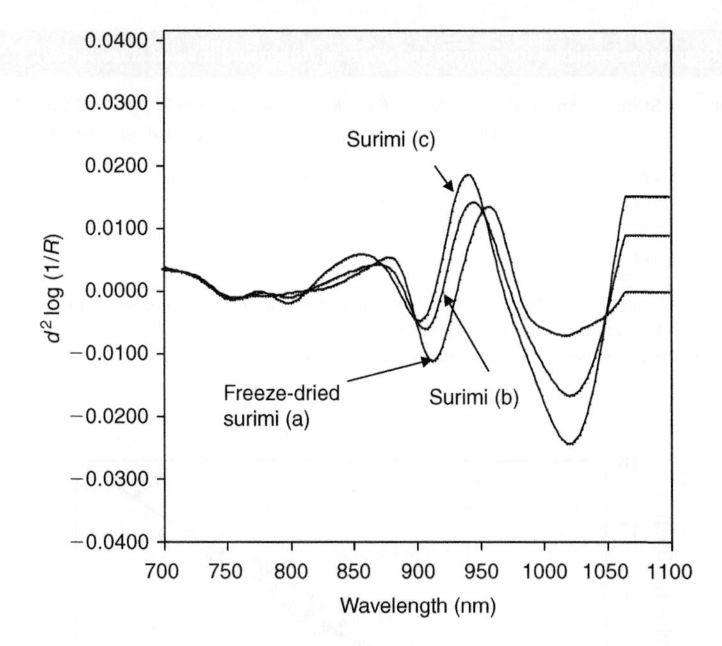

Figure 9.8 Second derivative spectra of (a) freeze-dried surimi, (b) surimi containing 40% water, and (c) surimi containing 75% water.

being 0.38 OD and 0.50 OD value, respectively. This phenomenon can be explained by the fact that in thawed samples the penetration depth of the light beam is deeper, so more light is absorbed, which results in increased OD values. This is in contrast to the frozen sample spectra, where the icy surface reflects more light. In addition, the higher difference and the increase of difference with wavelength indicate higher levels of scattering. Unlike frozen sample spectra, the major informative region is centered on the water band at approximately 982 nm. The spectra are shown in Figure 9.7b after scattering treatment and second derivative transformation. In these spectra, information related to protein is not visible at all. Only water shows two weaker and one stronger absorption bands near 752 nm, 846 nm, and 970 nm, respectively (Golic *et al.*, 2003). The first of these wavelengths appears at a smaller wavelength compared with frozen state surimi.

Regression results for crude protein and water in frozen and thawed surimi samples are summarized in Table 9.3. All results in this table were achieved using spectra after MSC and the second derivative treatment described previously. The number of factors in the models varies greatly depending on the constituents and the state of the surimi. Correlation coefficients are in the range of 0.90 to 0.98, indicating good linear relationships between spectra and chemical data, as indicated in Figures 9.9 and 9.10. Coefficients of determination values are more different. Since they indicate the variance explained by the model, they are a good measure of the goodness of the calibration. As can be seen, calibration with the frozen samples produced much higher values than the thawed samples, and protein calibrations explain more variance than calibrations for water. This means that these frozen models describe the relationship better, therefore, are more suitable for determination of protein and

Table 9.3 Regression statistics for protein and water in frozen and thawed state surimi

Constituent	State	Spectral range (nm)	*N*	*F*	*R*	R^2	SEC (g/100 g)	RMSECV (g/100 g)	RPD
Crude protein	FR	750–1050	52	9	0.98	0.96	0.16	0.22	3.50
	TH	800–1098	52	4	0.92	0.85	0.30	0.37	2.16
Water	FR	700–1000	52	6	0.96	0.92	0.18	0.22	3.09
	TH	750–1050	52	7	0.90	0.81	0.28	0.37	2.16

FR, frozen; TH, thawed; *N*, number of samples; *R*, correlation coefficient; R^2, coefficient of determination; SEC, standard error of calibration; RMSECV, root mean square error of cross-validation; RPD, ratio of standard deviation in the chemical data to RMSECV.

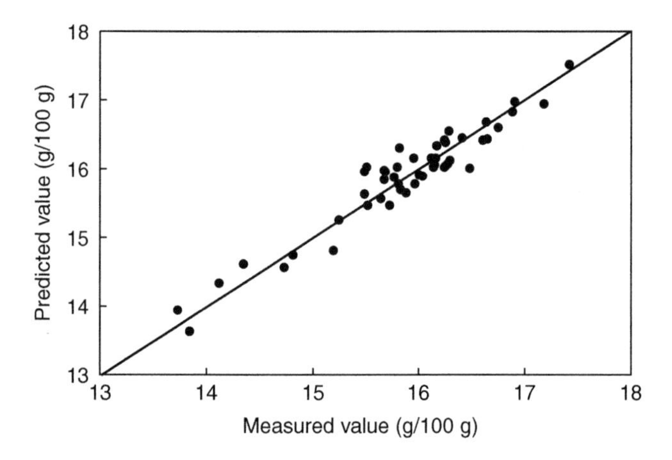

Figure 9.9 Scatter plot for crude protein in frozen state surimi in the validation phase.

water in commercial surimi samples. The performance advantage is clear from the SEC, RMSECV, and RPD values as well, the first two being smaller and the last one being higher for the frozen samples. Another important point is the difference between the calibration and the validation error. When frozen state surimi was measured the differences between the calibration and the validation errors were smaller. RPD values below 4 are generally considered as not so good, however, it is important to remark that this particular figure is heavily dependent on the standard deviation of the chemical data. As mentioned previously, the standard deviation is much smaller in these data sets compared to the previous work (Uddin *et al.*, 2006b), in which the standard deviation for protein and water was 1.35% and 2.90%, respectively. Moreover, the RMSECV for water was almost the same with 0.38. The RMSECV value in this experiment for water in the frozen state was 0.22, which is considerably smaller. With regards to this low RMSECV value, the results in this work are promising. The reason for the higher errors in thawed state surimi for water calibration and validation, besides the higher range, might be that the distribution of water across the measured sample is not uniform, resulting in local highs and lows randomly occurring in each sample. This could affect the spectra, which implies that

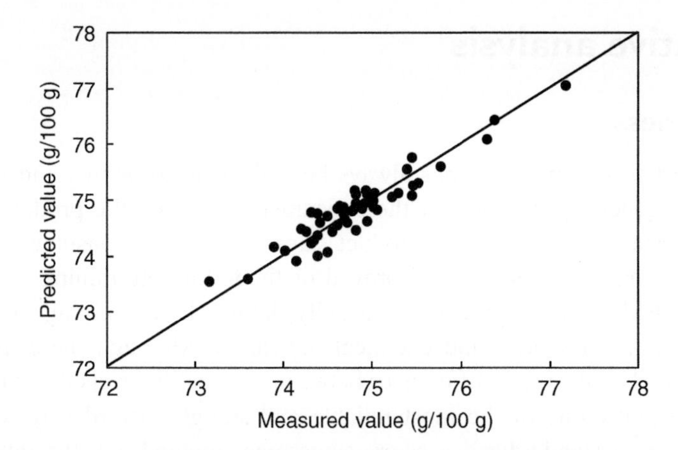

Figure 9.10 Scatter plot for water in frozen state surimi in the validation phase.

errors will increase. On the other hand, frozen samples are not affected by this problem, if good manufacturing processes are applied and fast freezing occurs. In the case of protein, which forms the matrix-like structure where water molecules are entrapped, the distribution is more even. Applicability of NIR spectroscopy to the determination of water and protein content in commercial surimi samples has been demonstrated. Results suggest that frozen samples are better media to use for these determinations.

NIR spectroscopy was also applied to determine the chemical composition in minced raw fish samples used to make fishmeal (Cozzolino *et al.*, 2002). Kaneko and Lawler (2006) investigated the utility of NIR spectroscopy as a means to quantify the diet of seals via analysis of feces. Five of the six calibrations could accurately and precisely quantify how much of a given dietary component the seal had eaten the previous day from an NIR scan of the feces. NIR spectroscopy is therefore potentially a viable way to quantify seal diets. Kaneko and Lawler (2006) discussed the logistical requirements for the development of calibration equations for application to a field study. The adoption of NIR spectroscopy may confer significant benefits for such studies. NIR spectroscopy is also successfully employed to determine moisture and sodium chloride in cured and cold smoked Atlantic salmon (Huang *et al.*, 2002, 2003). NIR reflectance spectroscopy in the spectral range of 1000–2500 nm, was also measured directly for brine from barrel-salted herring, to investigate the potential of NIR as a rapid method to determine the protein content. A PLS regression model between selected regions of the NIR spectra and the protein content yielded a correlation coefficient of 0.93 and a prediction error of 0.25 g/100 g (Svensson *et al.*, 2004). The findings may be used as an indicator for the ripening quality of barrel-salted herring. The oxidative and hydrolytic degradation of lipids in fish oil was monitored using PLS regression and NIR reflectance spectroscopy. Fish oil hydrolytic degradation of lipids which seriously affects oil use and storage under industrial conditions can be successfully monitored using PLS regression and NIR spectroscopy in the fishmeal industry (Cozzolino *et al.*, 2005).

Qualitative analysis

Fish freshness

The quality of fishery products has always been difficult to define, and is typically based on the general perception of the consumer evaluating the product. With the increasing globalization of fishery product sales, processors, consumers, and regulatory officials have been seeking improved methods for determining the freshness and quality. Quality measurements are usually defined by examining the microbial count, sensory panel scores, and chemical indicators. Although these methods all show some overlap, there are differences between the quality levels that each method indicates. Currently one of the most reliable and straightforward ways of describing freshness is a standardized sensory assessment method, i.e. the quality index method (QIM). The disadvantage of sensory panels is that they require highly trained personnel who can be expensive to train and employ. Microbial quality can show contamination in meat; however it is possible for meat to be spoiled or be unfit for consumption with no sign of microbial activity. Sensory panel scores, while often repeatable when using a trained panel, are time-consuming and expensive, making them cost prohibitive for most food manufacturers. Sensory panels are also impractical for use on a large scale, such as at a processing plant where many lots of food need to be tested. Chemical analysis, which measures the chemical breakdown in a food product, may not correlate with sensory scores. While chemical indicators provide a good overall measurement of food quality, the chemical makeup of each food material is different, so it is difficult to establish a standard chemical indicator (Dodd *et al.*, 2004). A rapid, non-destructive method to ascertain fish quality would be of great benefit to both the industry that is eager to provide its consumers with a fresh, safe product and to consumers who are increasingly looking for a reliable guarantee of food quality.

In order to evaluate new technologies that could improve quality determination of fishery products, several researchers have investigated the application of NIR spectroscopy as a possible sensing and rapid technique (Bechmann and Jørgensen, 1998; Lin *et al.*, 2006). Determination of freshness of fish, which is a very complex problem, is the prime concern in this regard. The freshness as storage time (two weeks) in ice of cod (*Gadus morhua*) and salmon (*Salmo salar*) was estimated by visible/NIR spectroscopy (Nilsen *et al.*, 2002). According to this study the best-fit model was found by using the visible wavelength range, giving a correlation of prediction value of 0.97 with an error value of 1.04 day, however, for salmon, NIR range giving a correlation of prediction value of 0.98 and an error value of 1.2 day. NIR spectroscopic measurements provided promising results for evaluation of freshness for thawed–chilled modified atmosphere packed (MAP) cod fillets (Bøknæs *et al.*, 2002). However, it is necessary to study for example the effect of sample preparation, season, fishing ground and cod size together with more sophisticated pre-treatments of NIR spectra before the NIR method can be integrated as a method for evaluation of thawed–chilled MAP cod fillets. NIR diffuse reflectance spectroscopy was also evaluated as a rapid technique to assess quality parameters (water-holding capacity, concentration of total volatile nitrogen bases, dimethylamine, and formaldehyde) of frozen cod (Bechmann and Jørgensen, 1998). A principal component analysis (PCA) showed that these four quality parameters were strongly correlated with each other. It was found that the high water content in fish is a major limitation in the use of NIR analysis for the determination of chemical quality parameters in fish tissue. However, NIR reflectance measurements provide an acceptable determination of these four quality parameters.

The possibility of using visible and short-wavelength NIR spectroscopy to detect the onset of spoilage and to quantify microbial loads in rainbow trout (*Oncorhynchus mykiss*) was investigated by Lin *et al.* (2006). Spectra were acquired for the skin and flesh side of intact trout fillet portions and for minced trout muscle samples stored at 4°C for up to 8 days or at room temperature (21°C) for 24 h. PCA and PLS regression chemometric models were developed to predict the onset and degree of spoilage. PCA results showed clear segregation between the control (day 1) and the samples held 4 days or longer at 4°C. Clear segregation was observed for samples stored 10 h or longer at 21°C compared with the control (0 h), indicating that the onset of spoilage could be detected with this method. Quantitative PLS prediction models for microbial loads were also established. This report demonstrated that NIR in combination with multivariate statistical methods can be used to detect and monitor the spoilage process in rainbow trout and quantify microbial loads rapidly and accurately.

Process verification

Inadequate cooking of food products and use of improper holding times are common causes of food-borne disease outbreaks (Walton and McCarthy, 1999; Uddin *et al.*, 2000). The NIR technique has been successfully applied to investigate heat treated fish and shellfish (Uddin *et al.*, 2002). Fish-meat gels are gaining popularity in recent years as a good protein source, therefore, non-destructive NIR spectroscopy would be beneficial to ensure proper heat treatment of fish-meat gels since the optimum heat treatment of food products is important not only for reducing the risk of infection by pathogens but also for improving the shelf-life, producing a palatable product, and maintaining the optimum food quality. To assess the heat treatment (end-point temperature, EPT) of fish-meat gels, i.e. kamaboko, using NIR spectroscopy, kamaboko gels were prepared by pressing salt-ground surimi into a polyvinylidene chloride casing measuring 48 mm in circumference and approximately 100 mm in length. Both ends of each tube casing were tied with cotton thread and incubated at 10°C intervals between 30 and 90°C for 30 min in water-bath incubators then cooled immediately in ice-cold water and kept at 4°C overnight before NIR spectroscopic measurement. Thermocouples were inserted through the side edge of a separate sample into the geometric center and the temperature was monitored using a recorder. NIR spectra were collected from 1100 to 2498 nm at 2-nm intervals. Only a single spectrum was taken from each individual sample therefore 70 spectra were used to develop a calibration and the remaining 70 spectra were used for validation. Multiple linear regression (MLR) and PLS regression techniques were used to develop the calibration model and validation set.

Spectral changes upon heat treatment might be related to the changes in the secondary structure due to the denaturation of proteins, and to changes in the state of water (Ellekjaer and Isaksson, 1992; Uddin *et al.*, 2006a). The second derivative spectra of kamaboko gels heated between 30 and 90°C showed systematic differences in absorbance related to the heat treatment at different wavelengths throughout the spectra and these differences were more evident between 1300 and 1600 nm where the main absorbance band of proteins occurs (Figure 9.11). It is well established that NIR spectra are affected by sample temperature. Since protein conformations, protein water interactions, or their combination might depend on the variation of water content, the heating process could also affect the NIR reflectance spectra of a heated sample. The kamaboko gels used in this study were heat-treated in a stirred-water bath. The water content was constant before and after the heat treatment which allowed minimizing the differences in water contents between samples heat-treated at different temperatures. In MLR analysis, the wavelengths selected by a step forward–step reverse regression provide the calibration equations with the lowest SEC and highest correlation coefficients of calibration. It is suggested that the wavelength region selected by MLR could be used as a good indicator for selection of the wavelength region in the PLS calibration (Saranwong *et al.*, 2003). Therefore, the wavelength region selected by MLR was used for PLS regression. It was found that the selected MLR wavelength region improved the accuracy of the PLS calibration (Table 9.4).

Figure 9.12 depicts scatter plots of NIR predicted end-point temperature (EPT), obtained by MLR and PLS calibration sets against the actual heating temperatures for the kamaboko gels (Uddin *et al.*, 2005c). A comparison of the NIR-predicted EPT with actual heating temperatures showed close agreement. The MLR and PLS

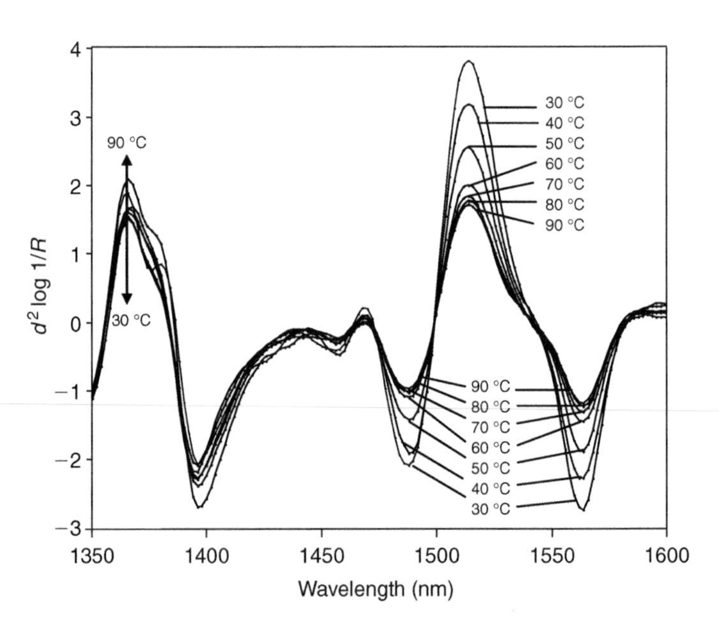

Figure 9.11 Second derivative magnified spectra of kamaboko gels between 30 and 90°C.

calibration had a similar performance for determining EPT of kamaboko gel if the appropriate wavelengths or wavelength region were selected. The *R* was greater than 0.98 indicating a good model structure. The results discussed above demonstrate the potential of NIR-reflectance spectroscopy for determining EPT of kamaboko gels in a rapid and non-invasive manner. Once perfected, this technique will have several advantages over other techniques in that it will take the least time for analysis and will not require any consumable or supporting equipment nor sample preparation. The most promising future use for this NIR spectroscopy application is for online processing control in the food industry.

Differentiation between fresh and frozen–thawed fish

Given the perishable nature of fish, extension of its shelf-life is beneficial for normal trading. However, frozen fish usually have a much lower market price than fresh fish therefore the substitution of frozen–thawed for fresh fish is a significant authenticity issue. In practise, a considerable number of frozen fish are thawed in fish shops, stored on ice and sold as unfrozen fish without being labeled as such prior to retail. To differentiate fresh and frozen–thawed fish, various techniques have been proposed

Table 9.4 Calibration and validation results of partial least squares regression using whole-spectrum and selected region obtained by multiple linear regression

Regression	Wavelength region	*F*	*R*	SEC (°C)	SEP (°C)	Bias (°C)
PLS	1100–2500	10	0.98	2.25	2.21	−0.31
	1300–1600	5	0.98	1.97	2.09	−0.17
MLR	1650–1580	4	0.98	2.04	2.16	−0.11

F, number of factors in the model; *R*, multiple correlation coefficient; SEC, standard error of calibration; SEP, bias-corrected standard error of prediction; Bias, the average of difference beteen actual value and NIR value; PLS, partial least squares; MLR, multiple linear regression.

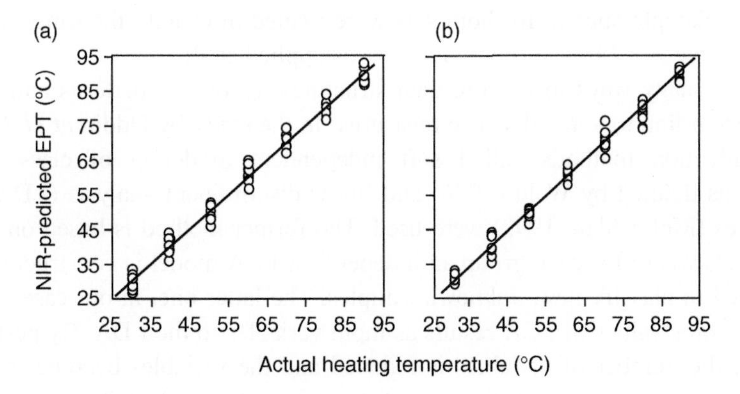

Figure 9.12 Near-infrared-predicted end-point temperatures (EPT) obtained by (a) multiple linear regression and (b) partial least squares calibration sets of kamaboko gels plotted against the actual heating temperatures.

by several researchers (Rehbein, 1992; Rehbein and Cakli, 2000; Uddin *et al.*, 2005a), however, all the methods reported are either time-consuming, destructive or have limitations for practical use. Therefore, the need exists for a method which is capable of differentiating between fresh and frozen–thawed fish or fillets. Ideally, any such method should be non-destructive and rapid as well as reliable.

Dry extract spectroscopy by infrared reflection (DESIR) of fresh and frozen–thawed fish was performed on extracted meat juices then fresh and frozen–thawed fish differentiated (Uddin and Okazaki, 2004).

Uddin *et al.* (2005b) proposed non-destructive visible/NIR spectroscopy to investigate whether fish have been frozen–thawed. Compared with DESIR, no extraction is needed and no wastes are produced in Vis/NIR spectroscopy using a fiber-optic probe, which would be an eco-friendly instrumental technique. In this study, 108 fresh red sea bream (*Pagrus major*) were transported in seawater to the National Research Institute of Fisheries Science, Yokohama, Japan. Fish used for this study were between 416 and 1307g and fork length between 23.2 and 34.9 cm. Fish were divided into two equal groups and used for further evaluation. For fresh and unfrozen fish, 54 samples were used soon after arrival while the second lot of 54 fish was kept at −40°C. After 30 days of frozen storage, the fish were removed and thawed overnight at 5°C then evaluated as frozen–thawed samples. The fish samples were scanned using a NIRSystems 6500 spectrophotometer equipped with a surface interactance fiber-optic accessory. The fish were measured at a location just behind the dorsal fin, midway on the epaxial part. Spectra were recorded in the wavelength range 400–1100 nm at 2-nm intervals. The spectra were stored in optical density units log(1/T), where T represents the percent of energy transmitted.

Among the total of 108 samples, 54 of them fresh and 54 of them frozen-then-thawed, the fish were then divided into a modeling set and a prediction set. The modeling set contained 35 samples for the fresh and 35 for the frozen fish. Twenty-seven of these samples were picked as each odd numbered sample in the order of recording, and the remaining 8 samples were selected randomly. Thus, a total of 38 samples for both fresh and frozen fish (19 samples each) were allocated to the prediction set. Sample spectra for both sets were treated in exactly the same way with second derivative or MSC, or no treatment was applied at all.

There are many ways to explore data structures, recognize patterns and classify samples according to some distance measures. In the study by Uddin *et al.* (2005b), the classification methods called soft independent modeling of class analogy (SIMCA) as defined by Wold (1976) and linear discriminant analysis (LDA) using PCA scores (McLachlan, 1992) were used. The former method is based on disjoint PCA models where for each group an independent PCA model is constructed which is then used to classify new, unknown samples. The latter one, in our case, uses the so-called scores values of PCA results as input variables to the LDA. By performing PCA first, the number of variables is reduced and the variables become independent. By doing so, only a small fraction of the information is lost. This is important, since LDA requires the number of samples to be considerably higher than number of variables to have a statistically meaningful classification.

For a classification to be successful, two things are needed. First, the number of samples belonging to the same group should be as similar as possible and second, the groups should be as far away from each other as possible. The major effect of freeze–thawing treatment involves a gross change in the total absorbance after freezing and thawing; this arises from changes in light scatter presumably arising from alterations in the physical structure of at least the surface layer of fish (Uddin *et al.*, 2005b). In Figure 9.13 the PCA score plot clearly shows that the fresh (right side) and the frozen–thawed (left side) samples are well separated. For this model, only one factor was enough to separate the two groups. As it can also be seen that the frozen–thawed samples have a more compact structure, i.e. data points are closer to each other while in the fresh samples, the group is not so well defined (larger spread of data points). A similar separation was also observed in DESIR analysis of fresh and frozen–thawed fish that was performed on the sample juices (Uddin and Okazaki, 2004). Using the results of this exploratory stage for all the spectral treatments applied, two independent PCA models (900–1098 nm wavelength range with original absorbance spectra) were generated with the modeling sets and then were used to build SIMCA models.

There are several powerful advantages of the SIMCA approach compared with methods such as cluster analysis. First, SIMCA is not restricted to situations in which the number of objects is significantly larger than the number or variables as is invariably the case with classical statistical techniques. This is not so with the present bilinear methods, which are stable with respect to any significant imbalance in the ratio of objects/variables, be it either many objects with respect to variables—or vice versa. Because of the score-loading outer product nature of bilinear models, the entire data structure in a particular data matrix will be well modeled even in the case where one dimension of the data matrix in much smaller than the other. Another advantage is that all the pertinent results can be displayed graphically, allowing exceptional insight regarding the specific data structure behind the modeled patterns.

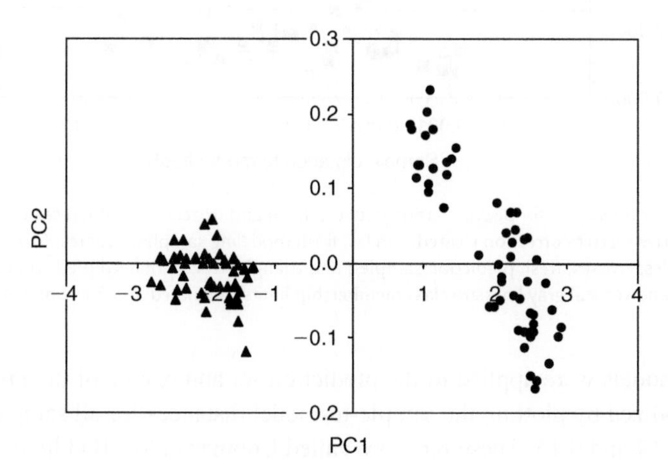

Figure 9.13 Two-dimensional principal components analysis score plot of all 108 red sea bream samples. Samples on the left side of the ordinate axis are frozen samples (triangles), while those on the right are fresh samples (circles).

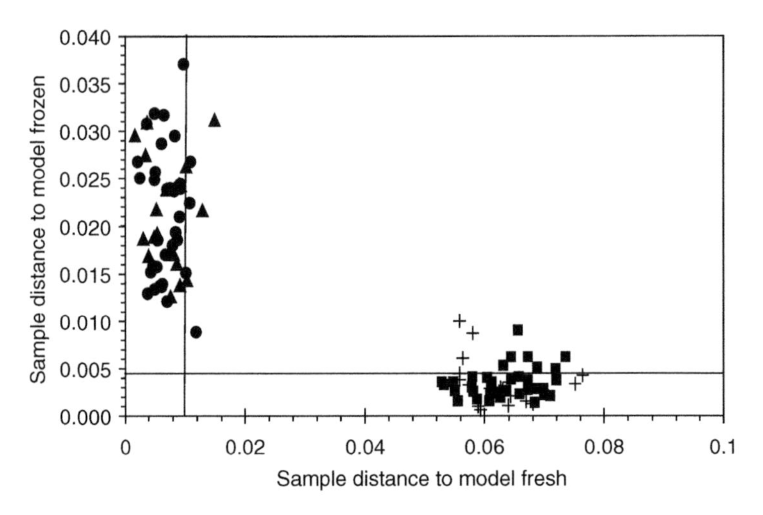

Figure 9.14 Coomans plot for discrimination between fresh and frozen-thawed red sea bream. Spectra were submitted to SIMCA without any treatment. Circles, fresh modeling samples; squares, frozen-thawed modeling samples; crosses, fresh prediction samples; and triangles frozen-thawed prediction samples. The horizontal and vertical gray lines are class membership limits calculated at a 5% confidence limit.

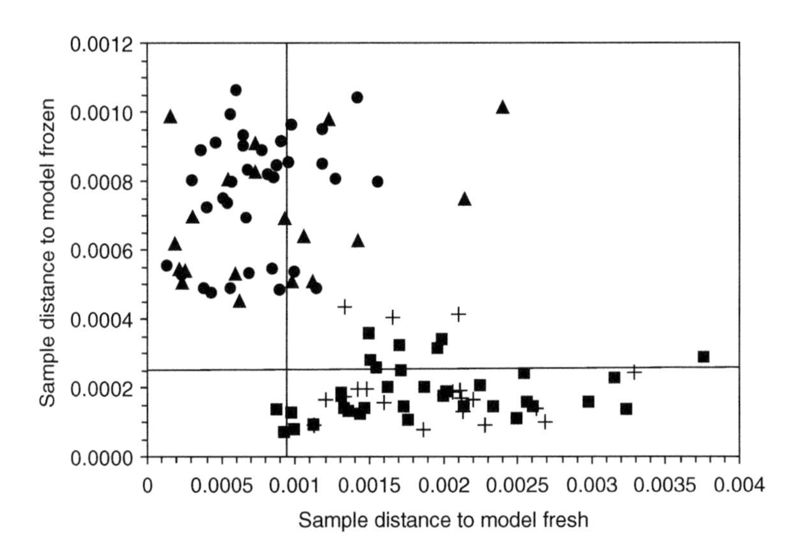

Figure 9.15 Coomans plot for discrimination between fresh and frozen-thawed red sea bream. Spectra were multiplicative scatter correction treated. Circles, fresh modeling samples; squares, frozen-thawed modeling samples; crosses, fresh prediction samples; and triangles frozen-thawed prediction samples. The horizontal and vertical gray lines are class membership limits calculated at a 5% confidence limit.

SIMCA models were applied to the prediction set and results of the prediction can be best visualized by plotting the sample-to-model distances for all samples as shown in Figures 9.14 and 9.15. These plots are called Coomans plots (Uddin *et al.*, 2005b), which show orthogonal (transverse) distances from all new objects (samples) to two selected models (classes) at the same time. In Figure 9.14, the two groups are well defined and separated. All prediction samples are close to the group that they should

Table 9.5 Discrimination results between fresh and frozen–thawed prediction for red sea bream samples using the SIMCA method

Spectral transformation	Kind of sample	Classified correctly[a]	Classified to none[b]	Classified to both[c]	No. of PCs[d]
None	Fresh	63	37	0	1
	Frozen	84	16	0	1
	Total[e]	73	27	0	–
MSC	Fresh	63	37	0	1
	Frozen	84	16	0	3
	Total[e]	73	27	0	–

MSC, multiplicative scatter correction.
[a]Proportion (%) of samples which were classified to the correct model at the 5% significance level.
[b]Proportion of samples which were classified to none of the models at the 5% significance level.
[c]Proportion of samples which were classified to both models at the 5% significance level.
[d]Number of PCs which were used for making class model.
[e]Proportion of fresh and frozen samples combined.

belong to. However, not every sample is within the membership limits for both the modeling and the prediction samples. As can be seen, some samples are located in the upper right quadrant, indicating that they are not included in the defined models. No sample is in the lower left quadrant, meaning that no sample was classified to both groups simultaneously. The upper left and lower right quadrants define samples which belong to a specific group. In Figure 9.15, however, where the sample spectra were subjected to MSC transformation, modeling and classification seem much less clear.

The two groups are very close; in fact, they almost overlap even at the modeling stage. This means that the MSC transformation removed information (i.e. scattering) on which the previous model is based, therefore models are not that far apart. However, the units in Figure 9.15 are by one to two orders of magnitude smaller than those of Figure 9.14, which explains why samples of the two groups are closer and more scattered. The sample distance, which is by an order of magnitude smaller compared to the previous model, between groups is important in terms of reliability or robustness of the model.

This does not necessarily mean that the classification accuracy is worse, as indicated in Table 9.5, where results are summarized for models with original absorbance and MSC-transformed spectra. The same proportions of samples to none or to both groups are classified correctly, meaning that the classification accuracy is the same for both models; however, the model using original absorbance spectra has a higher reliability. This is also an important model feature, since the model is more stable against random errors or interferences from any source.

With regard to LDA, the results are much more clear-cut as seen in Table 9.6. To perform modeling and classification in the same wavelength range (900–1098 nm), spectral transformation and prediction samples were used for LDA analysis as well. It is clear from Table 9.6 that the model using the original absorbance spectra achieved much better classification accuracy (100%) for the prediction samples, but the results obtained using the MSC-treated spectra are considerably worse, indicating again that

Table 9.6 Discrimination results between fresh and frozen–thawed prediction for red sea bream using linear discriminant analysis with principal component analysis scores as input variables

Spectral transformation	*N* correct in groups[a]		Group proportion correct		Overall proportion correct
	Type of fish		Type of fish		
	Fresh	Frozen	Fresh	Frozen	
None	19	19	100%	100%	100%
MSC	15	16	79%	84%	81%

MSC, multiplicative scatter correction.
[a]The number of correctly classified samples out of the 19 prediction samples for the fresh and frozen–thawed red sea bream groups, respectively.

scattering is the information that makes classification work. For fresh fish, as the cellular structure is intact when NIR light enters the fresh fish, cells not only absorb the light but change its direction until the light reaches the next cell. This process may continue until all light is absorbed or until the light emerges at the other side of the sample. This multiple change in the direction of light is called scattering, which increases the distance that the light travels from the entry point to the exit point of the sample. On the other hand, when freezing and thawing is done, the cell membrane may be damaged, causing the leakage of the intracellular contents into the extracellular space. Thus, there is a much smaller number of cells that can scatter light as it travels through the sample, reducing the distance that the light has to cover.

It is interesting to note that frozen–thawed samples were slightly better classified than fresh ones, but as the number of samples is limited the drawing of conclusions is as yet premature. This method maximizes the ratio of between-class variance to the within-class variance in any particular data set, thereby guaranteeing maximal separability. It is possible to achieve better accuracy and reliability by using a custom-made fiber-optic probe or improved sample presentation. Nevertheless, the results are promising that a rapid measurement method can be developed to detect practises such as the selling of frozen–thawed fish as fresh fish.

Conclusions

Although variable in results, studies dealing with the predicting ability of NIR spectroscopy to determine sample chemical properties show good potential to replace analytical procedures, which can be time-consuming, expensive and sometimes hazardous to operator health or the environment. NIR technology incorporates all the benefits brought by the evolution of related fields such as chemometrics, new materials for optical components, new sensors and sensor arrays, microcomputers and microelectronics. The development in the technology for providing better information on raw materials, manufacturing parameters and their impact on finished product quality will result in more robust processes, better products, more uniform results, and potential cost savings for the manufacturer. In spite of its great potential, the practical use of NIR spectroscopy however may be limited by the requirement for laborious calibration for each purpose. On the other hand, by considering the universal, non-invasive and non-destructive nature of the NIR spectroscopy, its speed, and the robustness of the NIR spectrophotometers commercially available today, the disadvantages indicated above may be overcome in the near future.

The number of scientific papers and the successes of international congresses on this theme is evidence of this fact especially in seafood analysis. As mentioned earlier, the safety and quality of seafood has been of particular concern in recent years, NIR spectroscopy could be a non-destructive testing method allowing reproducible and rapid assessment, something that has not been available in the past.

References

Bechmann IE, Jørgensen BM (1998) Rapid assessment of quality parameters for frozen cod using near infrared spectroscopy. *LWT-Food Science and Technology*, **31**, 648–652.

Bøknæs N, Jensen KN, Andersen CM, Martens H (2002) Freshness assessment of thawed and chilled cod fillets packed in modified atmosphere using near-infrared spectroscopy. *LWT-Food Science and Technology*, **35**, 628–634.

Christos AB, Anastasioss Z, Dimitrios P (2005) Production of fish-protein products (surimi) from small pelagic fish (*Sardinops pilchardusts*), underutilized by the industry. *Journal of Food Engineering*, **68**, 303–308.

Cozzolino D, Murray I, Scaife JR (2002) Near infrared reflectance spectroscopy in the prediction of chemical characteristics of minced raw fish. *Aquaculture Nutrition*, **8**, 1–6.

Cozzolino D, Murray I, Chree A, Scaife JR (2005) Multivariate determination of free fatty acids and moisture in fish oils by partial least-squares regression and near-infrared spectroscopy. *LWT-Food Science and Technology*, **38**, 821–828.

Dodd TH, Hale SA, Blanchard SM (2004) Electronic nose analysis of tilapia storage. *Transactions of the ASAE*, **47**, 135–140.

Downey G (1996) Non-invasive and non-destructive percutaneous analysis of farmed salmon flesh by near infra-red spectroscopy. *Food Chemistry*, **55**, 305–311.

Ellekjaer MR, Isaksson T (1992) Assessment of maximum cooking temperatures in previously heat-treated beef. Part 1: Near-infrared spectroscopy. *Journal of the Science of Food and Agriculture*, **59**, 335–343.

Endo Y, Tagri-Endo M, Kimura K (2005) Rapid determination of iodine value and saponification value of fish oils by near-infrared spectroscopy. *Journal of Food Science*, **70**, C127–C131.

Geladi P, McDougall D, Martens H (1985) Linearization and scatter correction for near infrared reflectance spectra of meat. *Applied Spectroscopy*, **39**, 491–500.

Golic M, Walsh K, Lawson P (2003) Short-wavelength near-infrared spectra of sucrose, glucose and fructose with respect to sugar concentration and temperature. *Applied Spectroscopy*, **57**, 139–145.

Huang Y, Cavinato AG, Mayes DS, Bledsoe GE, Rasco BA (2002) Nondestructive prediction of moisture and sodium chloride in cold smoked Atlantic salmon (*Salmo salar*). *Journal of Food Science*, **67**, 2543–2547.

Huang Y, Cavinato AG, Mayes DS, Kangas LJ, Bledsoe GE, Rasco BA (2003) Nondestructive determination of moisture and sodium chloride in cured Atlantic salmon (*Salmo salar*) (Teijin) using short-wavelength near-infrared spectroscopy (SW-NIR). *Journal of Food Science,* **68**, 482–486.

Isaksson T, Togersen G, Iversen A, Hildrum KI (1995) Non-destructive determination of fat and moisture and protein in salmon fillets by use of near-infrared diffuse spectroscopy. *Journal of the Science of Food and Agriculture,* **69**, 95–100.

Kaneko H, Lawler IR (2006) Can infrared spectroscopy be used to improve assessment of marine mammal diets via fecal analysis? *Marine Mammal Science*, **22**, 261–275.

Kays SE, Archibald DD, Shon M (2005) Prediction of fat in intact cereal food products using near-infrared reflectance spectroscopy. *Journal of the Science of Food and Agriculture*, **85**, 1596–1602.

Lin M, Mousavi M, Al-Holy M, Cavinato AG, Rasco BA (2006) Rapid near infrared spectroscopic method for the detection of spoilage in rainbow trout (*Oncorhynchus mykiss*) fillet. *Journal of Food Science*, **71**, S18–S23.

Luo YK, Kuwahara M, Kaneniwa Y, Murata Y (2001) Comparison of gel properties of surimi from Alaska Pollock and three freshwater fish species: effects of thermal processing and protein concentration. *Journal of Food Science*, **66**, 548–554.

McLachlan GJ (1992) *Discrimimant Analysis and Statistical Pattern Recognition.* Chichester, UK: John Wiley and Sons.

Nilsen H, Esaiassen M, Heia K, Sigernes F (2002) Visible/near-infrared spectroscopy: A new tool for the evaluation of fish freshness? *Journal of Food Science*, **67**, 1821–1826.

Okazaki E, Uddin M (2006a) The potentiality of NIR spectroscopy as a nondestructive, simple and rapid analytical tool to measure seafood qualities. *The Food Industry*, **49**, 26–35.

Okazaki E, Uddin M (2006b) Rapid determination of water and protein contents in commercial frozen surimi by NIR spectroscopy. *Proceedings of the European Seminar on Infrared Spectroscopy*, Lyon, France.

Osborne BG, Fearn T, Hindle PH (1993) *Practical NIR Spectroscopy with Applications in Food And Beverage Analysis.* London: Longman Scientific and Technical.

Peng L, Xiaoxue W, Ronald WH, Delbert M (2004) Nutritional value of fisheries by-catch and by-product meals in the diet of red drum (*Sciaenops ocellatus*). *Aquaculture*, **236**, 485–496.

Rehbein H (1992) Physical and biochemical methods for the differentiation between fresh and frozen–thawed fish or fillets. *Italian Journal of Food Science*, **2**, 75–86.

Rehbein H, Cakli S (2000) The lysosomal enzyme activities of fresh, cooled, frozen and smoked salmon fish (*Onchorhyncus keta* and *Salmo salar*). *Turkish Journal of Veterinary and Animal Sciences*, **24**, 103–108.

Saranwong S, Sornsrivichai J, Kawano S (2003) Performance of a portable near infrared instrument for Brix value determination of intact mango fruit. *Journal of Near Infrared Spectroscopy*, **11**, 175–181.

Savitzky A, Golay MJE (1964) Smoothing and differentiation of data simplified least squares procedures. *Analytical Chemistry*, **36**, 1627–1639.

Shimamoto J, Hiratsuka S, Hasegawa K, Sato M, Kawano S (2003a) Rapid non-destructive determination of fat content in frozen skipjack using a portable near infrared spectrophotometer. *Fisheries Science*, **69**, 856–860.

Shimamoto J, Hasegawa K, Hattori S, Hattori Y, Mizun T (2003b) Non-destructive determination of the fat content in glazed bigeye tuna by portable near infrared spectrophotometer. *Fisheries Science*, **69**, 1247–1256.

Shimamoto J, Hasegawa K, Sato M, Kawano S (2004) Non-destructive determination of fat content in frozen and thawed mackerel by near infrared spectroscopy. *Fisheries Science*, **70**, 345–347.

Solberg C, Saugen E, Swenson L, Bruun L, Isaksson T (2003) Determination of fat in live atlantic salmon using non-invasive NIR techniques. *Journal of the Science of Food and Agriculture*, **83**, 692–696.

Sollid HE, Solberg CH (1992) Salmon fat content estimation by near infrared transmission spectroscopy. *Journal of Food Science*, **57**, 792–793.

Svensson VT, Nielsen HH, Bro R (2004) Determination of the protein content in brine from salted herring using near-infrared spectroscopy. *LWT-Food Science and Technology*, **37**, 803–809.

Uddin M, Okazaki E (2004) Classification of fresh and frozen–thawed fish by near-infrared spectroscopy. *Journal of Food Science*, **69**, C665–C668.

Uddin M, Ishizaki S, Tanaka M (2000) Coagulation test for determining end-point temperature of heated blue marlin meat. *Fisheries Science*, **66**, 153–160.

Uddin M, Ishizaki S, Okazaki E, Tanaka M (2002) Near-infrared reflectance spectroscopy for determining end-point temperature of heated fish and shellfish meats. *Journal of the Science of Food and Agriculture*, **82**, 286–292.

Uddin M, Okazaki E, Fukuda Y (2005a) Classification of fresh and frozen–thawed fish by dry extract spectroscopy by infrared reflection. *Near Infrared News*, **16**, 4–7.

Uddin M, Okazaki E, Turza S, Yumiko Y, Fukuda Y, Tanaka M (2005b) Non-destructive visible/NIR spectroscopy for differentiation of fresh and frozen–thawed fish. *Journal of Food Science*, **70**, C506–C510.

Uddin M, Okazaki E, Ahmed MU, Fukuda Y, Tanaka M (2006a) NIR spectroscopy: A non-destructive fast technique to verify heat treatment of fish-meat gel. *Food Control*, **17**, 660–664.

Uddin M, Okazaki E, Fukushima H, Turza S, Yumiko Y, Fukuda Y (2006b) Nondestructive determination of water and protein in surimi by near-infrared spectroscopy. *Food Chemistry*, **69**, 491–495.

Uddin M, Turza S, Okazaki E (2007) Rapid determination of intact sardine fat by NIRS using surface interactance fibre probe. *International Journal of Food Engineering*, **3**(6), Art. 12.

Walton JH, McCarthy JM (1999) New method for determining internal temperature of cooking meat via NMR spectroscopy. *Journal of Muscle Foods*, **22**, 319–330.

Williams P (1996) Observations on the use, in prediction of functionality of cereals, of weights derived during development of partial least squares regression. *Journal of Near Infrared Spectroscopy*, **4**, 175–187.

Williams PC, Norris KH (2001) *Near Infrared Technology in the Agriculture and Food Industries*. MN, USA: American Association of Cereal Chemists.

Wold JP, Isaksson T (1997) Non-destructive determination of fat and moisture in whole Atlantic salmon by near-infrared diffuse spectroscopy. *Journal of Food Science*, **62**, 734–736.

Wold JP, Jakobsen T, Karne L (1996) Atlantic salmon average fat content estimated by near-infrared transmittance spectroscopy. *Journal of Food Science*, **61**, 74–77.

Wold S (1976) Pattern recognition by means of disjoint principal components models. *Pattern Recognition*, **8**, 127–139.

Milk and Dairy Products

10

CC Fagan, CP O'Donnell, L Rudzik, and E Wüst

Introduction

Acquiring information on the quantitative and qualitative properties of raw materials, intermediate and final products is currently gaining more and more importance. This is primarily due to the economic benefits which can be achieved by utilizing this information. Near-infrared (NIR) and mid-infrared (MIR) spectroscopy have both been investigated to determine their potential in a range of applications in the dairy industry. This has included process monitoring, determination of quality, geographical origin, and adulteration of dairy products in processes such as milk, milk powder, butter, and cheese production. Therefore infrared (IR) spectroscopy has a role to play

Infrared Spectroscopy for Food Quality Analysis and Control
ISBN: 978-0-12-374136-3

in producing dairy products of high quality and consistency from farm to final product. The vast majority of published reports, however, relate to studies carried out at laboratory or pilot scale.

Production facilities are growing in scale due to the merging of companies. By their nature the successful operation of large production lines, with high outputs, is critical. In such situations large economic losses could occur in a short period of time if the process runs "out of control" or the produced product does not conform to set specifications. IR spectroscopy allows rapid analysis of a large volume of samples in the production line. Due to the effort involved in calibrating an IR spectrometer, however, this technique is only worthwhile if there is a financial benefit. This benefit can include but is not limited to an increase in the product yield.

This chapter describes applications of IR spectroscopy in the dairy industry, including the application of IR spectroscopy in process control through to the analysis of the final product, and its use in helping to produce good food at reasonable production costs.

Milk production

Milk production involves a number of stages including herd management, milking, bulk storage, collection, transportation, reception, storage, and processing. The use of NIR spectroscopy to facilitate the production of high-quality milk has been investigated in applications ranging from monitoring rumen metabolism through to the standardization of milk in the milk-processing plant.

The by-products of fermentation in the rumen are precursors for milk production in dairy cows; hence rumen metabolism should be controlled in order to obtain production of a high quality and quantity of milk. Turza *et al.* (2002) suggested that this could be achieved by using NIR spectroscopy to determining the composition of the rumen fluid as it is an important indicator of the state of rumen metabolism. They developed a technique for online monitoring of rumen fluid, which involved inserting the head of a fiber-optic measuring device through a fistula into the rumen of the cow. Using transmittance spectra in the range 1100–1046 nm and 1550–1860 nm in conjunction with partial least squares (PLS) regression, Turza *et al.* (2002) predicted a number of rumen fluid constituents including acetic, butyric, propionic, and isovaleric acids.

IR spectroscopy has been widely used to determine the composition of milk products (Carl, 1991; Luinge *et al.*, 1993; Nathier-Dufour *et al.*, 1995; Lefier *et al.*, 1996; Albanell *et al.*, 1999). Carl (1991) used NIR spectra and PLS regression to predict the total fat content of milk with a relative standard deviation of 2% while Luinge *et al.* (1993) and Van de Voort (1992) reported on a Fourier transform infrared (FTIR) method that utilizes the MIR region of the electromagnetic spectrum, for an easy and fast determination of the total fat content in milk. FTIR spectroscopy has also been investigated for the compositional analysis of sweetened condensed milk (Nathier-Dufour *et al.*, 1995). Fat and total solids content were predicted with accuracy in the order of ±0.09 and ±0.55 for fat and total solids respectively. They concluded that with the standardization of the sample preparation protocol, the method produced good results and offered both ease of sample handling and rapid sample processing. Subsequently, predictions of the fat, crude protein, true protein, and lactose content of raw milk by FTIR spectroscopy and a traditional filter-based milk analyzer were assessed (Lefier *et al.*, 1996). They concluded that because the FTIR instrument provided more spectral information related to milk composition than the filter instrument, the single-calibration FTIR analysis of milk samples collected in different seasons was more accurate (Lefier *et al.*, 1996). Due to the success of the FTIR measuring principle for the analysis of milk this technology has been successfully commercialized with products such as the MilkoScan™ FT 120 (Foss, Denmark; www.foss.dk) which employs this principle in compliance with IDF and AOAC standards.

Milk composition is an important factor in all stages of milk production, therefore the use of IR technology in determining milk composition has been employed in a variety of situations. Determining milk composition is essential for the efficient management of dairy herds. It has been stated that milk fat content could be used to regulate the diet forage concentration ratio, protein content could be used as an indicator of adequate dietary energy supply, and that lactose content could be used to detect mastitis (Tsenkova *et al.*, 2000). The implementation of an online IR sensing system which would allow determination of milk composition of individual cows during milking at farm level would assist in the management of dairy herds. In recent years a number of studies have examined the potential of NIR spectroscopy in such an application (Tsenkova *et al.*, 1999, 2000; Woo *et al.*, 2002; Kawamura *et al.*, 2007). If NIR spectroscopy is to be employed for online determination of milk composition during milking then the technique must be suitable for determining the composition of unhomogenized milk. Tsenkova *et al.* (2000, 1999) investigated the potential of NIR spectroscopy to measure milk fat, total protein, and lactose content of unhomogenized milk samples taken during milking. Tsenkova *et al.* (1999) found that NIR spectroscopy was adequate for determining the composition of unhomogenized cows' milk over one lactation. They also found that the most important region for predicting composition was 1100–2400 nm, with fat, lactose, and total protein predicted with standard errors of 0.11%, 0.082%, and 0.096%, respectively.

Tsenkova *et al.* (2000) further investigated the 1100–2400 nm range to predict fat, total solids, and lactose contents in milk from three different cows. They found that prediction accuracy was improved when individual calibration models for each cow were developed.

Recently Kawamura *et al.* (2007) constructed a sensing system which consisted of an NIR instrument, a milk flow meter and a milk sampler, to monitor milk quality during milking. They installed the system between the teatcup cluster and the milk bucket of the milking machine, which allowed a continuous flow of milk by the NIR sensor during milking. Calibration models were then developed using PLS regression to predict fat, protein, lactose, somatic cell count, and milk urea nitrogen. They found the calibrations had R^2-values between 0.82 and 0.95 and standard error between 0.05 and 1.33 units (Kawamura *et al.*, 2007). The implementation of such IR-based systems will assist in the transition to precision dairy farming based on the data gathered from each individual cow.

Incoming product control

Considering a medium size dairy, with approximately 10 000 farmers, 10 000 samples have to be analyzed every day or every second day depending on the collection of the raw milk. The fat, protein, and lactose contents are determined for each sample using MIR instruments. These parameters are used for the payment of the farmers. The average fat content is approximately 4.3%, the protein around 3.3%, and the lactose around 4.7%. The accuracy is approximately ±0.03% abs. for these constituents. It is also possible to separate the total protein content into levels of casein and whey protein using IR spectroscopy. At present discussions are on-going as to whether citrate can also be determined using this technique.

Food manufacturers are required to demonstrate the authenticity of their products. The rights of consumers and genuine food processors in terms of food adulteration and fraudulent or deceptive practices in food processing are set out in a recent European Union regulation regarding food safety and traceability (European Commission, 2002). Detection of milk adulterated with synthetic milk using NIR spectroscopy has been studied by Jha and Matsuoka (2004). They found that NIR spectroscopy in the range 700–1124.8 nm was capable of predicting the level of adulterants such as urea, sodium hydroxide, oil, and shampoo in milk with low standard errors.

Some dairy products require additives (e.g. stabilizers in yoghurt or dessert products). Due to time constraints, testing of the incoming additives is often limited to an inspection of the loading paper and a sensory test. In some cases the additives are brought into the store and used in the production after further investigation. Both situations have drawbacks: in the first case the sensory test is not sufficient to detect specific problems with the additive. In the second situation if the load does not fulfil the requirements the load must be reshipped following the failed test. A robust method is necessary which is fast enough so that it can be performed during the time when the truck is standing at the manufacturing site and which can give more information than a sensory test alone. An application based on an NIR instrument with fiber optics, for instance, is able to do this task at the entrance to the plant. Having a spectral library of correct additives one compares the spectrum of the delivered additive with the spectra in the library. A hit-list will show the classification result. Using a spectral library for additives a classification report (first place in the hit-list) shows the following information:

- Name of the substance: sorbic acid
- Sum formula: $C_6H_8O_2$
- Molecule weight: 112
- CAS number: 110-44-1
- Hit quality: 915

Using products from different manufacturers qualitative information can be important for further processing: What is the dominant modification of the lactose because the different modifications (α, β, or amorphous) behave differently in the production process? In the case of fat, knowing the ratio of *cis*- and *trans*-fatty acids is worthwhile from a nutritional standpoint. There exists an international standard (ISO Standard 13884:2003, 2003) allowing the determination of *trans*-fatty acids by MIR spectroscopy.

One should be aware, however, that IR spectroscopy is not a method with which one can solve everything. The limitations of the method must be recognized. Detecting or identifying small concentrations of constituents in complex matrices is always a problem!

Process control

To control a process it is necessary to have the appropriate information just-in-time. Traditional chemical analysis of constituent concentrations is very time consuming. Therefore by the time the result is available it may not be possible to take suitable corrective action if the test results are out of specification. IR spectroscopy can potentially fulfill this task almost simultaneously to the production process.

There are three possibilities: off-line, at-line, and in-line analysis.

Off-line analysis

The IR analysis can be performed in the industrial laboratory. This means that a sample has to be taken out of the process and sent to the laboratory. There the IR analysis is performed and the result is sent back to the process operator. This procedure is called *off-line* analysis because it is carried out away from the processing line. Using IR spectroscopy in this way also results in a short lag period prior to the test results being available.

At-line analysis

In at-line analysis the IR spectrometers are standing beside the production lines. The process operator takes a sample and performs the analysis personally. Therefore the results are available more rapidly and the operator can quickly take corrective action if necessary. Furthermore, process operators can perform the analysis whenever they deem it necessary.

Table 10.1 shows a list of calibrations for the major constituents of dairy products which is continuously growing. The accuracy, defined as the difference between IR prediction and value according to the reference method, is approximately of the order of the repeatability limit of the reference method. The repeatability limit r of a reference method is defined in such a way that the absolute difference of two analyses under repeatability conditions is smaller than r with 95% probability. (Note: Repeatability conditions are that the same person performs the analysis several times using the same material, the same chemicals, and the same instruments shortly after another. r is determined by a proficiency test.) In principle, one cannot do better

Table 10.1 Constituents of dairy products determined by infrared spectroscopy

Product	Constituents
Liquid	
Raw milk	Fat, protein, casein, whey protein, lactose, dry matter
Skim milk	Dry matter, protein, casein
Market milk	Fat, protein, dry matter
Coffee cream	Dry matter, fat
Evaporated milk	Dry matter, fat
Whipped cream	Dry matter, fat
UHT-Cream	Dry matter, fat
Cocoa concentrate	Dry matter
Viscous	
Low fat curd cheese	Dry matter, protein
Modified curd cheese	Dry matter
Curd cheese	Dry matter, fat, protein
Fruit curd cheese	Dry matter, fat
Yoghurt	Dry matter, fat
Butter	Water, salt
Cheese	Dry matter, fat
Hard cheese	
Slicing cheese	
Semi-solid slicing cheese	
Soft cheese	
Processed cheese	
Powder	
Skim milk powder	Water, fat, protein
Milk powder	Water, fat, protein, lactose
Coffee creamer	Water, fat, protein
Capuccino	Water, fat
Yeast autolyzate	Water, salt
Creamer	Water, fat
Others	
Ice cream mix	Dry matter, fat

than this because the reference values that are needed for the calibration procedure include this error. On the other hand, comparing different methods (reference method with IR prediction) the reproducibility limit R is the measure for the judgment. (Note: R has the same meaning as r. The difference is that reproducibility conditions are applied. Beside the same material, these conditions are given if one or more of the repeatability conditions are violated or a different method has been used.) A good empirical approximation for R is $2*r$. The experience with the mentioned calibrations shows that 90–95% of the differences between reference value and the IR prediction of independent samples fall within R.

Several calibrations are not based on the reference method because some companies do not use the reference method for process control due to the lengthy time they take to perform. In these cases the calibration is based on the individual method. Checking the IR prediction with the result of the individual method from time to time has the advantage that all errors are handled on the same basis, which is a great help in the case of deviations.

Beside the quantitative predictions of the major constituents, qualitative aspects of the products are becoming more important. More insight into the changes of the microstructure during the processing is required to optimize the process. For example, the conformation of the protein (α-helix or β-sheet) influences the properties of the product very strongly. The amide I band at approximately $1650\,cm^{-1}$ is the best investigated structure for the determination of protein conformation.

It is also useful to be able to measure qualitative product characteristics at the end of shelf life during manufacturing so that action can be taken if the product does not fulfill the specifications. In the case of whipped cream the IR spectra during manufacture has been related to the product characteristic (volume increase by whipping, firmness) at the end of shelf life via discriminating analysis. Figure 10.1 shows that there are two distinct clusters. One cluster represents the samples where the properties are within specification, while the samples in the other cluster do not fulfill the set specification. If this is done during manufacturing it is possible to modify the recipe.

In-line analysis

IR spectroscopy also allows measurement of the quantity of interest directly in the process line (in-line measurement). Mixing and separating are important process operations in which the control of constituent concentrations influences product quality and economy. The following applications exist in the dairy industry:

- The water content of milk powder is measured directly after the drying tower with a NIR spectrometer. Having this information, one can regulate the concentrate feed to the tower.
- The water content of butter can be measured at the end of the butter-making machine with an NIR spectrometer. This can be used to influence the water content.
- To standardize fat and protein for cheese milk or fat for market milk a MIR spectrometer may be used to determine these quantities for process control.
- The dry matter content of curd cheese is determined by an NIR spectrometer with fiber optics. The feeding of the separator is modified according to the dry matter of the curd cheese.

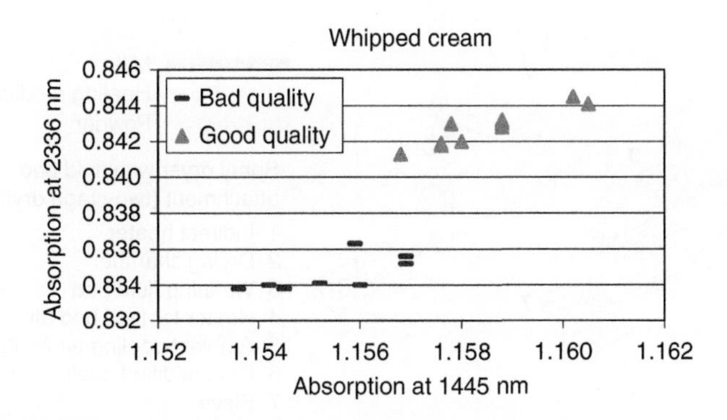

Figure 10.1 Absorption at wavelength 2336 nm vs. absorption at wavelength 1445 nm measured during the production of whipped cream to predict the quality at the end of the shelf life.

- The dry matter content of soft cheese curd is measured right before the filler to control the filling.
- The formation of the coagulum during the cheese-making process is monitored with NIR diffuse reflectance spectroscopy with fiber optics to optimize the curd cutting.
- Mozzarella cheese production is controlled by a NIR reflectance measurement to control the qualities of interest.

All in-line applications will be discussed in some detail because this is where the future of process control lies.

Milk powder production

The water content of powder (milk powder, skim milk powder, whey powder and other special powder) is important because the economy is dependent on the water content and usually there are specifications with respect to the maximum amount of water. Powder will mainly be produced by spray drying (Figure 10.2). The concentrate is pumped to the top of the spray dryer. A nozzle sprays the product, which then falls down. During the fall hot air, which is tangentially fed into the chamber, causes

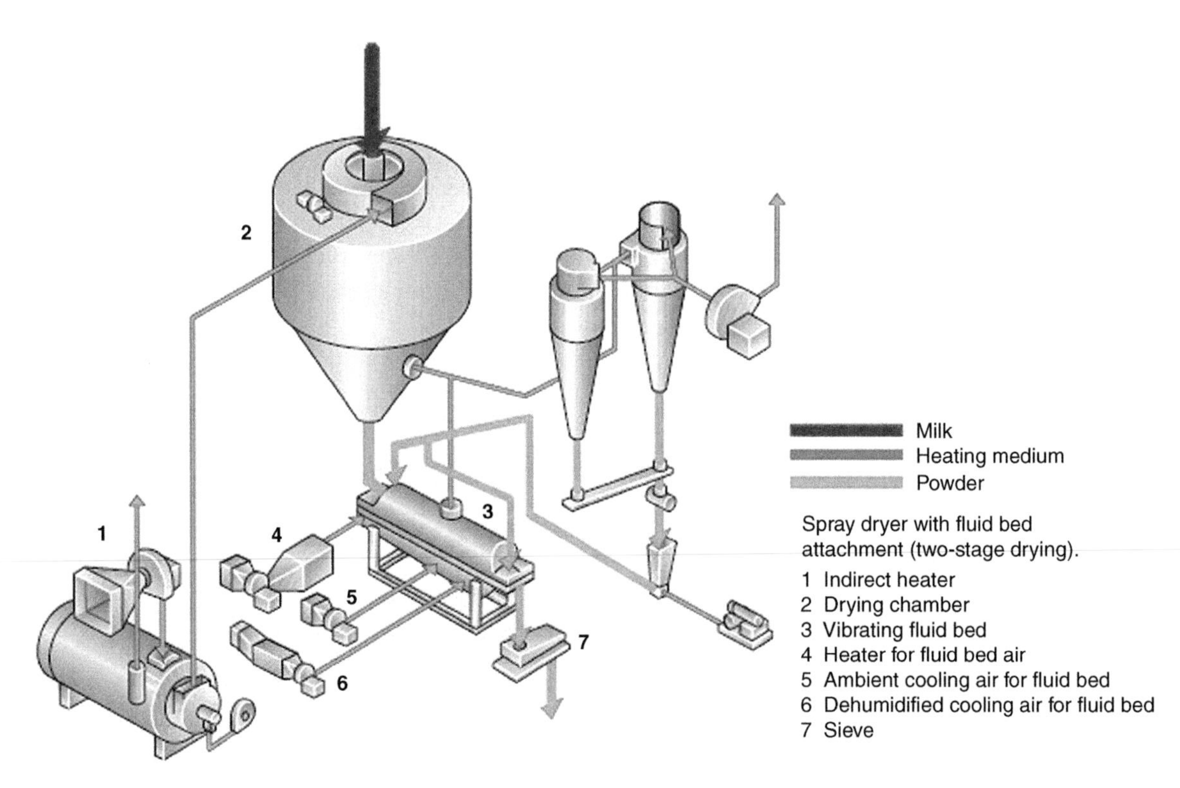

Figure 10.2 Powder production with a spray dryer with fluid bed. (Tetra Pak, 1995)

the product and air to circle. The product leaves the dryer with a water content of 5–10%.

Two positions exist to install an in-line device. One is at the outlet of the dryer, another is the end of the vibrating fluid bed. Figure 10.3 shows a realization based on a single reflectance probe inserted into an automatic sampler close to the outlet of the drying chamber. The sampler uses a pneumatic ram to take subsamples into a pipe in which the probe is located. The scanned sample is then returned into the product stream or can be pulled out for reference analysis which is a great advantage. Another advantage of the sampler is that the measuring conditions can be well defined so that the prediction accuracy is high. The same accuracy is obtained as with laboratory IR measurement. Having information about water content allows the drying conditions in the chamber to be changed by altering the feed of concentrate or modifying the air temperature.

Using an in-line measurement at the end of the vibrating fluid bed it is also possible to extract the mean diameter of the powder particles with IR spectroscopy. The basis for this is that different particle size distributions cause a shift in absorption due to the scattering of IR light. Knowing the mean diameter (Frake *et al.*, 1997; Goebel and Steffens, 1998) the particle size distribution can be influenced by modifying the drying conditions in the fluid bed.

Figure 10.3 In-line measuring device at the end of a spray dryer. (A Niemoeller, 2007, personal communication)

It must also be taken into account that process control is getting more complicated and the actions must be very smooth if the measuring place is further away from the component to be influenced. In the case of the drying chamber there is the additional problem that the component reacts very slowly with respect to changes.

Infrared spectroscopy has been studied as a rapid method to determine the composition and adulteration of milk powder. In fact milk powder is particularly suited to NIR spectroscopic analysis as the particles are uniform in size and shape and the small particle size negates any requirement for grinding.

Mendenhall and Brown (1991) investigated the potential of MIR spectra in the range 1200–1400 cm^{-1} to predict whey protein concentration in non-fat dry milk (NDM). Whey protein and NDM powder were mixed in varying proportions and reconstituted to a constant total solids content prior to analysis. They found that the PLS regression model successfully predicted the concentration of whey protein in adulterated samples ($R = 0.99$) and that accuracy was not affected by processing conditions, source of NDM, or origin of whey protein concentrate powder (Mendenhall and Brown, 1991). More recently, NIR spectroscopy also in conjunction with PLS regression successfully determined the protein content of milk powder with a root mean square error of prediction of 0.687% (Chang *et al.*, 2007).

As previously mentioned, the water content of milk powder is very important as it will influence the physico-chemical stability of milk powder during both storage and distribution. Reh *et al.* (2004) compared a number of different methods for determining the moisture content of milk powders, which included analyzing the powders using NIR spectroscopy (400–2500 nm). They found that there was a very good correlation between the moisture content of the milk powder and absorbance at 1940 nm with an R^2 of 0.94 and a standard deviation of differences of 0.07wt%. It was also stated that the major advantage of NIR absorbance at 1940 nm was that it was almost completely absence of interference from other food ingredients. Another study also investigated the potential of NIR spectroscopy to predict the moisture content of milk powder (Nagarajan *et al.*, 2006). They developed models for predicting moisture content using two spectral ranges (i.e. 1900–1950 nm and 1425–1475 nm) using PLS regression. They found that the 1900–1950 nm region was superior at predicting milk powder moisture content with the validation model, having an R^2 of 0.98.

Studies have also highlighted that IR spectroscopy may have a role in determining milk powder quality and process control during manufacture (Downey *et al.*, 1990; Koc *et al.*, 2002; Qin *et al.*, 2004; Deng *et al.*, 2005; Cen *et al.*, 2006; Zhou *et al.*, 2006; Huang *et al.*, 2007). While Downey *et al.* (1990) found it was possible to differentiate between milk powder based on heat treatment using NIR spectroscopy, Deng *et al.* (2005) utilized FTIR spectroscopy to identify 11 kinds of milk powder based on variations in the main nutritious components. Powdered infant milk formula is another important milk powder product which has been investigated using NIR spectroscopy. Cen *et al.* (2006) used NIR and visible spectroscopy (400–1000 nm) to successfully distinguish between nine different varieties of powdered infant milk formula ($R^2 = 0.98$). Using a similar spectral region Huang *et al.* (2007) developed a back propagation neural network, which with proper training could achieve a recognition accuracy of 100%.

NIR technology has also been used in a study in which fuzzy logic was applied to real-time control of a spray-dried whole milk powder processing system in order to achieve a product of consistent quality (Koc *et al.*, 2002). The NIR measurements were employed to determine product color and the study found that the developed algorithm controlled the process at ±0.074 kW of the desired power consumption and provided a whole milk product within ±3.0 units of the desired color value.

NIR spectroscopy has also been demonstrated as an effective technique for detecting milk powders adulterated with 0–5% vegetables protein ($R^2 = 0.99$, standard error of prediction = 0.23%) (Maraboli *et al.*, 2002).

Oil and fat production

One of the earliest studies examining the use of IR spectroscopy in the analysis of oils and fats was by Kliman and Pallansch (1965). They found that the water content of butter oil was relative to the absorption at 1900 nm. In more recent times IR spectroscopy has been utilized for the rapid assessment of authenticity and composition of butter.

Van de voort *et al.* (1992) used MIR spectroscopy to determine the fat and moisture content of butter samples. They found that the MIR technique produced compositional values comparable to those obtained by conventional wet chemical methods. Safar *et al.* (1994) studied 27 commercial samples of oils, butters, and margarines and determined that as MIR spectra contained information regarding carbonyl groups and double bonds, the samples were classified on the basis of their degree of esterification and their degree of unsaturation. MIR spectroscopy has also been used for the quantitative analysis of edible *trans* fats in butter (Dupuy *et al.*, 1996), and the characterization of butter, soybean oil, and lard (Yang and Irudayaraj, 2000).

Sato *et al.* (1990) studied the detection of foreign fat in butter using NIR spectroscopy and found that it was capable of detecting as little as 3% foreign fat in butter and margarine mixtures. NIR spectroscopy has also been used for determining the composition of butter, with Hermida *et al.* (2001) predicting moisture ($R^2 = 0.83$), solids-non-fat ($R^2 = 0.94$) and fat ($R^2 = 0.72$) in butter without any previous sample treatment.

The role of IR analyzers for the at-line analysis of butter during manufacture has also been recently examined by (Meagher *et al.*, 2007), who studied the potential of NIR spectroscopy to predict the solid fat content of milk fat extracted from butter during manufacture. The solid fat content was successfully predicted ($R^2 = 0.92$–0.98) using NIR spectroscopy and PLS regression and the method had the added advantage of not requiring a 16 h delay period for sample preparation as is currently required in the at-line determination of solid fat content by nuclear magnetic resonance spectroscopy.

Figure 10.4 shows schematically a butter-making machine. Cream is churned so that pieces of butter and buttermilk will be produced. In the following section pieces of butter (nearly pure fat) and the buttermilk are separated from each other. In the squeeze drying section some amount of buttermilk is worked into the pieces of butter to form butter. A moisture content limit for butter exists as a product called "butter" may not have a water content of more than 16%. Due to the fact that water is cheaper than fat the producer wants to keep the water content as high and close as possible to

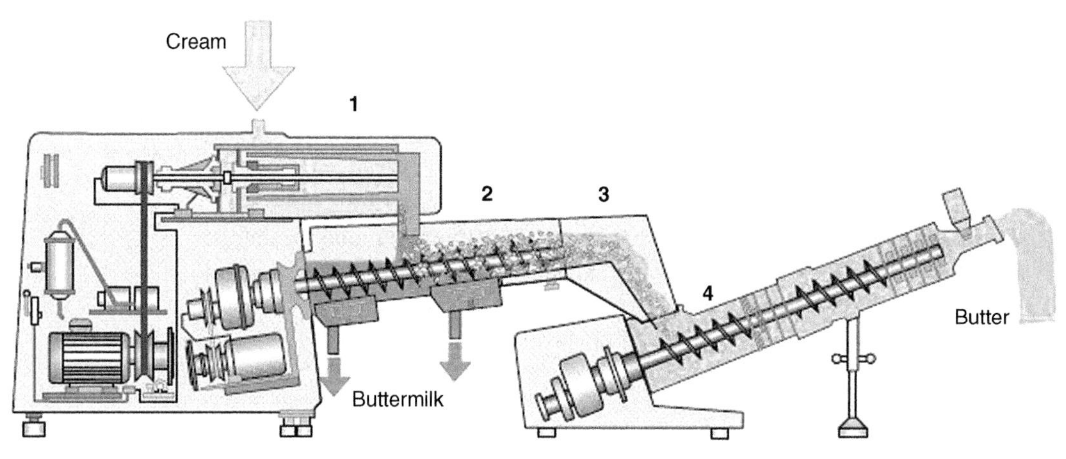

Figure 10.4 Butter-making machine. (Tetra Pak, 1995)

16%. The second working section is a new development which is not very often realized to soften the butter.

The water content is measured by IR spectroscopy in the squeeze drying section. There are two different applications: (a) One is based on a single reflectance probe a little bit downstream of this section. A sampling valve is located close by allowing the extraction of the scanned representative sample. (b) Another option (Figure 10.5) is to measure the water content with a transmission device directly in this section. The representative sample can be collected at the outlet. Both calibrations have accuracies close to the reference method because water content determination with IR spectroscopy is straightforward due to the large electrical dipole moment of the water molecule. If the separation or the squeezing are changed the water content of butter can be modified. Due to the location of the measuring device a direct control is possible, giving an almost constant water content.

Cheese production

Process milk is usually standardized in the following way (Figure 10.6): The milk is completely or partly separated into cream and skim milk which are then recombined to achieve the desired fat content. Furthermore, products such as fat and protein are often recovered from whey, a by-product of cheese production. Such products can be added to the process milk to ensure maximum cheese-making efficiency. These products normally change the concentration of other constituents. There are two possibilities to solve the problem: (1) The mixture can be poured into a process tank

Figure 10.5 Transmission measuring device at the end of the butter-making machine. (M Sievers, 2007, personal communication)

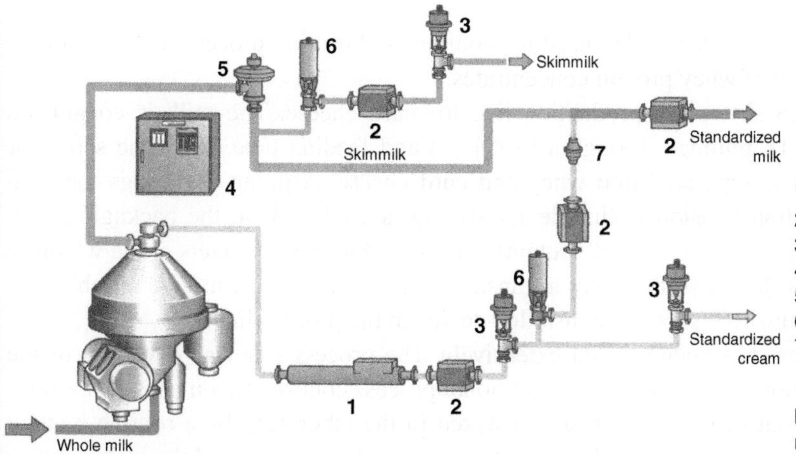

Figure 10.6 Traditional milk standardization by separation and mixing via density and flow measurement. (Tetra Pak, 1995)

and analyzed by the laboratory. The constituent concentration is then corrected by adding the appropriate component. This is very time consuming because the laboratory must analyze the mixture again after the corrective action. (2) The mixture can be analyzed at the inlet of the tank during the filling and the different flows adjusted to obtain the correct concentrations. This is a better solution. To achieve excellent IR predictions this analysis is done using an MIR spectrometer which can only be operated in bypass mode (Figure 10.7). In addition, protein and/or carbohydrate standardization is possible.

Figure 10.7 Mid-infrared measurement of fat, protein, lactose and dry matter content via a bypass. (M Sievers, 2007, personal communication)

The MIR spectrometer is also used to monitor the filtration processes, for example, the production of whey protein concentrates.

Figure 10.8 shows the production line for curd cheese. The milk is coagulated in the tank (1), pumped through a heater (2) and feeding pipe (3) to the separator (4), where it is separated into whey and curd cheese. A pump (5) brings the curd through the transmission in-line device (6) and a cooler (8) to the packaging unit. The protein-rich curd (12–13% protein) causes the building of layers at the windows of the in-line device so that only a transmission measurement can give reliable predictions. Figure 10.9 shows the installed device in the process line.

This process has been studied extensively. The process standard deviation of the dry matter content is 0.24% using traditional process control. This means that a sample is taken out of the process and analyzed in the laboratory by a routine method similar to the reference method (drying by 102°C). Having arranged the in-line measurement so that the process operator gets the IR predictions directly to their desk, the standard deviation is reduced to 0.12% (i.e. the standard deviation is reduced by a factor 2). The last step is to operate the separation process as a closed loop, meaning that the IR predictions are directly used to automatically control the feed of the separator. In this case the standard deviation is brought down to 0.07%. Standard low fat curd cheese has a minimum dry matter content of 18%. Producing this curd cheese using traditional process control the target dry matter is 18.3%. In the automatic case the target value will be 18.1%, saving a lot of raw material. In a medium-sized company the pay-back period of the investigation is approximately a year.

Figure 10.8 Schematic showing process line with in-line transmission device. (Pos. 6)

Figure 10.9 In-line transmission device in a quarg production line behind the separator.

Infrared spectroscopy has been employed in several stages of cheese production. These include the monitoring of milk coagulation, syneresis and ripening and the determination of composition, texture, and authenticity.

In cheese production the cheese milk is poured into a cheese vat (Figure 10.10) and enzyme is added. The splitting of casein, a protein component, causes the casein micelles to be destabilized, allowing them to be linked together and form the coagulum. Subsequently the coagulum is cut so that whey can be expelled and then removed. The correct cutting time is necessary to reduce the loss of constituents through the whey. In fact the importance of obtaining objective online measurements for monitoring gel time, coagulum firmness and cutting point during cheese manufacture, in order to obtain high-quality and consistent cheese products is well known (Payne *et al.*, 1993a). This has resulted in the development of a number of online sensors, which can be used to successfully monitor milk coagulation. O'Callaghan *et al.* (2002) comprehensively reviewed a number of systems (optical, thermal, mechanical, and vibrational) for monitoring curd setting during cheese making. McMahon *et al.* (1984) found that the absorbance of light during milk coagulation resulted from changes in the molecular weight, size, and number of colloidal casein micelle aggregates.

Developments in fiber optics have assisted the development of NIR reflectance and transmission sensors. Payne *et al.* (1993a) developed a method based on changes

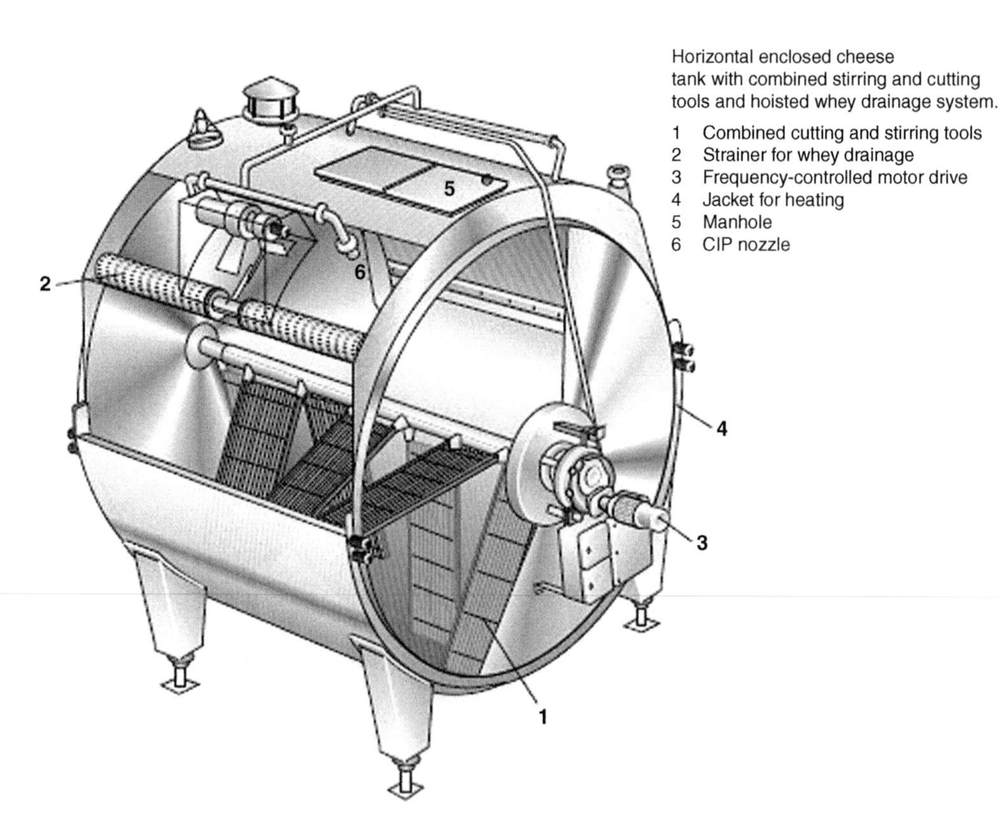

Figure 10.10 Cheese vat with cutting and stirring tools. (Tetra Pak, 1995)

in diffuse reflectance at 940 nm during milk coagulation. The time to the inflection point (t_{max}) was determined from the first derivative and was found to correlate well with Formograph cutting times. In further work by Payne *et al.* (1993b), a sensor operating at 950 nm was developed and t_{max} was again found to be well correlated with the observed cutting time. Linear prediction equations, which were considered to be of the form required for predicting cutting time, were also developed using t_{max}. These equations had standard errors of between 1.5 and 2.4 min.

These studies only monitored coagulation at one wavelength, however, and so Laporte *et al.* (1998) used full spectrum information and PLS regression. Reflectance was monitored during coagulation and spectra collected between 1100 and 2500 nm with a resolution of 2 nm. This method was found to be reliable in monitoring milk coagulation.

O'Callaghan *et al.* (2000) compared the response of three NIR sensors (two transmission and one reflectance) to the response of thermal and vibrational sensors where protein and enzyme concentration was varied. While they found that all sensors were sensitive to changes in enzyme concentration, the NIR sensors were more sensitive to changes in the rate of curd firming due to varying protein levels than the hot wire and torsional vibrational systems. Indeed the NIR sensors predicted the curd-firming time with a standard error less than 100 s. However O'Callaghan *et al.*

Figure 10.11 Infrared measuring device in a cheese vat. View through the manhole. (A Niemoeller, 2007, personal communication)

(2000) also stated that an algorithm combining the output of an NIR sensor as well as protein level was required to accurately predict curd-firming times.

Figure 10.11 shows an example of a measuring device that is put into the cheese vat through a manhole. In this example the sensor must be removed prior to cutting due to the action of cutting and stirring tools. However other NIR cutting time sensors, such as the CoAguLite™ (Reflectronics, Lexington KY, USA), which do not require removal prior to cutting are commercially available.

Development of a syneresis control technology is another emerging area of research due to its significant impact on cheese quality and yield. Currently only regions of the NIR spectra have been investigated as a technology for monitoring syneresis (Guillemin *et al.*, 2006; Fagan *et al.*, 2007b, 2007e).

Fagan *et al.* (2007b) proposed that a sensor detecting NIR light backscatter (300–1100 nm) in a cheese vat and with a large field of view (LFV) relative to curd particle size would have potential for monitoring both milk coagulation and curd syneresis. Fagan *et al.* (2007b) reported that the response of the prototype sensor was affected by temperature and that the sensor showed potential for predicting whey fat content, curd moisture content, and curd yield. Further work by Fagan *et al.* (2007e) found that the LFV optical sensor provided the information on gel assembly and curd shrinkage kinetics required for accurate predictions of whey fat losses and curd yield prediction as well as for curd moisture control. Fat losses, curd yield, and moisture content at 85 min from cutting were predicted using a combination of independent variables, milk compositional parameters and LFV light backscatter parameters with SEP of 2.37 g, 0.91% and 1.28%, respectively (Fagan *et al.*, 2007a). Curd moisture as a function of processing time was predicted with a SEP of 1.27% over the range of 50–90% curd moisture content. Guillemin *et al.* (2006) modified an NIR (700 nm) milk coagulation sensor to investigate its potential for the online determination of casein particle size distribution and of the volume fraction relative to the whey as a function of time. It was found that utilizing multiple thresholds of the optical signal associated with data processing using neural networks provided useful results. The casein particle volume fraction was estimated with a relative error of 23%, and the casein particle size distribution was estimated with a maximum relative error of 7.5% (Guillemin *et al.*, 2006).

The determination of the dry matter content in soft cheese before pouring into the cheese mold is another example. This is of interest because it is necessary to standardize the dry matter content of the product to be poured. The soft cheese is produced with a coagulator (Pos. 4.1–4.4 in Figure 10.12). The measurement device is installed on top of the whey removal drum (Pos. 6) before the filling machine (Pos. 7). Figure 10.13 shows the product (curd). The surface is very rough, deep, and structured so that one needs a good averaging. This can be achieved in two ways: (a) The IR reflectance measurement device (Figure 10.14) measures an area of approximately 200 cm^2. (b) During multiple IR scans the product is continuously moving.

The accuracy of the dry matter calibration is of the order of the repeatability of the reference method. The prediction is used to influence the stirring at the end of the coagulator as well as the amount of filling in the next process step.

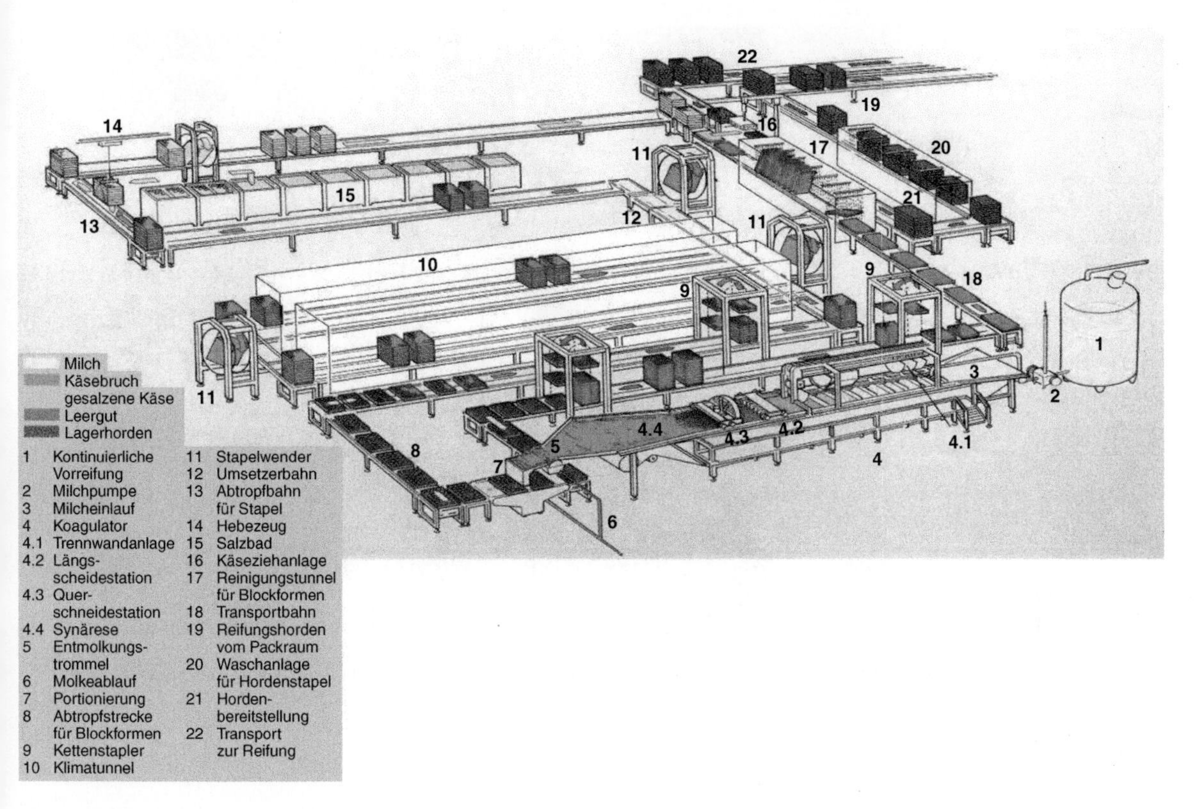

Figure 10.12 Soft cheese production process via coagulator (Pos. 4.1–4.4). (ALPMA, 1998)

Figure 10.13 Picture of the curd on top of the whey removal drum.

Figure 10.14 Infrared measuring device on top of the whey removal drum. Product is coming from the left side.

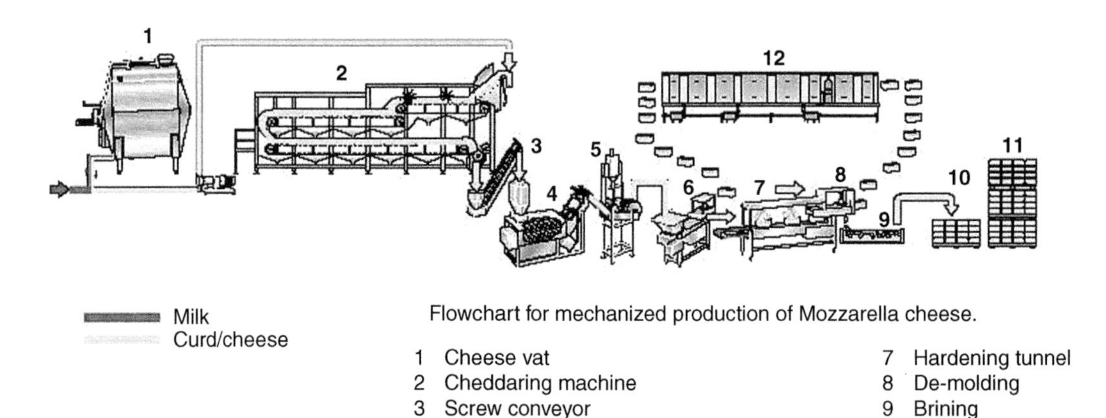

Figure 10.15 Mozzarella cheese production process with cocker/stretcher (Pos. 4). (Tetra Pak, 1995)

A reflection measurement with a single fiber has also been used to determine qualities of interest in mozzarella cheese after stretching. Figure 10.15 schematically shows the production process. After the cooker/stretcher the in-line device (Figure 10.16) is located to influence the previous production steps.

Irudayaraj *et al.* (1999) investigated the use of MIR spectroscopy to follow texture development in Cheddar cheese during ripening. They demonstrated that springiness could be successfully correlated with a number of bands in the MIR spectra. The development of cheese microflora during ripening is extremely important in the development of flavor and texture and Lefier *et al.* (2000) demonstrated that FTIR spectroscopy could be used as a rapid and robust method for the qualitative analysis of cheese flora, while Lucia *et al.* (2001) found significant differences during ripening in the region of the MIR spectra associated with amides I and II, of cheeses produced using different starter cultures. Further studies have also shown that MIR spectroscopy is a useful technique for characterizing changes in proteins during cheese ripening (Mazerolles *et al.*, 2001) as well as having the potential to differentiate between cheeses of different age (Dufour *et al.*, 2000). As the level of water-soluble nitrogen (WSN) is as an indicator of cheese ripening, some attempts have also been made to predict the level of WSN in cheese with some success ($R^2 = 0.80$) (Karoui *et al.*, 2006a, 2006b).

More recently a number of studies have assessed IR spectroscopy to predict cheese sensory and rheological properties. MIR spectroscopy has been applied to

Figure 10.16 In-line measuring device after the cocker/stretcher. (M Sievers, 2007, personal communication)

the prediction of processed cheese texture and meltability attributes (Fagan *et al.*, 2007c, 2007d). Models predicting hardness, springiness, massforming, and masscoating gave approximate quantitative results (R^2 = 0.66–0.81), models predicting cohesiveness, Olson and Price meltability, firmness, rubberiness, creaminess, and chewiness gave good prediction results (R^2 = 0.81–0.90), while only the fragmentable model provided excellent predictions ($R^2 > 0.91$). Blazquez *et al.* (2006) also predicted these attributes in processed cheese; however in this study NIR spectroscopy was utilized. In general, NIR spectroscopy predicted each of the attributes with a similar or better accuracy than MIR spectroscopy, with Blazquez *et al.* (2006) reporting models with excellent prediction accuracies for chewiness, meltability, creaminess, springiness and hardness attributes (R^2 = 0.94). Downey *et al.* (2005) also predicted crumbly, rubbery, and chewy attributes in Cheddar cheese using NIR spectroscopy and the reported models gave approximate quantitative results.

Analysis of the shelf-life period in which freshness is maintained in Crescenza cheese has also been examined using MIR spectroscopy (Cattaneo *et al.*, 2005). They found that using principal component analysis (PCA) of the spectra it was possible to detect the decrease of Crescenza freshness and to define the critical day during shelf life.

Many processes can be monitored by IR spectroscopy and their applications in the dairy industry, while in its early stages, is growing. Beside its quantitative application, qualitative in-line analysis will become more important in the future.

Final product control

Nearly all major constituents of dairy products (see Table 10.1) can be determined with IR spectroscopy with sufficient accuracy. The composition and texture of cheese is intrinsically linked with its quality. Infrared spectroscopy therefore has been considered for predicting both the composition and textural attributes of cheese. NIR spectroscopy (Rodriguez Otero *et al.*, 1994; Lee *et al.*, 1997; Wittrup and Nørgaard, 1998; McKenna, 2001; Pérez-Marín *et al.*, 2001; Blazquez *et al.*, 2004; Čurda and Kukačková, 2004; Karoui *et al.*, 2006c, 2007) has probably been more widely utilized than MIR spectroscopy for cheese composition determination (McQueen *et al.*, 1995; Chen and Irudayaraj, 1998). McQueen *et al.* (1995) used FTIR spectroscopy to predict protein, fat, and moisture content of cheese (R = 0.81–0.92). Chen and Irudayaraj (1998), also using FTIR spectroscopy, found that the intensity of certain bands in the MIR spectra increased with increasing fat and protein contents. McKenna (2001) stated that published data demonstrated that NIR transmittance spectroscopy is more accurate than NIR reflectance spectroscopy for the determination of cheese moisture content. However NIR reflectance spectroscopy is the more widely used mode of NIR spectroscopy in cheese compositional analysis (Rodriguez Otero *et al.*, 1994; Lee *et al.*, 1997; Pérez-Marín *et al.*, 2001; Blazquez *et al.*, 2004; da Costa Filho and Volery, 2005; Karoui *et al.*, 2006c; Skeie *et al.*, 2006). These authors report predictions with varying degrees of accuracy, but overall the results are extremely accurate.

Pillonel *et al.* (2003) used MIR spectroscopy in combination with chemometrics to investigate the potential for discriminating Emmental cheeses of various geographic origins. The normalized spectra were analyzed by PCA and linear discriminant analysis (LDA) of the PCA scores and Swiss Emmental was correctly classified 100% of the time. Karoui *et al.* (2004) also investigated the potential of MIR spectroscopy to discriminate between Emmental cheeses produced during the summer and winter and from five European countries: Germany, Austria, Finland, France, and Switzerland. Using PCA and PLS regression they found that it was possible to distinguish between cheeses according to the season of production, the treatment of the milk and the geographic origin. The same research group have also investigated the potential of combining MIR and fluorescence spectroscopy for determining the geographic origin of experimental French Jura hard cheeses and Swiss Gruyère and L'Etivaz cheeses (Karoui *et al.*, 2005a). Although it was possible to discriminate between the samples based on origin it was found that fluorescence spectra produced better results than the MIR spectra. Karoui *et al.* (2005b) also compared the potential of NIR, MIR, and front-face fluorescence spectroscopy to discriminate Emmental cheeses from different European geographic origins. Almost 90% of cheese samples were classified by factorial discriminant analysis using either MIR or NIR spectra. However in this case the classification obtained with the fluorescence spectra was considerably lower.

The microbiology of the product is a major issue because it takes a long time to perform traditional microbial analysis. Infrared spectroscopy has the potential to reduce the time required for analysis and therefore the product can be delivered earlier and hence more cost efficiently. Infrared spectroscopy is used to identify microorganisms in a routine laboratory. The method, based on MIR spectroscopy, has been developed by Naumann and co-workers (Naumann *et al.*, 1988; Helm *et al.*, 1991). After a cultivation step and extraction of a single colony the spectrum of the unknown species is compared with the spectra of a library. A dendrogram (Figure 10.17) helps to identify the unknown species. The advantage of the method is that it can reduce the time for analysis by 2–3 days. The microbiological department of the Wissenschaftszentrum Weihenstephan (Kummerle *et al.*, 1998; Oberreuter *et al.*, 2002) has developed huge libraries of microorganisms and applies this qualitative method to identify microorganisms in dairy samples. This method has also been developed as a routine method within large hospitals and other industrial facilities (e.g. pharmaceutical industry and breweries).

To ensure the correct operation of the calibrations one must arrange a monitoring system to check the performance of the IR method. In the following section the "Good Laboratory Practice" (GLP) procedure for IR calibrations will be described.

Good Laboratory Practice for IR calibrations

Like all chemical analysis methods one must check the performance of the method regularly. Due to the fact that the IR method is an indirect method (calibration step) one must establish a monitoring routine which tests for the three following types of potential problems:

- One must ensure that the instrument operates well. This can be done by taking the IR spectra of inert standards and comparing the spectra over time, setting appropriate limits.

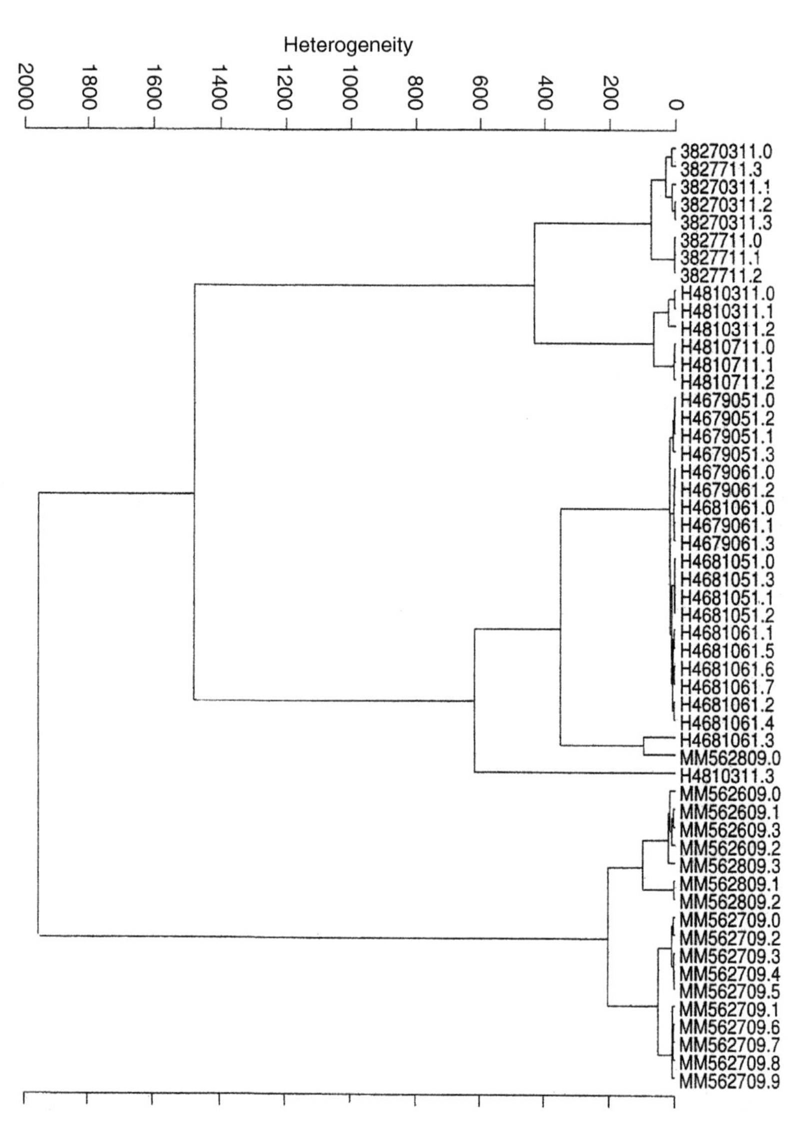

Figure 10.17 Infrared identification of microorganisms via dendrogram. A heterogeneity of zero means that the spectra of the microorganisms are identical.

- If sample preparation is a necessary step for the IR technique it must be tested whether the operating personnel fulfill the demands of the appropriate standard operating procedure. This can be evaluated by preparing the same material several times and predicting the constituent concentrations. Limits will help to clarify this step.
- The performance of the calibration must be monitored. This is not always possible because certified material may not be available in all cases (e.g. having a calibration for fat in yoghurt, i.e. no yoghurt exists which can be bought as certified material). The only way is to analyze the corresponding sample by reference

analysis and to compare the difference between reference value and IR prediction over time. Limits of the difference can help to define warning and action levels.

The last point is of special importance because changes in the recipe can influence the IR spectrum. If constant difference between the predictions and the reference values or a drift over a period of time is observed then the development of a new calibration may be necessary. To visualize the performance in an effective way a plot of the difference vs. time is the optimal presentation. These tasks can be done by control charts also showing the warning and action limits.

Having established such a system it can be shown that the IR method gives more constant values than the reference method when performing multiple analyses of the same inert material. The IR predictions have smaller variations than the reference values. It is believed that the IR predictions are more accurate. This is reasonable but cannot be proven because the reference methods are required to judge this.

To establish such a GLP system much experience is necessary. The quickest and most efficient way is to operate a network so that many instrument performances can contribute to the definition of the Good Laboratory Practice (GLP) arrangements.

Networking

For process control of multiple dairy plants, networking can be used so that NIR instruments in individual plants can be connected to headquarters. An example employed by the Ahlemer Institute of the Landwirtschaftskammer, Hannover in Germany is discussed below.

Service network

Since 1988 the Ahlemer Institute of the Landwirtschaftskammer, Hannover (now LUFA Nord-West, Agricultural Chamber Lower Saxony) has operated a service network which has been accredited by the German Accreditation Council (DAP). There are five dairies with 12 NIR instruments connected by phone and modem to the headquarters in Oldenburg (Lower Saxony, Germany). The institute performs feasibility studies and develops new applications as well as GLP procedures. The advantage is that the individual dairies do not need to employ highly qualified staff with specialized training in the area of IR spectroscopy. Furthermore staff members with different scientific backgrounds work together in the institute so that problems can be solved very efficiently. Today businesses tend to concentrate on their key areas of expertise and therefore are outsourcing their other activities.

Surveillance network

The institute has been given the responsibility by the government of Lower Saxony to check MIR instruments that are used for the prediction of various constituents in raw milk for the farmers' payment. This is done in the following manner:

- On a weekly basis specific samples are sent to the laboratories to test the performance of instrument and calibration. After approximately 200 raw milk

samples one of the specific samples has to be analyzed. This procedure is repeated after each 200 raw milk samples. The results of these specific samples are transferred by modem to the institute for further evaluation.
- Each month a series of samples are prepared artificially to check the calibration over a larger concentration range. These results are also transferred to the institute.

The advantages of this network are that the calibration is monitored more often and the inspection is completed without the necessity of traveling and is thus less time intensive. The controller is also able to use his or her experience to assist. The network also serves to build confidence between farmers and laboratories.

Harmonization network

Harmonization of analytical results is an important issue for two reasons:

- Large dairies with more production sites that ship milk or products from one production site to another would like all measurements performed on the same product at different locations to be the same or at least in good agreement. Otherwise the use of mass balance procedures creates problems.
- Exported products need to be analyzed in such a way that the results of different laboratories are in good agreement.

With respect to chemical methods of analysis, standard operating procedures are defined as well as precision parameters (used for the comparison of the results). However the results are strongly dependent upon the laboratory technician. Within a research and development project funded by the European Union it has been shown that this harmonization goal could be achieved for IR spectroscopy. The method is based on the concept of "matching instruments," where one instrument is used as the master instrument. Having determined the characteristic between master and the other instrument (Wang *et al.*, 1991), the spectra of the other instrument is transformed in a mathematical way. In doing so spectra are obtained which look as if they were recorded using the master instrument. Using the calibration of the master instrument the appropriate constituent concentration is predicted. This ensures that all predictions include the same information and therefore all instruments behave in the same manner.

Future trends

Infrared spectra are dominated by the water content of the product and nearly all dairy products, with the exception of powder products, have high moisture levels. Spectra of water and milk, for example, look very similar. Only statistical methods are able to extract the information out of the spectra. Usually the information is extracted on a direct and fundamental basis because experience shows that better prediction accuracy can be achieved. If Raman spectroscopy is applied, water hardly disturbs the measurement

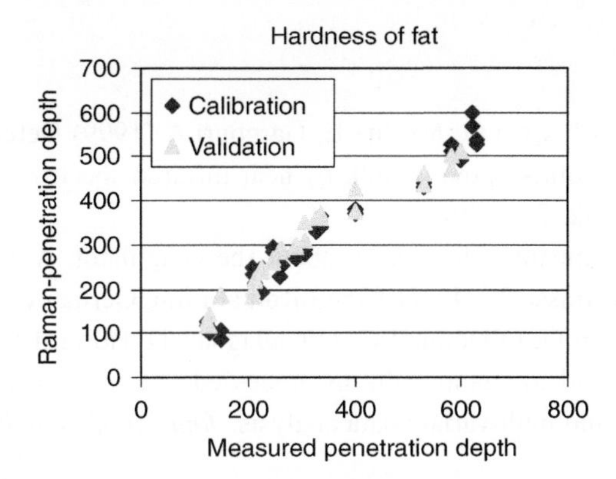

Figure 10.18 Hardness of fat predicted by Raman spectroscopy for calibration and validation (samples in arbitrary units).

results. Based on this complementary information in some cases a more precise determination of constituents may be possible. Figure 10.18 shows the result of a calibration where the hardness of fat is predicted by Raman spectroscopy. The hardness is measured by a penetration method in which a cone penetrates a block of fat. Fat is soft if it has a lot of unsaturated carbon–carbon bonds. Due to the fact that Raman spectroscopy is sensitive to non-polar molecular binding the unsaturated carbon–carbon bonds can be correlated to the hardness of fat. By measuring only a small amount of the product Raman spectroscopy still has to prove that the measurement can also be done quantitatively. Raman imaging may be the way to overcome this. A combination of both vibration and excitation mechanisms combined with the appropriate statistical methods and in-line devices will boost the industrial possibilities.

Conclusions

Infrared spectroscopy is a powerful tool to determine constituent concentrations and qualitative characteristics of dairy products. Many examples and applications show that the technique is accurate and fast and can be used for process control. To ensure proper performance it is necessary to establish GLP guidelines. Within these guidelines the main issue is monitoring of the calibration performance. This includes the adjustment of the existing calibrations or the generation of new calibrations. Some dairies use the services of a network, thus outsourcing the calibration and application work. This type of measurement is moving into the process line. With the appropriate process control strategy the payback period of the investment will be between one and two years, suggesting that it is a worthwhile investment.

Acknowledgments

The authors would like to thank A Fehrmann-Reese, Dr A Hoffmann, Dr A Niemoeller (Bruker Optik GmbH), and M Sievers (Foss GmbH) for their support and the material.

References

Albanell E, Caceres P, Caja G, Molina E, Gargouri A (1999) Determination of fat, protein, and total solids in ovine milk by near-infrared spectroscopy. *Journal of AOAC International*, **82**, 753–758.

ALPMA (1998) Company information about the coagulator, Alpenland Maschinenbau GmbH, Alpenstrasse 39–43, D-83543 Rott am Inn, Germany.

Blazquez C, Downey G, O'Donnell C, O'Callaghan D, Howard V (2004) Prediction of moisture, fat and inorganic salts in processed cheese by near infrared reflectance spectroscopy and multivariate data analysis. *Journal of Near Infrared Spectroscopy*, **12**, 149–157.

Blazquez C, Downey G, O'Callaghan D, Howard V, Delahunty C, Sheehan E, Everard C, O'Donnell CP (2006) Modelling of sensory and instrumental texture parameters in processed cheese by near infrared reflectance spectroscopy. *Journal of Dairy Research*, **73**, 58–69.

Carl RT (1991) Quantification of the fat-content of milk using a partial-least-squares method of data-analysis in the near-infrared. *Fresenius Journal of Analytical Chemistry*, **339**, 70–71.

Cattaneo TMP, Giardina C, Sinelli N, Riva M, Giangiacomo R (2005) Application of FT-NIR and FT-IR spectroscopy to study the shelf-life of Crescenza cheese. *International Dairy Journal*, **15**, 693–700.

Cen HY, Bao YD, Huang M, He Y (2006): Comparison of data pre-processing in pattern recognition of milk powder Vis/NIR spectra. *Advanced Data Mining and Applications, Proceedings*, pp. 1000–1007:

Chang M, Chu PJ, Xu KX (2007) Study on noninvasive detection using NIR diffuse reflectance spectrum for monitoring protein content in milk powder. *Spectroscopy and Spectral Analysis*, **27**, 43–45.

Chen M, Irudayaraj J (1998) Sampling technique for cheese analysis by FTIR spectroscopy. *Journal of Food Science*, **63**, 96–99.

Čurda L, Kukačková O (2004) NIR spectroscopy: a useful tool for rapid monitoring of processed cheese manufacture. *Journal of Food Engineering*, **61**, 557–560.

da Costa Filho PA, Volery P (2005) Broad-based versus specific NIRS calibration: Determination of total solids in fresh cheese. *Analytica Chimica Acta*, **554**, 82–88.

Deng YE, Zhou Q, Sun SQ (2005) Quality analysis of powdered milk via FTIR spectroscopy. *Spectroscopy and Spectral Analysis*, **25**, 1972–1974.

Downey G, Robert P, Bertrand D, Kelly PM (1990) Classification of commercial skim milk powders according to heat treatment using factorial discriminant analysis of near-infrared reflectance spectra. *Applied Spectroscopy*, **44**, 150–155.

Downey G, Sheehan E, Delahunty C, O'Callaghan D, Guinee T, Howard V (2005) Prediction of maturity and sensory attributes of Cheddar cheese using near-infrared spectroscopy. *International Dairy Journal*, **15**, 701–709.

Dufour E, Mazerolles G, Devaux MF, Duboz G, Duployer MH, Riou NM (2000) Phase transition of triglycerides during semi-hard cheese ripening. *International Dairy Journal*, **10**, 81–93.

Dupuy N, Duponchel L, Huvenne JP, Sombret B, Legrand P (1996) Classification of edible fats and oils by principal component analysis of Fourier transform infrared spectra. *Food Chemistry*, **57**, 245–251.

European Commission (2002) Laying down the general principles and requirements of food law, establishing the European Food Safety Authority and laying down procedures in matters of food safety. Article 8, Regulation (EC) No. 178/2002D.

Fagan CC, Castillo M, O'Donnell CP, O'Callaghan DJ, Payne FA (2007a) On-line prediction of cheese making indices using backscatter of near infrared light. *International Dairy Journal,* DOI: 10.1016/j.idairyj.2007.09.007.

Fagan CC, Castillo M, Payne FA, O'Donnell CP, Leedy M, O'Callaghan DJ (2007b) Novel online sensor technology for continuous monitoring of milk coagulation and whey separation in cheese making. *Journal of Agricultural and Food Chemistry*, **22**, 8836–8844.

Fagan CC, Everard C, O'Donnell CP, Downey G, Sheehan EM, Delahunty CM, O'Callaghan DJ (2007c) Evaluating mid-infrared spectroscopy as a new technique for predicting sensory texture attributes of processed cheese. *Journal of Dairy Science*, **90**, 1122–1132.

Fagan CC, Everard C, O'Donnell CP, Downey G, Sheehan EM, Delahunty CM, O'Callaghan DJ, Howard V (2007d) Prediction of processed cheese instrumental texture and meltability by mid-infrared spectroscopy coupled with chemometric tools. *Journal of Food Engineering*, **80**, 1068–1077.

Fagan CC, Leedy M, Castillo M, Payne FA, O'Donnell CP, O'Callaghan DJ (2007e) Development of a light scatter sensor technology for on-line monitoring of milk coagulation and whey separation. *Journal of Food Engineering*, **83**, 61–67.

Frake P, Greenhalgh D, Grierson SM, Hempenstell JM, Rudd DR (1997) Process control and end-point determination of a fluid bed granulation by application of near-infrared spectroscopy. *International Journal of Pharmaceutics*, **151**, 75–80.

Goebel SG, Steffens K-J (1998) Online-measurement of moisture and particle size in the fluidized-bed processing with the near-infrared spectroscopy. *Pharmazeutische Industrie*, **60**, 889–895.

Guillemin H, Trelea IC, Picque D, Perret B, Cattenoz T, Corrieu G (2006) An optical method to monitor casein particle size distribution in whey. *Lait*, **86**, 359–372.

Helm D, Labischinski H, Schallehn G, Naumann D (1991) Classification and identification of bacteria by Fourier-transform infrared-spectroscopy. *Journal of General Microbiology*, **137**, 69–79.

Hermida M, Gonzalez JM, Sanchez M, Rodriguez-Otero JL (2001) Moisture, solids-non-fat and fat analysis in butter by near infrared spectroscopy. *International Dairy Journal*, **11**, 93–98.

Huang M, He Y, Cen HY, Hu XY (2007) Fast discrimination of varieties of infant milk powder using near infrared spectra. *Spectroscopy and Spectral Analysis*, **27**, 916–919.

Irudayaraj J, Chen M, McMahon DJ (1999) Texture development in Cheddar cheese during ripening. *Canadian Agricultural Engineering*, **41**, 253–258.

ISO Standard 13884:2003 (2003) Animal and vegetable fats and oils—Determination of isolated trans isomers by infrared spectrometry. International Organization for Standardization, Geneva, Switzerland.

Jha SN, Matsuoka T (2004) Detection of adulterants in milk using near infrared spectroscopy. *Journal of Food Science and Technology-Mysore*, **41**, 313–316.

Karoui R, Dufour É, Pillonel L, Picque D, Cattenoz T, Bosset JO (2004) Determining the geographic origin of Emmental cheeses produced during winter and summer using a technique based on the concatenation of MIR and fluorescence spectroscopic data. *European Food Research and Technology*, **219**, 184–189.

Karoui R, Bosset J-O, Mazerolles G, Kulmyrzaev A, Dufour E (2005a) Monitoring the geographic origin of both experimental French Jura hard cheeses and Swiss Gruyere and L'Etivaz PDO cheeses using mid-infrared and fluorescence spectroscopies: a preliminary investigation. *International Dairy Journal*, **15**, 275–286.

Karoui R, Dufour É, Pillonel L, Schaller E, Picque D, Cattenoz T, Bosset JO (2005b) The potential of combined infrared and fluorescence spectroscopies as a method of determination of the geographic origin of Emmental cheeses. *International Dairy Journal*, **15**, 287–298.

Karoui R, Mouazen AM, Dufour É, Pillonel L, Picque D, De Baerdemaeker J, Bosset JO (2006a) Application of the MIR for the determination of some chemical parameters in European Emmental cheeses produced during summer. *European Food Research and Technology*, **222**, 165–170.

Karoui R, Mouazen AM, Dufour E, Pillonel L, Picque D, Bosset JO, De Baerdemaeker J (2006b) Mid-infrared spectrometry: A tool for the determination of chemical parameters in Emmental cheeses produced during winter. *Lait*, **86**, 83–97.

Karoui R, Mouazen AM, Dufour É, Pillonel L, Schaller E, De Baerdemaeker J, Bosset JO (2006c) Chemical characterisation of European Emmental cheeses by near infrared spectroscopy using chemometric tools. *International Dairy Journal*, **16**, 1211–1217.

Karoui R, Pillonel L, Schaller E, Bosset JO, De Baerdemaeker J (2007) Prediction of sensory attributes of European Emmental cheese using near-infrared spectroscopy: A feasibility study. *Food Chemistry*, **101**, 1121–1129.

Kawamura S, Kawasaki M, Nakatsuji H, Natsuga M (2007) Near-infrared spectroscopic sensing system for online monitoring of milk quality during milking. *Sensing and Instrumentation for Food Quality and Safety*, **1**, 37–43.

Kliman PG, Pallansch MJ (1965) Method for determination of water in butteroil by near-infrared spectrophotometry. *Journal of Dairy Science*, **48**, 859–862.

Koc AB, Heinemann PH, Ziegler GR, Roush WB (2002) Fuzzy logic control of whole milk powder processing. *Transactions of the ASAE*, **45**, 153–163.

Kummerle M, Scherer S, Seiler H (1998) Rapid and reliable identification of food-borne yeasts by Fourier-transform infrared spectroscopy. *Applied and Environmental Microbiology*, **64**, 2207–2214.

Laporte M-F, Martel R, Paquin P (1998) The near-infrared optic probe for monitoring rennet coagulation in cow's milk. *International Dairy Journal*, **8**, 659–666.

Lee SJ, Jeon IJ, Harbers LH (1997) Near infrared reflectance spectroscopy for rapid analysis of curds during cheddar cheese making. *Journal of Food Science*, **62**, 53–56.

Lefier D, Grappin R, Pochet S (1996) Determination of fat, protein, and lactose in raw milk by Fourier transform infrared spectroscopy and by analysis with a conventional filter-based milk analyzer. *Journal of AOAC International*, **79**, 711–717.

Lefier D, Lamprell H, Mazerolles G (2000) Evolution of *Lactococcus* strains during ripening in Brie cheese using Fourier transform infrared spectroscopy. *Lait*, **80**, 247–254.

Lucia V, Daniela B, Rosalba L (2001) Use of Fourier transform infrared spectroscopy to evaluate the proteolytic activity of *Yarrowia lipolytica* and its contribution to cheese ripening. *International Journal of Food Microbiology*, **69**, 113–123.

Luinge HJ, Hop E, Lutz ETG, Vanhemert JA, Dejong EAM (1993) Determination of the fat, protein and lactose content of milk using Fourier-transform infrared spectrometry. *Analytica Chimica Acta*, **284**, 419–433.

Maraboli A, Cattaneo TMP, Giangiacomo R (2002) Detection of vegetable proteins from soy, pea and wheat isolates in milk powder by near infrared spectroscopy. *Journal of Near Infrared Spectroscopy*, **10**, 63–69.

Mazerolles G, Devaux MF, Duboz G, Duployer MH, Mouhous Riou N, Dufour É (2001) Infrared and fluorescence spectroscopy for monitoring protein structure and interaction changes during cheese ripening. *Lait*, **81**, 509–527.

McKenna D (2001) Measuring moisture in cheese by near infrared absorption spectroscopy. *Journal of AOAC International,* **84**, 623–628.

McMahon DJ, Richardson GH, Brown RJ (1984) Enzymic milk coagulation: role of equations involving coagulation time and curd firmness in describing coagulation. *Journal of Dairy Science*, **67**, 1185–1193.

McQueen DH, Wilson R, Kinnunen A, Jensen EP (1995) Comparison of two infrared spectroscopic methods for cheese analysis. *Talanta*, **42**, 2007–2015.

Meagher LP, Holroyd SE, Illingworth D, van de ven F, Lane S (2007) At-line near-infrared spectroscopy for prediction of the solid fat content of milk fat from New Zealand butter. *Journal of Agricultural and Food Chemistry*, **55**, 2791–2796.

Mendenhall IV, Brown RJ (1991) Fourier transform infrared determination of whey powder in nonfat dry milk. *Journal of Dairy Science*, **74**, 2896–2900.

Nagarajan R, Singh P, Mehrotra R (2006) Direct determination of moisture in powder milk using near infrared spectroscopy. *Journal of Automated Methods & Management in Chemistry*, **2006**, 1–4.

Nathier-Dufour N, Sedman J, Voort FRvd (1995) A rapid ATR/FTIR quality control method for the determination of fat and solids in sweetened condensed milk. *Milchwissenschaft*, **50**, 462–466.

Naumann D, Fijala V, Labischinski H (1988) The differentiation and identification of pathogenic bacteria using FT-IR and multivariate statistical analysis. *Microchimica Acta*, **94**.

O'Callaghan DJ, O'Donnell CP, Payne FA (2000) On-line sensing techniques for coagulum setting in renneted milks. *Journal of Food Engineering*, **43**, 155–165.

O'Callaghan DJ, O'Donnell CP, Payne FA (2002) Review of systems for monitoring curd setting during cheesemaking. *International Journal of Dairy Technology*, **55**, 65–74.

Oberreuter H, Seiler H, Scherer S (2002) Identification of coryneform bacteria and related taxa by Fourier-transform infrared (FT-IR) spectroscopy. *International Journal of Systematic and Evolutionary Microbiology*, **52**, 91–100.

Payne FA, Hicks CL, Madangopal S, Shearer SA (1993a) Fiber optic sensor for predicting the cutting time of coagulating milk for cheese production. *Transactions of the ASAE*, **36**, 841–847.

Payne FA, Hicks CL, Shen PS (1993b) Predicting optimal cutting time of coagulating milk using diffuse reflectance. *Journal of Dairy Science*, **76**, 48–61.

Pérez-Marín MD, Garrido-Varo A, Serradilla JM, Núnez N, Ares JL, Sánchez J (2001) Chemical and microbial analysis of goat's milk, cheese and whey by near infrared spectroscopy. In: *Near Infrared Spectroscopy: Proceedings of the 10th International Conference* (Davies AMC, Cho RK, eds). West Sussex, UK: R Publications, pp. 225–228.

Pillonel L, Luginbühl W, Picque D, Schaller E, Tabacchi R, Bosset JO (2003) Analytical methods for the determination of the geographic origin of Emmental cheese: mid- and near-infrared spectroscopy. *European Food Research and Technology*, **216**, 174–178.

Qin Z, Xu CH, Zhou Q, Wang J, Fang X, Sun S (2004) Quality analysis of powdered milk by two dimensional correlation infrared spectroscopy. *Chinese Journal of Analytical Chemistry,* **32**, 1156–1160.

Reh C, Bhat SN, Berrut S (2004) Determination of water content in powdered milk. *Food Chemistry*, **86**, 457–464.

Rodriguez Otero JL, Hermida M, Cepeda A (1994) Determination of fat, protein, and total solids in cheese by near-infrared reflectance spectroscopy. *Journal of AOAC International*, **78**, 802–806.

Safar M, Bertrand D, Robert P, Devaux MF, Genot C (1994) Characterization of edible oils, butters and margarines by Fourier-transform infrared-spectroscopy with attenuated total reflectance. *Journal of the American Oil Chemists' Society*, **71**, 371–377.

Sato T, Kawano S, Iwamoto M (1990) Detection of foreign fat adulteration of milk fat by near infrared spectroscopic method. *Journal of Dairy Science*, **73**, 3408–3413.

Skeie S, Feten G, Almoy T, Ostlie H, Isaksson T (2006) The use of near infrared spectroscopy to predict selected free amino acids during cheese ripening. *International Dairy Journal*, **16**, 236–242.

Tetra Pak (1995) *Dairy Processing Handbook.* Lund, Sweden: Tetra Pak Processing Systems AB.

Tsenkova R, Atanassova S, Toyoda K, Ozaki Y, Itoh K, Fearn T (1999) Near-infrared spectroscopy for dairy management: Measurement of unhomogenized milk composition. *Journal of Dairy Science*, **82**, 2344–2351.

Tsenkova R, Atanassova S, Itoh K, Ozaki Y, Toyoda K (2000) Near infrared spectroscopy for biomonitoring: Cow milk composition measurement in a spectral region from 1,100 to 2,400 nanometers. *Journal of Animal Science*, **78**, 515–522.

Turza S, Chen JY, Terazawa Y, Takusari N, Amari M, Kawano S (2002) On-line monitoring of rumen fluid in milking cows by fibre optics in transmittance mode using the longer NIR region. *Journal of Near Infrared Spectroscopy*, **10**, 111–120.

van de Voort FR (1992) Fourier transform infrared spectroscopy applied to food analysis. *Food Research International*, **25**, 397–403.

van de Voort FR, Sedman J, Emo G, Ismail AA (1992) A rapid FTIR quality control method for fat and moisture determination in butter. *Food Research International*, **25**, 193–198.

Wang YD, Veltkamp DJ, Kowalski BR (1991) Multivariate instrument standardization. *Analytical Chemistry*, **63**, 2750–2756.

Wittrup C, Nørgaard L (1998) Rapid near infrared spectroscopic screening of chemical parameters in semi-hard cheese using chemometrics. *Journal of Dairy Science*, **81**, 1803–1809.

Woo YA, Terazawa Y, Chen JY, Iyo C, Terada F, Kawano S (2002) Development of a new measurement unit (MilkSpec-1) for rapid determination of fat, lactose, and protein in raw milk using near-infrared transmittance spectroscopy. *Applied Spectroscopy*, **56**, 599–604.

Yang H, Irudayaraj J (2000) Characterization of semisolid fats and edible oils by fourier transform infrared photoacoustic spectroscopy. *Journal of the American Oil Chemists' Society*, **77**, 291–295.

Zhou Q, Sun S-Q, Yu L, Xu C–H, Noda I, Zhang X-R (2006) Sequential changes of main components in different kinds of milk powders using two-dimensional infrared correlation analysis. *Journal of Molecular Structure*, **799**, 77–84.

Cereals and Cereal Products

11

Birthe Møller Jespersen and Lars Munck

Introduction

Although early in the twentieth century emission spectroscopy played a crucial role in implementing atomic theory, nobody could have forecast the present widespread practical use of diffuse near-infrared (NIR) spectroscopy for the identification of the more mundane components of foods. Spectra from NIR spectroscopy multi-meters

Infrared Spectroscopy for Food Quality Analysis and Control
ISBN: 978-0-12-374136-3

were chemically evaluated by spectral inspection. Spectral data were correlated to predict specific chemical analytes by classical statistics and pattern recognition data models (chemometrics) such as partial least squares (PLS) regression and artificial neural networks (ANN).

Nowadays, NIR spectroscopy analyses have replaced wet analyses in the industry to a great extent. There is a rich literature that is mainly focused on the prediction of specific analytes by more or less global calibration models that have great commercial and technical value as indicators for food functionality. Details can be found in textbooks by Osborne *et al.* (1993), Burns and Ciurczak (2001), Williams and Norris (2001), Siesler *et al.* (2002), and Ozaki *et al.* (2007). In order to widen the perspective on NIR spectroscopy applications, this chapter will mainly focus on other less exploited applications such as differential spectral analysis (Jacobsen *et al.*, 2005; Munck, 2005, 2006, 2008) and batch classification through discriminate principal component analysis (PCA) for food functionality (Mark, 2001; Munck, 2005; Williams, 2007), which is also used for industrial single seed NIR spectroscopy sorting for quality. These explorative applications, whereby unknown phenomena can be detected and physiochemically explored, are available for any owner of an NIR spectrometer without the need for specific commercial calibration software.

In a global context, cereals are the most important food crops in the world with a combined yearly production in excess of 2 billion tons of grains. Quality characteristics are important in both national and international trade and will become even more important in the future as markets become more competitive due to increasing demand and prices for food, non-food, and energy (Munck, 2004). After the end of World War II in 1945 the large-scale international industrialization of the cereal production chains for food and feed raised demands for screening for physical, chemical, and technical quality as well as for healthy grains free from microbiological contamination. As an example, a malting company in Canada (Pitz, 1990) producing 375 000 tons of malt a year will require about 500 000 tons of barley.

If the average size of a barley unload from one farmer is about 75 tons, the yearly intake would represent about 6700 permits. On average for three samples offered, two samples are rejected because they do not fulfill quality specifications. A grain inspector should therefore visually evaluate, test for germination energy, and chemically analyze about 20 000 submitted samples each year. There is therefore a need for almost instant physicochemical and functional at-line or online analyses, which in the case of barley should combine analyses of moisture, protein, starch (extract), germination energy, and beta-glucan (malting resistance). Prior to the 1980s it was beyond common belief that such a multi-tasking endeavor could ever be possible through just one kind of an instrument for all cereals.

Comparison of mid-infrared (MIR) and near-infrared (NIR) spectroscopy

In analyzing solid and semi-solid samples from cereal grain and grain products by diffuse reflection, near-infrared (NIR) spectroscopy has a clear advantage in sampling over mid-infrared (MIR) spectroscopy because of the more effective sample penetration by light at shorter wavelengths. Sampling is less important in liquid matrices such as milk and brewers' wort, which are collected in transmission cells as short as 0.025 mm for MIR spectroscopy, compared with in the order of 1 cm for NIR spectroscopy (Meurens and Yan, 2002). Samples are most conveniently recorded dry either finely milled in an attenuated total reflection diamond attachment (Tønning, 2007, Figure 11.1b) or analyzed on a barium mineral plate in an infrared transmission microscope (Chen *et al.*, 1998; Philippe *et al.*, 2006). The mass of the involved atom in infrared spectroscopy determines the response absorption frequency of light, including the fundamental functions of vibration, symmetric stretching, asymmetric stretching, bending, rocking, wagging, and twisting (Miller, 2001). MIR spectra contain bands of all the fundamental vibrations. However, NIR spectra include only pure overtones of the stretching vibrations. The remaining vibration types are represented here only as combination tones of the most anharmonic vibrations mediated by the C–H, N–H, O–H, and SH bonds.

In Figure 11.1a, b NIR is compared with MIR spectroscopy by measuring a wheat flour sample in the reflection mode (Tønning, 2007). The information from an MIR spectrum at 4000–400 cm^{-1} (2500 nm–25 μm) in Figure 11.1b is reflected as a hologram in the NIR spectrum at 700–2500 nm in Figure 11.1a. The 1st and 2nd overtones of the fundamental overlapping stretching vibration of O–H and N–H in the IR spectrum around 3300 cm^{-1} (peak 1, Figure 11.1b) correspond to the NIR bands at 1465 and 1000 nm (Figure 11.1a). The 1st, 2nd, and 3rd overtones of the fundamental stretching vibration of C–H in the MIR spectrum around 2927 cm^{-1} (peak 2, Figure 11.1b) are reflected in the NIR at 1780, 1200, and 920 nm (Figure 11.1a).

The fingerprint region in MIR (Figure 11.1b) from 1900 to 700 cm^{-1} contains the amide I and II bands (peaks 3 and 4) and the characteristic C–O and C–N stretching bands (peak 5). These bonds are represented in NIR spectroscopy (Figure 11.1a)

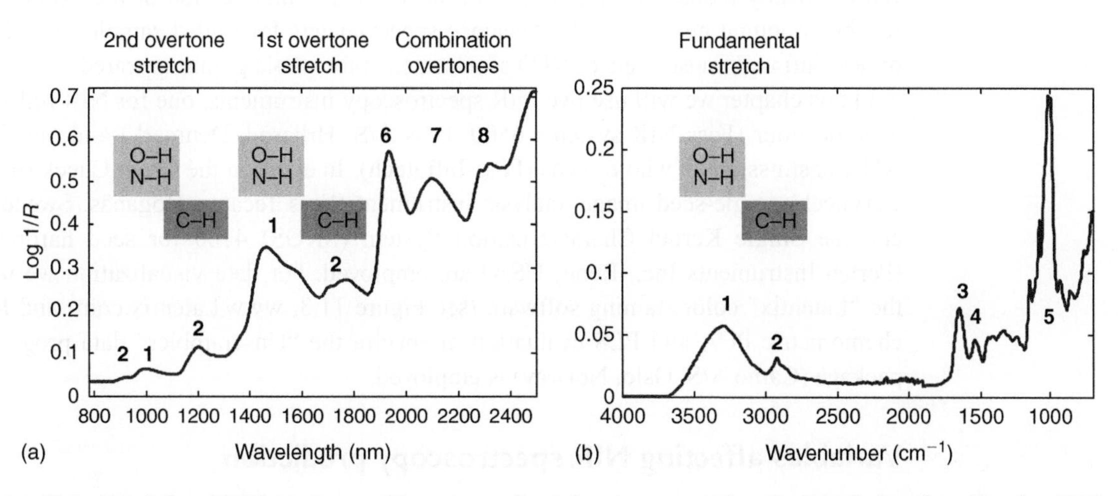

Figure 11.1 Near-infrared (NIR) spectrum and its corresponding infrared spectrum of a typical bread wheat flour (Tønning, 2007). (a) NIR reflectance spectrum (log(1/*R*) MSC) of typical bread wheat flour; (b) the corresponding IR spectrum of the same wheat flour sample. Selected vibrational bands in (a) and (b) assigned to: **1**: O-H and N-H stretch, **2**: C-H stretch, **3**: amide I at 1540–1660 cm^{-1} (C-N + C=O stretch), **4**: amide II at 1530–1540 cm^{-1} (N-H bend + C-N stretch), **5**: C-O and C-N stretch, **6**: O-H combinations and N-H combinations, **8**: C-H combinations.

as combination overtone bands such as for amide at 2100 nm (peak 7) and for C–H stretching from 2280 to 2330 nm (peak 8). The overtone bands dominate the NIR spectrum from 1900 nm and also include those of O–H and N–H at 1934 nm (peak 6).

Because of a more straightforward relationship between peaks and bonds in MIR compared with NIR, the first alternative is preferred for structural elucidation of purified components such as in the investigations on carbohydrates by Kacurakova and Wilson (2001) or those on arabinoxylanes from wheat endosperm by Robert *et al.* (2005). Chen *et al.* (1998) used an IR microscope in transmission mode to identify cell-wall mutants from lyophilized dry leaves of *Arabidopsis* and flax. However, Burns and Schultz (2001) found in a lignocellulose material that NIR spectroscopy had "an edge" over Fourier transform infrared (FTIR) spectroscopy on all counts: the correlation was better, the equations were more robust, and there were fewer errors.

Current practice in NIR spectroscopy for cereals and cereal products

Instrumentation

Modern NIR spectroscopy technology started in the 1950s when it was applied to cereals by its founding father, Karl Norris, at the USDA laboratory in Beltsville, USA (Hindle, 2001). Norris was probably the first to realize the practical potential of NIR spectroscopy, using the primitive instrumentation and computer technology available at that time, for developing moisture and protein determination in wheat. In the early 1970s reproducible NIR filter instruments became available (Williams, 1975). It is now possible to apply full-scale NIR spectroscopy, so that in principle the wet laboratory could be replaced, making dry instant analyses available in an instrument box, which can drastically reduce the time taken. In the 1980s the first version of the extremely reliable scanning near-infrared reflectance spectrometers for milled samples and that of near-infrared transmittance (NIT) spectrometers for whole grains appeared.

In this chapter we will use two NIR spectroscopy instruments, one for NIR reflectance on flour (Foss-NIR-systems 6500, Foss A/S, Hillerød, Denmark) and one for NIT transmission on whole seeds (Foss-Infratech). In addition the Grain Check (now Cervitech) single-seed image analysis instrument (Foss Tecator, Höganäs, Sweden) and the Single Kernel Characterization System (SKGS) 4100 for seed hardness (Perten Instruments Inc., Reno, USA) are employed. For data visualization we use the "Latentix" color staining software (see Figure 11.3, www.Latentix.com) and for chemometric PCA and PLS evaluation of spectra the "Unscrambler" data program package (Camo A/S, Oslo, Norway) is employed.

Variables affecting NIR spectroscopy prediction

As outlined by Williams and Norris (2001) there is a need for a strictly defined standardization routine for NIR spectroscopy measurements to be successful. The measurements are dependent on the reproducibility of the instrument with regard to slit, wavelength, and photometric scale, the instrument temperature control and the relative humidity. Williams (2001) identified over 50 variables that influenced NIR spectra in a complex way and documented 24 of them in spectral plots (Williams, 2007).

Over a period of 10 years we have had positive experience with the very high reproducibility over time with our Near Infrared Reflection Instrument NIR-Systems 6500 (Foss A/S, Hillerød, Denmark) that is used in applications in this chapter. The finely milled seed samples are measured through a quartz window with a rotary cup at 16 references and 32 sample scans with the reference setting at 0.09691. Temperature in the laboratory is between 23.5 and 24.5°C and every second wavelength in a scan 400–2500 nm is retained. There is no spectral smoothing. To check the quality of the light source to indicate when a change of lamp is necessary, a pectin standard is measured before each measurement at 1934 nm and the height of the peak is noted. Deviations on the third decimal are typically $\log(1/R)$ 0.516–0.518. In addition a measurement cell with a permanent barley sample is measured regularly several times during each experiment. After measurement the reproducibility of the barley standard is checked in a PCA score plot for sample classification together with the other measurements. The standard measurements should then occupy a narrow overlapping circle in the plot where the standards overlap.

In our laboratory we condition all the whole grain samples to the ambient relative humidity of the air in the laboratory by open storage for a few days. In this way moisture in a sample within an experiment is a specific expression of its water activity, which is of great importance in interpreting NIR data. The samples have a water content of typically 9–11% for barley.

Spectral pre-treatment

A minor part of the stochastic noise in NIR data is due to instrumental noise. The major part is due to physical scatter (Dahm and Dahm, 2001). Variables related to the hardness of the seed influence both whole seed transmission (NIT) spectra and reflection (NIR) through the particle distribution of the milled seed sample. Thus NIR allows effective calibrations for hardness in milled wheat samples (Williams, 2002). As shown earlier it is not a problem to reduce the stochastic variation in $\log(1/R)$ NIR spectra by the classical first and second derivative pre-treatment techniques, or by chemometric methods such as multiple scatter correction (MSC) (Martens and Næs, 2001) and the extended inverted signal correction (EISC) used on NIT spectra of single wheat seeds by Pedersen *et al.* (2002). It is possible to separate the chemical information from the physical by the difference between raw and pre-treated (e.g. MSC) spectra (Jacobsen *et al.*, 2005). The extended multivariate scatter correction (EMSC) (Martens *et al.*, 2003) allows a further more sharp separation and quantification of the chemical and physical influences in NIR spectra. EMSC works through prior knowledge of the specific patterns of spectra from the pure chemical components.

NIR spectra as a reproducible physiochemical fingerprint

In order to empirically demonstrate the relative specificity of chemical information in NIR spectra, it is important to span the variation in chemical composition.

Table 11.1 Chemical composition (% dry matter of the barley mutant material) (Munck *et al.*, 2004) used in the near-infrared (NIR) spectral studies in Figures 11.2–11.4 and 11.10–11.17. C = carbohydrate mutants; P = protein, high-lysine mutants; N = normal barley cultivars. The mutants are listed in Figure 11.14

		Dry matter	Protein	Amide	A/P	BG	Fat	Starch
All	*n*	92	84	81	81	88	20	35
	Mean	90.66 ± 0.97	16.08 ± 2.22	0.4 ± 0.2	15.19 ± 2.22	8.06 ± 4.79	2.48 ± 0.76	45.18 ± 8.78
	Max	93.02	9.7	0.2	9.46	2.2	1.66	27.3
	Min	88.91	22.28	0.6	18.58	20	3.77	60.4
Greenhouse	*n*	69	69	69	69	69	15	25
	Mean	9.71 ± 0.996	16.85 ± 1.48	0.42 ± 0.07	15.44 ± 2.20	8.45 ± 4.91	2.36 ± 0.68	42.85 ± 8.24
Field	*n*	23	15	12	12	23	6	10
	Mean	90.42 ± 0.86	12.53 ± 1.43	0.28 ± 0.05	13.71 ± 1.78	6.95 ± 4.36	2.77 ± 0.92	51.01 ± 7.55
C	*n*	29	26	25	25	29	9	7
	Mean	91.62 ± 0.75	16.98 ± 2.13	0.42 ± 0.06	15.32 ± 0.81	14.21 ± 2.9	2.71 ± 0.76	32.1 ± 5.8
P	*n*	23	23	18	18	23	3	15
	Mean	90.18 ± 0.49	15.77 ± 2.33	0.29 ± 0.05	11.67 ± 1.54	3.79 ± 1.36	3.5 ± 0.06	44.75 ± 4.81
N	*n*	40	40	37	37	40	9	14
	Mean	90.22 ± 0.79	15.64 ± 2.09	0.42 ± 0.07	16.77 ± 0.78	5.72 ± 0.99	1.91 ± 0.16	52.16 ± 4.25

An example on barley endosperm mutants is described in Table 11.1 (Munck *et al.*, 2004), where all chemical variables except dry matter have a very wide range. The variation is caused by both genetics and environment. Carbohydrate C (high beta-glucan, low starch) and protein P (low prolamine/amide, high lysine, low beta-glucan) endosperm mutants and normal N barley controls are grown in two extreme environments: field F (low protein, high starch) and greenhouse G (high protein and beta-glucan, low starch). In Figure 11.2a log(1/*R*) MSC corrected NIR spectra of milled seeds from four barley genotypes selected from Table 11.1 grown in the field, are displayed for the complex C–H part of the first overtone region 1680–1810 nm. They represent genotypes from the classes C (mutant *lys5f*), N (Bomi), and P (*lys3a*). Spectra from the latter genotype are replicated from separate plots to check reproducibility. The 1680–1810 nm region was selected because it could perfectly classify the genotypes (C, N, P) as described above. The positive and negative standard correlation coefficients (*r*) for every second wavelength to the protein, amide, starch and beta-glucan analyses for the whole material listed in Table 11.1 are displayed as curves in Figure 11.2b at 1680–1810 nm. The wavelength-dependent correlation curves verify empirically the finely tuned specific chemical information that is available in NIR spectroscopy.

Spectral assignments selected from the literature (Osborne *et al.*, 1993; Williams, 2001) are displayed in Figure 11.2b. Starch gives high correlations in the 1750–1780 nm range while those for protein peak at 1685–1705 nm and at 1710 and 1734 nm, for amide at 1690 and 1745 nm and finally for beta-glucan at 1683, 1705, and 1735 nm. The protein band at 1734 nm (Williams, 2001) was confirmed. The C and the P mutants are high in oil (Table 11.1). The oil peaks at 1724 and 1762 nm are confirmed in Figure 11.2a for the *lys5f* sample that has 50% more oil than Bomi.

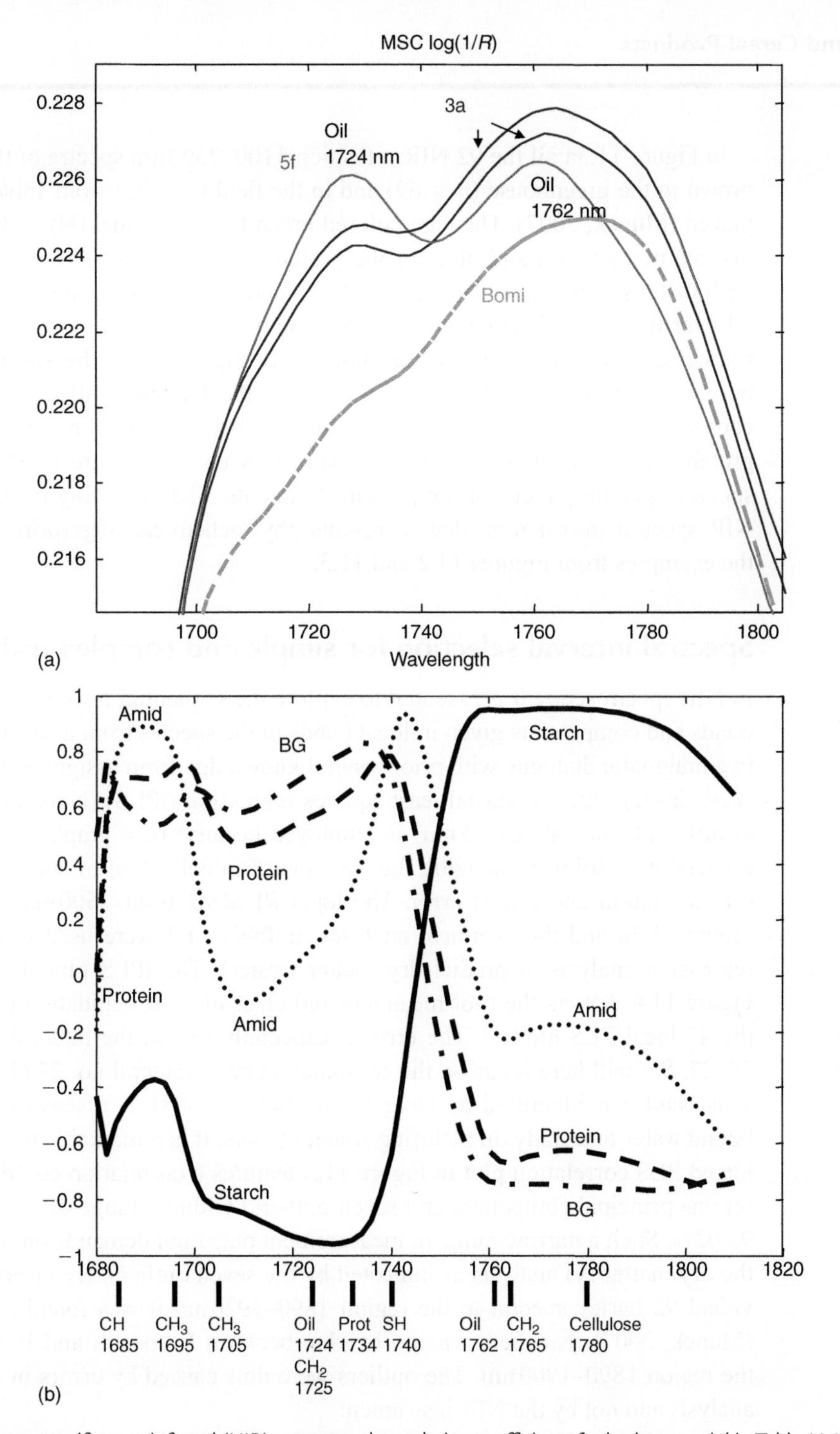

Figure 11.2 Mutant-specific near-infrared (NIR) spectra and correlation coefficients for barley material in Table 11.1. (a) Examples of mutant-specific NIR spectra (log(1/*R*) MSC) 1700–1810 nm from the barley material in Table 11.1 (field): Normal N Barley, cv. Bomi and its P (protein) mutant 3a ($n = 2$, *lys3a*) and C (carbohydrate) mutant 5f (*lys5f*). The peaks at 1724 and 1762 nm of the 5f and 3a spectra indicates the high oil (fat) content in the C and P mutants described in Table 11.1. (b) Simple correlation coefficients (*r*) between every second NIR wavelength in the 1680–1820 nm area and four chemical analyses from the whole barley material in Table 11.1. BG = beta-glucan. Assignments of wavelengths according to the literature (Osborne *et al.*, 1993; Williams, 2001) for comparison to the mutant-specific spectra in Figure 11.2a are given at the bottom of the figure. (see Plate 11.1 for colour version)

In Figure 11.3a all the 92 NIR reflection 1100–2500 nm spectra of the barley lines grown in the greenhouse ($n = 69$) and in the field ($n = 23$) from Table 11.1 are displayed (Munck, 2007). They are colored green for the normal (N), red for the carbohydrate mutants (C) and blue for the protein mutants (P) genotypes. It is clear that each of these spectral categories displays genotype specific patterns for N, C and P all over the spectral range and that the genetic differences dominate over the environmental ones (field versus greenhouse). In Figure 11.3e the spectra are stained by a beta-glucan color gradient (2–19%) by the Latentix software visualizing the carbohydrate mutant category that is high in beta-glucan. Similarly a low amide/protein (A/P) index (9–19 units) is visualized by color in Figure 11.3f, indicating the specific spectral pattern of the protein P mutants. The versatility of the finely tuned NIR spectral information that represents physiochemical fingerprints is evident by the examples from Figures 11.2 and 11.3.

Spectral interval selection for simple and complex traits

In NIR spectroscopy it is essential to explore the sequential information on chemical bonds and components given in local bands in the spectra from a specific experiment in a pragmatic dialogue with prior general knowledge from assignment tables (Shenk *et al.*, 2001). Interval partial least squares regression (iPLS) (Nørgaard *et al.*, 2000) at different intervals (5–35 nm) is employed because of a graphic interface that is especially useful in visualizing the "hot spots" in collection of spectra for the highest correlation and lowest error. The log(1/*R*) MSC 1100–2500 nm NIR spectra in Figure 11.3a and the chemical analyses in Table 11.1 were used in a 30-nm iPLS regression analysis to predict dry matter (water). The iPLS plot for dry matter in Figure 11.4 depicts the root mean squared error of cross-validation (RMSECV) for the 47 local PLS models. The error is especially low in the intervals of 8, 10, and 25–27. We will here focus on the combination band interval no. 27 (1890–1920 nm). This band was identified by Gergely and Salgo (2003) as a sensitive indicator for bound water in a study on maturing wheat kernels. It is remarkable that the cross-validated PLS correlation plot in Figure 11.5 features a correlation coefficient $r = 0.96$ (at one principal component and seven outliers) within a range of dry matter 88.91–93.02%. Such a narrow range in measurement puts high demands on the precision of the dry matter (y) analysis as indicated by the seven outliers. By inspecting the individual 92 barley spectra in the region 1890–1920 nm it was found in Figure 11.3c (Munck, 2007) that there was no overlap between C spectra and P + N spectra in the region 1890–1900 nm. The outliers were thus caused by errors in the dry matter analysis and not by the NIR instrument.

The spectral classification for dry matter is verified by the Latentix software in a dry matter gradient from 89 to 93% in Figure 11.3d. The feasibility of the visual classification of spectra explains that there is only need for a simple PLS model with one principal component (PC). It is amazing to find by inspection (see insert in Figure 11.5) that the total mean spectral response C versus P + N spectra at 1905 nm is only 4×10^{-4} log(1/*R*) and that the mean difference in dry matter analysis between the C samples on one side and the P + N samples on the other is as low

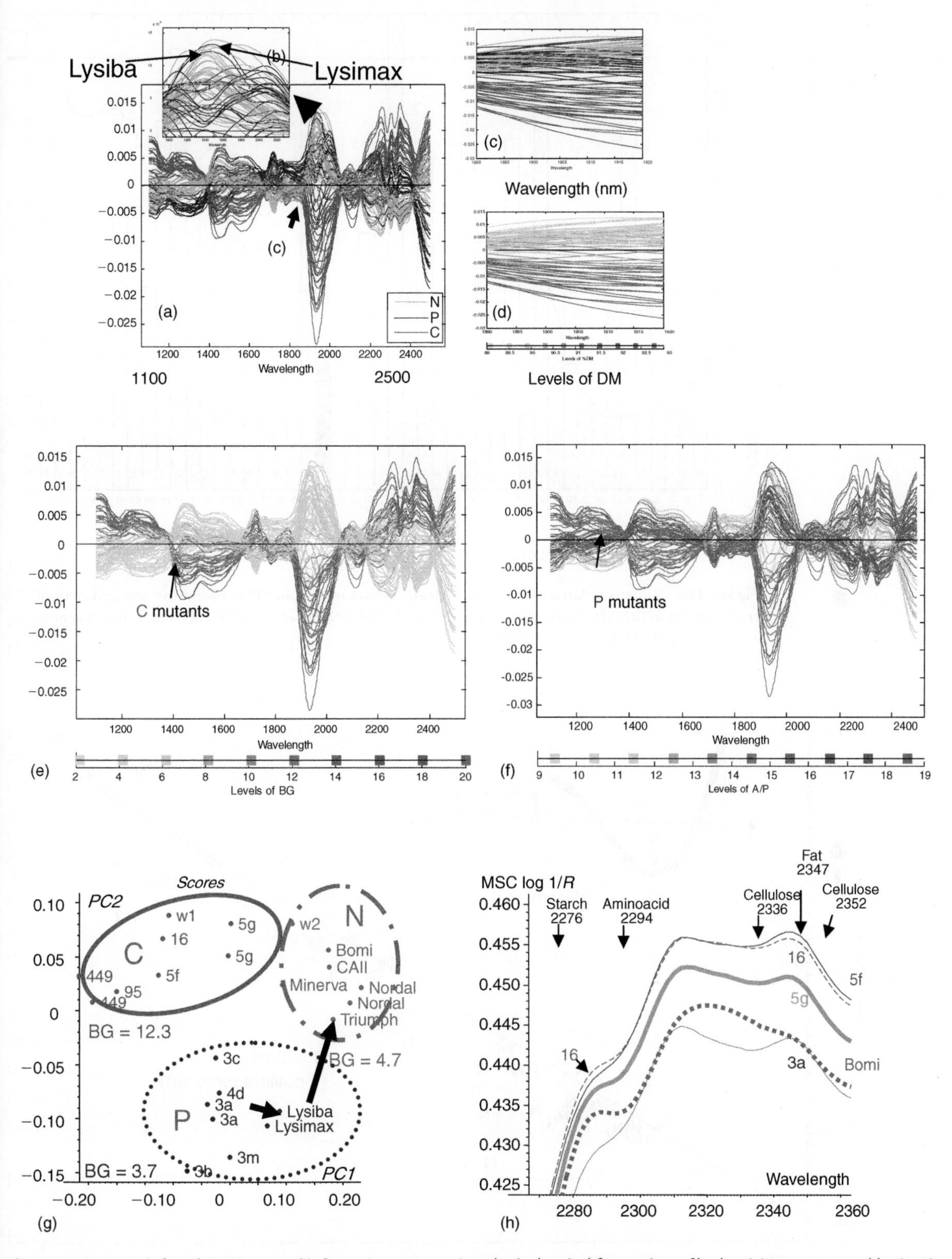

Figure 11.3 Near-infrared (NIR) spectral information representing physiochemical fingerprints of barley. (a) Mean centered log(1/*R*) from 92 barley seed samples. Green = normal barley (N), blue = protein (high lysine, P), and red = carbohydrate (C) mutants. (b) Enlargement of peak at 1935 nm; P outliers Lysimax and Lysiba. (c) Interval 1890–1920 nm from (a). (d) Marking of spectra from (c) for dry matter 89–93%. (e) Coloring for beta-glucan (2–20%) of the spectra in (a). (f) The same for amide/protein index (9–19). (g) Principal component analysis (PCA) classification of 23 field barley NIR spectra (1100–2500 nm) for normal barley (N) and protein (P) and carbohydrate mutants (C). BG, β-glucan % dry matter. (h) NIR spectra (2270–2360 nm) of Bomi (N) and its P *lys3a*, and C mutants *lys5f*, *lys5g* and 16 grown in greenhouse. a–d and g–h. (From Munck, 2007 with permission from Wiley & Sons Ltd.) (see Plate 11.7 for colour version)

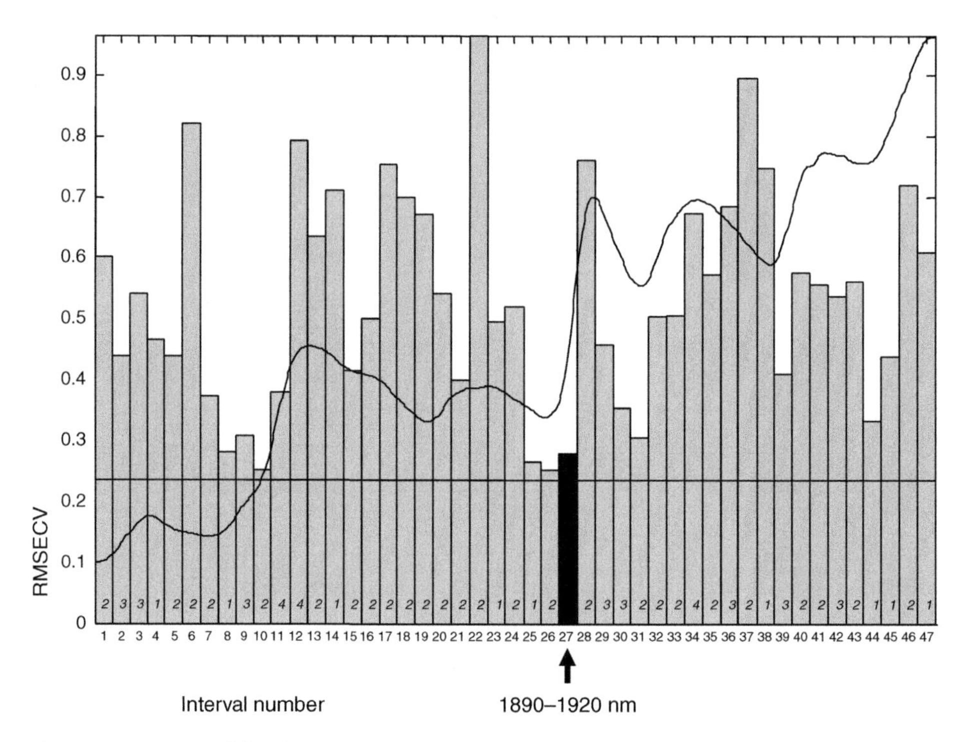

Figure 11.4 Cross-validated interval partial least squares regression (iPLS) root mean squared error of cross-validation (RMSECV) plot for dry matter on 69 barley NIR spectra 1100–2500 nm from a greenhouse as listed in Table 11.1 (Munck, 2006).

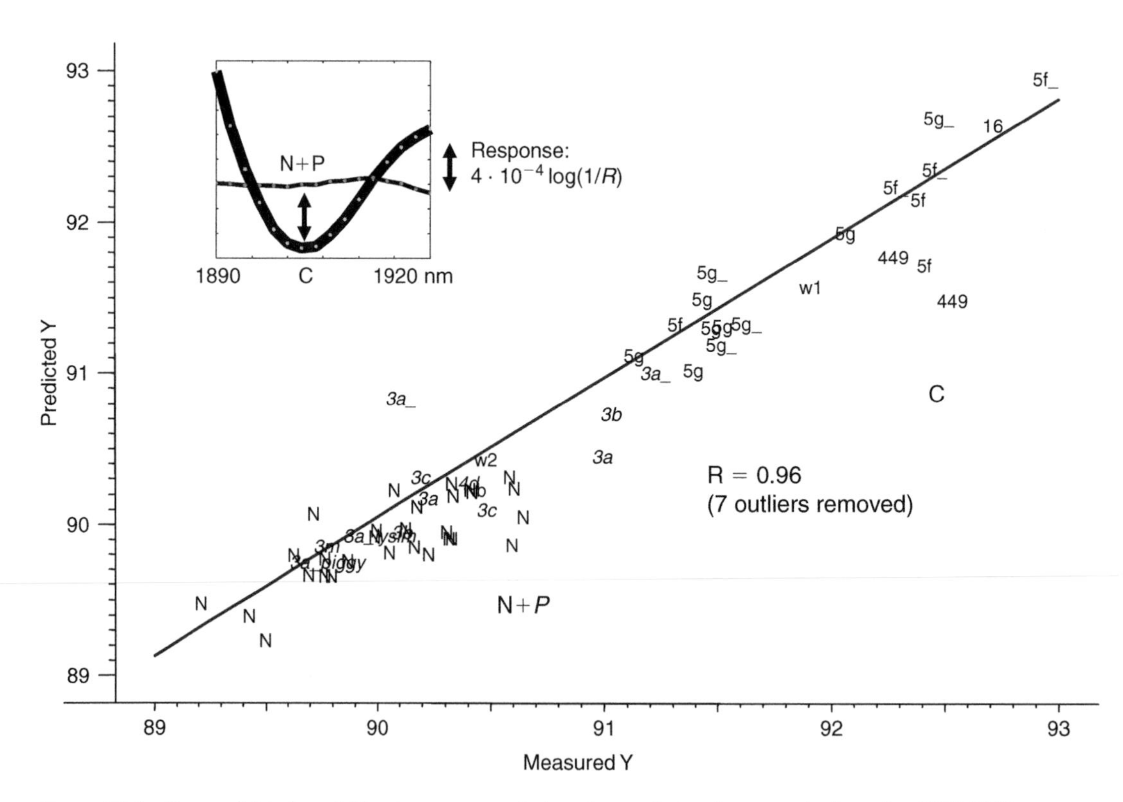

Figure 11.5 Cross-validated partial least squares correlation plot predicting dry matter in the region 1890–1920 nm selected by interval partial least squares regression (iPLS) from the material in Figure 11.3 (Munck, 2006).

as an absolute percentage of 1.5. This example verifies that water can be sensitively detected by NIR due to its dipolar qualities (Miller, 2001) that also play a fundamental role in the chemistry of biological networks. Water binding, detected and studied by NIR spectroscopy was thus named aqua-photomics by Tsenkova (2007).

Spectral ranges in the barley mutant material can, in analogy with iPLS, be classified by interval extended canonical variate analysis (iECVA) (Nørgaard *et al.*, 2006; Munck, 2007) for genotype (C, N, and P) and for environment greenhouse (G) and field (F) as demonstrated in the iECVAs in Figure 11.6a, b. The *y*-axis depicts the number of misclassifications at 35 nm spectral intervals along the *x*-axis 1100–2500 nm. It can be seen that there are broad areas from zero to a few misclassifications of the genotype, which are all associated in a remarkable way to vibrations from C–H bonds. First, around 2280–2330 nm in Figure 11.6a the combination overtone area (Figure 11.1a) represents the fingerprinting region in MIR 1900–700 cm^{-1} (Figure 11.1b). Second, the second overtone stretch area 1130–1300 nm and the second part of the first overtone stretch region 1675–1800 nm in Figure 11.6a reflect the fundamental C–H stretch vibrations around 2927 cm^{-1} in MIRS as shown in Figure 11.1b. The "hot" area for genetic classification 1680–1810 nm was chemically validated with success by the correlations in Figure 11.2b. The finely tuned genotype specific spectra in Figures 11.2a and 11.3a thus have a solid chemical significance.

The iECVA classification in Figure 11.6b for the environmental differences (G, F) demonstrates large differences between intervals. The area with the best result is six intervals from four to zero misclassification around 2200 nm. The iECVA approach can also be used for spectral classification of complex traits other than mutant genotypes such as food functionality, e.g. baking volume to identify the spectral "hot spots."

NIR seed batch analytical applications

Prediction of single analytes

Wheat

NIR spectroscopy application pioneer Phil Williams at the Canadian Grain Commission introduced large-scale analysis of ground wheat and soybeans for moisture, protein and oil in the early 1970s (Williams, 1975). The commercial market in the world trade of grain raw materials needs global calibration models for specific analytes such as water (%) and protein (%) that are included in trade contracts. However, the disadvantage is that there is a need for a separate calibration for each commodity and constituent in a calibration set of thousands of samples. This is an expensive and tedious affair (Buchmann *et al.*, 2001; Williams, 2007). Calibration models are therefore often developed by the instrument manufacturers (Buchmann *et al.*, 2001) and included as an essential, however, quite costly part of NIR spectroscopy instrument packages that are tailored to the specific needs of traders, plant breeders, and the cereal industry. Alternatively the instrument maker can associate with a network of users such as in the Danish Near-Infrared Transmission (NIT) calibration network for wheat and barley (Buchmann *et al.*, 2001) for providing the calibration models.

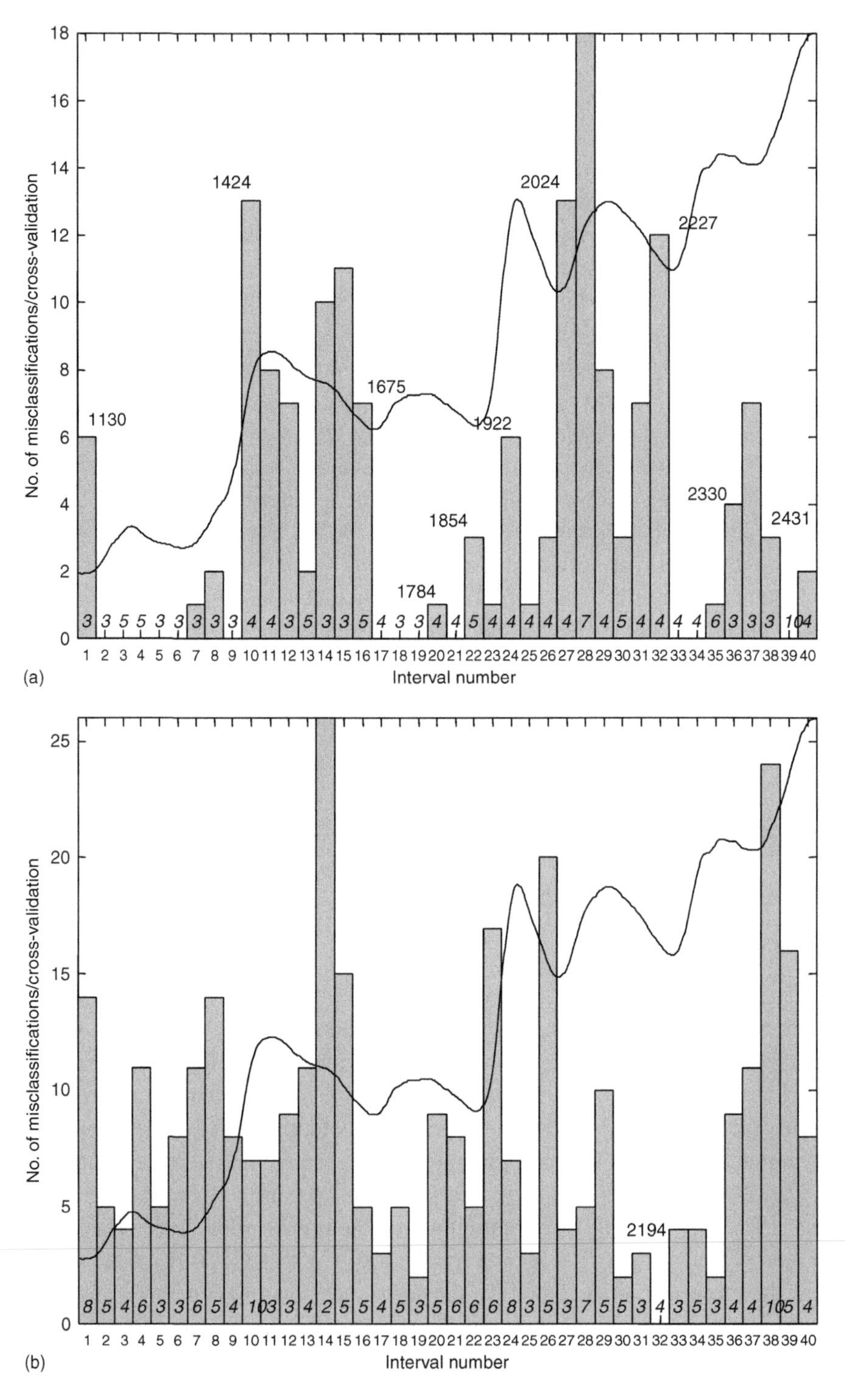

Figure 11.6 Cross-validated 35 nm interval extended canonical variate analyses (iECVAs) for 1100–2500 nm (log(1/*R*) MSC) on the 92 barley samples in Table 11.1 (Nørgaard *et al.*, 2006). x = interval number; y = number of misclassifications. (a) Classification of genotypes N, C, and P. (b) Classification of environment G (greenhouse) and F (field).

For wheat processing, the Australian scientist Brian Osborne (Osborne *et al.*, 1993) has pioneered many NIR spectroscopy applications summarized in reviews on wheat milling (Osborne, 2007) and bread making (Osborne, 2001). The ability of NIR spectroscopy to measure protein level, hardness and starch damage by estimating free and bound hydroxyl groups in starch and water in flour may explain the ability of NIR spectroscopy to predict water absorption in flour (Delwiche and Weaver, 1994). NIR spectroscopy is used to discriminate between amylopectin (waxy) starch and amylose (Delwiche and Graybosch, 2002). An in-depth NIR spectroscopic study on the chemical aspects of gluten in relation to water uptake has recently been published by Bruun *et al.* (2007a, 2007b).

Barley

Meurens and Yan (2002) has given a thorough overview of the use of NIR reflectance and NIT spectroscopy in the brewing industry with regard to barley, malt, hops, and yeast raw materials, and also in the mashing and fermentation processes. The hardness, protein, starch, and beta-glucan parameters are important parameters to predict malt quality, e.g. in pilot maltings where extract yield and wort viscosity (high beta-glucan) are most important (Pram Nielsen, 2002; Møller, 2004b). Malting quality is greatly influenced by the physiological aspects of germination such as the viability (live or dead) and vigor (potential growth rate) of the germ, which are difficult to predict by NIR spectroscopy from the ungerminated kernel. Speed of germination is an essential marker for enzyme production triggered by the excretion of plant hormones from the germ. Beta-glucanases break down the endosperm cell walls in the process of malt modification (Nielsen and Munck, 2003; Munck and Møller, 2004) and alpha amylase is produced to be exploited in the brewhouse in breaking down starch to produce wort with a high extract yield.

Rice

Rice is mainly used for cooking after de-hulling the seed into brown rice that is polished to white rice, with a yield of about 67% (Bergman *et al.*, 2003). Rice fat from bran and testa may produce taste problems with rancidity. Starch constitutes about 90% of the milled rice. The amylopectin (glutinous) and amylose (firm and non-sticky) components in starch determine cooking quality and gelatinization temperature. The sensory attributes of cooked rice such as smell, mouth feel, and flavor are of great practical importance. NIR spectroscopy is used in rice in general, e.g. for moisture and protein (Natsuga, 1999), to predict rice lipid acitivity and surface lipid content (Chen *et al.*, 1997; Li and Shaw, 1997) as well as grain cooking quality. Prediction of the important amylose component for cooking quality has been reported (Delwiche *et al.*, 1995; Shimizu *et al.*, 1999) with varying degree of success.

According to Bergman *et al.* (2003), NIR spectroscopy for selection of rice by cooking quality is practised by rice breeders with a defined breeding material, however, large-scale genetically broad evaluation programs by NIR in the US and Australia have stopped. In the Japanese rice industry (Tanaka *et al.*, 1999) special NIR spectroscopy rice "taste analyzers" have been developed to produce test scores related to a combination of the parameters of amylose, protein, moisture, and lipid activity. However, these NIR spectroscopy "tasters" are designed to measure the rice preference of Japanese consumers and are thus unsuited for use in other markets (Champagne *et al.*, 1996).

Maize

NIT and NIR reflectance spectrometers are widely used in the industry to measure moisture, protein, starch, and fiber content in whole maize and in maize products (Paulsen *et al.*, 2003). NIR spectroscopy prediction models are available for the estimation of extractable starch for the wet milling industry and for predicting true density, seed breakage susceptibility and kernel density in maize. Quality in maize is mainly associated with the handling of the seed to avoid cracks during drying, which influence the ability to store, and to avoid mold and insects that are related to seed size and hardness. There are a wide range of starch and protein mutants in the diploid maize species in analogy with barley, some of which are exploited in the industry (e.g. waxy and high amylose maize).

Use of NIR spectrometers independent of commercial calibration models

It is remarkable that the overwhelming majority of NIR spectrometer owners are not aware of that their instruments can be used in a great many applications for both prediction and classification (Williams, 2007; Munck, 2008) without commercial software. The instrument industry has been reluctant in promoting such an opportunity, probably because they do not want to distract their customers from using the commercial calibration models. However, there should be no conflict but instead an expansion in the use of NIR instruments for the new classification applications. As pointed out by Williams (2007), the manufacturers should benefit in sales by marketing the whole technology instead of focusing on instruments and calibration software for limited uses.

Prediction models for food functionality in local cereal laboratories

The Kjeldahl protein analysis has had a dominating economic importance in trade contracts in the last 100 years as an indicator for high gluten content in wheat and for low extract in malting barley. However, high protein content in a wheat variety does not always imply high baking performance because baking functionality depends on many other factors besides protein. Plant breeders have now developed environmentally improved wheat varieties that need less nitrogen fertilizers in spite of having an excellent baking performance and yield at low protein content. However, remarkably, these varieties cannot be officially acknowledged (e.g. in Germany) because they do not live up to the required standard protein level of premium wheat that is needed in official testing and by the market.

Wet gluten, dough performance, and baking volume can now be estimated by NIR spectroscopy (Williams, 2002, 2007). These functional analyses by NIR spectroscopy, where wavelengths for the protein contribution are included as an integrated part in the spectral signatures for complex quality should now substitute for crude protein as a standard for quality in wheat. There are no established global calibrations available yet for the more complex quality parameters to be used in trade. Until such calibrations are obtained the individual industrial laboratory should be trained in making local calibrations that are continuously improved (Munck, 2008) to be used to upgrade quality control within the company.

Spectral classification with reference samples independent on commercial calibration data

The French chemometrician Dominique Bertrand from Nantes and his group probably are the first to apply the explorative NIR classification (discriminant analysis) for wheat variety identification (Bertrand *et al.*, 1985). Kim *et al.* (2003) used a modified least squares regression analysis of NIR spectroscopy to check the authenticity of Korean grown rice versus imported rice for short and medium long grains.

Strategy in the physiochemical validation of classification While many representative samples in calibrations and in test sets are needed to get reliable predictions of single analytes by PLS and ANN, there is another situation if one just wants to verify differences between (seed) samples. For classification of NIR spectra, a PCA score plot (Martens and Martens, 2000) will be able to differentiate between classes of samples with different chemical composition as demonstrated in the following examples. There are two options for validation that can be used separately or combined. The first is to introduce extreme control samples with known quality in the material to be tested. They should all be grown together in the same field. In selecting unknown samples, the low- and high-quality variants will then appear near to the high and low calibration sample respectively in a PCA score plot. As a second alternative, validation of NIR spectroscopy data from a PCA score plot is made in a separate PCA bi-plot, on physical and chemical laboratory analyses from the same samples. The discrimination ability of the two PCA plots is then compared, followed by integration in a PLS-2 NIR spectroscopy prediction (x) of all the physiochemical parameters as dependent variable (y). Here the aim is to explore the combined data-structure in a PLS score plot and to validate the significance of the y-parameters by Jack-knifing (Martens and Martens, 2000) after the most obvious outliers have been identified and removed in leverage and residual plots (Munck and Møller, 2004; Jacobsen *et al.*, 2005).

Classification example by NIR spectroscopy for understanding quality concept in malting barley In malting barley production, climatic differences between years influence malting quality as much as variety. This was investigated by Møller (2004a, 2004b) in the PCAs in Figures 11.7 and 11.8 comprising intact seed samples of the barley varieties A (Alexis), B (Blenheim), and M (Meltan) grown in 1993–1998 marked **3** to **8** (Figures 11.7a, b and 11.8). A PCA (Figure 11.7a) on the 63 NIT (850–1150 nm) for whole seed spectra by the Foss Infratech 1255 Analyzer is clearly discriminating between two groups of barley grown in 1996, 1997, 1998, and 1993 versus those grown in 1994 and 1995. There was a general consensus in Denmark that the malting barley harvest in 1994–1995 was difficult to modify into

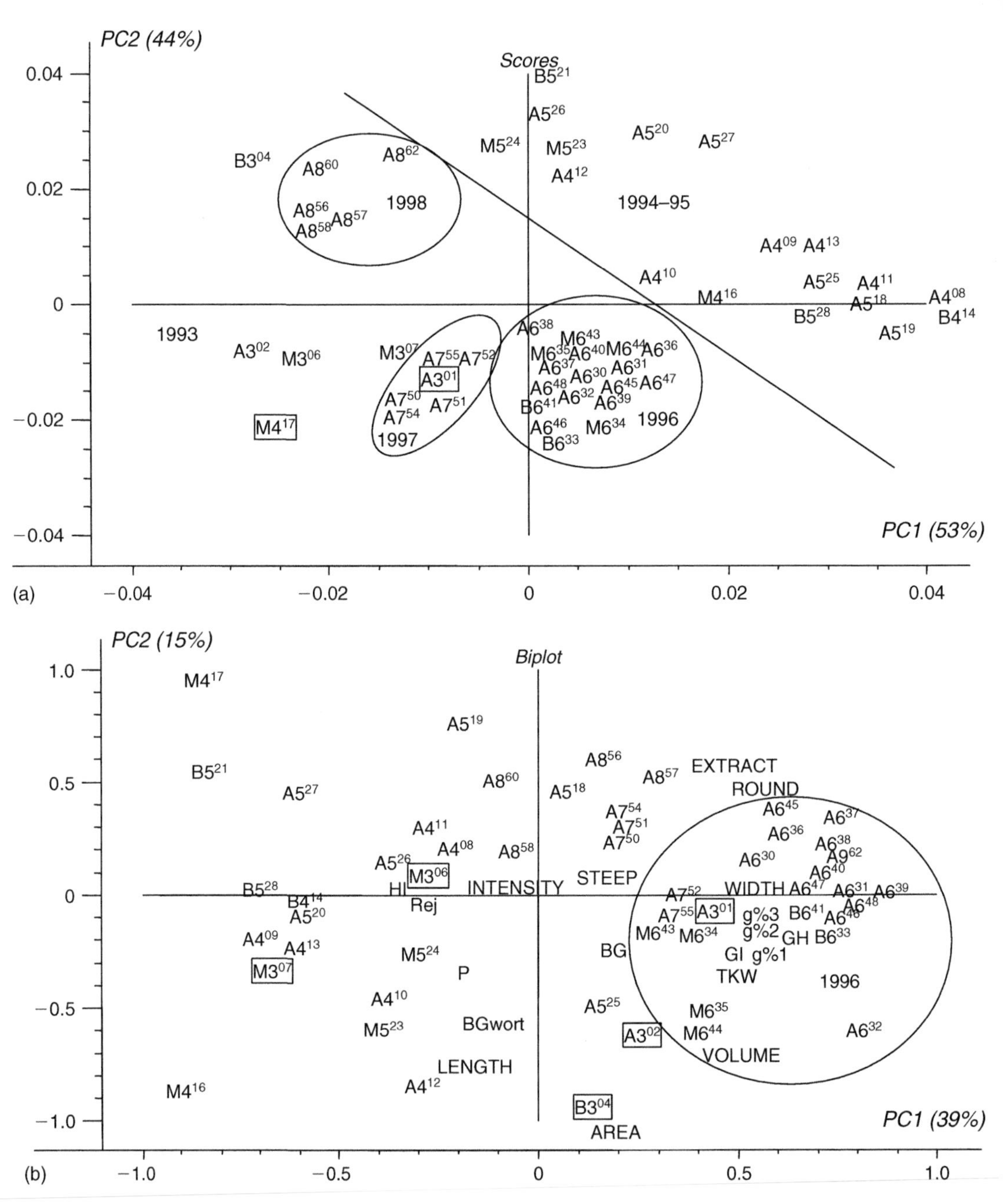

Figure 11.7 Cross-validated principal component analysis (PCA) score plot and bi-plot (Møller, 2004b). (a) Cross-validated PCA score plot for whole seed batch near-infrared transmittance (NIT) (first derivate) spectra (850–1050 nm) on 63 malting barley varieties (A = Alexis, B = Blenheim, M = Meltan) grown in different years 1993 to 1998 (3–8). (b) Cross-validated PCA bi-plot for physiochemical analyses of the material in Figure 11.7a involving nine grain check parameters (single seed (mean): area, length, volume, width, round (ness), (color) intensity). SKCS 4100 single seed hardness index (mean) (HI); rejected seeds (Rej); water uptake during steeping (STEEP); germination % day 1 (g%1), day 2 (g%2), day 3 (g%3); pilot malting data (EXTRACT); % beta-glucan in wort (BGwort); beta-glucan in barley (BG), protein (P) and thousands kernel weight (TKW).

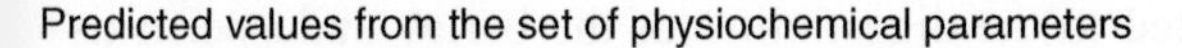

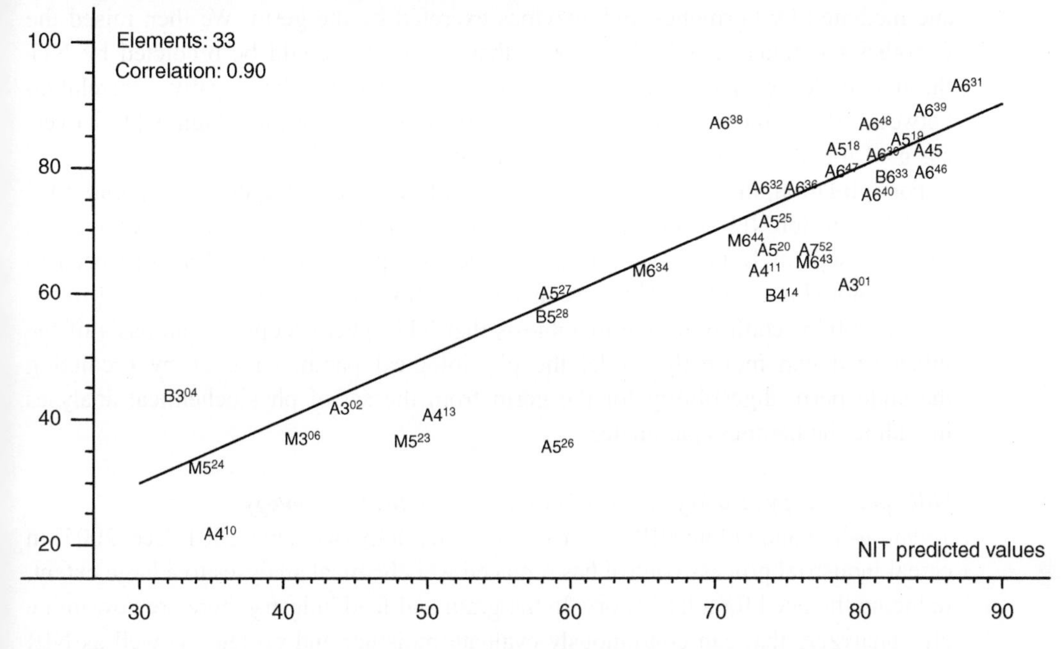

Figure 11.8 Near-infrared transmittance (NIT) prediction of "vigor" 1 day germination (g%1) (*x*-axis) correlated to the same prediction (*y*-axis) by physiochemical analyses in the material from Figure 11.6 (Møller, 2004b).

malts. The NIR results in Figure 11.7a were now validated as suggested above by 19 physiochemical variables including 10 seed form and hardness (HI) parameters from the single seed grain analyzers as well as germination percentage for 1–3 days (i.e. g%1–g%3), germination index (GI) and homogeneity (GH), water uptake (steep) and by extract percentage and beta-glucan in wort (BGwort) and beta-glucan in barley (BG) that require pilot maltings.

From the cross-validated PCA bi-plot on these 19 parameters displayed in Figure 11.7b it can be seen that samples from the favorable malting quality year 1998 form a cluster (encircled) standing out as an independent verification of the pattern of NIR spectroscopy data displayed in the separate PCA in Figure 11.7a. Near to the encircled 1998 samples to the right in Figure 11.7b we find the parameters ROUND, VOLUME, WIDTH, TKW, and EXTRACT, all associated with fast malting verified by the adjacent germination parameters g%1–g%3, GI, and GH. The slower germinating barleys from 1994–1995 to the left are associated with the negative parameters longer (LENGTH), harder seeds (HI) and Bgwort, all inversely correlated to the favorable quality characters to the right.

It was surprising to find that NIT spectroscopy could predict germination rate g%1 by PLS (Møller, 2004a, 2004b). A physiological variable would not be expected to be predicted by NIT data from the ungerminated kernel, bearing in mind that germ weight constitutes only 2% of the seed. Vigor as well as malt modification is dependent on the substrate availability that is needed for the emerging plant to grow. However, seed hardness and beta-glucan resistance to break down (BG wort) are variables that should be related to the digestibility of the barley endosperm tissue mediated by hormones and enzymes excreted by the germ. We then raised the hypothesis (Munck and Møller, 2004) that vigor g%1 could be predicted by NIT through endosperm digestibility. This was indicated by the negative correlation between HI together with BGwort to the left in the PCA bi-plot (Figure 11.7b) versus g%1 together with EXTRACT to the right as discussed above.

For validation, the g%1 result for each sample was predicted by two separate PLS models, one for NIT spectroscopy and the other for the physiochemical parameters. The two separately predicted values for each sample were plotted against each in an x–y plot (Figure 11.8). This reveals a correlation coefficient of $r = 0.90$ (Møller, 2004a, 2004b) confirming the hypothesis that NIT spectroscopy on the basis of the intact seed can indirectly model the physiological parameter g%1 by predicting the endosperm digestibility for the germ from the set of physiochemical analyses including the hardness parameter.

NIR spectroscopy as a key element in process analytical technology

Today, at-line and online NIR spectroscopy technology (Kemeny, 2001; Lee, 2007) in cereal industrial process control has replaced wet chemical analyses to a large extent, reducing the need for a laboratory. In the grain and feed industry there are now inline NIT analyzers that can continuously evaluate moisture and protein, as well as NIR spectroscopy equipment for online control of, for example, extract and alcohol during brewing and fermentation. In principle, the development of a spectral fingerprint from a window in a process can be followed as a trajectory by the self-modeling PCA algorithm without a standard (Munck *et al.*, 1998). The integration among NIR spectroscopy instruments, chemometrics, and knowledge about processes is now taking place as "process analytical technology" (PAT) with multi-way (Smilde *et al.*, 2004) chemometric models including the time dimension built on n-way PLS (Bro, 1996) and PARAllel FACtor analysis (PARAFAC). Allosio *et al.* (1997) demonstrated the use of NIR spectroscopy in the reflectance mode (1100–2500 nm) evaluated by PARAFAC in the transformation from barley to malt.

As an example, the process control in a wheat flourmill by multi-block modeling of NIR spectroscopy data is discussed below. The wheat flour data set (Nielsen *et al.*, 2001) is modeled in a multi-block PLS analysis (Berg, 2001; Bro *et al.*, 2002) as displayed in Figure 11.9a, b. Flour samples from six different positions in the mill marked 1–6 were sieved into six fractions marked with a–f, making a total of 42 flour fractions including the origin. The samples were analyzed by NIR at 1100–2500 nm. The NIR information (Figure 11.9a) was included in two blocks: the original information in the upper block to the left and in a second SNV block below, to emphasize chemical information over scatter by the standard normal variates (SNV) transformation. These two NIR blocks are now used as predictors for the two blocks to the right in Figure 11.9a, i.e. chemical composition including damaged starch (five variables upper block) and for particle distribution as analyzed in the laboratory by laser-scatter size distribution (lower block).

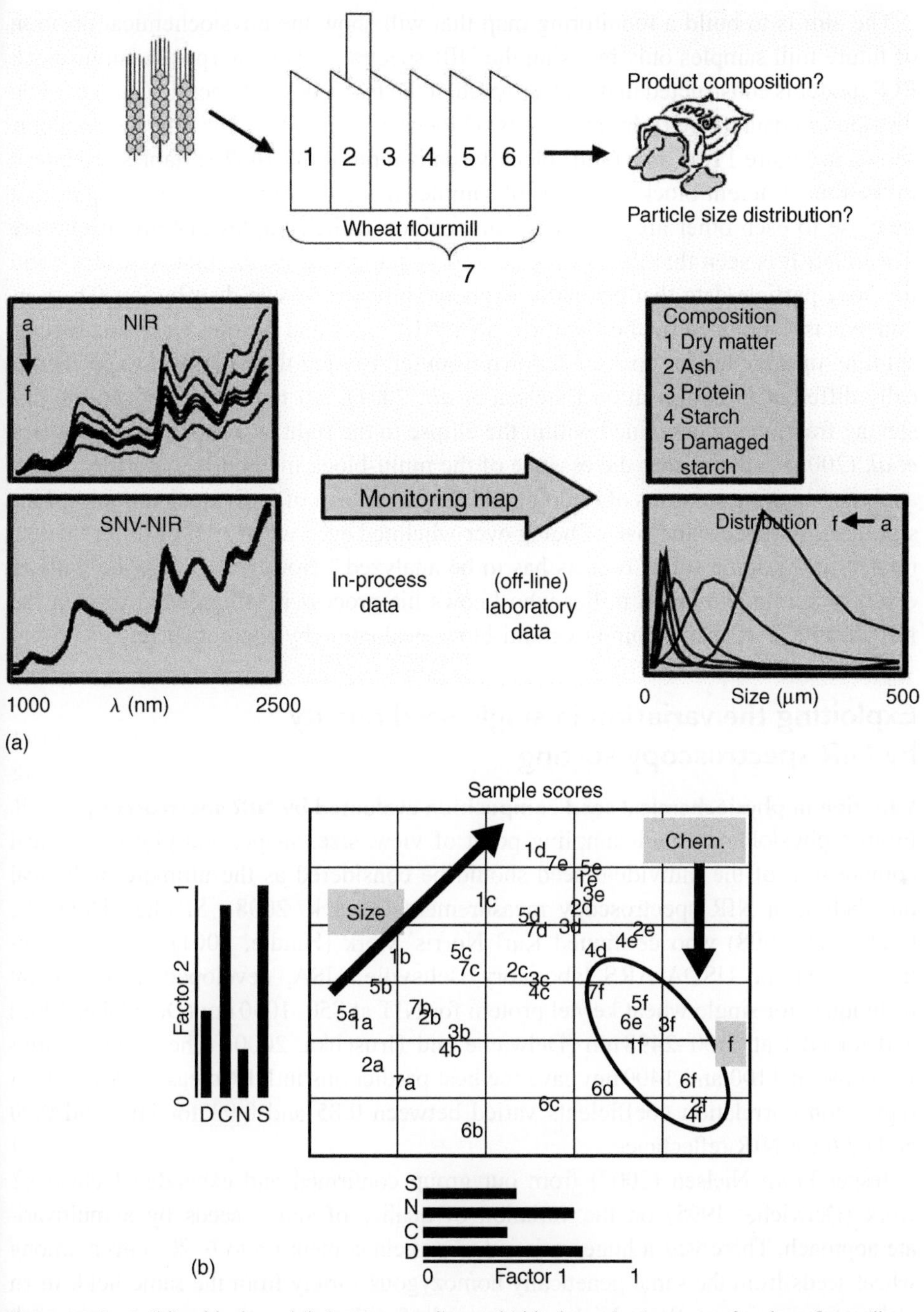

Figure 11.9 Building blocks and their corresponding multi-block monitoring-map of a wheat flourmill data set (Berg, 2001; Bro *et al.*, 2002). (a) Building blocks of a wheat flourmill data set (Nielsen *et al.*, 2001) (see explanation in the text). (b) Partial least squares multi-block monitoring map for the wheat flourmill data set in Figure 11.9a. S = SNV-NIR, N = NIR, C = chemical composition, and D = particle size distribution.

The aim is to build a monitoring map that will show the physiochemical position of future mill samples only by using the NIR spectra. For this purpose a multi-block PLS model is constructed that seeks to predict the two blocks of chemical and particle distribution data to the right from the two blocks of NIR data to the left. The model is shown in Figure 11.9b as a multi-block visualization of the 1437 variables organized in the four different blocks in terms of samples marked by their labels. Samples that lie close to each other are similar in composition while samples that are distant are dissimilar. It is seen that the horizontal axis is dominated by the first NIR block and the laser particle data that primarily explains the particle size distribution. The second axis is dominated by the chemical SNV-NIR block and the chemical data blocks, splitting up samples for position 6 down from left to right that is known to be chemically different in composition (Nielsen *et al.*, 2001). All the samples from the "f" sieving fraction are contained within the ellipse to the right as a separate cluster. Bro *et al.* (2002) explains that the essence of the multi-block method is to perform "data analysis, thinking in terms of building blocks rather than of individual variables. This significantly reduces the risk of being over-whelmed even when a lot of different data related to the same set of objects has to be analyzed." However, a dialogue with an experienced open-minded miller who knows his process is still needed to gain the full advantage of such a complex multi-block evaluation by chemometrics.

Exploiting the variation in single seed quality by NIR spectroscopy sorting

Variation in physiochemical seed composition evaluated by NIR spectroscopy

From a physiological and sampling point of view, size, shape, and physiochemical composition of the individual seed should be considered as the ultimate biological unit behind a NIR spectroscopy measurement (Munck, 2008). Stephen Delwiche (Delwiche, 1998) who continued Karl Norris' work (Hindle, 2001) on NIR spectroscopy at the USDA-ARS laboratory Beltsville, USA, developed measurement techniques for single wheat kernel protein for NIT at 850–1050 nm (Delwiche, 1995) and for NIR at 1100–2498 nm (Delwiche and Hruschka, 2000). The spectral interval between 1100 and 1400 nm gave the best predictions in NIR measurements. PLS regression correlation coefficients varied between 0.85 and 0.93 for NIT and 0.90 and 0.96 for NIR reflectance.

Jesper Pram Nielsen (2002) from our group confirmed and expanded Delwiche's work (Delwiche, 1995) on the variation of quality of single seeds by a multivariate approach. There was a huge variation in protein content up to 6–20% even among wheat seeds from the same genetically homozygous variety from the same field. In an elaborate experiment, Pram Nielsen measured 15 different quality parameters on each of 523 wheat kernels collected from 43 different wheats grown at two locations in Denmark (Pram Nielsen *et al.*, 2003). Single seed laboratory Kjeldahl protein and density measurements for reference were combined with single seed measurements with the Infratech instrument (850–1050 nm), the Grain Check single seed imaging analyzer, and the Perten SKCS 4100 single seed hardness meter. There was a great variation in single seed hardness (from 28.8 to 101.5 hardness units), protein (6.8–17.0%), and density (0.99–1.25 g/cm^3). The PLS regression correlation coefficients were 0.98 for protein, 0.76 for virtuousness, 0.70 for density, and 0.59 for hardness.

Development of automated single seed NIR spectroscopy sorters

A commercial laboratory single seed sorter Luminar 3076 ("Seed Meister" NIR analyzer, Brimrose, MD, USA) for plant breeding use was launched in 1996. NIR devices for laboratory single seed sorting work for moisture and protein in maize and soybean (Armstrong, 2006) and in rice (Rittiron *et al.*, 2004) have also been developed. Recently Dowell *et al.* (2006) have published data on sorting by another automated NIR system for selecting individual kernels in a laboratory scale, based on PLS calibrations. The seed-sorting system was designed for plant breeders. It is able to differentiate and sort between 100% amylopectin (waxy)- and amylose-containing kernels in a segregating seed population for the waxy gene in proso millet. The protein content in a batch of wheat seeds could be increased by 3.1 absolute percentage points higher than the low protein fraction. In sorting for hardness the hardness index was 29.4 hardness units higher in the hardest fraction than in the fraction with the softest kernels. Delwiche *et al.* (2005) in single grains confirmed Williams (2002) indication in bulk wheat that it is possible to obtain a considerable reduction of deoxynivalenol by sorting *Fusarium*-infested grains.

It should be attractive in plant breeding and industry to exploit the extreme single seed variation in quality for sorting that has been shown for univariate variables in the NIR spectroscopy literature by PLS prediction. However, it is now possible to advance a step further by the pilot scale NIR/NIT TriQ single seed sorter (Bomill AB, Lund, Sweden) (Löfqvist and Pram Nielsen, 2003; Munck, 2008). The TriQ machine sorts directly for complex functional traits such as baking quality in wheat by classification without the elaborate data model development needed for univariate parameters such as protein. The pilot machine with a capacity of 2–500 kg/h shown in Figure 11.10a is based on a cylinder indent machine with pockets (Figure 11.10b) that position the seeds for individual NIR analysis. The NIR spectroscopy information is then classified according to the functional trait by the data program and connected for sorting to an air jet ejector.

Results for a batch of seeds from the spring wheat variety Vinjett are shown in Table 11.2. These were fractionated on a single seed basis into these fractions and were analyzed for flour dough and baking quality parameters (Tønning *et al.*, 2007). The fractionation difference in quality from F1 (yield: 30.4%), F2 (33.3%) to F3 (36.3%) demonstrates a significant systematic increase in falling number, water absorption, Zeleny value, wet gluten, flour protein, bulk grain density, and baking volume, confirming the data from another single seed wheat fractionation experiment by the same manufacturer published by Munck (2008). There is now industrial-scale single seed quality classification capacity (Figure 11.10c) (Pram Nielsen and Löfqvist, 2006) for complex functional characteristics such as malting quality, baking quality, and *Fusarium* infection with the full-scale TriQ-20 single seed NIR/NIT (2 tons per hour) sorter module from Bomill A/B as shown in Figure 11.10b. By integrating five TriQ-20 units in one machine a stunning single seed NIR sorting capacity of 10 tons per hour can be obtained.

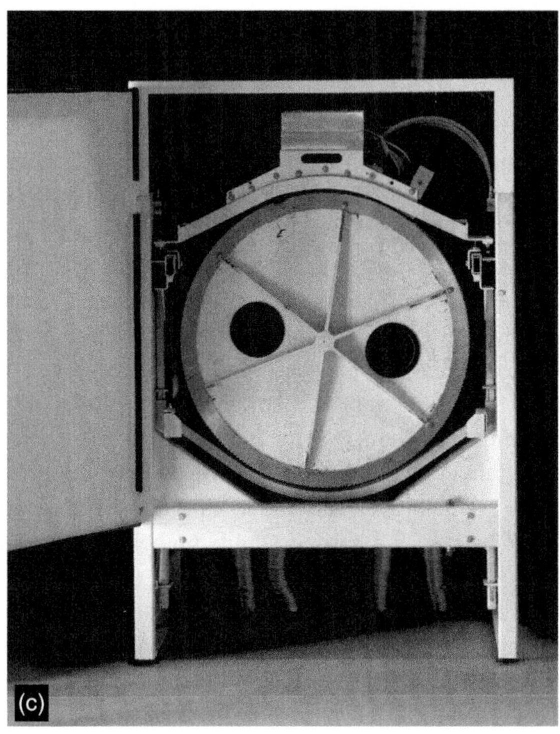

Figure 11.10 Single seed near-infrared (NIR) spectroscopy sorters. (a) Bomill pilot (2–500 kg h^{-1}) TriQ single seed NIR spectroscopy sorting machine (Bomill AB, Lund, Sweden) used in single seed sorting in wheat in Table 11.2. (b) Close-up of cylinder of machine in Figure 11.10a showing pockets to position the seeds before NIR measurement. (c) The industrial-scale 2 tons h^{-1} TriQ-20 single seed NIR spectroscopy sorter unit (Bomill AB, Lund, Sweden). The door is open to show the cylinder. (see Plate 11.2 for colour version)

Table 11.2 The quality of three fractions (F1–F3) from single seed sorting of wheat by the pilot near-infrared (NIR) TriQ sorter (Bomill AB, Lund, Sweden) shown in Figure 11.10 (Tønning *et al.*, 2007). Flour quality analyses are based on 14% moisture

	F1	F2	F3
Bread volume	2.6	2.7	3.1
Bulk grain density	80.2	81.1	81.3
Flour protein (dry matter)	9.5	10.4	11.9
Wet gluten (%)	20.2	21.3	26.5
Zeleny number	26.0	28.8	30.5
Water absorption (%)	51.1	51.5	53.5
Falling number seconds	372	404	404

Reproducibility of NIR spectra from multi-grain samples as influenced by genotype and environment

It is not yet known how precisely one can select for complex traits in early single seed generation breeding with the new equipment. Careful seed sampling is of paramount importance (Tønning *et al.*, 2006) in order to obtain a representative sample to estimate the correct protein value for the batch. However, sampling is getting another edge when moving from the univariate (protein) to the multivariate (e.g. baking quality) perspectives on single kernels compared with those on seed batches. A whole NIR spectroscopy fingerprint of a seed sample is much better in representing the intrinsic quality of identity (the origin = genetics + environment) than a single chemical component. Thus a NIR spectroscopy fingerprint can be used to ascertain homogeneity regarding origin when delivering from a silo that is supposed to contain many loads from one variety grown in the same area.

In measuring seed batches with NIR reflectance spectroscopy, the environment tends to influence spectral offset to a higher degree than genetics that seems to be more pattern-specific (Munck *et al.*, 2001). The single seeds within an advanced barley or wheat line that are naturally inbred should be homozygous and almost 100% genetically uniform. As shown above, the seed batches from the barley mutant material in Table 11.1 can be classified by 35 nm iECVAs on NIR 1100–2500 nm both for genetic (Figure 11.6a) and for environmental differences (Figure 11.6b). The underlying single seed composition within each batch must be extremely variable. The crucial step to be taken to relate single seed NIR pattern with those from the bulk is to define the consistent spectral patterns for origin (genotype + environment) in single seeds from a seed sample and to prove how they are reproduced in bulk. It was shown that contamination of normal seeds in a mutant line detected in a PCA on NIR spectroscopy data from bulk could be verified by visual inspection of the spectra and seeds (Munck *et al.*, 2001).

As demonstrated in Figure 11.11a, b, we were impressed by the extremely high reproducibility of NIR reflectance spectra (1680–1810 nm) from milled batches of seeds, here exemplified by two lines 0404 and 1105 of the homozygous barley mutant *lys5g* grown in the greenhouse and in the field. The two lines were propagated separately in over 10 generations. The two overlapping spectra to the left in Figure 11.11a (enlarged in Figure 11.11b) demonstrate the extreme reproducibility of the greenhouse environment displaying line 0404 grown in 1998 and in 1999. The two lines of mutant *lys5g* 0404 and 1105 (below to the right) grown outdoors in the same field in 2000 also closely reproduce each other's spectrum. Because of uncontrolled climate, it is impossible to reproduce spectra from different years in field trials as for greenhouse. In spite of that, as seen in the color graph in Figure 11.3a (Munck, 2007) of all 92 barley spectra (1100–2500 nm) from the material in Table 11.1, the barley genotype N, P, C specific patterns dominate over the field/greenhouse environment effect. To demonstrate the genotypic reproducibility of spectra from mutants *lys3a* (P mutant), *lys5f* (C mutant) and parent variety Bomi (N normal) over three greenhouse cultivations in 1998, 1999 and 2000, Figure 11.12 focuses on the spectral region 2260–2380nm (Munck, 2006). The spectra are normalized to the control genotype *lys5g* set equal to zero. The high reproducibility of the spectra from the three barley genotypes is evident.

It is now clear that spectral patterns in bulk samples grown in a controlled environment are specific for each homozygote barley genotype and mutant. The complex morphological and physiochemical seed emergence that is behind each population of seeds seems to have an almost computational significance as documented by the high

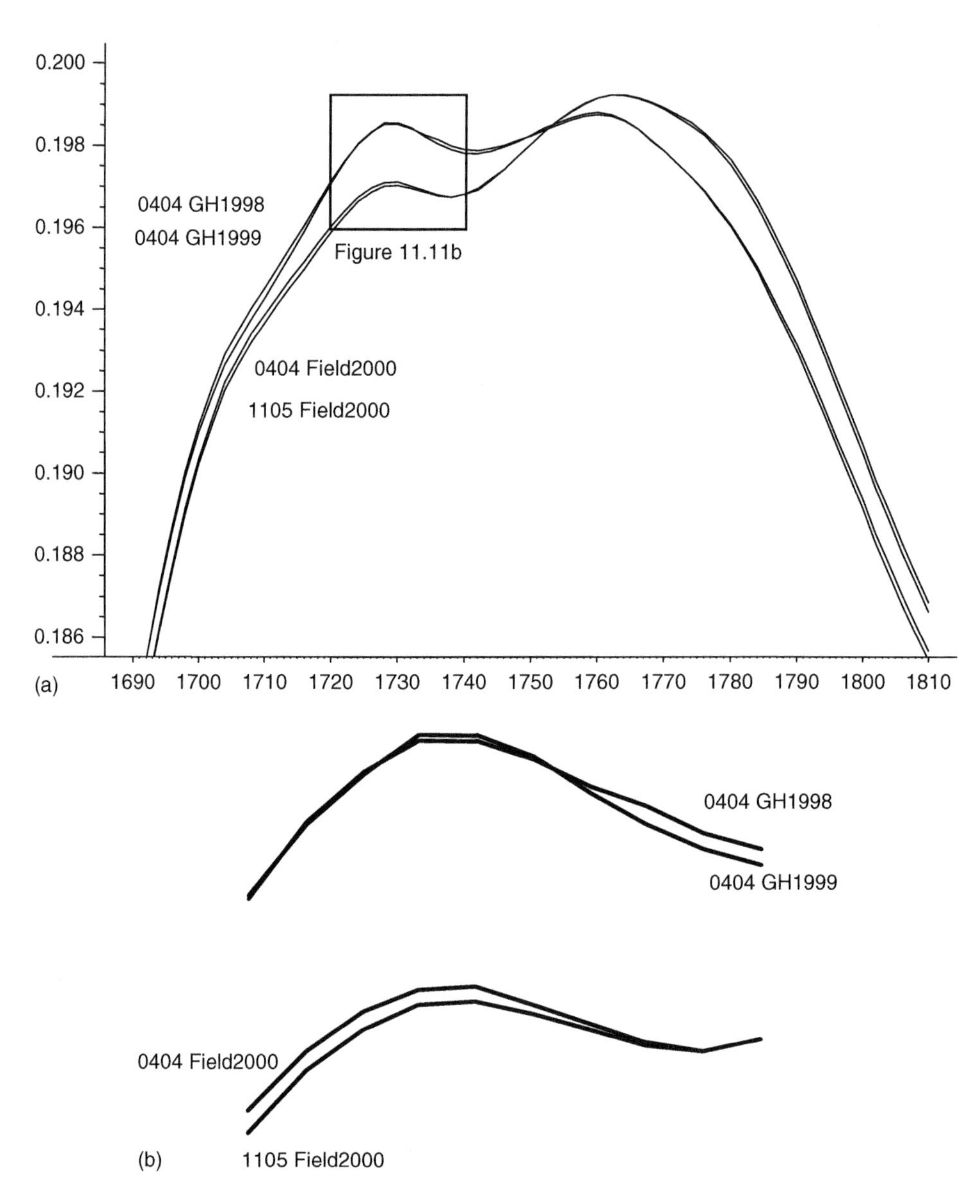

Figure 11.11 Four near-infrared (NIR) spectra 1690–1810 nm (log(1/*R*) MSC) for two parallel lines (0404 and 1105) with the *lys5g* carbohydrate C mutant included in the barley material in Table 11.1. To the left above line 0404 grown in greenhouse (GH) in 1998 and 1999. To the left below line 0404 and 1105 grown in the field in 2000. (a) Whole spectra; (b) enlargement of the four spectra within the square in Figure 11.11a to show reproducibility within environment.

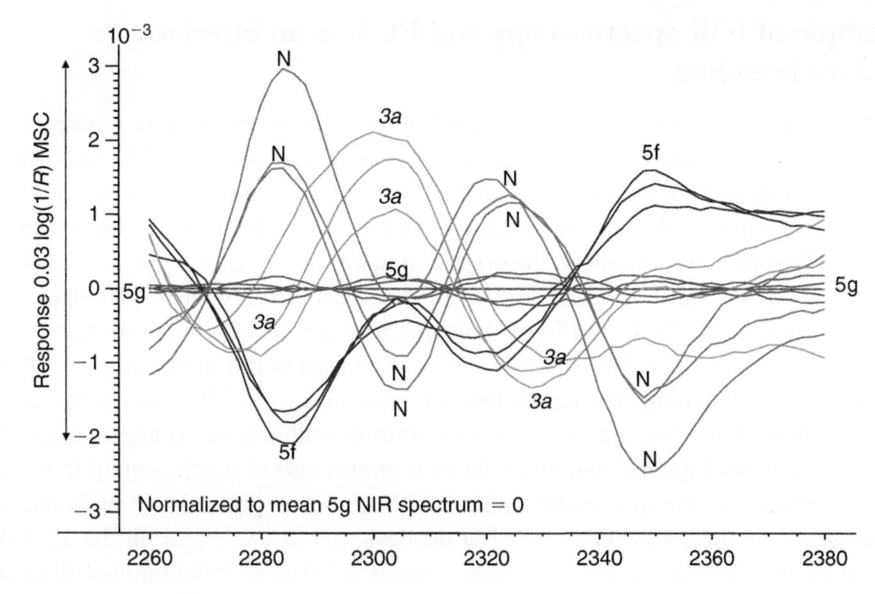

Figure 11.12 Differential (log(1/*R*) MSC) spectra 2260–2380 nm featuring a normal (N) control (cv. Bomi) and three barley endosperm mutants 3a (*lys3a*), 5f (*lys5f*) and 5 g (*lys5g*) grown in three years (1998–2000) in the greenhouse environment (Munck, 2006). The mean spectrum of 5g spectra is the reference equal to zero in the plot subtracted from the other spectra. (see Plate 11.3 for colour version)

reproducibility of the NIR spectra from the genotypes in Figures 11.2a, 11.3a, 11.11 and 11.12. In analysis it requires that sampling is correct.

Classification of NIR spectroscopy data by principal component analysis

Explorative classification of NIR spectroscopy data

The traditional role of the plant breeder in selecting cereal lines is to optimize the performance pattern of the whole plant, using the tools of visual observation. However, in the last 50 years science, technology, and trade have increasingly focused on causal relationships for economic value, with univariate variables as indicators for quality as discussed for plant breeding by Osborne (2006). In the NIR literature before 2001 there were few publications that used classification. Delwiche *et al.* (1995) used NIR reflectance spectroscopy and PCA to classify amylose content in rice and Campbell *et al.* (2000) employed the same approach by NIT for maize endosperm mutants. However, two-way causal analysis by PLS was used by Wang *et al.* (1999) to predict by NIR the number of dominant *R* alleles for color in wheat, and by Delwiche *et al.* (1999) to identify wheat translocations from NIR spectra, as well as by Kim *et al.* (2003) for the authentication of Korean rice. The PLS correlation coefficients were used for discrimination between categories. The informative score plot option in PCA is also available in PLS for classification as shown in Figure 11.9b on wheat milling. This option was not used, however, in the examples discussed above.

Example of NIR spectroscopy and PCA as an interface by data breeding

As described by Munck (2008) a range of barley endosperm mutants (Table 11.1) were selected for lysine by a dye-binding method in the 1960s and 1970s in order to improve nutritional quality in barley. One of those, the P mutant *lys3a* (Risø mutant 1508), had an improved amino acid (lysine) composition with a nutritive value near to that of animal proteins. However, the original mutant was lower in yield, seed quality, and starch content but had a higher oil percentage compared with its parent variety Bomi (Munck *et al.*, 2004). A 15-year breeding program was started at the Carlsberg Research Center in 1975. This material will here be used to test the efficiency of NIR spectroscopy in breeding. The barley lines in Table 11.3 ($n = 15$) grown in the same field are divided into four groups + normal controls with respect to improvement in starch content, seed quality, and yield. There is an increase in starch content from the original mutants in group 4 to the commercial varieties in group 1, which is due to selection for plump seeds. Figure 11.13a displays a PCA score plot on the 15 NIR spectra (1100–2500 nm) using Foss-NIR systems 6500 from whole-milled flour of the samples in Table 11.3. The original *lys3* mutants (3a, 3 m, group 4, 48.7% starch) in the third quadrant have been improved in starch as represented by the movement of the recombinants Lysimax and Lysiba (group 1, 52.6% starch) in the plot towards the high starch variety Triumph down to the left (normal group 54.6% starch).

The PCA classification of NIR spectroscopy data (Figure 11.13a) is verified by a separate PCA bi-plot (Figure 11.13b) on the chemical data from Table 11.3 as suggested earlier. The variable "starch" that is situated near to the Triumph and Lysiba and Lysimax samples in Figure 11.13b indicates a relatively high starch content in these samples. In a second validation step the NIR spectroscopy (x) and starch (y) data are combined in a cross-validated PLS starch prediction plot as shown in Figure 11.14a. The starch-improved Lysimax and Lysiba in the circle are displayed as intermediate in starch between the *lys3* genotypes to the left and the normal controls to the right, including a Triumph outlier. In Figure 11.14b the advantage of a PLS score plot is demonstrated. Here the starch and NIR spectroscopy information is integrated to obtain a more subtle classification of the samples than in the correlation

Table 11.3 Average and standard deviation of chemical data from 15 lines from the Carlsberg *lys3a* starch improvement breeding programs discussed in conjunction with Figures 11.13 and 11.14 (Møller, 2004b)

	Normal ($n = 6$)	Group 1	Group 2	Group 3	Group 4
Protein (P)	11.3 ± 0.4	11.7 ± 0.1	11.7 ± 0.1	12.6 ± 0.2	12.5 ± 0.2
Amide (A)	0.28 ± 0.03	0.21 ± 0.007	0.21 ± 0.007	0.22 ± 0.02	0.23
A/P	15.5 ± 0.9	11.0 ± 0.3	10.9 ± 0.4	10.7 ± 0.8	11.4
Starch	54.6 ± 2.5	52.6 ± 0.5	50.0 ± 0.1	49.4 ± 1.5	48.7 ± 0.2
β-glucan (BG)	4.7 ± 1.1	3.1 ± 0.1	3.1 ± 0.2	3.1 ± 0.3	2.8 ± 0.5
Rest (100 − P + S + BG)	29.5 ± 1.8	32.7 ± 0.5	35.3 ± 0.3	34.9 ± 1.8	36.1 ± 0.5

Group 1 = Lysiba, Lysimax; Group 2 = 502, 556; Group 3 = 505, 531, 538; Group 4 = *lys3a*, *lys3m*.

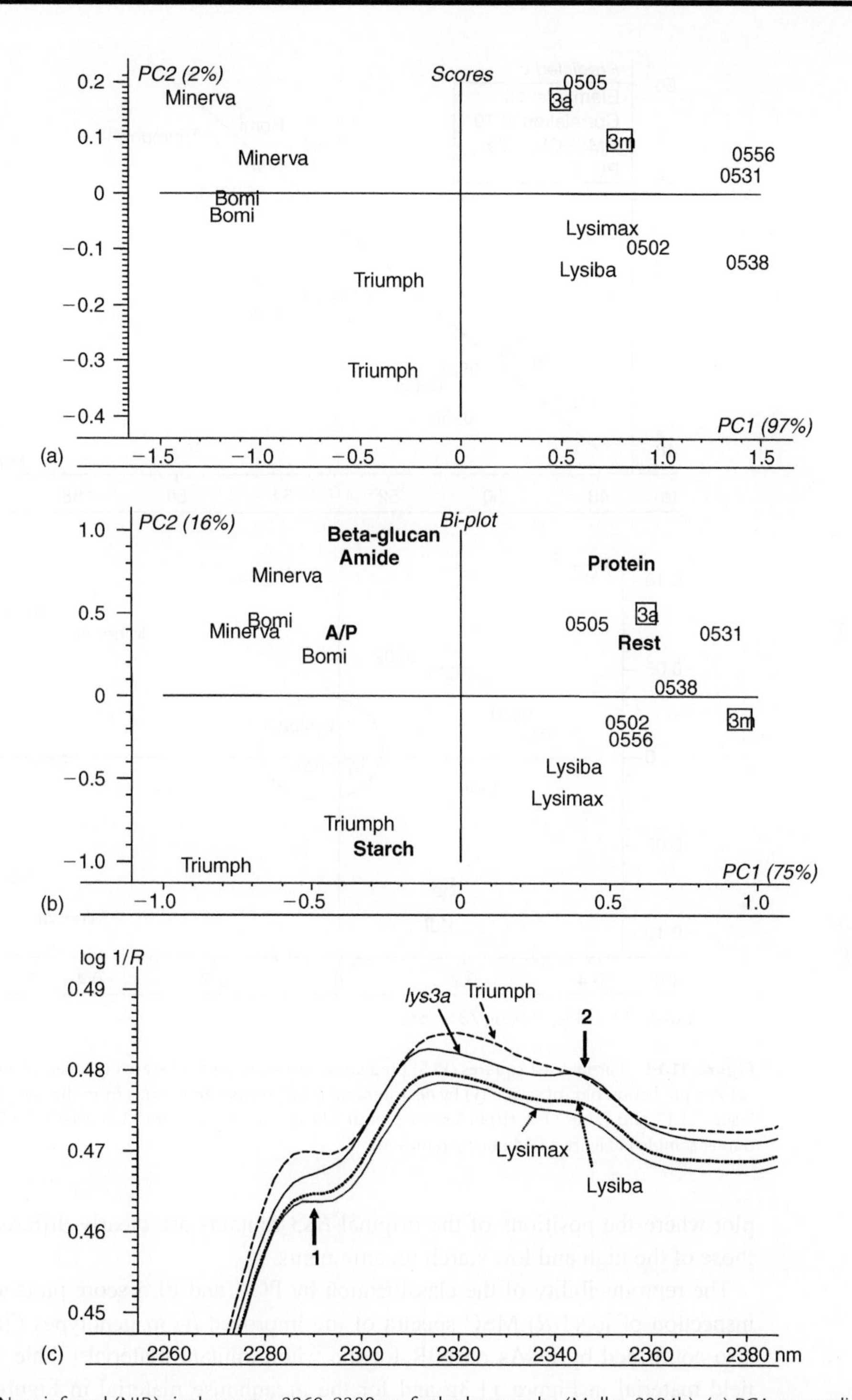

Figure 11.13 Near-infrared (NIR) single spectra 2260–2380 nm for barley samples (Møller, 2004b). (a) PCA cross-validated score plot of 15 NIR spectra (400–2500 nm, log(1/*R*) MSC) from the material in Table 11.3 featuring normal barley (Bomi, Minerva, Triumph), original mutants (*lys3a*, *lys3m*) as well as high-lysine recombinant lines from the Carlsberg material (0502, 0505, 0531, 0538, 0556, Lysiba, Lysimax); (b) a PCA cross-validated bi-plot on chemical data (protein, beta-glucan, amide (A), protein (P), A/P index, starch) from Table 11.3; (c) log(1/*R*) MSC NIR single spectra 2260–2380 nm for barley samples Triumph N, the P mutant *lys3a*, and its starch-improved recombinants by cross-breeding Lysiba and Lysimax from Table 11.3.

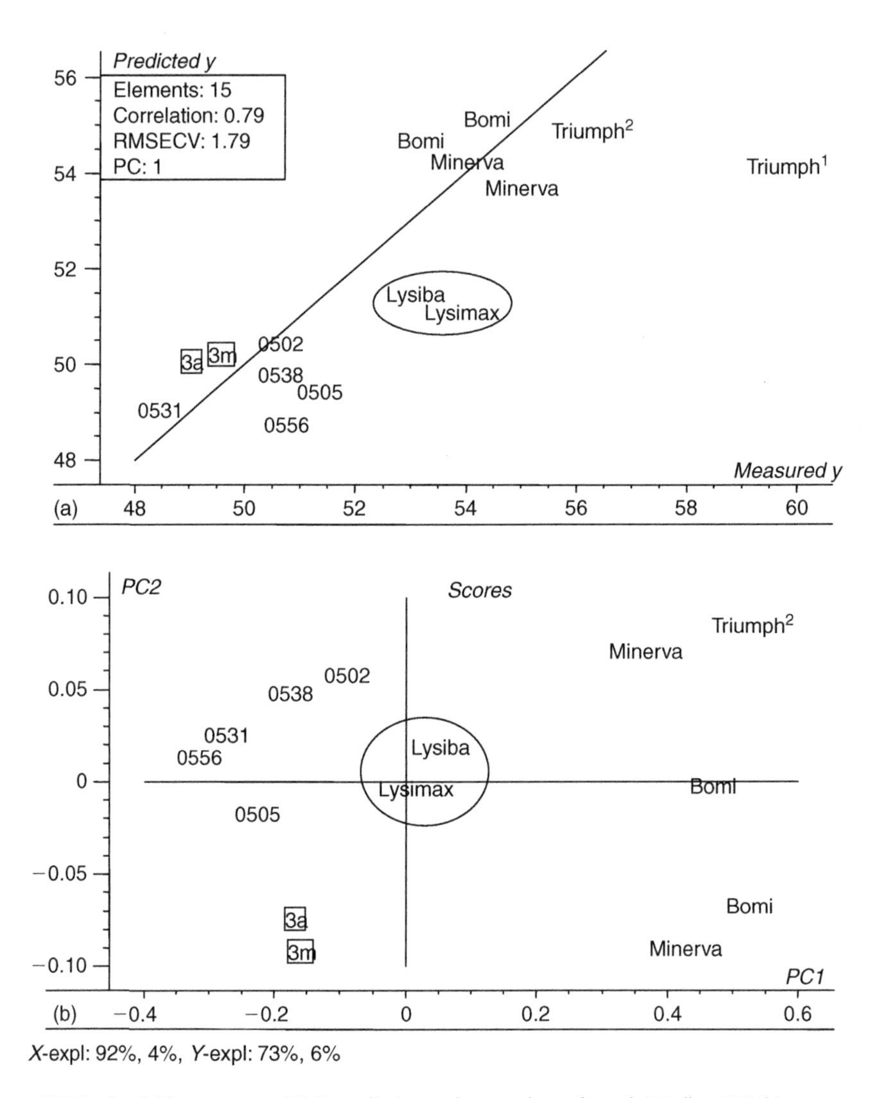

Figure 11.14 Partial least squares (PLS) prediction and score plots of starch (Møller, 2004b). (a) PLS prediction plot of starch (y) by near-infrared (NIR) measurements (x) from the samples set in Figure 11.13 and Table 11.2. (b) PLS score plot of NIR spectra from Figure 11.14a with the Triumph outlier sample in Figure 11.14a being removed.

plot where the positions of the original *lys3* mutants are clearly differentiated from those of the high and low starch recombinants.

The reproducibility of the classification by PCA and PLS score plots and by direct inspection of log(1/*R*) MSC spectra of the improved *lys3a* genotypes (Table 11.3) is also confirmed by PCAs on NIR for the whole mutant material (Table 11.1) for the field material in Figure 11.3g and for the greenhouse material in Figure 11.15. The improved *lys3a* recombinants Piggy, Lysimax, and Lysiba have also moved in these PCAs according to the arrows from the original *lys3a* position in the P category towards the normal barley's (N). However, as demonstrated for four genotypes in Figure 11.13c in the log(1/*R*) MSC-treated interval 2270–2380 nm, a spectral inspection

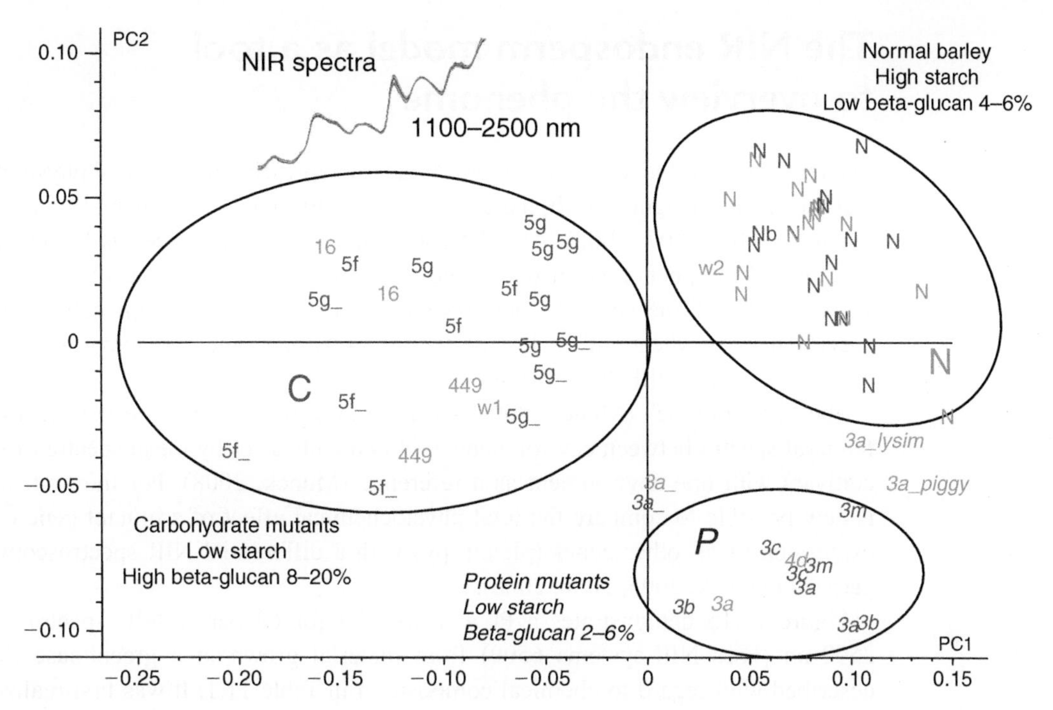

Figure 11.15 A cross-validated PCA score plot 1100–2500 nm for the 69 barley samples grown in the greenhouse from Table 11.1 featuring normal lines N, high-lysine protein mutants P, and carbohydrate mutants C (Munck, 2006). Mutant crosses are indicated with a line and it can be seen that the starch-improved recombinants of *lys3a* Lysimax and Piggy are moving towards the normal N barley class. (see Plate 11.4 for colour version)

is necessary to give the finely tuned reproducible spectra full justice. This is not possible by data compression by a PCA score plot, even if selection of wavelengths from the PCA loadings plot is a valuable compliment. The plateau at 2290 nm (arrow 1, Figure 11.13c) in the spectrum from the high starch cultivar Triumph is indicative for starch. This plateau is a slope in the low-starch *lys3a* mutant that is approaching the plateau of the starch-rich Triumph control when transferred to the improved gene backgrounds of Lysiba and Lysimax by breeding. The *lys3a* mutant spectrum brings interpretable information on fat content (Munck *et al.*, 2004). There is an increase in oil content in mutant *lys3a* compared to Triumph from about 2 to 3 absolute percentage that is reflected in the bulb in the *lys3a* spectrum at 2347 nm (Figure 11.3h) indicative for fat (Figure 11.13c, arrow 2). This bulb is significantly reduced (as well as the fat content) in the two improved *lys3a* lines Lysimax and Lysiba as shown in the same figure.

It is impressive that the small relative changes in physiochemical composition can be traced back to subtle but quite reproducible changes in NIR spectra. The use of existing NIR spectroscopy instruments can now be expanded by classification and selection for complex functional traits such as baking and malting quality through "data breeding" without the need for specific commercial calibrations (Munck *et al.*, 2000; Munck and Møller, 2005; Munck, 2008).

The NIR endosperm model as a tool to overview the phenome

The realization of the fact that NIR spectroscopy represents finely tuned physiochemical fingerprints (Figure 11.2b) has a profound genetic and environmental significance (Munck, 2005, 2007, 2008). An NIR spectrum from a cereal seed is highly representative of the phenotype of one tissue—the endosperm—that constitutes between 70 and 90% of the seed. In homozygous seeds from lines from self-fertilizing species such as wheat and barley the (endosperm) phenome and "environome" can now be defined by NIR fingerprints.

The "environome" (Munck, 2008) of a homozygous line can be defined by differential spectra between environments within a cultivar or by mean spectra of many cultivars with one environment as a reference (Munck, 2008). For the first time it is now possible to compare the total physiochemical effect of a mutant gene on the expression of all other genes (pleiotropy) with a differential NIR spectroscopy fingerprint (Munck, 2005, 2006, 2008).

Figure 11.15 demonstrates a PCA score plot for 69 barley NIR spectra 1100–2500 nm (Foss NIR-systems 6500) from material grown in a greenhouse that is described with regard to chemical composition in Table 11.1. It was first realized in 2004 (Munck *et al.*, 2004), that barley mutants (and their cross-bread genotypes, indicated with a line in Figure 11.15) were classified into two groups: regulative protein P mutants (large change in lysine/amino acids with moderate decrease in starch) and structural carbohydrate C mutants (small change in lysine/amino acids with large to moderate decrease in starch). The C mutants *lys5f*, *lys5g*, and mutant 16 have successfully been used by molecular biologists to investigate the starch synthesis pathway (see review by Rudi *et al.*, 2006). The function of the mutant genes is precisely documented in the literature as defaults for specific isoenzymes in the synthesis and transport of ADP-glucose. It was surprising when it was found (Figures 11.3g and 11.15) that nearly all the C mutants had compensated for the decrease in starch with beta-glucan that reached up to 20% d.m. for mutant *lys5f*. A new regulative pathway involving beta-glucan was detected (Munck *et al.*, 2004).

The differential spectra 1100–2500 nm between the spectrum from the isogenic background (Bomi) and those of the P mutants *lys4d* and *lys3* and the C mutants *lys5f* and mutant 16 are outlined in Figure 11.16a. The two finely tuned differential spectra from each of the mutant pairs follow their respective P and C patterns with small but chemically interpretable differences. Such an interpretation is demonstrated for the *lys3* locus in Figure 11.16b, which shows the analog differential spectra at 2200–2500 nm for the three mutant alleles in Bomi—the *lys3* locus: *lys3a*, *lys3b*, and *lys3c*. These closely follow each other except for *lys3c*, which is an outlier above in the 2425–2500 nm area indicative of beta-glucan. It was verified (Munck *et al.*, 2004) that the *lys3c* mutant has a normal content of beta-glucan (6.1%), compared with *lys3a* (4.7%) and *lys3b* (3.1%), which have lower values than normal lines.

The effect of the changed background of the *lys3a* gene obtained by breeding for plump seeds (starch) is shown in Figure 11.16b as the mean differential spectrum

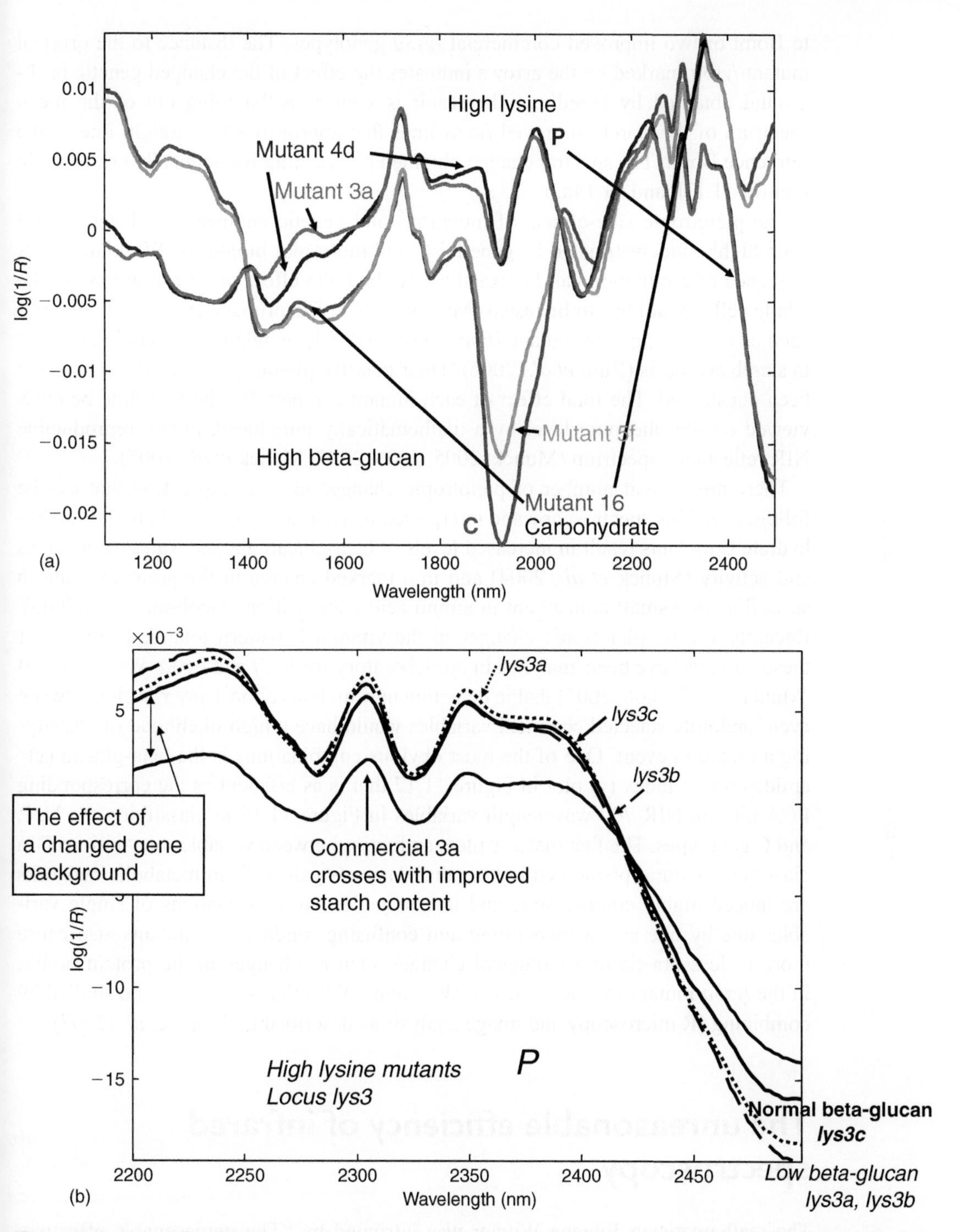

Figure 11.16 Near-infrared (NIR) spectral representation of gene interaction (pleiotropy) (Munck, 2005). (a) NIR spectral representation of gene interaction (pleiotropy): Mean differential NIR spectra log(1/*R*) MSC 1100–2500 nm to the parental genetic background of cv. Bomi of protein P mutants *lys3a* and *lys4d* compared with those of the carbohydrate C mutants *lys5* and Risø mutant 16. (b) NIR log(1/*R*) MSC 2200–2500 nm spectral representations of pleiotropy for the three *lys3* alleles *a*, *b*, and *c* and by the change in gene background for the spectral mean of the two starch-improved *lys3a* breeding cultivars Lysimax and Piggy. All spectra in (a) and (b) are differential spectra to cv. Bomi isogenic to the original mutants. (see Plate 11.5 for colour version)

to Bomi of two improved commercial *lys3a* genotypes. The distance to the original mutant *lys3a* marked by the arrows indicates the effect of the changed genetic background obtained by breeding. The result is seen as a flattening out of the mean spectrum of the starch-improved *lys3a* lines that approaches the straight line of the reference Bomi. The spectral change obtained by breeding is chemically evaluated in Figures 11.13b and 11.14a,b.

The pleiotropic side-effects of mutations and genetic engineering of genes have been highly underestimated by geneticists and molecular biologists. While the DNA sequence of a mutation can be exactly described, the effect of its expression on the whole cell (tissue) has to be tested (Munck, 2007). The primary causes of C mutants such as *lys5f*, *lys5g*, and mutant 16 are described only as related to specific enzymes in starch synthesis (Rudi *et al.*, 2006). Their massive pleiotropic side-effects have not been considered. The total effect of each mutant can now for the first time be overviewed on the phenome level by a mathematically unreduced, highly reproducible NIR reflectance spectrum (Munck, 2005, 2006, 2007; Munck *et al.*, 2007).

There are a great number of pleiotropic changes in gene expression that can be followed by NIR spectroscopy and interpreted by chemical analyses. In barley, carbohydrate C mutants result in increased levels of beta-glucan, fat and water percentages and activity (Munck *et al.*, 2004) and in a marked change in the proteome pattern as well as in a small adjustment in amino acid composition (Jacobsen *et al.*, 2005). Recently, drastic pleiotropic changes in the vitamin E pattern relative to Bomi for these mutants have been analyzed in our laboratory for *lys5f* and *lys3a*. We concluded (Munck, 2005, 2006, 2007) that in detecting new mutants almost any *x*–*y* plot between even randomly selected chemical variables would have a high likelihood of classifying a mutation event. One of the most obvious combinations is the beta-glucan (*x*)–amide/protein index (*y*) plot in Figure 11.17 that is as efficient as the corresponding PCA plot on NIR 700 wavelength variables in Figure 11.15 in classifying the N, P, and C genotypes. The fact that *x*–*y* plots and ratios between variables are "rational" in classifying natural phenomen reveals that data sets gained from metabolic networks are indeed highly compressible and interpretable, while evaluations of single variables one by one are overwhelming and confusing. Endosperm mutants also cause more or less drastic morphological changes such as changes in the protein bodies in the *lys3a* mutant (Munck and von Wettstein, 1976) that should now be studied by combining IR microscopy and image analysis as described by Lewis *et al.* (2007).

The unreasonable efficiency of infrared spectroscopy

The mathematician Eugene Wigner was intrigued by "The unreasonable effectiveness of mathematics in natural sciences" (Wigner, 1960). The reproducible patterns of chemical bonds from seeds in approximately identical initial conditions as read by NIR spectroscopy discussed above creates highly structured data material. Our spectral and mathematical interpretation of these data supports Wigner's statement

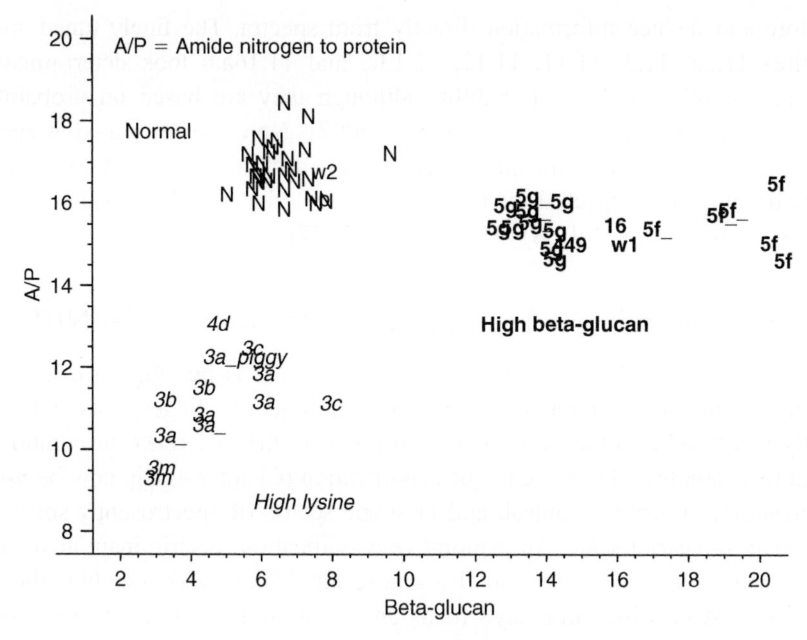

Figure 11.17 An *x–y* correlation plot of barley greenhouse material (n = 69, Table 11.1) with beta-glucan as abscissa and amide to protein (A/P) index as ordinate (Munck, 2006).

on "unreasonable efficiency," although it is now attributed to the NIR spectra and explained by self-organization.

Finely tuned reproducible NIR spectra from seeds

The amazing reproducibility of the finely tuned NIR spectra between replicated experiments on genetics and environment demonstrated in Figures 11.2a, 11.3a, 11.11, and 11.12 gives associations to a biological networking computer (Munck, 2005, 2006, 2007) that introduces a stabilizing performance due to self-organization that is reflected in the NIR measurements. DNA and environment is the input to the dynamic "plant–seed (endosperm) computer" and the final output is recorded after "computation" as patterns of chemical bonds of the ripe endosperm that is read by NIR spectroscopy as a strip code. The stunning ability of the integrated plant–seed–spectrograph system to reproduce NIR spectra under identical conditions (Munck, 2005, 2006, 2007, 2008) must also involve the dynamic development behind seed populations from each genotype. This involves parameters such as earliness, plant leaf area, number of tillers, number of seeds per spike, seed weight, form, and hardness and chemical composition. The finely tuned changes in the spectra from the individual samples that make genetic and chemical sense (e.g. in Figure 11.13c) can neither be modeled in detail by the PCA on NIR data in Figure 11.13a nor by the PLS NIR predictions for starch in Figures 11.14a,b. One has then to revive the skills of the classical spectroscopists supplemented by new data visualization programs in color (e.g. Latentix at www.latentix.com as in Figures 11.3a–f) to explore and deduce information directly from spectra. The finely tuned spectra in Figures 11.2a, 11.3, 11.11, 11.12, 11.13c, and 11.16a,b look deterministic in their reproducibility and interpretability although they are based on probabilistic self-organizing chemical reactions (Munck, 2007). Now the terrain—the spectral phenome—is equal to the map and new gentle, less destructive mathematical models must be developed for spectral interpretation such as fine-tuned interval versions of PLS regression analysis (ftiPLS) (Munck *et al.*, 2007).

Self-organization in seed synthesis for classification by NIR

It is generally acknowledged in forensic science that it is possible to differentiate between two human individuals by just two high-quality fingerprints. Only two carefully measured spectra are needed to investigate the difference in composition between two samples. This principle of classification (Chapter 4) can now be used in plant breeding, in process control, and in single seed NIR spectroscopy sorting for quality at an industrial scale. An explorative classification (discriminant analysis) of NIR spectra is able to give easy and clear-cut results without elaborate hypotheses to find unexpected surprises. In today's focus on causation this option has been too little used in industry and science up till now.

The sharp view of the DNA sequence of a gene as the primary cause in genetics is now dominating in biology. The secondary (pleiotropic) causes of a gene in the whole biological network (phenotype) have hereto been highly underestimated due to the lack of methods for an overview. As demonstrated here, NIR spectroscopy can now supply such an overview. Figure 11.18 (Munck, 2008) is an attempt to visualize how the endosperm tissue is working as a "biological computer" in "calculating" the physiochemical composition in the synthesis of endosperm of the *lys5f* mutation from the Bomi parent line (Munck, 2007). The mutant has a lesion in the DNA sequence for one of the ADP-glucose transporters needed for starch synthesis. This primary cause (position **1** to the left in Figure 11.18) that reduces starch with 50%, now starts a cascade of secondary metabolic events, leading to a regulative compensation in beta-glucan (Munck *et al.*, 2004) that increases from 6% (in Bomi) to 20% in *lys5f* (**2** Figure 11.18). Our research group has recently verified that the increased synthesis of beta-glucan in *lys5f* endosperms peaks at day 20 post anthesis and that the synthesis is combined with a rise in water content from 68.2% in the normal control to 73.1% in *lys5f*. During seed synthesis water content in the seeds of the *lys5f* mutant relative to the mutant control increases by 10%. Water activity is related to moisture content in a non-linear relationship known as a moisture sorption isotherm curve, which was studied in dry seeds in the passive moisture conditioning experiment described above. It is not possible to study water activity in the same way as in an actively growing tissue. The chemical structure of crystalline starch versus beta-glucan and the above-verified difference in water content and in beta-glucan due to the *lys5f* mutation makes it likely that a drastic change in water activity in the endosperm cells takes place at day 20. This change should have the potential to change the performance of any enzyme (**2** Figure 11.18) that is active in the cell during seed synthesis. It may partly explain

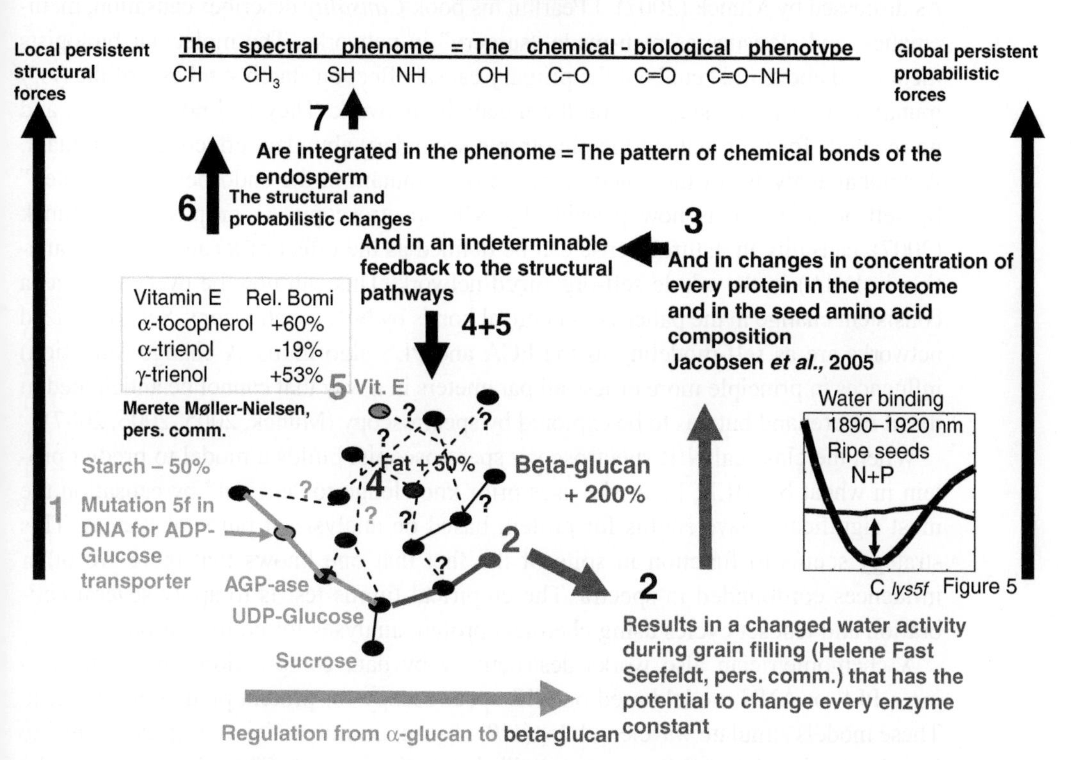

Figure 11.18 Overview graph of the primary and secondary effects on endosperm synthesis of the expression of the carbohydrate C mutant *lys5f* (Munck, 2007). The effect of the cascade of gene expressions numbered **1** to **7** is described in the text. (see Plate 11.6 for colour version)

the multifaceted change (relative to Bomi) in the pattern of the water-soluble proteome (including the total amino acid composition), which was observed in the ripe seeds of *lys5f* by Jacobsen *et al.* (2005) (**3** Figure 11.18). The likely change in water activity in the cell will also give a feedback to the enzyme-driven structural pathways in the metabolome and may be involved in the increases in oil content in *lys5f* by 50% (**4** Figure 11.18, Munck *et al.*, 2004) and in the drastically changed vitamin E pattern (**5** Figure 11.18).

The described change in the internal environment of the cell has a probabilistic character including a component of biological indeterminacy (Munck, 2007) that cannot be amended by causal path modeling. However, the probabilistic effect of changed water activity in the cell due to beta-glucan produced by the mutation is included in the final physiochemical composition of the seed (**6** Figure 11.18) as a whole integrated reproducible spectral response (Figure 11.16a) together with information on structural changes in pathways (e.g. increase in beta-glucan). The total effect of the mutation on the physiochemical composition as measured by NIR spectroscopy (**7** Figure 11.18) is reproducible and supports the genetic and environmental classification by PCA of NIR spectra demonstrated in Figures 11.3g, 11.6a,b, and 11.15.

Causality as destructive surgery in networks

As discussed by Munck (2007), J Pearl in his book *Causality* describes causation, mathematics, and statistics as man-made "surgery" in networks. The molecular biologists mentioned above did "cut" out the primary causal effect on starch synthesis of the *lys5f* mutation by way of "surgery" on the metabolic network. They had no interest in and no methods for overviewing the pleiotropic secondary abundant effects of the mutant. A "global analysis" of the outcome of the *lys5f* mutant in the endosperm "computed" by self-organization is now possible by NIR spectroscopy. As proposed by Munck (2007), causality in nature's sense can be defined as the effect of a cause (e.g. a mutation in DNA) in the whole self-organized network. This can now be overviewed as a consistent change in the pattern of chemical bonds by NIR spectroscopy. Self-organized networks are as self-modeling as the PCA and PLS algorithms. A change (mutation) influences in principle more or less all parameters in a way that cannot be anticipated in detail beforehand but has to be explored by spectroscopy (Munck, 2005, 2006, 2007).

When the classical NIR spectroscopy spectroscopist builds a model to predict protein in wheat by MLR, he or she uses prior knowledge to "cut out" by causation the most significant wavelengths for protein based on analysis of pure substances. This strategy seems to function in spite of the fact that one knows that there are other influences confounded in spectra. The empirical litmus test is to apply several calibration and test set cycles using chemical protein analyses for confirmation.

A chemometrician also works destructively by data compression when establishing a PLS or ANN model based on NIR spectroscopy for protein prediction in wheat. These models simulate wavelength $\log(1/R)$ absorption interaction (covariance) and are less destructive than MLR, however, still destructive enough. The advantage with the self-modeling chemometric algorithms is that it is possible with a minimum of prior spectroscopic knowledge to manage a global overview of the pattern of thousands of variables that is not amendable in classical statistics because of distributional assumptions. The chemometric model is 100% empirical. It completely ignores whether or not the wavelengths that are used in the calibrations have a direct causal relation to the protein that was assigned by measurement of pure substances. It just cuts out "quick and dirty" patterns of covariance, including the wavelengths that directly or indirectly give any strong positive or negative correlations for the best prediction of protein as in Figure 11.2b. The correlations are therefore highly influenced by the "biological networking computer" guided by genetics and environment that has "computed" the physiochemical composition during plant growth and seed synthesis beforehand. This information is now included as physiochemical fingerprints in the NIR spectroscopy library of wheat. Consequently one has to respect nature's way of "computing" in a different way when including new unknown samples that are registered as outliers in NIR spectroscopy and contemplate whether this new "experience" should be included in the global prediction model to serve as a representative source of artificial intelligence.

Physiochemical fingerprint observable by NIR spectroscopy

As discussed by Miller (2001), the physical NIR spectral theory explains why "vibrational spectroscopy made complicated is NIR made possible." However, we can now conclude that the built in "biological computer" is assisting more than complicating. In understanding the theory behind the efficacy of NIR spectroscopy in cereals one has to consider the physical/measurement/biological/physiochemical/self-organized/plant/endosperm system as a whole dynamic entity. The output should be preliminarily evaluated by spectral inspection and self-modeling chemometric algorithms and supported by prior physiochemical and biological knowledge. This also explains why "there is no definitive theory for diffuse reflectance NIR spectroscopy" as claimed by Karl Norris in 1996 (Davis, 1996).

This notion is still more valid when applying NIR spectroscopy to complex biological systems such as cereals. In the quest for a theory, the answer is that the result is already computed out there in the form of a physiochemical fingerprint. All the details of the computation are not and will never be available to us because of the probabilistic indeterminacy that is intrinsic to the system (Munck, 2007), which also applies to quantum physics (Dahm and Dahm, 2001; Miller, 2001).

Causal deterministic practice based on "surgery in networks" either in physics, chemistry, molecular biology, or in engineering has severe limitations in a self-organized world that is built on networks. The latter are driven by probability in self-organization. The reproducible results look deterministic to us, far from the arbitrary probabilistic world of pure chance (Munck, 2007). However, so far spectroscopy is the only way to achieve the coarse-grained physiochemical overview that is an absolutely necessary compliment to the causal strategy. It is needed to explore new surprising elements (Munck *et al.*, 1998). The advice from nature's own calculations should always be requested as when "global" specific chemometric NIR spectroscopy calibrations have to be upgraded as cereal varieties, climate, and food processes change. Self-organization and amplification of gene expression also explains why NIR spectroscopy and classification software is effective in exploring genetic manipulated plants, as well as in finding human cell mutations that cause cancer (Munck, 2007).

Future development

Today there is a major increase in global cereal commodity prices that is favoring analysis for quality. The increasing purchase power of the Asian developing economies is leading to increased meat consumption that supports the demand for cereals for feed. This request has now to compete with the use of cereals for non-food purposes, including energy, due to the very high world market prices of fuel. The result is the rapidly increasing prices that make quality diversification by NIR spectroscopy economic both for sorting in bulk and on a single seed basis. This is because it is now possible to sell the low-quality fraction for a high price for energy production.

The variation in quality between single seeds even from the same field is stunning. It will always be sensible to sort toxic fungal-infected seeds from non-infected and use them for energy (e.g. ethanol) and feed respectively. It will not, however, be possible to use a major portion of the world cereal seed production for non-food purposes (Munck, 2004) when the present world storage supply of cereals for food security is almost down to zero. New worldwide trade regulations for securing cereals for food and feed will therefore be necessary if the current trends are sustained, which seems to be likely.

The sorting of seeds for optimal food and feed quality should be feasible in the future using the new industrial-scale NIR seed sorters. We can envisage a new area of single seed research and technological development based on NIR spectroscopy technology, which is presently at an embryonic stage. In the future the quality sorting profile of a cereal seed batch for optimized diversified use will probably have a greater impact on the price than its quality measurement in the non-separated bulk.

Current NIR spectroscopy quality control technology including chemometrics to "cut out" correlations for prediction of specific quality criteria of economical importance such as protein in wheat is now at its almost fully developed stage. Its motivation is the causal relation between the price and protein content established by the market. A less-destructive NIR spectroscopy fingerprint should provide a much better summary of functional criteria as a whole, evaluated by visual inspection and discriminative analysis, and should represent a much better indicator for baking quality and economic value than the old indication by protein. The deciding question that has not yet been solved is how a multivariate NIR spectroscopy calibration to baking quality could be standardized as a basis for trade contracts. While waiting for the problem to be solved, it should be economic to upgrade the internal management of quality in a cereal company by local calibrations, e.g. to baking and malting quality or even less destructively by sample classification and selection for physiochemical composition with high- and low-quality cultivars as standards ("data breeding").

The present rapid change in environment is to a large extent due to human selection on a focused, limited cause–effect, cost–benefit basis, resulting in unexpected secondary environmental side-effects. Causality in nature's sense that not only involves the primary cause but also the effects on the whole network must be acknowledged in an exploratory strategy with any global screening method (e.g. NIR spectroscopy measuring first and hypothesizing afterwards) (Munck *et al.*, 1998; Munck, 2007). The necessary overview given by the endosperm/NIR spectroscopy/chemometric model creates a unique window in understanding gene–environment interaction and the possibilities and limits of human intervention by technology. The supremacy of self-organization in nature has to be respected in a dialogue through exploratory inventories by NIR spectroscopy to find and define the new surprising elements that cannot be predicted by limited causal path modeling. Global tools have to be developed to explore the total effects of human causal interventions in biological and environmental networks.

The endosperm is not only relevant as the most important food source for the human population, which amounts to more than 1.6 billion tons per year. It is also unique as a cast global model for the phenome in systems biology to be used in molecular biology with the potential to inspire medical research. The holistic spectral overview by chemometrics and spectral inspection also provides a unique opportunity in using human blood serum as a mirror for modeling health as a whole concept for individual patients. Spectroscopy and chemometrics have great potential in serving as a training ground for students to develop a new science in a "Global Data Modeling University," which is urgently needed.

Conclusions

NIR spectroscopy technology is an extremely successful tool in the cereal industry, allowing fast screening methods for the prediction of specific analytes. NIR spectroscopy analyses make quality control instant and move it from the laboratory to the production line. This review focuses on the next step in NIR spectroscopy technology—spectral classification—which promises new, challenging options in identifying functional factors in plant breeding, single seed sorting, process control as well as in food production and design. Single seed industrial-scale sorting by NIR spectroscopy will have a future great potential in using the huge variation in quality within the same seed batch for added value. The ability of NIR spectra under controlled conditions to represent the effects of genes and environment of cereal seeds at batch level is stunning. The high reproducibility can be interpreted as the output from a built-in networking endosperm computer based on self-organization that is read like a strip code by NIR. Mathematical models cannot do justice to the finely tuned specific NIR spectra by data reduction. A transfer of the skills in spectral chemical evaluation by inspection from the classical spectroscopists to biologists has great potential in realizing the spectral phenome in molecular biology. In the near future the Vis/NIR–IR–Fluorescence–NMR spectroscopic tools will prove to have significant theoretical and practical impact on the biological, molecular, and medical sciences as that of emission spectroscopy in the history of atomic physics and astronomy.

Acknowledgments

This work has been supported by the Danish Research Councils through the project "BygDinMad" to Professor Søren Balling Engelsen. The assistance from Lars Nørgaard (Latentix and ECVA computation), Helene Fast Seefeldt (analysis of the developing endosperm), Merete Møller Nielsen (tocopherol in kernels), Frans van den Berg, and Rasmus Bro in completing this manuscript is gratefully acknowledged. We thank Lisbeth Hansen for skillful analyses.

References

Allosio N, Boivin P, Bertrand D, Courcoux P (1997) Characterization of barley transformation into malt by three-way factor analysis of near infrared spectra. *Journal of Near Infrared Spectroscopy*, **5**, 157–166.

Armstrong PR (2006) Rapid single-kernel NIR measurement of grain and oil seed attributes. *The Chemical Engineer*, **238**, 108–112.

Berg van den F (2001) Multi-blockPLSR models in food technology. In: *Proceedings on the 2nd International Symposium on PLS and Related Methods held in Anacapri (Naples), Italy, October 1–3, 2001*, pp. 385–394.

Bergman CJ, Bhattacharya KR, Ohtsubo K (2003) Rice and end use quality analysis. In: *Rice: Chemistry and Technology* (Champagne ET, ed.). St. Paul, MN, USA: American Association of Cereal Chemists, Inc., pp. 415–472.

Bertrand D, Robert P, Loisel F (1985) Identification of some wheat varieties by near infrared reflectance spectroscopy. *Journal of the Science of Food and Agriculture*, **36**, 71–78.

Bro R (1996) Multiway Calibration. Multilinear PLS. *Chemometrics and Intelligent Laboratory Systems*, **10**(1), 47–61.

Bro R, Berg van den F, Thybo A, Andersen CM, Jørgensen BM, Andersen H (2002) Multivariate data analysis as a tool in advanced quality monitoring in the food production chain. *Trends in Food Science & Technology*, **13**, 235–244.

Bruun SW, Søndergaard I, Jacobsen S (2007a) Analysis of protein structures and interactions in complex food by near-infrared spectroscopy. Part 1: Gluten powder. *Journal of Agriculture and Food Chemistry*, **55**, 7234–7243.

Bruun SW, Søndergaard I, Jacobsen S (2007b) Analysis of protein structures and interactions in complex food by Near-infrared spectroscopy. Part 2: Hydrated gluten. *Journal of Agriculture and Food Chemistry*, **55**, 7244–7251.

Buchmann NN, Josefsson H, Cowe IA (2001) Performance of European artificial neural network (ANN) calibrations for moisture and protein in cereals using the Danish near-infrared transmission network. *Cereal Chemistry*, **78**(5), 572–577.

Burns DA, Ciurczak EW (eds) (2001) *Handbook of Near Infrared Analysis*. New York: Marcel Dekker Inc.

Burns DA, Schultz TP (2001) FT/IR versus NIR—A study with lignocellulose. In: *Handbook of Near Infrared Analysis* (Burns DA, Ciurczak EW, eds). New York: Marcel Dekker Inc., pp. 563–571.

Campbell MR, Sykes J, Glover DV (2000) Classification of single- and double-mutant corn endosperm genotypes by near-infrared transmittance spectroscopy. *Cereal Chemistry*, **77**(6), 774–778.

Champagne ET, Richard OA, Bett KL, Grimm GC, Vinyard BT, Webb BD, McClung AM, Barton FE, Lyon BG, Moldenhauer K, Linscombe S, Mohindra R, Kolwey D (1996) Quality evaluation of US medium-grain rice using a Japanese taste analyzer. *Cereal Chemistry*, **73**, 290–294.

Chen H, Marks BP, Siebenmorgen TJ (1997) Quantifying surface lipid content on milled rice via visible/near-infrared spectroscopy. *Cereal Chemistry*, **74**, 826–831.

Chen L, Carpita NC, Reiter WD, Wilson RH, Jeffries C, McCann MC (1998) A rapid method to screening for cell-wall mutants using discriminant analysis of Fourier transform infrared spectra. *The Plant Journal*, **16**(3), 385–392.

Dahm DJ, Dahm KD (2001) The physics of near infrared scattering. In: *Near Infrared Technology in the Agricultural and Food Industries* (Williams P, Norris K, eds). St. Paul, MN: American Association of Cereal Chemists, pp. 1–17.

Davis AMC (1996) Some highlights from the 8th International Diffuse Reflection Conference. *Spectroscopy Europe*, **8**(5), 22–24.

Delwiche SR (1995) Single wheat kernel analysis by near-infrared transmittance: protein content. *Cereal Chemistry*, **72**(1), 11–16.

Delwiche SR (1998) Protein content of single kernels of wheat by near-infrared reflectance spectroscopy. *Journal of Cereal Science*, **27**, 241–254.

Delwiche SR, Graybosch RA (2002) Identification of waxy wheat by near infrared reflectance spectroscopy. *Journal of Cereal Science*, **35**, 29–38.

Delwiche SR, Weaver M (1994) Bread quality of wheat flour by near-infrared spectroscopy: feasibility of modeling. *Journal of Food Science*, **59**(2), 410–415.

Delwiche SR, Bean MM, Muller RE, Webb BD, Williams PC (1995) Apparent amylose content of milled rice by near-infrared reflectance spectrophotometry. *Cereal Chemistry*, **72**, 182–187.

Delwiche SR, Graybosch RA, Peterson J (1999) Identification of wheat lines possessing the 1AL.1RS or 1BL.1RS wheat-rye translocation by near-infrared spectroscopy. *Cereal Chemistry*, **76**(2), 255–266.

Delwiche SR, Hruschka WR (2000) Protein content of bulk wheat from near-infrared reflectance of individual kernels. *Cereal Chemistry*, **77**(1), 86–88.

Delwiche SR, Pearson TC, Brabec DL (2005) High speed optical sorting of soft wheat for reduction of deoxynivalenol. *Plant Disease*, **89**(11), 1214–1219.

Dowell FE, Maghirang EB, Graybosch RA, Baenziger PS, Baltensperger DD, Hansen LE (2006) An automated near-infrared system for selecting individual kernels based on specific quality characteristics. *Cereal Chemistry*, **83**(5), 537–543.

Gergely S, Salgo A (2003) Changes in moisture content during wheat maturation—what is measured by near infrared spectroscopy. *Journal of Near Infrared Spectroscopy*, **11**, 17–26.

Hindle PH (2001) Historical Development. In: *Handbook of Near Infrared Analysis* (Burns DA, Ciurczak EW, eds). New York: Marcel Dekker Inc., pp. 1–5.

Jacobsen S, Søndergaard I, Møller B, Desler T, Munck L (2005) A chemometric evaluation of the underlying physical and chemical patterns that support near infrared spectroscopy of barley seeds as a tool for explorative classification of endosperm genes and gene combinations. *Journal of Cereal Science*, **42**(3), 281–299.

Kacurakova M, Wilson RH (2001) Developments in mid-infrared FT-IR spectroscopy of selected carbohydrates. *Carbohydrate Polymers*, **44**, 291–303.

Kemeny GJ (2001) Process analysis. In: *Handbook of Near Infrared Analysis* (Burns DA, Ciurczak EW, eds). New York: Marcel Dekker Inc., pp. 729–782.

Kim SS, Mee-Ra R, Kim MJ, Lee S-H (2003) Authentication of rice using near-infrared reflectance spectroscopy. *Cereal Chemistry*, **80**(3), 346–349.

Lee AK (2007) On analysis in food engineering. In: *Near Infrared Spectroscopy in Food Science and Technology* (Ozaki Y, McClure WF, Christy AA, eds). Hoboken, NJ: Wiley-Interscience, pp. 361–378.

Lewis EN, Dubois J, Kidder LH (2007) NIR Imaging and its applications to Agriculture and Food Engineering. In: *Near Infrared Spectroscopy in Food Science and Technology* (Ozaki Y, McClure WF, Christy AA, eds). New Jersey, USA: Wiley-Interscience, Hoboken, pp. 121–131.

Li WS, Shaw JT (1997) Determining the fat acidity of rough rice by near-infrared spectroscopy. *Cereal Chemistry*, **74**, 556–560.

Löfqvist B, Pram Nielsen J (2003) Method of sorting objects comprising organic material, European Patent, EC B07C5/34; G01N2/35G.

Mark H (2001) Qualitative discriminant analysis. In: *Handbook of Near Infrared Analysis* (Burns DA, Ciurczak EW, eds). New York: Marcel Dekker Inc., pp. 363–399.

Martens H, Martens M (2000) *Multivariate Analysis of Quality—An Introduction.* Chichester: John Wiley & Sons, pp. 93–109, 235–256.

Martens H, Næs T (2001) Multivariate calibration by data compression. In: *Near Infrared Technology in the Agricultural and Food Industries* (Williams P, Norris K, eds). St. Paul, MN: American Association of Cereal Chemists, pp. 59–100.

Martens H, Nielsen JP, Engelsen SB (2003) Light scattering and light absorbance separated by extended multiplicative signal correction. Application to near-infrared transmission analysis of powder mixtures. *Analytical Chemistry*, **75**(3), 394–404.

Meurens M, Yan SH (2002) Applications of vibrational spectroscopy in brewing. In: *Handbook of Vibrational Spectroscopy* (Chalmers JM, Griffiths PR, eds). Chichester: John Wiley & Sons Ltd, Vol. 5, pp. 3663–3671.

Miller CE (2001) Chemical principles of near-infrared technology. In: *Near Infrared Technology in the Agricultural and Food Industries* (Williams P, Norris K, eds). St. Paul, MN: American Association of Cereal Chemists, pp. 19–37.

Møller B (2004a) Near infrared transmission spectra of barley of malting grade represent a physical-chemical fingerprint of the sample that is able to predict germinative vigour in a multivariate data evaluation model. *Journal of the Institute of Brewing*, **110**(1), 18–33.

Møller B (2004b) Screening analyses for quality criteria in barley. PhD thesis, Department of Food Science, the Royal Veterinary and Agricultural University, Rolighedsvej 30, DK1958 Frederiksberg C, Denmark, p. 205.

Munck L (2004) Whole plant utilization. In: *Encyclopedia of Grain Science*. Amsterdam: Elsevier, pp. 459–466.

Munck L (2005) The Revolutionary Aspect of Exploratory Chemometric Technology. The Royal and Veterinary University of Denmark Narayana Press, Gylling, Denmark, p. 352. Available at www.models.life.ku.dk and lmu@life.ku.dk

Munck L (2006) Conceptual validation of self-organization studied by spectroscopy in an endosperm gene model as a data driven logistic strategy in chemometrics. *Chemometrics and Intelligent Laboratory Systems*, **84**, 26–32.

Munck L (2007) A new holistic exploratory approach to systems biology by near infrared spectroscopy evaluated by chemometrics and data inspection. *Journal of Chemometrics*, **21**, 406–426.

Munck L (2008) Breeding for quality traits in cereals—a revised outlook on old and new tools for integrated breeding. In: *Cereals* (Carena MJ, ed.). New York: Springer Publishers.

Munck L, Møller B (2004) A new germinative classification model of barley for prediction of malt quality amplified by a near infrared transmission spectroscopy calibration for vigour "on-line" both implemented by multivariate data analysis. *Journal of the Institute of Brewing*, **110**(1), 3–17.

Munck L, Møller B (2005) Principal component analysis of near infrared spectra as a tool of endosperm mutant characterization and in barley breeding for quality. *Czech Journal for Genetics and Plant Breeding*, **41**(3), 89–95.

Munck L, von Wettstein D (1976) Effects of genes that change the amino acid composition of barley endosperm. In: *Proceedings on a Workshop "Genetic improvement of seed proteins." March 18–20, 1974*. Washington, D.C.: National Academy of Sciences, pp. 71–82.

Munck L, Nørgaard L, Engelsen SB, Bro R, Anderson SA (1998) Chemometrics in food science—a demonstration of the feasibility of a highly exploratory, inductive evaluation strategy of fundamental scientific significance. *Journal of Chemometrics and Intelligent Laboratory Systems*, **44**, 31–60.

Munck L, Møller B, Pram Nielsen J (2000) From plant breeding to data breeding—How new multivariate screening methods could bridge the information gap between the genotype and the phenotype. In: *Proceedings from Barely Genetics VIII* (Logue S, ed.). Adelaide: Adelaide University, Vol. I, pp. 179–182.

Munck L, Pram Nielsen J, Møller B, Jacobsen S, Søndergaard I, Engelsen SB, Nørgaard L, Bro R (2001) Exploring the phenotypic expression of a regulatory proteome-altering gene by spectroscopy and chemometrics. *Analytica Chimica Acta*, **446**, 171–186.

Munck L, Møller B, Jacobsen S, Søndergaard I (2004) Near infrared spectra indicate specific mutant endosperm genes and reveal a new mechanism for substituting starch with (1–3, 1–4)-β-glucan in barley. *Journal of Cereal Science*, **40**, 213–222.

Munck L, Nørgaard L, Møller Jespersen B (2007) Mathematical models and data visualization programmes adapted to explore causality in the finely tuned reproducible spectra from biological networks. In: *PLS-07—5th International Symposium on PLS: Causalities Explored by Indirect Observation*, Matforsk, Ås, Norway, September 5–7, 2007, pp. 12–16. Available at www.pls07.org.

Natsuga M (1999) Grain quality determination by near-infrared spectroscopy. *Memoirs of the Faculty of Agriculture, Hokkaido University*, **22**, 127–168.

Nielsen JP, Munck L (2003) Evaluation of malting quality using explorative data analysis I. Extraction of information from micro-malting data of spring and winter barley. *Journal of Cereal Science*, **80**(3), 274–280.

Nielsen JP, Bertrand D, Micklander E, Courcoux P, Munck L (2001) Study of NIR spectra, particle size distributions and chemical parameters of wheat flours: a multi-way approach. *Journal of Near Infrared Spectroscopy*, **9**, 275–285.

Nørgaard LA, Saudland J, Wagner J, Nielsen JP, Munck L, Engelsen SB (2000) Interval partial least squares regression (iPLS): A comparative chemometric study with an example from near infrared spectroscopy. *Applied Spectroscopy*, **54**(3), 413–419.

Nørgaard L, Bro R, Westad F, Engelsen SB (2006) A modification of canonical variates analysis to handle highly collinear multivariate data. *Journal of Chemometrics*, **20**, 425–435.

Osborne BG (2001) NIR analysis of baked products. In: *Handbook of Near Infrared Analysis* (Burns DA, Ciurczak EW, eds). New York: Marcel Dekker Inc., pp. 475–497.

Osborne BG (2006) Applications of near infrared spectroscopy in quality screening of early-generation material in cereal breeding programmes. *Journal of Near Infrared Spectroscopy*, **14**, 93–101.

Osborne BG (2007) Flours and breads. In: *Near Infrared Spectroscopy in Food Science and Technology* (Ozaki Y, McClure WF, Christy AA, eds). Hoboken, NJ: Wiley-Interscience, pp. 281–295.

Osborne BG, Fearn T, Hindle PH (1993) *Spectroscopy with Applications in Food and Beverage Analysis*. Harlow: Longman Scientific & Technical, p. 230.

Ozaki Y, McClure WF, Christy AA (eds) (2007) *Near Infrared Spectroscopy in Food Science and Technology*. Hoboken, NJ: Wiley-Interscience.

Paulsen MR, Watson SA, Singh M (2003) Measurement of corn quality. In: *Corn: Chemistry and Technology* (White PJ, Johnson LA, eds). St. Paul, MN: American Association of Cereal Chemists, pp. 159–212.

Pedersen DK, Martens H, Pram Nielsen J, Balling Engelsen S (2002) Near-infrared absorption and scattering separated by extended inverted signal correction (EISC): Analysis of near-infrared spectra of single wheat seeds. *Applied Spectroscopy*, **56**(9), 1206–1214.

Philippe S, Robert P, Barron C, Saulnier L, Guillon F (2006) Deposition of cell wall polysaccharides in wheat endosperm during grain development: Fourier transform-infrared micro-spectroscopy study. *Journal of Agricultural and Food Chemistry*, **54**, 2303–2308.

Pitz WJ (1990) An analysis of malting research. *Journal of American Society of Brewing Chemists*, **48**, 33–48.

Pram Nielsen J (2002) Fast quality assessment of barley and wheat. PhD thesis, Department of Food Science, The Royal Veterinary and Agricultural University, Rolighedsvej 30, DK1958 Frederiksberg C, Denmark, p. 212.

Pram Nielsen J, Pedersen DK, Munck L (2003) Development of non-destructive screening methods for single kernel characterization of wheat. *Cereal Chemistry*, **80**(3), 274–280.

Pram Nielsen J, Löfqvist B (2006) Method and device for sorting objects. European Patent, EC B07C5/36C1;B075/34.

Rittiron R, Saranwong S, Kawano S (2004) Useful tips for constructing a near-infrared based quality sorting systems for single brown rice kernels. *Journal of Near Infrared Spectroscopy*, **12**(2), 133–139.

Robert P, Marquis M, Barron C, Guillon F, Saulnier L (2005) FT-IR investigation of cell wall polysaccharides from cereal grains. Arabioxylan infrared assignement. *Journal of Agricultural and Food Chemistry*, **53**, 7014–7018.

Rudi H, Uhlen AK, Harstad OM, Munck L (2006) Genetic variability in cereal carbohydrate compositions and potential for improving nutritional value. *Animal Feed Science and Technology*, **130**(1–2), 55–65.

Shenk JR, Workman Jr JJ, Westerhaus MO (2001) Application of NIR Spectroscopy to agricultural products. In: *Handbook of Near Infrared Analysis* (Burns DA, Ciurczak EW, eds). New York: Marcel Dekker Inc, pp. 419–473.

Shimizu N, Kimura T, Yanagisawa T, Inoue S, Withey RP, Cowe IA, Eddison CG, Blakeney AB, Kimura T, Yoshizaki S, Okadome H, Toyoshima H, Ohtsubo K (1999) Determination of apparent amylose content in Japanese milled rice using near-infrared transmittance spectroscopy. *Food Science and Technology Research*, **5**, 337–342.

Siesler HW, Ozaki Y, Katawata S, Heise HM (eds) (2002) *Near Infrared Spectroscopy: Principles, Instruments, Applications*. Weinheim: Wiley-VCH Verlag GmbH.

Smilde A, Bro R, Geladi P (2004) *Multi-way Analysis*. Chichester: John Wiley & Sons, p. 381.

Tanaka K, Jinnouchi N, Harada K (1999) How to use NIR spectrophotometer for evaluating rice palatability. *Kyushu Agriculture Research*, **61**, 9.

Tønning E (2007) Wheat baking quality in a process analytical perspective—Sampling, diversification, prediction and chemometric method development, PhD thesis, Quality and Technology Research Section, Department of Food Science, Faculty of Life

Sciences, University of Copenhagen, Rolighedsvej 30, DK-1958 Frederiksberg C, Denmark, pp. 35–37. Available at www.models.life.ku.

Tønning EL, Nørgaard L, Engelsen SB, Pedersen L, Esbensen KH (2006) Protein heterogeneity in wheat lots using single-seed NIT—a theory of sampling (TOS) breakdown of all sampling and analytical errors. *Journal of Chemometrics and Intelligent Laboratory Systems*, **84**, 142–152.

Tønning E, Thybo AK, Pedersen L, Munck L, Hansen Å, Engelsen SB, Nørgaard L (2007) Bulk quality diversification of organic wheat by single-kernel near-infrared (SKNIR) sorting. *Journal of Cereal Science* (submitted).

Tsenkova R (2007) Aquaphotomics: water absorbance pattern as a biological marker for disease diagnosis and disease understanding. *NIR News*, **18**(2), 14–16.

Wang D, Dowell FE, Lacey RE (1999) Predicting the number of dominant R alleles in single wheat kernels using visible and near-infrared reflectance spectra. *Cereal Chemistry*, **76**(1), 6–8.

Wigner E (1960) On the unreasonable effectiveness of mathematics in the natural sciences. *Communications in Pure and Applied Mathematics*, **13**, 1–14.

Williams PC (1975) Application of near infrared reflectance spectroscopy to analysis of cereal grains and oilseeds. *Cereal Chemistry*, **52**, 561–576.

Williams PC (2001) Implementation of near infrared technology. In: *Near Infrared Technology in the Agricultural and Food Industries* (Williams P, Norris K, eds). St. Paul, MN: American Association of Cereal Chemists, pp. 145–170.

Williams PC (2002) Near infrared spectroscopy of cereals. In: Handbook of Vibrational Spectroscopy (Chalmers JM, Griffiths VR, eds). Chichester: John Wiley Ltd, Vol. 5, pp. 3693–3719.

Williams PC (2007) Grains and seeds. In: *Near Infrared Spectroscopy in Food Science and Technology* (Ozaki Y, McClure WF, Christy AA, eds). Hoboken, NJ: Wiley-Interscience, pp. 165–217.

Williams PC, Norris K (2001) Variables affecting near infrared spectroscopic analysis. In: *Near Infrared Technology in the Agricultural and Food Industries* (Williams P, Norris K, eds). St. Paul, MN: American Association of Cereal Chemists, pp. 171–198.

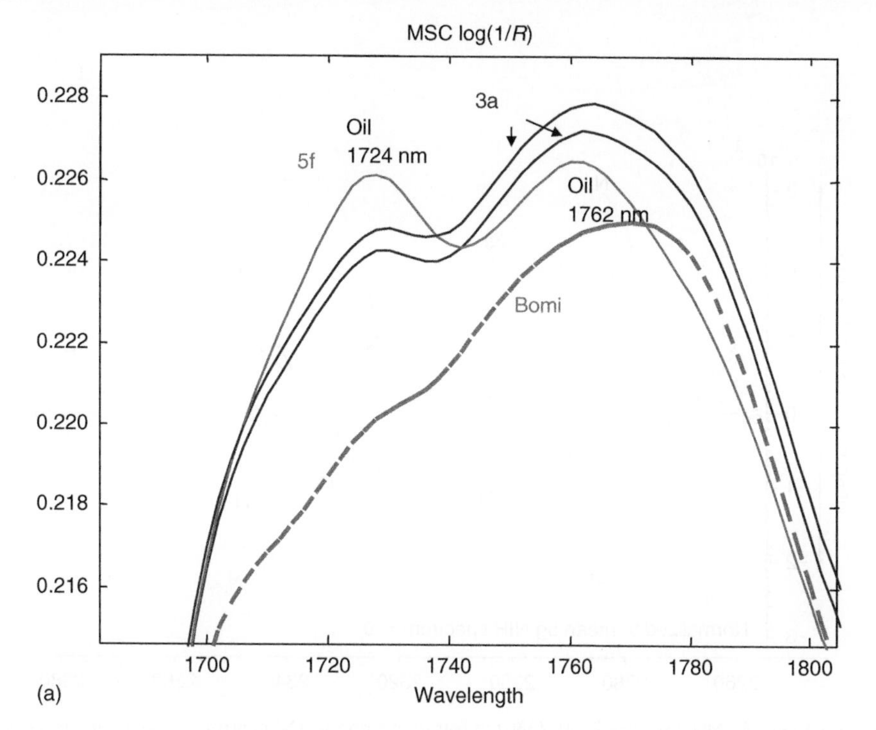

Plate 11.1 Mutant-specific near-infrared (NIR) spectra and correlation coefficients for barley material in Table 11.1. (a) Examples of mutant-specific NIR spectra (log(1/*R*) MSC) 1700–1810 nm from the barley material in Table 11.1 (field): Normal N Barley, cv. Bomi and its P (protein) mutant 3a ($n = 2$, *lys3a*) and C (carbohydrate) mutant 5f (*lys5f*). The peaks at 1724 and 1762 nm of the 5f and 3a spectra indicates the high oil (fat) content in the C and P mutants described in Table 11.1.

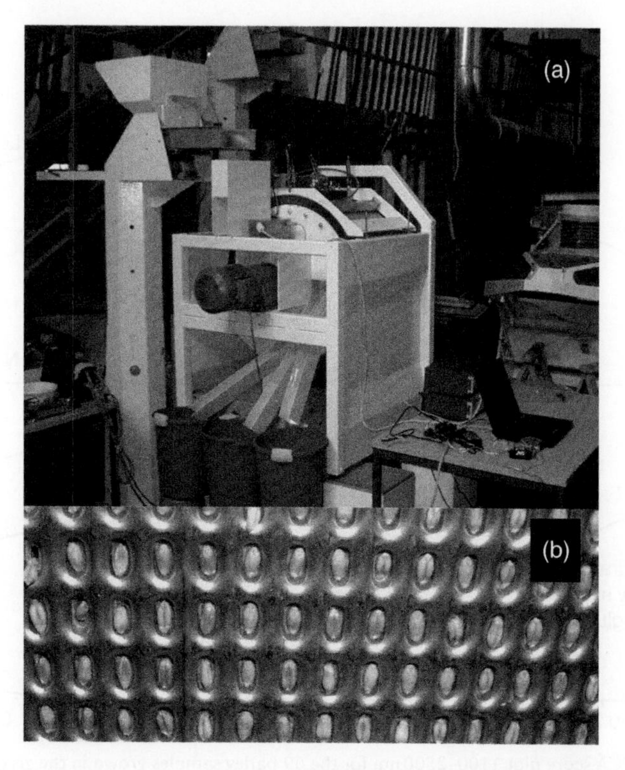

Plate 11.2 Single seed near-infrared (NIR) spectroscopy sorters. (a) Bomill pilot (2–500 kg h^{-1}) TriQ single seed NIR spectroscopy sorting machine (Bomill AB, Lund, Sweden) used in single seed sorting in wheat in Table 11.3. (b) Close-up of cylinder of machine in Figure 11.10a showing pockets to position the seeds before NIR measurement.

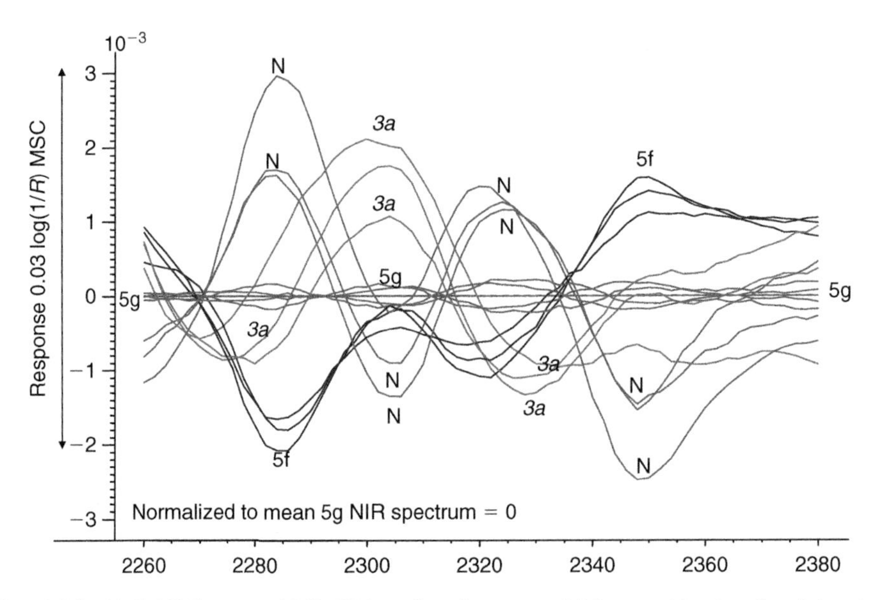

Plate 11.3 Differential (log(1/*R*) MSC) spectra 2260–2380 nm featuring a normal (N) control (cv. Bomi) and three barley endosperm mutants 3a (*lys3a*), 5f (*lys5f*) and 5 g (*lys5g*) grown in three years (1999–2000) in the greenhouse environment (Munck, 2006). The mean spectrum of 5g spectra is the reference equal to zero in the plot subtracted from the other spectra.

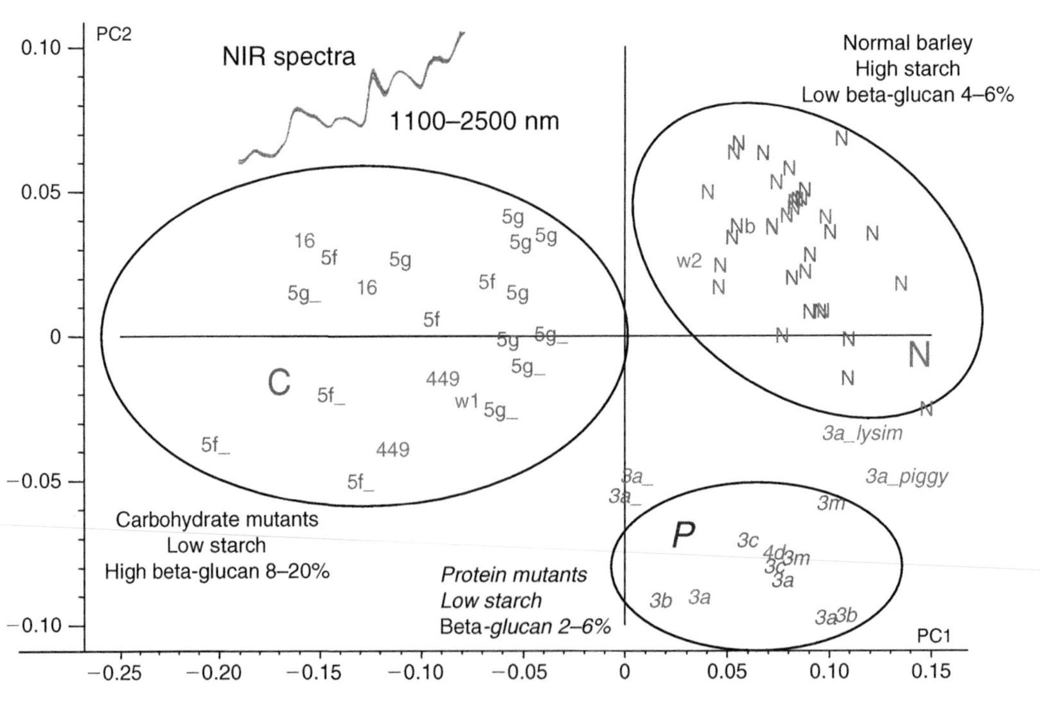

Plate 11.4 A cross-validated PCA score plot 1100–2500 nm for the 69 barley samples grown in the greenhouse from Table 11.1 featuring normal lines N, high-lysine protein mutants P, and carbohydrate mutants C (Munck, 2006). Mutant crosses are underlined and it can be seen that the starch-improved recombinants of *lys3a* Lysimax and Piggy are moving towards the normal N barley class.

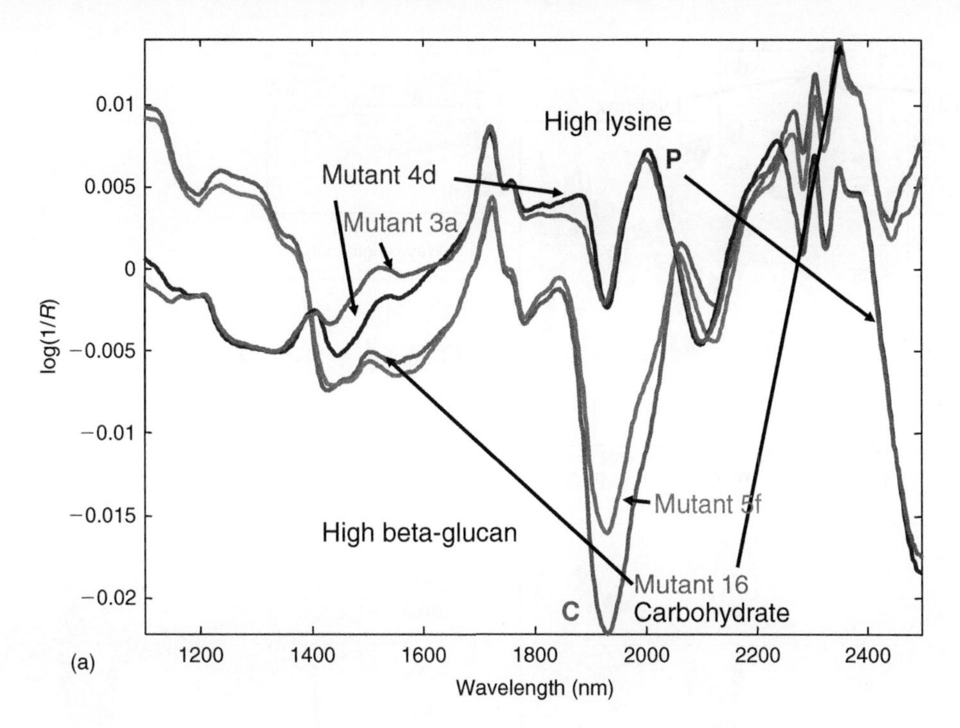

Plate 11.5 Near-infrared (NIR) spectral representation of gene interaction (pleiotropy) (Munck, 2005). (a) NIR spectral representation of gene interaction (pleiotropy): Mean differential NIR spectra log(1/*R*) MSC 1100–2500 nm to the parental genetic background of cv. Bomi of protein P mutants *lys3a* and *lys4d* compared with those of the carbohydrate C mutants *lys5* and Risø mutant 16.

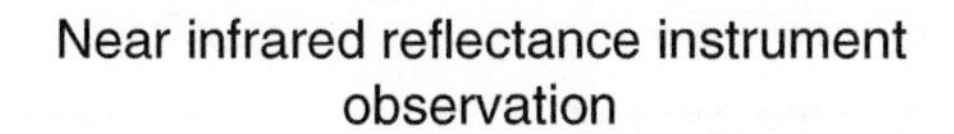

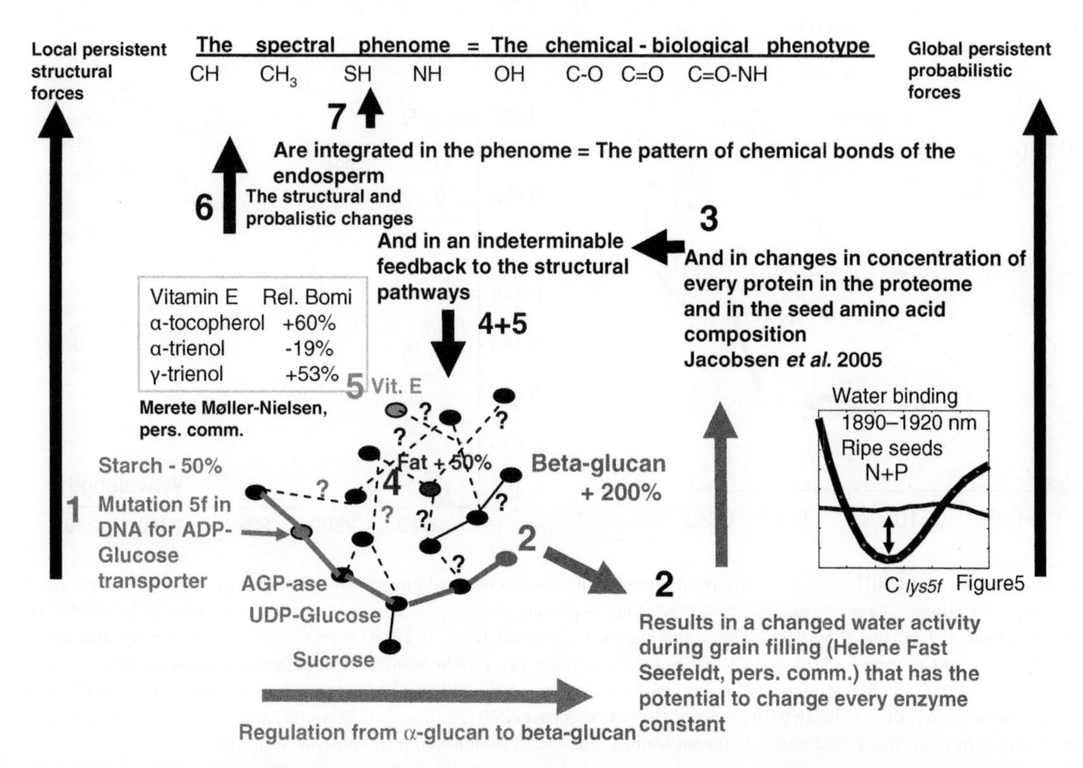

Plate 11.6 Overview graph of the primary and secondary effects on endosperm synthesis of the expression of the carbohydrate C mutant *lys5f* (Munck, 2007). The effect of the cascade of gene expressions numbered **1** to **7** is described in the text.

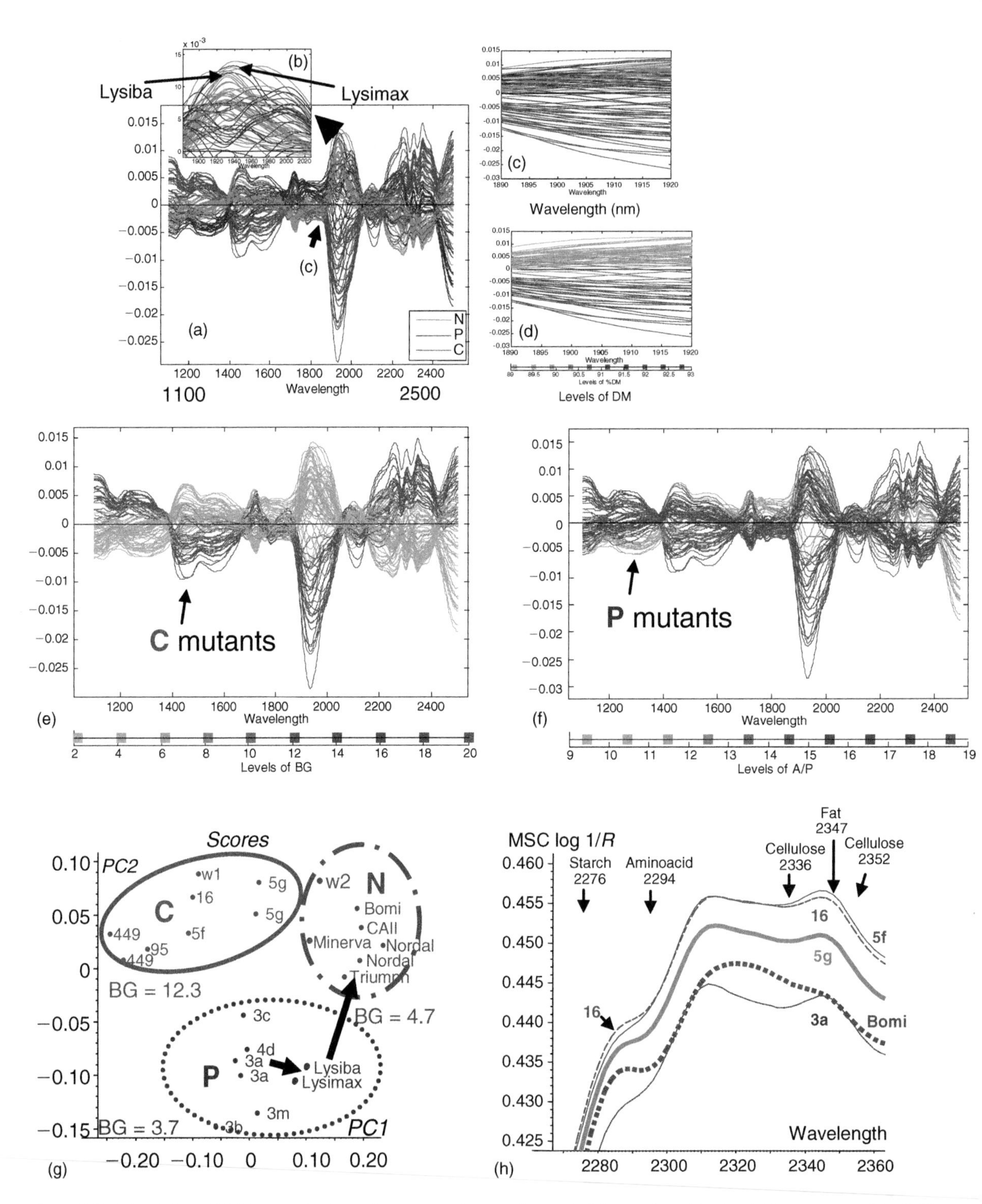

Plate 11.7 Near-infrared (NIR) spectral information representing physiochemical fingerprints of barley. (a) Mean centered log(1/*R*) from 92 barley seed samples. Green = normal barley (N), blue = protein (high lysine, P), and red = carbohydrate (C) mutants. (b) Enlargement of peak at 1935 nm; P outliers Lysimax and Lysiba. (c) Interval 1890–1920 nm from (a). (d) Marking of spectra from (c) for dry matter 89–93%. (e) Coloring for β-glucan (2–20%) of the spectra in (a). (f) The same for amide/protein index (9–19). (g) Principal component analysis (PCA) classification of 23 field barley NIR spectra (1100–2500 nm) for normal barley (N) and protein (P) and carbohydrate mutants (C). BG, β-glucan % dry matter. (h) NIR spectra (2270–2360 nm) of Bomi (N) and its P *lys3a*, and C mutants *lys5f*, *lys5g* and 16 grown in greenhouse. a–d and g–h. (From Munck, 2007 with permission from Wiley & Sons Ltd.)

Fruits and Vegetables

12

Hartwig Schulz and Malgorzata Baranska

Introduction

In comparison to near-infrared (NIR) reflectance spectroscopy, which has been used extensively for food authentication and quantification of various components in fruits and vegetables as well as medicinal and spice plants (Kawano, 2002; Slaughter and Abbott, 2004; Schulz, 2004), the use of mid-infrared (MIR) spectroscopy for the same purposes has been more limited. Until recently, MIR spectroscopy was primarily used in agricultural research as a qualitative technique for identification and verification of unknown pure substances isolated from extracts or distillates (Colthup *et al.*, 1990).

Infrared Spectroscopy for Food Quality Analysis and Control
ISBN: 978-0-12-374136-3

Usually infrared (IR) spectra obtained from plant samples are very complex because each functional group in a molecule contributes more or less to the spectral output. The net result is a spectrum in which band assignments may be difficult due to the fact that overlapping and mixing of various vibrational modes occur. Early attempts to combine IR spectroscopy with gas chromatography (GC) allowing volatile flavor substances released from fruits or vegetables to be analyzed directly were only partly successful. First developments using dispersive IR units suffered mainly because of their comparatively low sensitivity. The introduction of gas chromatography combined with Fourier transform infrared spectroscopy (GC-FTIR) provided better signal-to-noise ratio and spectra could be obtained in a shorter time. Applications of these techniques for the analysis of plant volatiles have been reviewed by Herres (1984) as well as David and Sandra (1992).

More recently, a matrix isolation interface has been developed for IR measurements. Here one part of the chromatographic effluent is trapped in an argon matrix onto a cylinder. This approach leads to significantly higher sensivity, sometimes rivaling that of gas chromatography-mass spectrometry (GC-MS) measurements. Due to the fact that band widths are usually very sharp for matrix-isolated substances, high spectral resolution can be obtained using this technique. In general, the additional spectral information obtained by GC-FTIR provides very important data, especially for a reliable characterization of complex flavor mixtures and discrimination of isomeric compounds. Nevertheless, it should be noted that spectra measured in the cryogenic state are different from those obtained from the condensed or vapor phase. Today GC-FTIR systems use a heated gas cell for measuring the spectra in the vapor state.

The comparatively new approach to use MIR spectra for the analysis of plant samples in the same fashion as NIR spectra brought the added advantage of spectral interpretability. It has recently been shown that the application of MIR leads to analysis results with an accuracy equal to or better than that found using NIR spectroscopy (Reeves, 1994, 1996; Briandet *et al.*, 1996; Downey *et al.*, 1997).

FTIR spectroscopy has become a powerful tool for elucidating the structure, physical properties, and interactions of various carbohydrates, including commercial sugars, cellulose, pectins, starch, hemicellulose, carrageenans and others. Applications in the area of systematic fingerprinting, quantification, as well as IR microspectroscopy to monitor cell wall constituents such as pectins, proteins, aromatic phenols, cellulose, and hemicellulose have been reviewed by Kačuraková *et al.* (2001).

Several attempts have also been made to correlate MIR spectra with NIR spectra to improve the interpretation of NIR data (Barton and Himmelsbach, 1993; Noda and Ozaki, 2004; Schulz *et al.*, 2007; Westad *et al.*, 2007).

Fruits

Apple

Soluble solid content (SSC) is a major characteristic used for assessing apple fruit quality and normally is performed destructively on juice. Non-destructive determination of SSC in apple fruits has been performed by the use of a portable fiber-optic NIR spectrometer (Ventura *et al.*, 1998). A total of 340 apples of cv. Golden Delicious and cv. Jonagol were analyzed and a multiple linear regression (MLR) equation was applied to calculate the °Brix value in the prediction data set. The most significant prediction quality ($R^2 = 0.56$) was found with the first derivative of $\log(1/R)$ (where R is reflectance), but when MLR was carried out separately on each cultivar, the reliability of the NIR method could be improved for "Golden Delicious" ($R^2 = 0.65$).

Recently, important applications of NIR spectroscopy in determining quality parameters of apples such as acids, sugars, moisture, soluble solids, nitrogen, maturity and firmness have been reviewed by Slaughter and Abbott (2004).

A hyperspectral imaging system was developed by Nicolaï *et al.* (2006) to identify a bitter pit lesion on apples. A PLS calibration model was constructed to discriminate between pixels of unaffected and bitter pit lesions. The obtained calibration model was successfully validated on different apples. The system was able to identify bitter pit lesions even when they were not visible, but could not discriminate between bitter pit lesion and corky tissue.

The surface of apples has been characterized by FTIR-photoacoustic spectroscopy (FTIR-PAS) in order to identify contamination by food microorganisms (Yang *et al.*, 2001). In the experiment described by the authors, suspensions containing test microorganisms (*Saccharomyces cerevisiae*, *Lactobacillus casei*, *Escherichia coli*, *Staphylococcus aureus*) were placed at the surface of the apple skin and dried at room temperature for 12 h before spectroscopic measurements. Five regions in the spectrum were found to be relevant for discrimination and classification of the individual microorganisms (3050–2800 cm^{-1} corresponding to CH_2 and CH_3 groups; 1750–1500 cm^{-1} due to protein and peptide bands; 1500–1200 cm^{-1} presenting bands of fatty acids, proteins and polysaccharides; 1200–900 cm^{-1} dominated by polysaccharide peaks; 900–700 cm^{-1} showing mainly deformation vibrational modes). In general, this new approach demonstrates that the FTIR-PAS technique can be effectively used to distinguish the coating surface of fruit species and to detect the presence of microorganisms. In this context two-dimensional correlation spectroscopy was extremely helpful in identifying coinciding vibrational bands.

Citrus fruits

The essential oils of various citrus species was studied by attenuated total reflectance (ATR)-FTIR spectroscopy by Schulz *et al.* (2002). They reported the application of this technique for classification and quantitative analysis of orange, grapefruit, mandarin, lemon, and lime oils. The most relevant monoterpene components occurring in these oils are limonene and γ-terpinene, but α- and β-pinene, myrcene, sabinene, octanal, decanal, citral, sinensal, and nootkatone can also be present. In grapefruit, orange, and bitter orange oils limonene occurs at levels of approximately 95% and in other citrus oils at 50–78%. It is therefore not surprising that the IR spectra of these oils are mainly characterized by limonene vibrational modes to be seen at 886 cm^{-1} (out-of-plane bending of the terminal methylene group), 1436/1453 cm^{-1} (δ_{CH2}) and 1644 cm^{-1} ($\nu_{C=C}$) (Lin-Vien *et al.*, 1991; Schulz and Baranksa, 2005,

2007). The stretching mode of cyclohexene C=C can be hardly recognized in the IR spectrum, but demonstrates a strong Raman band. Myrcene, which is the only acyclic substance of the six main monoterpene compounds, presents specific absorption bands of the vinyl substituent at 1637 cm^{-1} and two characteristic out-of-plane C–H bending vibrations at 989 and 890 cm^{-1}. The narrow band seen at 1595 cm^{-1} is assigned to the vibration of the double bond, which is conjugated with the terminal methylene group of the molecule. The spectra of α- and β-pinene differ from each other, especially in the region of 750–900 cm^{-1}. Whereas α-pinene presents the characteristic signal of the ω(C–H) at 787 cm^{-1}, β-pinene shows the absorption band of the terminal methylene group at 873 cm^{-1} and of the cyclohexane ring at 853 cm^{-1}. The corresponding cyclohexane vibration of α-pinene occurs at 886 cm^{-1}. The individual ν (C=C) stretching vibrations of both monoterpenes are found at 1658 cm^{-1} (α-pinene) and 1640 cm^{-1} (β-pinene), respectively. The IR spectra of most important monoterpene substances occurring in citrus oils are presented in Figure 12.1.

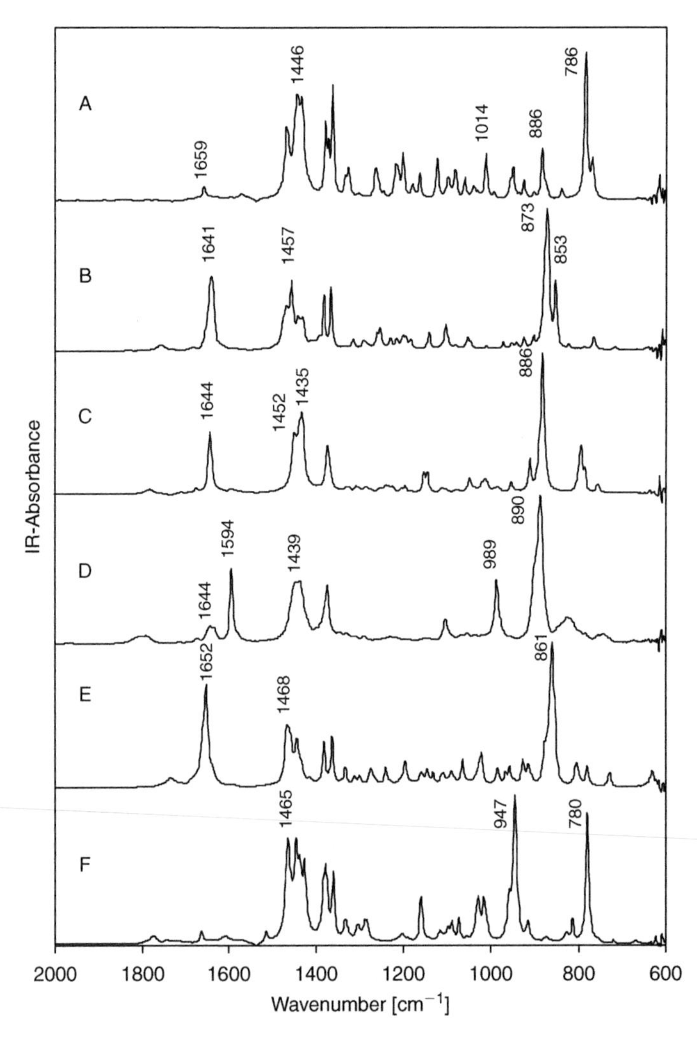

Figure 12.1 Mid-infrared spectra of pure monoterpene substances representing the main components in commercially used citrus oils. A: α-pinene, B: β-pinene, C: limonene, D: myrcene, E: sabinene, F: γ-terpinene.

In general, most of the discussed compounds show characteristic signals in the ATR-IR spectrum due to wagging vibrations of CH and CH_2 groups between 800 and 950 cm^{-1} (see Table 12.1), but by using Raman spectroscopy the differentiation between these groups is more clear (Daferera *et al.*, 2002; Schulz *et al.*, 2004, 2005a; Baranska *et al.*, 2005). Among monoterpenes, numerous aldehyde derivatives can be well recognized by IR spectroscopy where the intense IR band due to the C=O stretching mode is seen in the area of 1740–1750 cm^{-1}. Schulz *et al.* (2002) have reported that this key signal can be used to discriminate between cold-pressed and distilled lime oil. Whereas cold-pressed lime oil shows a relatively intense carbonyl signal at 1744 cm^{-1}, this band is completely missing in the IR spectrum of distilled oil.

In the range between 2000 and 3000 cm^{-1} bands associated with the C–H stretching vibrations of the volatile terpenoids as well as wax esters of the citrus peel can be found. Other significant signals resulting from ν(CH=CH) vibrations can be observed between 3020 and 3090 cm^{-1}.

Quantitative analysis of citrus oils based on ATR-IR spectroscopy achieved a better prediction quality than NIR reflectance spectroscopy, in terms of higher R^2 values for some parameters (e.g. myrcene and aldehyde) (Steuer *et al.*, 2001; Schulz *et al.*, 2002). Also the standard error of cross-validation (SECV) values, calculated from

Table 12.1 Assignment for the most characteristic infrared bands of some terpene compounds occurring in citrus oils

Terpene	Wavenumber (cm^{-1})	Assignment
Myrcene	1637	ν(C=C)
	1595	ν(C–C)
	989	$\omega(CH_2)$
	890	ω(C–H)
p-Cymene	1515	ν(ring)
	813	ω(C–H)
Limonene	1678[a]	ν(cyclohexene C=C)
	1644	ν(ethylene C=C)
	886	ω(C–H)
α-Terpinene	823	ω(C–H)
γ-Terpinene	947	$\omega(CH_2)$
	781	ω(C–H)
α-Pinene	1658	ν(C=C)
	886	$\omega(CH_2)$
	787	ω(C–H)
β-Pinene	1640	ν(C=C)
	873	$\omega(CH_2)$
	853	ω(C–H)
Sabinene	1653	ν(C=C)
	861	$\omega(CH_2)$

From Baranska *et al.* (2005), Schulz *et al.* (2004, 2005b), Schulz and Baranska (2005, 2007).
[a]Weak band.

the PLS calibrations of the IR validation sets, are in most cases a little bit lower in comparison to the related NIR chemometrical results. This is mainly due to the fact that the IR spectra generally have larger signal intensities and higher resolution of the characteristic (fundamental) vibrations than NIR transflection spectroscopy.

IR spectroscopy can also be used for investigation of the aging of citrus oils by following the (per)oxidation of γ-terpinene and formation of *p*-cymene (Lösing *et al.*, 1998). Because of its characteristic wavelength in the IR spectrum, the occurrence of *p*-cymene is easy to observe. Its two intense IR bands at 1515 and 815 cm^{-1} can be assigned to the aromatic skeleton resonance and C–H out-of-plane vibration of *para*-disubstituted benzene ring, respectively. These bands do not overlap with those of other essential oil components. The concentration of *p*-cymene has been used as an indicator for the advancing aging process of citrus oils and the formation of off-flavors.

Some flavones occurring in green tangerine peel have been isolated and characterized using spectroscopic technique (Dandan *et al.*, 2007). The system of bonded rings characteristic of flavones can be identified in the IR spectrum between 1570 and 1600 cm^{-1}.

It is well known that FTIR spectroscopy is an excellent tool for structural and quantitative analysis of polysaccharides, including pectins (Kačuraková *et al.*, 2001; Schulz and Baranska, 2007) (see Table 12.2 for details). Citrus pectin and its various carboxylic forms (i.e. pectinic acid, potassium pectinate, pectic acid, potassium pectate and pectinamides) were analyzed by Synytsya *et al.* (2003) by the use of diffuse reflectance FTIR. Spectra of these compounds demonstrated strong signals in the region of carbonyl stretching vibrations (1500–1800 cm^{-1}), which were sensitive to any chemical changes (e.g. bands of methyl ester group near 1750 cm^{-1}) and were decreased after alkali hydrolysis.

Steuer *et al.* (2001) have analyzed various citrus oils by the use of NIR spectroscopy. The measured citrus oils obtained from grapefruit, orange, mandarin, lemon, bitter orange, and lime could be distinguished on the basis of their individual spectral data using principal component analysis (PCA) by the first three PCA factors (98.3%

Table 12.2 Assignment for the most characteristic infrared bands of pectin

Wavenumber (cm^{-1})	Assignment
1745	$\nu(C=O)$
1605	$\nu_{as}(COO^-)$
1444	$\delta(CH)$
1419	$\nu_s(COO^-)$
1368	$\delta(CH_2)$, $\nu(CC)$
1335	$\delta(CH)$, ring
1150	$\nu(C\text{–}O\text{–}C)$, ring
1107	$\nu(CO)$, $\nu(CC)$, ring
1055	$\nu(CO)$, $\nu(CC)$, $\delta(OCH)$
1033	$\nu(CO)$, $\nu(CC)$, $\nu(CCO)$
1018	$\nu(CO)$, $\nu(CC)$, $\delta(OCH)$, ring
1008	$\nu(CO)$, $\nu(CC)$, $\delta(OCH)$, ring
972	OCH_3
963	$\delta(C=O)$

According to Schulz and Baranska (2007).

of the variation explained). The highest influence was found for factor 1 in accordance to high (orange, grapefruit) or low (lime) limonene contents. Discrimination between cold-pressed and distilled lime oil was also achieved. Since the optical rotation is strongly correlated with the limonene content, this physical parameter was also predicted precisely in all citrus oils. Not only were the individual terpenoid components (e.g. limonene, myrcene, α-pinene, β-pinene, sabinene, γ-terpinene and terpinolene) determined, but also the total aldehyde content (ranging from 0.3 to 3.2 g 100 g^{-1}).

In order to develop calibration equations for the two key flavor substances decanal and nootkatone, synthetic mixtures have been produced by diluting both components in orange oil terpenes and grapefruit oil, respectively. A high weight is observed at 2202 nm for decanal by the combination of the C=O and C–H stretching frequencies of the aldehyde group. Negative bands in the first loadings are observed according to the influence of limonene in the synthetic mixture.

Banana

The principal component of green bananas is starch that changes during fruit ripening. The retrogradation of starch isolated from banana has been studied using diverse techniques, including FTIR spectroscopy. Short-range order measurement has been performed with ATR-IR using a Golden Gate cell sealed with a sapphire anvil from the atmosphere. The applied vibrational method showed the structural changes in the starch during banana storage at the molecular level (Bello-Pérez, 2005).

To analyze the carbohydrate region of the starch spectra (800–1200 cm^{-1}), baseline correction was applied using a single point at 1900 cm^{-1}. The spectra were then deconvoluted in the above-mentioned region, where the assumed line shape was Lorentzian, with a deconvolution factor of 750 and a noise reduction factor of 0.2. After deconvolution, the 800–1200 cm^{-1} region consisted of a series of bands, mostly of C–O and C–C stretching vibrations that were reported to be very sensitive to the physical state of carbohydrates (Ottenhof *et al.*, 2003). The bands at 1045 and 1022 cm^{-1} are sensitive to the amounts of ordered and amorphous starch forms, respectively. The band at 1151 cm^{-1} was used as an internal correction standard. The ratio of the bands at 1045 and 1151 cm^{-1} was taken to follow the change in a short-range order during the banana starch retrogradation. This ratio increases with increasing storage times, reaching a plateau after approximately 11 h, a value that agrees with the results obtained by differential scanning calorimetry measurements (Bello-Pérez, 2005). To monitor the moisture loss occurring during the experiment, the ratio of the bands at 1635 and 1151 cm^{-1} from the deconvoluted spectra were correlated against time. The band at 1635 cm^{-1} was assigned to the O–H deformation and therefore it could directly be used to determine the moisture content in the sample. It has been found that a linear correlation exists between the moisture content of a retrograded starch sample and the band ratio of 1635:1151 cm^{-1}. The decrease of this ratio corresponds to the loss in moisture from the sample that would decrease the retrogradation kinetics, as was also observed to be the case for waxy maize starch (Farhart *et al.*, 2000).

Pear

The sun-dried pear possesses a unique organoleptic characteristic. In order to understand the modifications that occur during the drying process, cell wall extracts from dried and fresh tissue were prepared and measured by vibrational spectroscopy. The polysaccharides present in different extracts were characterized by FTIR spectroscopy in the region between 1200 and 850 cm^{-1}. The analysis of the obtained spectra using different chemometric algorithms allowed fresh and sun-dried pear extracts to be distinguished. Components that contribute to this distinction are the pectic and hemicellulosic polysaccharides rich in galacturonic acid (GalA) and xylose (Xyl), respectively (Ferreira *et al.*, 2001). Hemicellulosic extracts were characterized mainly by the peak located at 1041 cm^{-1} (Coimbra *et al.*, 1999) whereas pectic polysaccharide extracts showed absorbances at 1106, 1014, and 914 cm^{-1}. The absorbance bands detected at 1106 and 1014 cm^{-1} are due to the galacturonic acid (Coimbra *et al.*, 1998, 1999) and the peak located at 914 cm^{-1} is related to the non-dialyzable 1,2-*trans*-diaminocyclohexan-*N*,*N*,*N'*,N'-tetraacetate (CDTA) salt (the extracts were precipitated in CDTA).

Papaya

During the post-harvest ripening of fruits significant textural changes occur due to modification of the structure and composition of cell wall polysaccharides, mainly pectins and hemicelluloses (Melford and Prakash, 1986; Fry, 1995). It has been demonstrated that during fruit softening pectins undergo solubilization and depolymerization (Brady, 1987; Fischer and Bennett, 1991). It has been suggested that the methyl esterification degree (MED) of pectins in particular could be a parameter that would be useful to study some aspects of the fruit-softening mechanism (Bartley and Knee, 1982; Melford and Prakash, 1986). Several methods are available for measuring pectin MED, including the titration of carboxyl groups, the measurement of galacturonic acid in native pectin or measurement of galactose derived from selective esterified galacturonic acid reduction. The MED of pectins can also be determined by means of instrumental methods, such as high-performance liquid chromatography (HPLC) (Plöger, 1992) or ^{1}H-NMR spectroscopy (Grasdalen *et al.*, 1988).

A direct method using FTIR spectroscopy for the determination of MED of papaya pectins was applied by Guillermo and Lajolo (2002). The method was used to measure the methylation level of different pectin fractions isolated from papaya fruit at three ripening stages as well as of bulk pectin without isolation from the cell wall. The carbohydrates show high absorbances between 1200 and 950 cm^{-1} constituting the "fingerprint" region, specific for each polysaccharide. However, the assignment of bands in this region to a specific atom group vibration is ambiguous. MED values were calculated from absorbance spectra of the samples, using a relationship of the intensities for 1630 and 1740 cm^{-1} bands ($A_{1740}/(A_{1740} + A_{1630})$), that are intense and well separated. The signal at 1630 cm^{-1} has been assigned to the stretching frequency for the carbonyl groups of galacturonic acid whereas the band at 1745 cm^{-1} has been associated with its methyl ester. The obtained results were in agreement with those achieved using an established method.

Peach

As already mentioned, FTIR spectroscopy is a method suitable for monitoring chemical properties of cell walls and more specifically changes in the degree of esterification. The degree of esterification, defined as "number of esterified carboxylic groups/number of total carboxylic groups × 100," is one of the most important properties for characterization of the pectic molecules, which represent macromolecular constituents occurring in the cell wall.

FTIR spectra obtained from the cell walls of cv. Redhaven peaches immediately after harvest demonstrate several bands, which occur in a few wavenumber regions. The region between 3500 and 1800 cm^{-1} presents two major peaks centered at 3455 cm^{-1} (due to stretching of the hydroxyl groups) and at 2920 cm^{-1} (corresponding to the C–H stretching of CH_2 groups). The region between 1800 and 1500 cm^{-1} is of special interest with regard to the evaluation of the degree of esterification, since it allows the observation of IR absorption by the carboxylic acid and the carboxylic ester groups of the pectin molecules (Stewart and Morison, 1992). FTIR spectra of peach cell walls in the above-mentioned range revealed the existence of two peaks absorbing at 1749 and 1630 cm^{-1} assigned to the esterified and non-esterified carboxyl groups of pectin molecules, respectively (Chatjigakis *et al.*, 1998). A linear relationship between the degree of esterification and the ratio of the area underneath the peak at 1749 cm^{-1} over the sum of the areas underneath the two peaks at 1749 and 1630 cm^{-1} was established. The analysis was performed by the use of the second derivative and curve-fitting technique that allowed the elimination of spectral interferences from other cell wall components.

Strawberry

Different chemical components and their location in strawberry achene, vascular bundles, and cortical cell walls were studied by means of FTIR spectroscopy (Suutarinen *et al.*, 1998). First, the spectra of commercial pectin, protein, lignin, and cellulose were measured in the 700–4000 cm^{-1} range and compared with the spectra obtained from the different strawberry tissues. Spectra of the commercial compounds were obtained using the DRIFT (diffuse reflectance infrared Fourier transform) technique, whereas the sections of frozen strawberries cut in a cryostat were analyzed with FTIR microspectroscopy. Lignin is well known as an important component of achene and vascular bundles, whereas the cortical cell walls contain mainly pectin and cellulose in the middle lamella, and protein as deposits in the outer layer. The structure of all the walls was complicated and consisted of several compounds. The same samples have been examined by bright-field microscopy using different staining systems and the two methods gave comparable results.

A low-resolution gas phase FTIR analyzer was applied to the analysis of volatile compounds present in strawberry (Hakala *et al.*, 2001). The frozen fruit was thawed, mashed, and transformed into glass mounted to a sampling system. Volatile samples were collected by gently heating in a vacuum and then removed into the sample cell of the gas analyzer at atmospheric pressure. The quantitative analyses of 14 compounds were performed by multicomponent analysis. The main volatiles were esters, alcohols, and aldehydes. The highest proportions were measured for acetone and 2,5-dimethyl-4-hydroxy-3(2H)-furanone. Significant differences between strawberry varieties were noticed.

FTIR spectroscopy and chemometrics have been combined to detect adulteration in strawberry purées (Holland *et al.*, 1998). The MIR spectra of 983 fruit purées were used as raw data for PLS regression. After chemometric calculation, 94.3% of the samples were correctly classified by the applied model.

Grape

Phenols contribute in an important manner to the taste, bitterness, and bacteriological effects of grape wines. These compounds include catechins, leucoanthocyanidins, flavonols, flavonol glycosides, tannins, proanthocyanidins and anthocyanidins, phenolic benzoic and phenolic cinnamic acids (Spanos and Wrolstad, 1990; Cartoni *et al.*, 1991). High levels of polyphenols increase susceptibility to oxidation, leading to decreased visual and organoleptic qualities. To study the effect of wine processing on phenolic composition spectroscopic analysis was used. Various classes of phenolic compounds were detected and characterized by IR spectroscopy in white grapes of cv. Sauvignon Blanc and cv. French Colombard, as well as in wines prepared from these grape cultivars (Gorinstein *et al.*, 1993).

Comparisons of the polyphenol compositions of wines made from the same grape variety grown in different locations of the same vintage and between two vintages were reported. In general, the phenolic O–H stretch is observed at 3705–3125 cm^{-1} for phenols, catechols and resorcinols, and at 3335–2500 cm^{-1} for aromatic carboxylic acids. The IR spectra of some investigated polyphenols resulted in the following peaks: 2643, 2746, 2849, 2937, 3106 cm^{-1} for catechin, 2541, 3210 cm^{-1} for caffeic acid, 2875, 3132, 3544 cm^{-1} for gallic acid and 3232, 3441 cm^{-1} for tannic acid. FTIR spectra of standards and wine samples were totally consistent with one another in the O–H stretch region and the following bands were observed in samples of wine treated with bentonite and egg albumin at 2515, 3081, 3389, 3492, 3544 cm^{-1} and for samples of wine treated with bentonite, egg albumin, and polyclar at 2566, 2721, 2926, 3184, 3284, 3441, 3698 cm^{-1}.

Furthermore, IR spectroscopy has been successfully used to monitor wine fermentation (Urtubia *et al.*, 2004) as well as the quality of the final product, since many compounds can be measured quickly from a single sample without prior treatment (Coimbra *et al.*, 2002; Patz *et al.* 2004). The calibration model obtained by Urtubia *et al.* (2004) with a multivariable PLS algorithm proved to be effective for analyzing cv. Cabernet Sauvignon fermentation for glucose, fructose, glycerol, and ethanol as well as malic acid, succinic acid, lactic acid, acetic acid, and citric acid. Upon external validation an average relative predictive error of 4.8% has been found; in this context malic acid showed the largest relative predictive error (8.7%). A good calibration has been obtained for all investigated compounds, with the lowest value of R^2 for acids (about 0.98) and the highest for glucose and fructose (0.994).

Sugars in grapes were also determined by using NIR technology (Jarén *et al.*, 2001). NIR (800–2500 nm) reflectance spectra of 30 hand-harvested samples (cv. Garnacha and cv. Viura) were registered and analyzed using SPSS and SAS. It was possible to find a correlation between NIR data and sugar content in the analyzed grapes (°Brix). The calibration model obtained for cv. Viura ($R^2 = 0.925$, SEE = 1.0446) was slightly better than for cv. Garnacha ($R^2 = 0.89$, SEE = 1.0508).

NIR spectroscopy and the semi-parametric modeling technique least-squared support vector machine (LS-SVM), were used to predict the acidity of different grape varieties (Chauchard *et al.*, 2004). The performances of LS-SVM was found to be better than that of the classical linear methods (i.e. partial least square regression (PLSR) and multivariate linear regression (MLR)).

An innovative FTIR method coupled with thin layer chromatography (TLC) for the identification of pigments extracted from different red wine cultivars has been presented by Cserháti *et al.* (2000). The measurements prove that off-line TLC-FTIR can be successfully used for the determination of individual substances occurring in the main pigment fraction, whereas on-line TLC-FTIR failed to provide suitable results because of the strong background absorbance caused by the stationary TLC phase.

Paprika and chilli fruits

The widespread use of capsainoid extracts in flavorings make the quantitative analysis of these compounds of great importance. Five different capsainoids compounds have so far been found in paprika and chilli fruits, of which capsaicin is one of the most important (Schulz, 2004).

A new NIR method has been established for the determination of capsaicin content in red paprika and related extracts (Iwamoto *et al.*, 1984). The spectra of the solvent organic residues were recorded in the region between 1100 and 2500 nm using a single-beam spectrophotometer. The strong bands observed in the second derivative spectra between 2250 and 2350, 1700 and 1760 and at 1200 nm were assigned to the combination vibration, the first overtone and the second overtone vibrations of the individual C–H groups, respectively. The bands near 1950 and 2200 nm were interpreted as combination vibrations of N–H and O–H groups in the capsaicin molecule. A stepwise MLR analysis was performed to select the five most suitable absorption wavenumbers. The chemometric results ($R^2 = 0.993$, SEE = $0.0036\,g\,100\,g^{-1}$, calibration range $0.05–0.13\,g\,100\,g^{-1}$) led to the conclusion that NIR reflectance spectroscopy is applicable to the determination of capsaicin content in red paprika.

Melon

Different melon genotypes have been distinguished by Seregely *et al.* (2004) using NIR spectroscopy. First, a hybrid cultivar with its two parent lines were examined and bulk-seed samples of four muskmelon and five watermelon varieties were also examined. As a quantitative evaluation method, the polar qualification system (PQS) was used. For the hybrid lines, the value of sensitivity expressing the effectiveness of the classification was found to be 12.32. For classification of the muskmelon varieties, the sensitivity was between 1.35 and 3.37 among the different varieties. For watermelon samples, slightly lower values were received for sensitivity. The results achieved by PQS were comparable with the results obtained by classical evaluation methods (PCA, linear discrimination analysis).

Vegetables

Potato and cassava

The main signals to be observed in the IR spectra of potato and cassava are related to starch, representing a mixture of amylose and amylopectin (Figure 12.2). Amylose is essentially linear, whereas amylopectin is a highly branched polymer. Santha *et al.* (1990) measured the IR spectra of sweet potato and cassava in order to find key bands for amylose and its branched counterpart. In the fingerprint region the signal at 1263 cm^{-1} is assigned as a complex mode involving the CH_2OH side-chain in amylose. A band near 946 cm^{-1} is interpreted as a skeletal mode indicating the α-1,4 linkage of glucose molecules in potato amylose. Another signal observed at 943 cm^{-1} has been found to represent linkages in both amylose and amylopectin.

Two absorption bands at 861 and 840 cm^{-1}, formerly identified in potato (Cael *et al.*, 1974), do not occur in the IR spectra of sweet potato and cassava starch; only a small band around 870 cm^{-1} can be seen. It is assumed that this effect is mainly caused

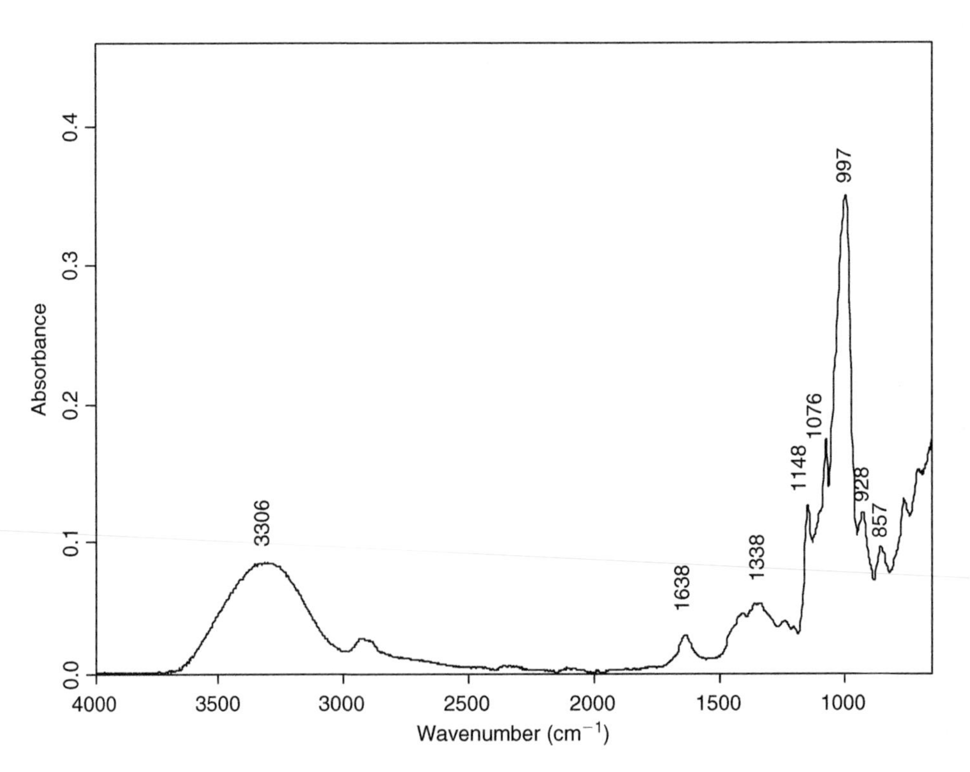

Figure 12.2 Attenuated total reflectance infrared spectrum obtained from pure amylose.

by lack of crystallinity in the measured starch samples. Similar experiments studying the change from semi-crystalline to amorphous starch by ATR-IR spectroscopy have been performed by van Soest *et al.* (1995). They found that the IR absorbance band at $1047\,cm^{-1}$ is sensitive to the amount of crystalline starch whereas the band at $1022\,cm^{-1}$ is characteristic of amorphous starch. The IR spectra obtained from native potato starch showed intensive CH stretching vibrations in the region $2900–3000\,cm^{-1}$ and at 1150, 1124, and $1103\,cm^{-1}$ (CO, CC stretching with some COH contributions), 1077, 1047, 1022, 994, and $928\,cm^{-1}$ (COH bending and CH_2-related modes) and $861\,cm^{-1}$ (COC symmetrical stretching and CH deformation). Signal assignment was generally limited due to overlapping and poor resolution of the detected bands. In some cases enhanced resolution of overlapped bands could be achieved by deconvolution.

Capron *et al.* (2007) studied the variation in powders of starch materials at various controlled hydrations using ATR-FTIR spectroscopy. An intense IR absorption at $1000\,cm^{-1}$ was assigned to hydrated crystalline domains whereas the signal at $1022\,cm^{-1}$ revealed the spectral contribution of amorphous starch. A sample set varying in structure, crystalline type, amylose content, and botanical origin presented a major contribution of the $1000/1022\,cm^{-1}$ ratio.

These studies illustrate that FTIR spectroscopic data can be used to describe phase transitions on biopolymers such as starch. The kinetics of conformational changes due to retrogradation during storage of potato and maize starch was monitored by ATR-FTIR (van Soest *et al.*, 1994). In this context a starch–water system (10% w/w gel) was used as a model to distinguish differences on a structural level in various starch materials. The authors showed that major contributions in the IR spectra ($800–1300\,cm^{-1}$) were obtained on gelatinization and subsequent retrogradation of waxy maize and potato starch/water systems.

It has also been shown that FTIR spectroscopy can be successfully applied to follow the gelatinization of starch granules *in situ* under high pressure (Rubens *et al.*, 1999). During this process the spectra showed characteristic changes in bandwidth, increasing absorption intensities, and frequency shifts in the wavenumber range $900–1300\,cm^{-1}$. These spectral changes could be successfully used to determine the individual midpoint of gelatinization for various starch materials (Table 12.3).

Table 12.3 Determination of midpoint of gelatinization $p½$ based on infrared absorption ratio and frequency shift

Starch type	$p½$[a]	Frequency shift[b]
Rice	430 ± 6	390 ± 10
Waxy corn	440 ± 10	420 ± 20
Corn	520 ± 2	510 ± 4
Tapioca	440 ± 6	410 ± 9
Pea	460 ± 20	370 ± 10
Potato	650 ± 9	480 ± 30

From Stute *et al.* (1996).
[a]Calculated from the absorbance ratio $1017/1047\,cm^{-1}$.
[b]Calculated from the frequency shift of the $1080\,cm^{-1}$ band.

ATR-FTIR spectroscopy combined with procedures for spectrum deconvolution has also been used to investigate the external regions of starch granules. The IR spectra of potato and amylomaize starches were found to be closer to those of highly ordered acid-hydrolyzed starch than the spectra obtained from wheat, maize, and waxy maize (Sevenou *et al.*, 2002).

The absorptions at 1047, 1022, and 995 cm^{-1} were recorded from the individual starch samples and the ratio of absorbances 1047/1022 and 1022/995 cm^{-1} was calculated. Based on these data a clear discrimination between potato, wheat, and different maize starches could be achieved. It has been found that the band at 1022 cm^{-1} was generally less pronounced in potato and amylomaize than in wheat maize and waxy maize. From these results the authors reason that potato and amylomaize exhibit a higher level of organization in their external region compared with the other starch types; this also illustrates that potato and amylomaize starch granules show better resistance to amylase hydrolysis.

Various chemical modifications of starch are known to improve its technological properties such as mechanical stability, crystallinity, or water absorption. Several approaches exist to follow the chemical modification of starch by using IR spectroscopy. Acetylation and maleinization reactions have been investigated by Cyras *et al.* (2006). They found that the signal at 2933 cm^{-1} corresponding to the absorption of the CH_2 groups remains unchanged during chemical modification, which is why they used this band as internal standard. Peaks observed at 1740 cm^{-1} (C=O stretching band) and 1240 cm^{-1} (C–O stretching band) confirmed the acetylating of starch. Similar results could be seen during maleinization of starch: a peak at 1703 cm^{-1} (C=O) indicates the presence of an ester maleic group and allows following the modification of starch.

Cassava starch samples including native, fermented, and sun-dried, chemically treated with lactic acid, and native starch oxidized in presence of $KMnO_4$ solution were also characterized by FTIR spectroscopy and subsequent chemometric interpretation (Demiate *et al.*, 2000). In order to reduce influences of the starch's inhomogeneity, the first derivative of the spectra in the region 1800–1540 cm^{-1} was measured. Attempts to predict the baking behavior from the IR data were only partly successful. The best correlation to baking properties was observed in the region around 1600 cm^{-1}, which is associated with the appearance of carboxylate groups.

Because of the use of flavorings in the food industry there is increasing interest in the study of the interaction between polysaccharides and aroma substances. In this context the formation of complexes of homologs of γ- and δ-lactones with polysaccharides of corn and potato starches has been analyzed by IR spectroscopy (Misharina *et al.*, 2002). It could be observed, that in the IR spectrum of γ-decalactone adsorbed by starch, the intensity of CH_3 stretching bands (2931–2959 cm^{-1}) decreased significantly in comparison to those occurring in the IR spectrum of pure lactone standard. Also CH_2 deformation bands decreased by 5–11 times, indicating that the conformational mobility of aroma substances is reduced after their adsorption by starch polysaccharides.

The effect of deep-freezing on the surface of potato starch granule was also evaluated by FTIR spectroscopy (Szymońska *et al.*, 2000). Several changes occur in the IR spectra when samples of oven-dried, air-dried, moisturized potato starch are frozen in liquid nitrogen. It can be observed that some peaks shift to other wavenumbers or they are overlapped by adjacent signals which proves that deep-freezing somehow affects the internal structure of the potato starch granules.

Important results concerning the IR characterization of neutral polysaccharides (pullulan and dextran) have been presented by Shingel (2002). In particular, the mobility and conformation of the carbohydrate chains as well as the molecular interactions of these polysaccharides could be successfully characterized by means of IR spectroscopy. The main signals observed in the deconvoluted spectra of pullulan and dextrans at 1155, 1107, 1080, 1020, and 1000 cm^{-1} are assigned as C–O and C–C stretching vibrational modes and deformational vibrations of CCH, COH, and HCO bonds. Variations in the profile of certain IR bands at 1040, 1020 and in the case of pullutan also at 996 cm^{-1} were found to correlate with changes in conformation and short-range interactions of the polysaccharides. It was also shown that the band at 1080 cm^{-1} indirectly indicates a number of α-(1,6) linkages in the polysaccharide structure.

In the past, numerous studies have been performed to describe oil uptake during deep-fat frying of potatoes, but so far no satisfactory quantitative predictive model exists for this process. In order to get a better understanding of the chemical changes occurring on the surface of cut potatoes a special IR microspectroscopy apparatus has been applied (Bouchon *et al.*, 2001). Two distinct spectroscopic features were used to monitor oil penetration into the freshly fried potato samples. Based on the stretching vibration band at 1745 cm^{-1} (due to ester group of triglycerides) a clear identification of oil location across sections of the fried potato cylinders could be performed. In addition, the core region of the potato sample was measured to get a reference spectrum for the oil-free area. The total oil adsorbed was estimated by summing the contribution of each analyzed layer. The results of the study prove that the oil uptake is mainly a surface phenomenon and that the oil distribution is predominantly determined by the developed crust microstructure.

Vicentini *et al.* (2005) compared the FTIR spectra of raw cassava starch with edible cassava films prepared by casting technique. In general, they did not find significant differences in the spectral pattern. However, by applying PCA in the fingerprint region (1300–800 cm^{-1}) a separation into three different groups was successfully achieved, which can be correlated with the starch concentration in the filmogenic solution. This chemometric analysis proves that the functional properties of cassava starch are mainly related to spectral contributions observed around 1000 cm^{-1} (change from crystalline to amorphous state). Furthermore, a comparatively large influence of the band at 1240 cm^{-1} was seen, representing the intermolecular cohesion of cassava films.

The potential of FTIR spectroscopy in combination with photoacoustic detection for the rapid analytical characterization of potato chips has been demonstrated by Sivakesava and Irudayaraj (2000). The authors measured the samples in sealed cells and purged with helium. IR radiation striking the sample surface generated acoustic waves which could be detected by a very sensitive microphone. Beside moisture bands (3700–3100 cm^{-1} and 1640 cm^{-1}), amide A (3380 cm^{-1}), amide I (1645 cm^{-1}), and amide II (1450 cm^{-1}) absorption bands were identified. In addition, the spectra show key bands of sugars, oils, and fats in the range between 950 and 1153 cm^{-1}. A strong peak at 1748 cm^{-1} (ester carbonyl group of triglycerides) and a small peak at 3005 cm^{-1} (*cis*-double bond stretching) represent marker bands for oil to be found in potato chips. Hydroperoxides formed during the oxidation processes were detected by a peak around 3380 cm^{-1} due to OH stretching vibration of this molecule. At the same time, a decrease of the *cis*-C=C stretching vibration at 3005 cm^{-1} was observed. Non-oxidized chips also show a very weak band at 1652 cm^{-1} which is associated with *cis*-C=C of olefins. The authors also found that the intensity of the band at 1099 cm^{-1} decreased with proceeding oxidation.

Transmission IR microspectroscopy was successfully applied to study the infection of potato tubers by *Erwinia carotovora* ssp. *carotovora* (a widespread bacterium causing soft rot in potato tubers) in aerobic and anaerobic conditions (Steward, 1996). This sophisticated technique generally permits the study of localized changes occurring in cell wall composition and structure. According to present knowledge the bacteria degrade the starch granules, which is associated with a significant decrease of polysaccharide C–O stretching vibrations in the region 1150–980 cm^{-1}. Furthermore, a slight increase of protein bands can be seen which may be due to cell wall hydrolases secreted by the pathogenic bacteria. The author (Steward, 1996) observed that IR spectra obtained from potato tissue infected in anaerobic conditions show clearly higher degradation of cell walls than that infected in aerobic conditions. This is related to the fact that increased absorbances at 1650 cm^{-1} (protein) and 1710 cm^{-1} (fatty acid) occur, indicating higher degradation during anaerobic infection (Steward *et al.*, 1994).

A first attempt tentatively to analyze contamination with certain pesticides of potato tubers using ATR-IR was made in 1975 (Copin *et al.*, 1975). This method clearly lacks sensitivity and selectivity, however, and therefore has only limited application in practise.

Tomato

ATR-IR spectroscopy has been used for *in situ* analysis of naturally occurring carotenoids in tomato fruits and tomato products (Baranska *et al.*, 2006). The principal pigments of red tomato fruits are lycopene and β-carotene, which are 11- and 9-conjugated carotenes, respectively. IR spectra obtained from tomato samples show no visual evidence of the presence of lycopene and β-carotene (Figure 12.3). Most intense signals observed in the spectra of isolated carotenoids (Figure 12.3) due to wagging vibration ((RH)–C=C–(RH)) can be seen at 965 and 960 cm^{-1} for β-carotene and lycopene, respectively. An additional satellite band at 950 cm^{-1} indicates that the standard of β-carotene used for measurement was not pure and contained small amounts of another carotenoid. At about 1370 cm^{-1} both investigated pigments show a vibrational mode due to symmetric deformation δ_{sym} (CH_3), whereas near 1450 cm^{-1} a deformation vibration δ (CH_2) is observed. None of these bands can be noticed in the spectrum obtained from tomato purée (Figure 12.3), which is dominated by signals to be seen at 1635 and 1060 cm^{-1} that are due to vibrational modes of water. This is not surprising, since water molecules have high dipole moment and their IR bands coincide with signals from other plant constituents.

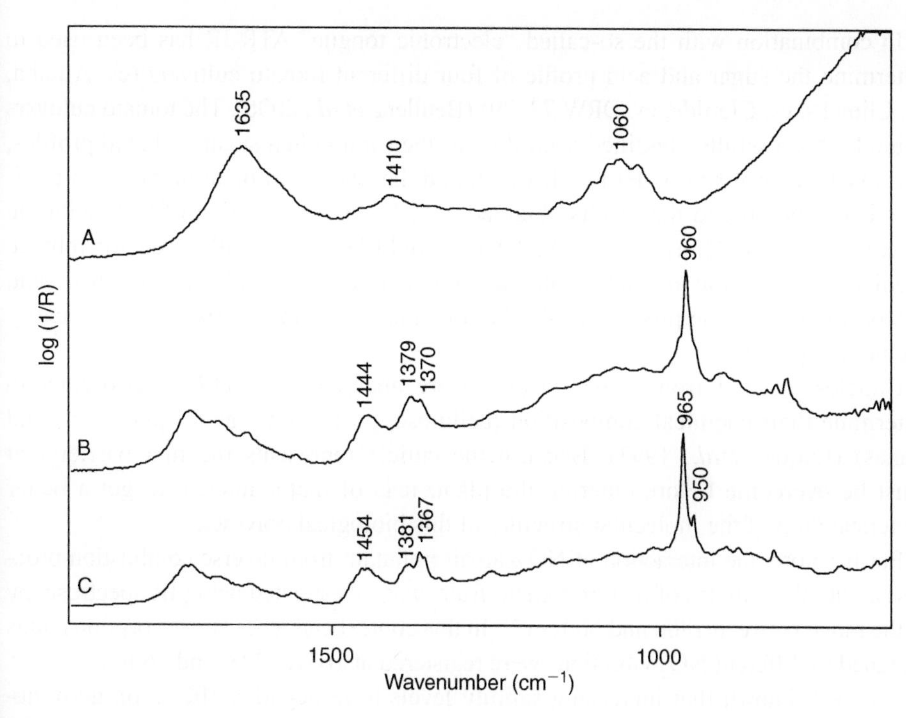

Figure 12.3 Attenuated total reflectance infrared spectra of tomato purée (A), lycopene (B) and β-carotene (C).

For IR spectra, vector normalization and baseline correction followed by mean centering was applied. A reasonable model was obtained by using the full wavenumber range ($R^2 = 0.95$, SECV = 0.21 for β-carotene and 0.97, SECV = 37.23 for lycopene), however the best model was obtained when the spectral range was limited to 650–1800 cm^{-1} ($R^2 = 0.97$, SECV = 0.16 for β-carotene and 0.98, SECV = 33.20 for lycopene). This narrow wavenumber range was more convenient since above 1800 cm^{-1} IR spectra are disturbed by CO_2 absorption. Some other methods of pre-processing data such as multiple scatter correction (MSC) also resulted in satisfactory prediction quality. In contrast, second-derivative pre-processing contributed considerable noise to the spectra and consequently gave very poor calibration models. Similar observations were made when MSC was applied to the calibration of NIR spectra obtained from tomato products (Pedro and Ferreira, 2005). NIR spectroscopy, which has been used for quantification purposes in the agricultural sector for several decades, showed the worst prediction quality ($R^2 = 0.85$ and 0.80, SECV = 91.19 and 0.41 for lycopene and β-carotene, respectively) in the study by Baranska *et al.* (2006).

A similar approach to detect and to quantify lycopene in tomatoes was performed by Halim and Schwartz (2006). They used the CH deformation vibration at 957 cm^{-1} as a marker band and successfully developed calibration models for the accurate prediction of unknown lycopene levels in tomatoes and tomato products by ATR-IR spectroscopy ($R = 0.96$, SECV $< 0.80\,mg\,100\,g^{-1}$).

In combination with the so-called "electronic tongue" ATR-IR has been used to determine the sugar and acid profile of four different tomato cultivars (cv. Aranka, cv. Climaks, cv. Clotilde, cv. DRW 73–29) (Beullens *et al.*, 2006). The tomato cultivars could be successfully classified according to their individual sugar and acid profiles, previously determined by HPLC. It was found that the discrimination between cultivars is comparable to the results obtained with HPLC (correlation of 98% with the reference method). Furthermore ATR-FTIR could be successfully used for reliable prediction of fumaric acid and malic acid content, as well as glucose, fructose, and sucrose. However, the prediction of other organic acids did not result in satisfactory prediction quality.

Cuticles isolated from tomato leaves were analyzed by FTIR spectroscopy to determine their chemical composition (cellulose, pectin, fatty acids, phenolics, and waxes) (Luque *et al.*, 1993). Because the cuticle represents the first barrier that must be overcome before entering the plant, it is of major interest to get a better understanding of the molecular structure of this biological polymer.

Furthermore, the interaction of NO_2, an air pollutant from diverse combustion processes, with the cuticles of mature tomato fruits was investigated using IR spectroscopy in the range between 1800 and 600 cm^{-1}. In this context characteristic absorption bands assigned to different NO_2 vibrations were registered at 1631, 1278, and 860 cm^{-1}.

It is well known that increasing salinity levels have negative effects on germination, plant growth, and fruit yield of tomatoes. Furthermore, salinization also results in an increase of various physiological disorders such as "blossom end rot" (Brown and Ho, 1993), which is related to a decrease in the absorption and translocation of calcium ions to the fruit (Franco *et al.*, 1994). In order to get a rapid classification of tomato fruits grown in saline conditions GC-MS is usually applied to obtain metabolic fingerprints of the analyzed plant tissues (Fiehn, 2001).

Applying FTIR spectroscopy, the interaction between cellulose and non-cellulosic polysaccharides in tomato fruits during cell elongation has been investigated in detail (Wilson *et al.*, 2000). Based on these experiments the authors found that pectin chains respond faster to oscillation than the more rigid cellulose. Another research group analyzed whole fruit flesh extracts obtained from salt-grown tomatoes using FTIR spectroscopy (Johnson *et al.*, 2003). Applying PCA, no discrimination between control samples and the individual salt-treated tomato varieties could be achieved. However discriminant function analysis (DFA) was able to classify the control group as well as salt-treated fruits correctly. Furthermore, a genetic algorithm (GA) was used as a variable selection method prior to discriminant multiple linear regression (D-MLR) to deconvolve hyperspectral data sets in order to identify discriminatory biomarkers for susceptible and salt-tolerant tomato varieties. Based on GA results two regions (2300–2100 cm^{-1} and 900–800 cm^{-1}) were selected for discrimination between control and salt-treated tomatoes. The first region corresponds to saturated and unsaturated nitrile compounds. The increased occurrence of these substances is attributed to the detoxification of hydrogen cyanide, a by-product produced during the biosynthesis of ethylene, which is enhanced in response to stress conditions, such as salinity (Mizrahi, 1982). The other key region at lower wavenumbers is mainly associated with absorption of amino radicals and other nitrogen compounds.

The structure of tomato cell walls comprising two independent but co-extensive networks (cellulose/hemicellulose and pectin) has also been characterized by FTIR microspectroscopy (Wells *et al.*, 1994). Applying the cellulose-synthesis inhibitor 2,6-dichlorobenzonitrile (DCB), the formation of the pectin network could be studied independently from the cellulose/hemicellulose network. The FTIR spectra obtained from cell walls of DCB-nonadapted cells were found to be very similar to those from onion parenchyma cell walls, showing most intense absorbances at 1550 and 1650 cm^{-1} (amide stretches), indicating a much larger protein content than in onion cells. In contrast, spectra obtained from walls of DCB-adapted tomato cells are very similar to purified pectin or a commercially available polygalacturonic acid standard.

Sugar beet

For several years NIR and FTIR spectroscopy have been presented as suitable for the analysis of various quality parameters in sugar beet extracts or pressed juice (Vaccari *et al.*, 1987; Huijbregts *et al.*, 1996). Reliable NIR spectroscopy calibration equations have been developed for the prediction of total sugars, potassium, sodium, and glucose (Huijbregts *et al.*, 2006). Comparable results were obtained for FTIR calibration models using the same sample set. A rapid ATR-FTIR method for the direct determination of sucrose in beet root has been proposed by Garrigues *et al.* (2000). The characteristic sugar band at 1056 cm^{-1} was selected for analytical measurements, with a baseline correction established between 1187 and 887 cm^{-1}. For measurements the cooked red beet root samples were directly crushed and placed on the ZnSe crystal of the ATR cell. It has been proved that the described method is free from matrix effects and from interferences with other sugar components occurring at lower concentration levels.

Other applications of IR spectroscopy methods are focused on the characterization of hemicellulose and cellulose fraction obtained from sugar beet pulp (Sun and Hughes, 1998). Furthermore, the degradation of beet pulps in the rumens of fistulated goats has been investigated (Robert *et al.*, 1989). In this context the absorption band at 1740 cm^{-1} decreased when the pulps were degraded, indicating that pectins were highly digestible. At the same time the protein content increased visibly by the evolution of the amide II band.

ATR-IR spectroscopy has also been used to detect beet invert sugar adulteration in various honey varieties (Sivakesava and Irudayaraj, 2001). Using the data in the wavenumber range 950–1500 cm^{-1}, best predictive values for adulterated honey samples were achieved with canonical variate analysis, which successfully classified 88–94% of the validation set.

The same authors also successfully developed predictive models for adulteration with glucose, fructose, sucrose as well as cane invert sugar (Sivakesava and Irudayaraj, 2002). The spectral region between 800 and 1500 cm^{-1} corresponding to sugar vibrational modes was selected for chemometric calculations and discriminant analysis was applied to identify the adulterated honey samples. A similar approach has been performed to classify adulterants in maple syrup (Paradkar *et al.*, 2002). Both, NIR and FTIR spectroscopic techniques have been used to detect adulterants such as cane and beet invert sugars in maple syrup. In general, FTIR led to more accuracy in predicting adulterations compared with NIR analyses. The absorption band at 991 cm^{-1} (C–O stretching in the C–OH group) was found to be the dominant signal in the spectra obtained from maple syrup, beet sugar, and cane sugar, whereas this peak is mostly absent in invert syrups.

Two-dimensional spectroscopy has been applied to assign overtones and combination bands of NIR absorbances (Maalouly *et al.*, 2004). Thus characteristic peaks of sugar in the MIR region (1056 and 995 cm^{-1}) were positively correlated with sugar bands in the NIR region (4808 cm^{-1}). In the same way, the water peak at 1610 cm^{-1} was related to corresponding bands occurring in the NIR region (5150 cm^{-1} and 7200 cm^{-1}).

Legume species

Whereas most legume species (e.g. soya and pea beans) are cultivated for animal nutrition, these raw materials are also used to isolate protein, starch, and fiber for food production. In order to check the chemical composition of legume species as well as related isolates, several vibrational spectroscopy methods have been developed within recent years (Jones *et al.*, 1995). FTIR in conjunction with photoacoustic detection and PLS interpretation has proved to be a reliable tool to predict major components (starch, protein, lipids) occurring in single pea seeds (Letzelter and Wilson, 1995). Linear regression of protein content calculated on the individual peak height measured at 1666 cm^{-1} provided a correlation coefficient of 0.99. Furthermore, it was also possible to achieve reliable predictions of the lipid content using the signal intensity at 1744 cm^{-1}.

Microspectroscopy using the light from a synchrotron, which is 2–3 orders of magnitude brighter than conventional IR light sources, provided valuable chemical information on the composition of the root zone of mung beans (Raab and Martin, 2001). The studies demonstrate that the roots of mung bean plants exposed to either low-phosphorus or nutrient-sufficient conditions can be clearly discriminated. Nutrient-sufficient plants showed very strong absorbance features at 1627, 1399, and 1269 cm^{-1} and weaker signals at 1003 and 775 cm^{-1}. In contrast, mung beans in low-phosphorus conditions had stronger aliphatic bands in the wavenumber range between 2950 and 2850 cm^{-1} and an isolated absorbance band at 1729 cm^{-1}.

Modifications occurring in the cell walls of beans (*Phaseolus vulgaris*) because of a habituation program to the herbicide dichlobenil have been characterized by FTIR spectroscopy (Alonso-Simón *et al.*, 2004). Applying various multivariate analyses to the obtained spectral data some changes in polysaccharide composition could be measured, which is mainly related to the interaction with the herbicide.

Other studies performed with common beans aimed to describe the so-called "hard-to-cook" (HTC) phenomenon, which occurs when the seeds are stored under adverse conditions of high temperature and high humidity (Maurer *et al.*, 2004). The authors showed that HTC beans contain higher concentrations of phenolic compounds (indicated by higher absorbances in the region between 1708 and 1581 cm^{-1}) which may have an effect on cell wall separation and cause HTC beans to take longer to cook to the same tenderness.

The secondary structure of legumin, a globular protein present in pea seeds, was also successfully characterized on the basis of FTIR data. Furthermore, the location of the structured regions along the primary sequence of legumin has been examined by cluster analysis (Subirade *et al.*, 1994). More recently, FTIR spectroscopy was used to study the conformation of red bean globulin under the influence of changing pH values, protein structure pertubants, and various heating treatments (Meng and Ma, 2001). The authors proved that FTIR spectroscopy, similar to differential scanning calorimetry and Raman spectroscopy, is an appropriate technique for studying conformational changes of plant proteins.

Carrots

Some attempts have been made to measure the carotenoids, sugars, and dry matter content of carrot roots using various NIR and IR methods (Schulz *et al.*, 1998; Quilitzsch *et al.*, 2005a). Whereas the calibrations for dry matter as well as α- and β-carotene content provide reliable data, the prediction quality of sugars occurring in carrot roots (fructose, glucose, sucrose) is comparatively low. That is the reason why the same authors applied ATR-IR spectroscopy in the wavenumber range from 850 to 2000 cm^{-1} in order to develop precise calibration models for these important quality parameters. As can be seen from Table 12.4, apart from sucrose the individual sugar substances can be successfully predicted resulting in high determination coefficients.

In general, Schulz *et al.* (1998) and Quilitzsch *et al.* (2005a) found that compared with NIR spectroscopy measurements, the MIR range clearly provides more spectral information of the analyte. For MIR analyses a portable FTIR spectrometer with a fixed mounted diamond-ATR unit accessory was used, allowing a very good interaction with the sample in the interface of the crystal to be obtained.

FTIR spectroscopy was also used to measure changes in the chemical composition of carrot cell walls during treatment with auxin, causing a significant cell elongation (McCann *et al.*, 1993). The described studies show that the walls of round carrot cells contain more protein, esters, and phenolics in a given area than the walls of elongated carrot cells.

In situ FTIR measurements have been used to study the heat stability of proteins and properties of glassy matrix in slowly dried, desiccation-tolerant and rapidly

Table 12.4 Cross-validation results for the prediction of sugar contents in carrot roots based on infrared spectra obtained from 260 different carrot juice samples

Component	Range ($g\,100\,g^{-1}$)	R^2	RMSECV ($g\,100\,g^{-1}$)
Fructose	0.00–7.72	0.94	0.39
Glucose	0.00–6.90	0.94	0.38
Sucrose	0.07–2.60	0.62	0.30
Total sugars	0.74–16.17	0.91	0.95

From Quilitzsch *et al.* (2005a).

dried, desiccation-sensitive carrot somatic embryos. Slight but consistent differences were observed between the amide I bands of both carrot types. The amide I band of slowly dried embryos was seen around 1654 cm^{-1} whereas in rapidly dried carrots this band was shifted to lower wavenumbers (around 1632 cm^{-1}).

Brassica species

Applying NIR reflectance spectroscopy, a number of species belonging to the *Brassicaceae* family have been analyzed for their total glucosinolates content (Salgo *et al.*, 1992; Velasco and Becker, 1998). Recently, even the individual glucosinolates occurring in *Brassica juncea* (sinigrin), *Raphanus sativus* (glucoerysolin, glucobrassicin, glucoraphenin), and *Brassica oleracea* (glucobrassicin, sinigrin) have been successfully predicted using reliable NIR correlation equations (Quilitzsch *et al.*, 2005b). Enzymatic degradation of glucosinolates results in the formation of a number of substances such as isothiocyanate, which can be used as substitutes for several synthetic organic pesticides. IR spectroscopy was found to be a convenient method to measure the release of isothiocyanate from plant glucosinolates in the soil (Brown *et al.*, 1991). The characteristic –N=C=S functional group in the region between 2174 and 2041 cm^{-1} could be successfully used for detection of isothiocyanates extracted from soil which has been amended with winter rapeseed meal before. In order to select *Brassica* genotypes with exceptionally high contents of individual glucosinolates a special FTIR method has been developed by Schütze *et al.* (2004). The authors applied the freeze-dried powder obtained from the freshly harvested plants directly to a diamond-ATR accessory and got acceptable predictions for individual glucosinolates and total glucosinolates content. Based on the results of this IR analysis special genotypes with high glucosinolates content could be clearly identified and these selected plants are now available as natural soil fumigants to reduce damage from pests and diseases in both agricultural and horticultural crops.

Cucumber

A non-destructive method to evaluate the quality parameters of pickling cucumbers was developed by Miller *et al.* (1995). In this context the relative amount of Vis-IR light passing through the longitudinal midsection of whole cucumber fruit was quantified on a unitless sigmoid scale from 1 to 10. Cucumbers measured directly after harvest exhibited transmission values between 2 and 3, regardless of the individual cultivar. Mechanical-stress treatment (e.g. bruising of the fruits) resulted in an increase in light transmission to a value of 6. Light transmission values also increased as fruit diameter decreased, but values within a particular size class of undamaged, hand-harvested fruit were found to be consistent.

FTIR spectroscopy was used to investigate the secondary structure of cucumber mosaic virus (CMV) which occurs worldwide in an array of host organisms and infects a diverse group of plants, including vegetables such as tomatoes and cucumbers as well as some other agricultural crops (Renugopalakrishnan *et al.*, 1998).

In order to isolate the IR bands arising from the protein backbone of CMV, the FTIR spectra of purified and precipitated RNA was obtained separately and digitally subtracted from the intact CMV spectra. The resulting spectra show two bands at $1682\,cm^{-1}$ and $1644\,cm^{-1}$ which are assigned to amide I vibrational modes (β-sheet structure). Amide II absorption band can be seen at $1546\,cm^{-1}$, whereas amide III band in the coat protein is registered at $1239\,cm^{-1}$, reaffirming the presence of β-sheet as the major conformational feature of the coat protein occurring in CMV.

Allium species

Onion oil, obtained by steam distillation of onion bulbs (*Allium cepa*), is well known as an important raw material used for various purposes in the flavor industry. Because of its high price there exists substantial incentive for adulteration with nature-identical flavoring substances or to offer a nature-identical substitute as genuine onion oil on the market. Usually, such adulterations are identified by GC-MS analysis, but IR spectroscopy has also been successfully applied to guarantee the authenticity of commercial onion oil (Lösing, 1999). Beside various dialk(en)yl(poly)sulfides, genuine onion oils contain 2-*n*-hexyl-5-methyl-3(2 H)furanone, which is well known in the literature as so-called "onion-furanone" (Thomas and Damm, 1986). This substance, which occurs in higher amounts in authentic onion oils, presents an intensive absorption band at $1720\,cm^{-1}$ in the IR spectrum (Figure 12.4). Nature-identical substitutes

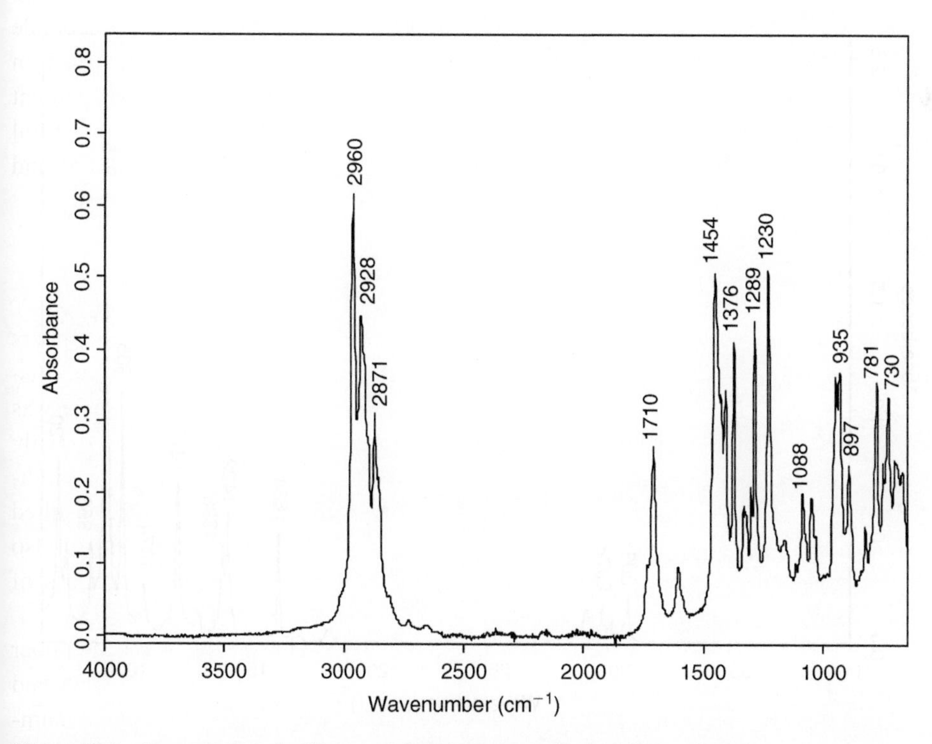

Figure 12.4 Attenuated total reflectance infrared spectrum obtained from onion oil.

or adulterated oils consist mainly of dialk(en)ylsulfides and consequently the furanone marker band is completely missing or shows significantly lower intensity.

The volatile sulfur components that occur in the essential oil of garlic (*Allium sativum*) have been characterized by Jirovetz *et al.* (1992) using GC-IR. Diallyltrisulfide was found to be the main constituent of the oil; the corresponding IR spectrum shows characteristic deformation vibration of the sulfur bridge (Figure 12.5).

Onion bulbs have also been used as a model to perform detailed studies of parenchyma cell walls using IR microspectroscopy. Most absorption bands were found in the region between 2000 and 900 cm^{-1}: at approximately 1740 cm^{-1} carboxylic ester groups were detected, amide-stretching bands of protein occur at 1650 and 1550 cm^{-1}, carboxylic acid groups related to pectin absorb at 1610 cm^{-1}, phenolics are observed at 1600 and 1500 cm^{-1}, and carbohydrate bands are seen between 1200 and 900 cm^{-1} (McCann *et al.*, 1992). The authors demonstrated that FTIR microspectroscopy could be successfully used to detect large conformational changes in pectic polymers on removal from the cell walls and on drying. A decrease in the proportion of unesterified pectin between maximal cell division and maximal cell elongation was found in every cell wall. By using polarized light, IR spectroscopy was also used to investigate single onion cells under different hydration conditions (Chen *et al.*, 1997). In this context it was observed that bands associated with pectin were stronger with polarization perpendicular to the direction of cell elongation. On the other hand, bands associated with cellulose were more intense with polarization parallel to the direction of cell elongation. Therefore, in contrast to the

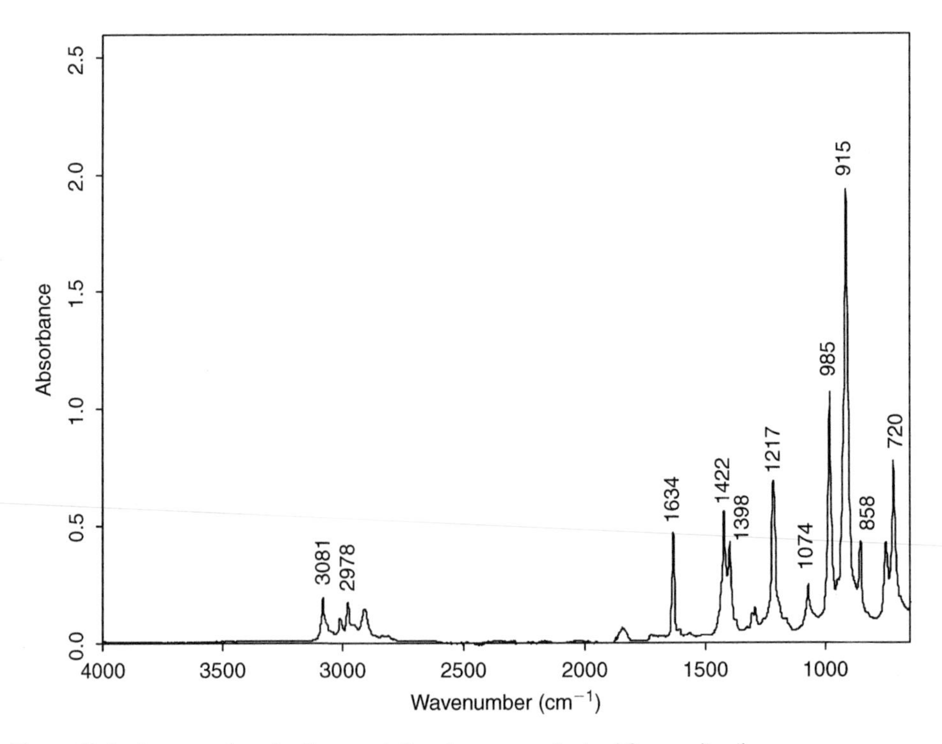

Figure 12.5 Attenuated total reflectance infrared spectrum obtained from garlic oil.

earlier opinion that pectin does not play any structural role in cell walls, it can be assumed that it contributes to the mechanical and structural properties of the cell network.

Conclusions

In recent years, vibrational spectroscopy methods have become attractive and promising analytical tools extensively applied in basic research, product development, and quality control for various agricultural crops and related products. Whereas the first NIR reflectance spectroscopy applications in the agricultural section appeared over 40 years ago, the use of MIR spectroscopy has been more limited. A general advantage of MIR over NIR spectroscopy is that absorption bands are usually well resolved and can be assigned to vibrational modes of specific chemical groups. Furthermore, the possibility of coupling NIR spectrometers with suitable, comparatively low-cost fiber optics opens numerous "in-line" and "on-line" commercial applications. Although special fiber-optics materials have been developed for MIR (e.g. ZrF_4 and AgCl), because of the signal attenuation and high costs at present their use is restricted to lengths of 2 m.

Today NIR and MIR spectroscopies are widely used to determine fats, proteins, and carbohydrates, but numerous secondary plant substances occurring in agricultural products can also be reliably predicted. For authentication purposes both spectroscopy techniques can provide in a few minutes valuable data allowing the discrimination of different agro-food samples. Quantitative analyses need accurate development of calibration equations based on reference data, therefore the worldwide trend (especially for NIR spectroscopy) is to use more and more private and public networks in this context.

In the past 20 years the development of NIR and MIR microscopes has considerably extended the field of application. These sophisticated techniques allow point by point measurements (mapping) to be performed or simultaneous spectra (imaging) from a small sample area to be obtained. Thus MIR and NIR spectra can be obtained nondestructively from the surface of plant samples, allowing the detection of, for example, the distribution of waxes, pesticide residues, or contamination by microorganisms. Moreover, sections of different parts of plants can be analyzed regarding their chemical composition on the cellular level.

References

Alonso-Simón A, Encina AE, García-Angulo P, Álvarez JM, Acebes JL (2004) FTIR spectroscopy monitoring of cell wall modifications during habituation of bean (*Phaseolus vulgaris* L.) callus cultures to dichlobenil. *Plant Science*, **167**, 1273–1281.

Baranska M, Schulz H, Krüger H, Quilitzsch R (2005) Chemotaxonomy of aromatic plants of the genus *Origanum* via vibrational spectroscopy. *Analytical and Bioanalytical Chemistry*, **381**, 1241–1247.

Baranska M, Schütze W, Schulz H (2006) Determination of lycopene and β-carotene content in tomato fruits and related products: comparison of FT-Raman, ATR-IR and NIR spectroscopy. *Analytical Chemistry*, **78**, 8456–8461.

Bartley IM, Knee M (1982) The chemistry of textural changes in fruit during storage. *Food Chemistry*, **9**, 47–58.

Barton FE, Himmelsbach DS (1993) Two-dimensional vibrational spectroscopy II: Correlation of the absorptions of lignins in the mid- and near-infrared. *Applied Spectroscopy*, **46**, 420–429.

Bello-Pérez LA (2005) Effect of storage time on the retrogradation of banana starch extrudate. *Journal of Agriculture and Food Chemistry*, **53**, 1081–1086.

Beullens K, Kirsanov D, Irudayaraj J, Rudnitskaya A, Legin A, Nicolai BM, Lammertyn J (2006) The electronic tongue and ATR-FTIR for rapid detection of sugars and acids in tomatoes. *Sensors and Actuators*, **B116**, 107–115.

Bouchon P, Hollins P, Pearson M, Pyle DL, Tobin MJ (2001) Oil distribution in fried potatoes monitored by infrared microspectroscopy. *Journal of Food Science*, **66**, 918–923.

Brady CJ (1987) Fruit ripening. *Annual Review of Plant Physiology*, **38**, 155–178.

Briandet R, Kemsley EK, Wilson RH (1996) Discrimination of Arabica and Robusta in instant coffee by Fourier transform infrared spectroscopy and chemometrics. *Journal of Agricultural and Food Chemistry*, **44**, 170–174.

Brown MM, Ho LC (1993) Factors affecting calcium transport and basipetal IAA movement in tomato fruit in relation to blossom-end rot. *Journal of Experimental Botany*, **44**, 1111–1117.

Brown PD, Morra MJ, Mc Cafferey JP, Auld DL, Williams L (1991) Allelochemicals produced during glucosinolate degradation in soil. *Journal of Chemical Ecology*, **17**, 2021–2034.

Cael JJ, Koenig JL, Blackwell J (1974) Infrared and Raman spectroscopy of carbohydrates. Part VI: Normal coordinate analysis of V-amylose. *Biopolymers*, **14**, 1885–1903.

Capron I, Robert P, Colonna P, Brogly M, Planchot V (2007) Starch in rubbery and glassy states by FT IR spectroscopy. *Carbohydrate Polymers*, **68**, 249–259.

Cartoni GP, Coccoili F, Quattrucci E (1991) Separation and identification of free phenolic acids in wines by high-performance liquid chromatography. *Journal of Chromatography*, **537**, 93–99.

Chatjigakis AK, Pappas C, Proxenia N, Kalantzi O, Rodis P, Polissiou M (1998) FT-IR spectroscopic determination of the degree of esterification of cell wall pectins from stored peaches and correlation to textural changes. *Carbohydrate Polymers*, **37**, 395–408.

Chauchard F, Cogdill R, Roussel S, Roger JM, Bellon-Maurel V (2004) Application of LS-SVM to non-linear phenomena in NIR spectroscopy: development of a robust and portable sensor for acidity prediction in grapes. *Chemometrics and Intelligent Laboratory Systems*, **71**, 141–150.

Chen L, Wilson RH, McCann MC (1997) Investigation of macromolecule orientation in dry and hydrated walls of single onion epidermal cells by FTIR microspectroscopy. *Journal of Molecular Structure*, **408/409**, 257–260.

Coimbra MA, Barros A, Barros M, Rutledge DN, Delgadillo I (1998) Multivariate analysis of uronic acid neutral sugars in whole pectic samples by FT-IR. *Carbohydrate Polymers*, **37**, 241–248.

Coimbra MA, Barros A, Rutledge DN, Delgadillo I (1999) FTIR spectroscopy as a tool for the analysis of olive pulp cell-wall polysaccharide extracts. *Carbohydrate Polymers*, **317**, 145–154.

Coimbra MA, Goncalves F, Barros A, Delgadillo I (2002) Fourier transform infrared spectroscopy and chemometric analysis of white wine polysaccharide extracts. *Journal of Agricultural and Food Chemistry*, **50**, 3405–3411.

Colthup NB, Daly LH, Wiberly SE (1990) *Introduction to Infrared and Raman Spectroscopy.* San Diego, CA: Academic Press.

Copin A, Martens PH, Kettmann R, Closset JL, Duculot C (1975) Application of spectrometry by attenuated total reflection in infrared and visible spectrometry to study antigerminative residues (Propham and Chlorpropham) on potato tubers. *Analytica Chimica Acta*, **74**, 437–440.

Cserháti T, Forgács E, Candeias M, Vilas-Boas L, Bronze R, Spranger I (2000) Separation and tentative identification of the main pigment fraction of raisins by thin-layer chromatography-Fourier transform infrared and high-performance liquid chromatography-ultraviolet detection. *Journal of Chromatographic Science*, **38**, 145–149.

Cyras VP, Zenklusen MCT, Vazquez A (2006) Relationship between structure and properties of modified potato starch biodegradable films. *Journal of Applied Polymer Science*, **101**, 4313–4319.

Daferera DJ, Tarantilis PA, Polissiou MG (2002) Characterization of essential oils from Lamiaceae by Fourier transform Raman spectroscopy. *Journal of Agricultural and Food Chemistry*, **50**, 5503–5507.

Dandan W, Jian W, Xuehui H, Ying T, Kunyi N (2007) Identification of polymethoxylated flavones from green tangerine peel (Pericarpium Citri Reticulatae Viride) by chromatographic and spectroscopic techniques. *Journal of Pharmaceutical and Biomedical Analysis*, **44**, 63–69.

David F, Sandra P (1992) Capillary gas chromatography-spectroscopic techniques in natural product analysis. *Phytochemical Analysis*, **3**, 145–152.

Demiate IM, Dupuy N, Huvenne JP, Cereda MP, Wosiacki G (2000) Relationship between baking behaviour of modified cassava starches and starch chemical structure determined by FTIR spectroscopy. *Carbohydrate Polymers*, **42**, 149–158.

Downey G, Robert P, Bertrand D, Wilson RH, Kemsley EK (1997) Near- and mid-infrared spectroscopies in food authentication: Coffee varietal identification. *Journal of Agricultural and Food Chemistry*, **45**, 4357–4361.

Farhart IA, Blanshard JMV, Mitchell JR (2000) The retrogradation of waxy maize starch extrudates: effects of storage temperature and water content. *Biopolymers*, **53**, 411–422.

Ferreira D, Barros A, Coimbra MA, Delgadillo I (2001) Use of FT-IR spectroscopy to follow the effect of processing in cell wall polysaccharide extracts of a sun-dried pear. *Carbohydrate Polymers*, **45**, 175–182.

Fiehn O (2001) Combining genomics, metabolome analysis and biochemical modelling to understand metabolic networks. *Comparative and Functional Genomics*, **2**, 155–168.

Fischer RL, Bennett AB (1991) Role of wall hydrolases in fruit ripening. *Annual Review of Plant Physiology and Plant Molecular Biology*, **42**, 675–703.

Franco JA, Banon S, Madrid R (1994) Effects of protein hydrolysate applied by fertigation on the effectiveness of calcium as a corrector of blossom-end ort in tomato cultivated under saline conditions. *Scientia Horticulturae*, **57**, 283–292.

Fry SC (1995) Polysaccharide-modifying enzymes in the plant cell wall. *Annual Review of Plant Physiology and Plant Molecular Biology*, **46**, 497–520.

Garrigues JM, Akssira M, Rambla FJ, Garrigues S, de la Guardia M (2000) Direct ATR-FTIR determination of sucrose in beet root. *Talanta*, **51**, 247–255.

Gorinstein S, Weisz M, Zemser M, Tilis K, Stiller A, Flam I, Gat Y (1993) Spectroscopic analysis of polyphenols in white wines. *Journal of Fermentation and Bioengineering*, **75**, 115–120.

Grasdalen H, Bakoy OE, Larsen B (1988) Determination of the degree of esterification and the distribution of methylated and free carboxyl groups in pectins by ^{1}H-NMR spectroscopy. *Carbohydrate Research*, **184**, 183–191.

Guillermo DM, Lajolo FM (2002) FT-IR spectroscopy as a tool for measuring degree of methyl esterification in pectins isolated from ripening papaya fruit. *Postharvest Biology and Technology*, **25**, 99–107.

Hakala M, Ahro M, Kauppinen J, Kallio H (2001) Determination of strawberry volatiles with low resolution gas phase FT-IR analyzer. *European Food Research and Technology*, **212**, 505–510.

Halim Y, Schwartz SJ*(2006) Direct determination of lycopene content in tomatoes (*Lycopersicon esculentum*) by attenuated total reflectance infrared spectroscopy and multivariate analysis. *Journal of AOAC International*, **89**, 1257–1262.

Herres W, GC-FTIR (1984) State-of-the-art in analysis of volatiles. In: *Analysis of Volatiles* (Schreier P, ed.). Berlin, New York: de Gruyter, pp. 183–217.

Holland JK, Kemsley EK, Wilson RH (1998) Use of Fourier transform infrared spectroscopy and partial least squares regression for the detection of adulteration of strawberry purees. *Journal of the Science of Food and Agriculture*, **76**, 263–269.

Huijbregts AWM, De Regt AH, Gijssel PD (1996) Determination of some quality parameters in sugar beet by near infrared spectrometry (NIRS). *Communication in Soil Science and Plant Analysis*, **27**, 1549–1560.

Huijbregts AWM, Heijuen CJ, Moulin B, Noé B (2006) Assessment of internal beet quality by NIRS (Near Infrared Spectroscopy) and FTIRS (Fourier transform mid infrared spectroscopy). *Sugar Industry/Zuckerindustrie*, **131**, 16–20.

Iwamoto M, Cho RK, Uozumi J, Iino K (1984) Near infrared reflectance spectrum of red pepper and its applicability to determination of capsaicin content. *Nippon Shokuhin Kogyo Gakkaishi*, **31**, 120–125.

Jarén C, Ortuno JC, Arazuri S, Arana JI, Salvadores MC (2001) Sugar determination in grapes using NIR technology. *International Journal of Infrared and Millimeter Waves*, **22**, 1521–1530.

Jirovetz L, Jäger W, Koch HP, Remberg G (1992) Investigations of volatile constituents of the essential oil of Egyptian garlic (*Allium sativum* L.) by means of GC-MS and GC-FTIR. *Zeitschrift für Lebensmittel-Untersuchung und-Forschung*, **194**, 363–365.

Johnson HE, Broadhurst D, Goodacre R, Smith AR (2003) Metabolic fingerprinting of salt-stressed tomatoes. *Phytochemistry*, **62**, 919–928.

Jones DA, Barber LM, Arthur AD, Hedley CL (1995) Assessing variations for fatty acid content by use of a non-destructive technique for single seed analysis. *Plant Breeding*, **114**, 81–83.

Kačuraková M, Wilson RH, Belton PS (2001) Development in mid-infrared FT-IR spectroscopy of selected carbohydrates. *Carbohydrate Polymers*, **44**, 291–303.

Kawano S (2002) Applications to agricultural products and foodstuffs. In: *Near-Infrared Spectroscopy—Principles, Instruments, Applications* (Siesler HW, Ozaki Y, Kawata S, Heise HM, eds). Weinheim: Wiley-VCH, pp. 115–124.

Letzelter NS, Wilson RH (1995) Quantitative determination of the composition of individual pea seeds by Fourier transform infrared photoacoustic spectroscopy. *Journal of the Science of Food and Agriculture*, **67**, 239–245.

Lin-Vien D, Colthup NB, Fateley WG, Grasselli JG (1991) *The Handbook of Infrared and Raman Characteristic Frequencies of Organic Molecules*. San Diego: Academic Press Inc.

Lösing G (1999) Einfache Authentizitätsprüfung bei Zwiebelölen. *Deutsche Lebensmittel-Rundschau*, **95**, 234–236.

Lösing G, Degener M, Günter M (1998) Investigation of the aging of citrus oils using IR spectroscopy. *Dragoco Report*, **4**, 181–187.

Luque P, Heredia A, Ramfrez FJ, Bukovac MJ (1993) Characterization of NO_2 bounds to the plant cuticle by FT-IR spectroscopy. *Zeitschrift für Naturforschung*, **48c**, 666–668.

Maalouly J, Eveleigh L, Rutledge DN, Ducauze CJ (2004) Application of 2D correlation spectroscopy and outer product analysis to infrared spectra of sugar beets. *Vibrational Spectroscopy*, **36**, 279–285.

McCann MC, Hammouri M, Wilson R, Belton P, Roberts K (1992) Fourier transform infrared microspectroscopy is a new way to look at plant cell walls. *Plant Physiology*, **100**, 1940–1947.

McCann MC, Stacey NJ, Wilson R, Roberts K (1993) Orientation of macromolecules in the walls of elongating carrot cells. *Journal of Cell Science*, **106**, 1347–1356.

Maurer GA, Ozen BF, Mauer LJ, Nielsen SS (2004) Analysis of hard-to-cook red and black common beans using Fourier transform infrared spectroscopy. *Journal of Agriculture and Food Chemistry*, **52**, 1470–1477.

Melford AJ, Prakash MD (1986) Postharvest changes in fruit cell wall. *Advanced Food Research*, **30**, 139–193.

Meng GT, Ma CY (2001) Fourier-transform infrared spectroscopic study of globulin from *Phaseolus angularis* (red bean). *International Journal of Biological Macromolecules*, **29**, 287–294.

Miller AR, Kelley TJ, White BD (1995) Nondestructive evaluation of pickling cucumbers using visible-infrared light transmission. *Journal of the American Society for Horticultural Science*, **120**, 1063–1068.

Misharina TA, Terenina MB, Krikunova NI (2002) Binding of lactones by polysaccharides of corn and potato starches. *Applied Biochemistry and Microbiology*, **38**, 583–587.

Mizrahi Y (1982) Effect of salinity on tomato fruit ripening. *Plant Physiology*, **69**, 966–970.

Nicolaï BM, Lotze E, Peirs A, Scheerlinck N, Theron KI (2006) Non-destructive measurement of bitter pit in apple fruit using NIR hyperspectral imaging. *Postharvest Biology and Technology*, **40**, 1–6.

Noda I, Ozaki Y (2004) *Two-Dimensional Correlation Spectroscopy: Applications in Vibrational and Optical Spectroscopy.* Chichester, UK: John Wiley and Sons Ltd.

Ottenhof M-A, MacNaughtan W, Farhart IA (2003) FTIR study of phase and state transitions of low moisture sucrose and lactose. *Carbohydrate Research*, **338**, 2195–2202.

Paradkar MM, Sivakesava S, Irudayaraj J (2002) Discrimination and classification of adulterants in maple syrup with the use of infrared spectroscopic techniques. *Journal of the Science of Food and Agriculture*, **82**, 497–504.

Patz C-D, Blieke A, Ristow R, Dietrich H (2004) Application of FT-MIR spectrometry in wine analysis. *Analytica Chimica Acta*, **513**, 81–89.

Pedro AMK, Ferreira MMC (2005) Non-destructive determination of solids and carotenoids in tomato products by near infrared spectroscopy and multivariate calibration. *Analytical Chemistry*, **77**, 2505–2511.

Plöger A (1992) Conductivity detection of pectin. A rapid HPLC method to analyze degree of esterification. *Journal of Food Science*, **57**, 1185–1186.

Quilitzsch R, Baranska M, Schulz H, Hoberg E (2005a) Fast determination of carrot quality by spectroscopy methods in the UV-VIS, NIR and IR range. *Journal of Applied Botany and Food Quality*, **79**, 163–167.

Quilitzsch R, Schulz H, Schütze W (2005b) Evaluation of glucosinolates in leaves and stems of various *Brassiaca* species by near infrared spectroscopy. In: *Proceedings of the 12th International Conference on Near Infrared Spectroscopy* (Burling GR, Holroyd SE, Sumner RMW, eds). Hamilton, New Zealand: Auckland: Print House Ltd, pp. 477–479.

Raab TK, Martin MC (2001) Visualizing rhizoshere chemistry of legumes with mid-infrared synchrotron radiation. *Planta*, **213**, 881–887.

Reeves JB (1994) Near-versus mid-infrared diffuse reflectance spectroscopy for the quantitative determination of the composition of forages and byproducts. *Journal of Near Infrared Spectroscopy*, **2**, 49–57.

Reeves JB (1996) Improvement in Fourier near- and mid-infrared diffuse reflectance spectroscopic calibrations through the use of a sample transport device. *Applied Spectroscopy*, **50**, 965–969.

Renugopalakrishnan V, Piazzolla P, Tamburro AM, Lamba OP (1998) Structural studies of cucumber mosaic virus: Fourier transform infrared spectroscopic studies. *Biochemistry and Molecular Biology International*, **46**, 747–754.

Robert P, Bertin C, Bertrand D (1989) Rumen microbial degradation of beet root pulps. Application of infrared spectroscopy to the study of protein and pectin. *Journal of Agricultural and Food Chemistry*, **37**, 624–627.

Rubens P, Snauwaert J, Heremans K, Stute R (1999) In situ observation of pressure-induced gelation of starches studied with FT IR in the diamond anvil cell. *Carbohydrate Polymers*, **39**, 231–235.

Salgo A, Weinbrenner-Varga Z, Fabian Z, Ungar E (1992) Determination of anti-nutritive factors by near infrared techniques in rapeseed. In: *Making Light Work: Advances in Near Infrared Spectroscopy* (Murray I, Cowe IA, eds). Weinheim, New York, Basel, Cambridge: VCH Publishers, pp. 328–341.

Santha N, Sudha KG, Vijayakumari KP, Nayar VU, Moorthy SN (1990) Raman and infrared spectra of starch samples of sweet potato and cassava. *Indian Academy of Sciences (Chemical Sciences)*, **102**, 705–712.

Schulz H (2004) Analysis of coffee, tea, cocoa, tobacco, spices, medicinal and aromatic plants and related products. In: *Near-infrared Spectroscopy in Agriculture* (Roberts C, Workman J, Reeves J, eds). Madison: American Society of Agronomy—Crop Science of America—Soil Science Society of America, Agronomy Monograph No. 44, pp. 345–375.

Schulz H, Baranska M (2005) Application of vibrational spectroscopy methods in essential oil analysis. *Perfumer and Flavorist*, **30**, 28–44.

Schulz H, Baranska M (2007) Identification and quantification of valuable plant substances by IR and Raman spectroscopy. *Vibrational Spectroscopy*, **43**, 13–25.

Schulz H, Drews HH, Quilitzsch R, Krüger H (1998) Application of near-infrared spectroscopy for the quantification of quality parameters in selected vegetables and essential oil plants. *Journal of Near-infrared Spectroscopy*, **6**, A125–A130.

Schulz H, Schrader B, Quilitzsch R, Steuer B (2002) Quantitative analysis of various citrus oils by ATR/FT-IR and NIR-FT Raman spectroscopy. *Journal of Applied Spectroscopy*, **56**, 117–124.

Schulz H, Pfeffer S, Straka P, Nothnagel T (2003) Rapid determination of alkaloids in poppy (*Papaver somniferum* L.) by vibrational spectroscopy methods. In: *Proceedings of the 11th International Conference on NIRS* (Davies AMC, Garrido-Varo A, eds). Chichester, UK: NIR Publications, pp. 883–885.

Schulz H, Baranska M, Belz HH, Rösch P, Strehle MA, Popp J (2004) Chemotaxonomic characterisation of essential oil plants by vibrational spectroscopy measurements. *Vibrational Spectroscopy*, **35**, 81–86.

Schulz H, Baranska M, Quilitzsch R, Schütze W, Lösing G (2005a) Characterization of peppercorn, pepper oil and pepper oleoresin by vibrational spectroscopy methods. *Journal of Agricultural and Food Chemistry*, **53**, 3358–3363.

Schulz H, Özkan G, Baranska M, Krüger H, Özcan M (2005b) Characterisation of essential oil plants from Turkey by IR and Raman spectroscopy. *Vibrational Spectroscopy*, **39**, 249–256.

Schulz H, Quilitzsch R, Baranska M (2007) Near infrared spectroscopy analysis of essential oils—Can statistical correlation to mid infrared and Raman spectra assist the interpretation?. In: *Proceedings of the 12th International Conference on NIRS* (Burling-Claridge GR, Holroyd SE, Sumner RMW, eds). Hamilton: Print House Ltd, pp. 250–255.

Schütze W, Quilitzsch R, Schlathölter M (2004) Glucosinolate testing of leaves and stems in brassicas with HPLC and mid IR spectroscopy. *Agroindustria*, **3**, 399–401.

Seregely Z, Deak T, Bisztray GD (2004) Distinguishing melon genotypes using NIR spectroscopy. *Chemometrics and Intelligent Laboratory Systems*, **72**, 195–203.

Sevenou O, Hill SE, Farhat IA, Mitchell JR (2002) Organisation of the external region of the starch granule as determined by infrared spectroscopy. *International Journal of Biological Macromolecules*, **31**, 79–85.

Shingel KI (2002) Determination of structural peculiarities of dextran, pullulan and gamma-irradiated pullulan by Fourier-transform IR spectroscopy. *Carbohydrate Research*, **337**, 1445–1451.

Sivakesava S, Irudayaraj J (2000) Analysis of potato chips using FTIR photoacoustic spectroscopy. *Journal of the Science of Food and Agriculture*, **80**, 1805–1810.

Sivakesava S, Irudayaraj J (2001) Detection of inverted beet sugar adulteration of honey by FTIR spectroscopy. *Journal of the Science of Food and Agriculture*, **81**, 683–690.

Sivakesava S, Irudayaraj J (2002) Classification of simple and complex sugar adulterants in honey by mid-infrared spectroscopy. *International Journal of Food Science and Technology*, **37**, 351–360.

Slaughter DC, Abbott JA (2004) Analysis of Fruits and Vegetables. In: *Near-infrared Spectroscopy in Agriculture* (Roberts C, Workman J, Reeves J, eds). Madison: American Society of Agronomy—Crop Science of America—Soil Science Society of America, Agronomy Monograph No. 44, pp. 377–398.

Spanos GA, Wrolstad RE (1990) Influence of processing and storage on the phenolic composition of Thompson seedless grape juice. *Journal of Agricultural and Food Chemistry*, **38**, 1565–1571.

Steuer B, Schulz H, Läger E (2001) Classification and analysis of citrus oils by NIR spectroscopy. *Food Chemistry*, **72**, 113–117.

Steward D (1996) Fourier transform infrared microscopy of plant tissues. *Applied Spectroscopy*, **50**, 357–365.

Steward D, Lyon GD, Tucker EJB (1994) A Fourier-transform infrared spectroscopic and microscopic study of the infection of potato tubers by *Erwinia carotovora* ssp. *carotovora* in aerobic and anaerobic conditions. *Journal of the Science of Food and Agriculture*, **66**, 145–154.

Stewart D, Morison IM (1992) FT-IR spectroscopy as a tool for the study of biological and chemical treatment of barley straw. *Journal of the Science of Food and Agriculture*, **60**, 431–436.

Stute R, Klinger RW, Boguslawski S, Eshtiaghi MN, Knorr D (1996) Effects of high pressure treatment on starches. *Starch*, **48**, 399–408.

Subirade M, Gueguen J, Pézolet M (1994) Conformational changes upon dissociation of a globular protein from pea: a Fourier transform infrared spectroscopy study. *Biochimica et Biophysica Acta*, **1205**, 239–247.

Sun R, Hughes S (1998) Fractional extraction and physico-chemical characterization of hemicelluloses and cellulose from sugar beet pulp. *Carbohydrate Polymers*, **36**, 293–299.

Suutarinen J, Anakainen L, Autio K (1998) Comparison of light microscopy and spatial resolved Fourier transformed infrared (FT-IR) microscopy in the examination of cell wall components of strawberries. *Lebensmittel-Wissenschaft-Technologie*, **31**, 595–601.

Synytsya A, Čopíková J, Matêjka P, Machovic V (2003) Fourier transform Raman and infrared spectroscopy of pectins. *Carbohydrate Polymers*, **54**, 97–106.

Szymońska J, Krok F, Tomasik P (2000) Deep-freezing of potato starch. *International Journal of Biological Macromolecules*, **27**, 307–314.

Thomas AG, Damm H (1986) A new synthesis of acetylenes, the addition of methanol to 5-hydroxyhex-3-yn-2-one. Synthesis of the onion furanone, 2-hexyl-5-methyl-3(2H)-furanone. *Tetrahedron Letters*, **27**, 505–506.

Urtubia A, Pérez-Correa JR, Meurens M, Agosin E (2004) Monitoring large scale wine fermentation with infrared spectroscopy. *Talanta*, **64**, 778–784.

Vaccari G, Mantovani G, Sgualdino G, Gogerti P (1987) Near infrared spectroscopy utilization for sugar products analytical control. *Sugar Industry/Zuckerindustrie*, **112**, 800–807.

van Soest JJG, deWit D, Tournois H (1994) Retrogradation of potato starch as studied by Fourier Transform infrared spectroscopy. *Starch/Stärke*, **46**, 453–457.

van Soest JJG, Tournois H, de Wit D, Vliegenthart JFG (1995) Short-range structure in (partially) crystalline potato starch determined with attenuated total reflectance Fourier-transform IR spectroscopy. *Carbohydrate Research*, **279**, 201–214.

Velasco L, Becker HC (1998) Analysis of total glucosinolates content and individual glucosinolates in *Brassica* spp. by near-infrared spectroscopy. *Plant Breeding*, **117**, 97–102.

Ventura M, Jager A, Putter H, Roelofs FPMM (1998) Non-destructive determination of soluble solids in apple fruit by near infrared spectroscopy (NIRS). *Postharvest Biology and Technology*, **14**, 21–27.

Vicentini NM, Dupuy N, Leitzelman M, Cereda MP, Sobral PJA (2005) Prediction of cassava starch edible film properties by chemometric analysis of infrared spectra. *Spectroscopy Letters*, **38**, 749–767.

Wells B, McCann MC, Shedletzky E, Delmer D, Roberts K (1994) Structural features of cell walls from tomato cells adapted to grow on the herbicide 2,6-dichlorobenzonitrile. *Journal of Microscopy*, **173**, 155–164.

Westad F, Afseth NK, Bro R (2007) Finding relevant spectral regions between spectroscopic techniques by use of cross model validation and partial least squares regression. *Analytica Chimica Acta*, **595**, 323–327.

Wilson RH, Smith AC, Kačuraková M, Saunders PK, Wellner N, Waldron KW (2000) The mechanical properties and molecular dynamics of plant cell wall polysaccharides studied by Fourier-transform infrared spectroscopy. *Plant Physiology*, **124**, 397–405.

Yang H, Irudayaraj J, Sakhamuri S (2001) Characterization of edible coatings and microorganisms on food surfaces using Fourier transform infrared photoacoustic spectroscopy. *Journal of Applied Spectroscopy*, **55**, 571–583.

Fruit Juices

13

Yiqun Huang, Barbara A Rasco, and Anna G Cavinato

Introduction

Fruit juices are an important food category in the USA, with an average consumption of over 8 gallons per capita since 1992, not including the use of juice as an ingredient in other food items (Pollack and Perez, 2006). Orange and apple juices are the primary fruit juices consumed in the USA, accounting for 57% and 25% of total consumption in 2005/2006 (Pollack and Perez, 2006). As the consumption of non-citrus fruit juices has slowly increased from 2.51 gallons since 1997/1998 to 2.91 gallons per capita in 2005/2006, the consumption of total citrus juices has continuously declined from 6.48 gallons in 1997/1998 to 4.67 gallons per capita in 2005/2006 (Figure 13.1). Although orange juice still accounts for more than half of the total fruit juice consumed, particularly as fresh pasteurized juices have gained popularity, grape juice has replaced grapefruit juice in popularity as of 2003/2004 and has become the third most widely consumed juice. In recent years, mixtures of tropical fruit juices, including blends with pineapple, mango, papaya, and banana, and more recently with carrot (although it is not a tropical fruit), are gaining popularity around the world. Juice blends of white grape juice, which has replaced apple juice as a base in many juice blends, are increasing in volume. These bases

Infrared Spectroscopy for Food Quality Analysis and Control
ISBN: 978-0-12-374136-3

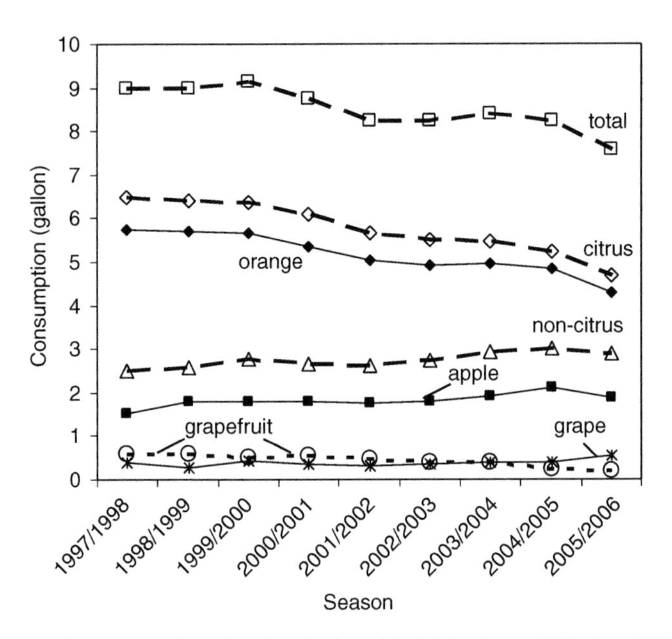

Figure 13.1 Per capita consumption of total and selected fruit juices in the USA since 1997/1998 (Pollack and Perez, 2006).

are blended with juices, juice concentrates or flavors (e.g. cranberry, strawberry, raspberry, tropical extracts) to make juice drinks directed towards the children's market.

Fruit juice is fluid extracted from fruits and is not fermented. Its basic process involves pre-treatment, juice extraction and post-pressing treatment. Pre-treatment mainly includes sorting, cleaning, and inspection. Pre-treatment with pectolytic enzymes may be required to help release juice from some fruits, such as strawberries and raspberries. Post-pressing involves clarification, adjustment and standardizing of solids content, sweetness and acid levels, addition of vitamins, antioxidants, extracts, flavors and/or colors, pasteurization, and packaging.

Fruit juices may be clarified, or not clarified (cloudy and pulpy), although clear juices are the most popular. The exception to this would be citrus juices that are now being produced with varying levels of pulp depending upon market demand. Also popular are juice products called "floats" in Asian and Middle Eastern markets, which are juices or juice blend with suspended large particles of fruit.

Fruit juices are often concentrated to reduce the cost of transportation as well as to stabilize the product. Concentrates are often a more convenient form for products that use juice as an ingredient, since less water is added, providing greater flexibility in product formulations. Concentrated juice can be diluted back to reconstitute the initial single strength juice (100% juice). Market and regulatory standards are set to control the amount of water added to constitute single strength juice, but these standards vary in different countries. For example, according to US regulations, a minimum of 11.8° Brix (soluble solids concentration) is required for reconstituted orange juice and 10.5° Brix for pasteurized orange juice (Food and Drug Administration, 2002; 21 CFR 146.140), but the

European Union allows a minimum of 11° Brix and 10° Brix for reconstituted and fresh-pressed orange juices, respectively (Fry, 1990). Different juices may also have different minimum requirements for soluble solids concentration (Table 13.1). For example, the minimum requirement of soluble solids content in reconstituted fruit juices could be as low as 4.5° Brix, such as in lemon and lime juice, and up to 22° Brix as in banana juice.

There are various juice products on the market that are standardized based upon the type of fruit from which the juice is made, such as apple, orange, and mango juice, and the processing methods. For example, orange juice products are defined and standardized based upon processing methods under the Federal Food, Drug, and Cosmetic Act in the USA as follows: pasteurized orange juice (21 CFR 146.140), canned orange juice (21 CFR 146.141), orange juice from concentrate (21 CFR 146.145), frozen concentrated orange juice (21 CFR 146.146), reduced acid frozen concentrated orange juice (21 CFR 146.148), canned concentrated orange juice (21 CFR 146.150), and concentrated orange juice for manufacturing (21 CFR 146.153 and 21 CFR 146.154) (USDA, 1982). It is therefore a topic of continuing importance to develop rapid and inexpensive analytical methods for online monitoring and control of the safety and quality of a large variety of juice products during processing, transportation, and storage. The use of spectroscopic methods is particularly important for online and continuous quality monitoring applications.

Table 13.1 Minimum brix levels for single strength fruit juices in the USA

Fruit	Juice	°Brix[a]
Citrus fruit	Orange	11.8
	Grapefruit	10.0
	Lemon	4.5
	Lime	4.5
	Tangerine	11.8
Pome fruit	Apple	11.5
	Pear	12.0
	Quince	13.3
Tropical fruit	Banana	22.0
	Kiwi	15.4
	Mango	13.0
	Papaya	11.5
	Passion fruit	14.0
	Pineapple	12.8
Other juices	Grape	16.0
	Peach	10.5
	Black currant	11.0
	Red currant	10.5
	Strawberry	8.0
	Blueberry	10.0
	Raspberry (black)	11.1

[a]Data were from 21 CFR 101.30 (FDA, 2002).

Basic components of fruit juices

Water

Water is the primary component of fruit juice and represents about 90% of the total weight of juice (Table 13.1). Water content is generally not directly used as a quality indicator for fruit juice. Instead, the total soluble solids content of juice is used. In clear fruit juice, the sum of water and soluble solids content is 100%. The use of total soluble solids as quality indicator is partially to control the water content. The amount of soluble solids in juice is normally referred as a Brix value and is determined by a Brix hydrometer or a refractometer. The measurement of soluble solids with a refractometer is based upon the assumption that all soluble ingredients have the same refractive index as sucrose. The refractometric solids content is normally corrected based upon acidity and temperature.

Sugars

Sugars in juice provide a sweet sensation and a viscous mouth-feel. The presence of sugars helps to increase the shelf-life of fruit juices, particularly concentrated fruit juices, by increasing the osmotic pressure of the aqueous environment. Since sugars contribute to the high caloric density of juice, there is a tendency to reduce the total content of sugars, and produce reduced calorie or "lite" juices by substituting non-nutritive sweeteners such as aspartame or sucralose for some of the soluble solids in the juice. Nevertheless, the functional properties, particularly the sensory properties of the sugars, must be considered to produce acceptable fruit juice.

Sugars account for about 75–85% of total soluble solids in fruit juices. Sugars, particularly sucrose, glucose, and fructose, are the major soluble solids in fruit juice. The ratio of each sugar varies in different fruit juices. Therefore, determining the ratio of sugars in a juice is a way to determine whether sugar has been added to adulterate and dilute expensive fruit juices and whether the composition of a juice is as it is represented. For example, fructose is the primary sugar in apple juice, and normally accounts for more than 50% of the total sugar. In orange juice, the ratio of sucrose, glucose and fructose is 1:1:1, as found in juice from Israel and Brazil, or 2:1:1, as found in juice from North America (Fry, 1990). The sugars in grape juice are mainly fructose and glucose, and the content of sucrose is less than 10% of total sugar (McLellan and Race, 1990).

Although sucrose, glucose, and fructose are major sugars found in fruit, a large variety of other sugars and sugar alcohols, such as maltose, xylose, and sorbitol, are also present in fruit juices. Some juices, such as pear juice, contain significant amounts of sorbitol, a sugar alcohol. One method commonly used to determine if more expensive apple juice has been "cut" with pear juice is to assay for sorbitol, since this component is not present in apple juice.

Organic acids

Organic acids are the second most abundant soluble solids component in fruit juices, and are typically present at about 1% of the total weight of a fruit juice. Citric and malic acids are the primary organic acids found in fruit juices. Citric acid accounts for about 90% of the total organic acids in orange juice, while malic acid is the main organic acid in apple juice. However, the tartness of grape juice is mostly caused by tartaric acid. A small amount of other acids including ascorbic, isocitric, citramalic, galacturonic, shikimic, lactic, quinic, succinic, and fumaric acids are also present in fruit juices.

Organic acids provide tartness, and the amount of organic acids is commonly represented as total acidity. Acidity, as a key quality indicator for fruit juice, is generally determined by a titration method using a standard sodium hydroxide solution and phenolphthalein indicator. The result is calculated as citric acid for orange juice, malic acid for apple juice, or tartaric acid for grape juice. The amount of acid and/or the brix-acid ratio is standardized for commercial juice, and must be within a certain range. Too high or too low an acid content or brix-acid ratio results in unacceptable product. Juice can have a relatively high concentration of organic acids, or a high titratable acidity, but not have a low pH ($= -\log_{10}[H^+]$). The pH of fruit juice is generally tied to the safety of fruit juice during storage, since certain types of microflora including pathogens may survive in juice of a lower pH but fail to grow unless the pH of the juice is slightly higher. Pasteurization conditions, such as heating temperature and time, are dependent upon the pH and juice solids concentration.

The types of acids as well as the amount of acids in fruit juices are rather variable, depending upon variety, season, processing methods, and storage time. The profiles of organic acids can be used for juice authentication, juice classification, in addition to predicting the safety and quality of fruit juices. For example, the ratio of citric acid to isocitric acid has been used as an indicator for orange juice authenticity to identify whether citric acid has been added to the juice (Fry, 1990). The ratio of citric acid to isocitric acid is around 100 for orange juice. Since isocitric acid is very costly and is unlikely to be added to adulterated fruit juice, a high ratio of citric acid to isocitric acid in orange juice indicates addition of citric acid, a common method for fruit juice fraud. As discussed later in this chapter another method, carbon isotope ratio of a specific organic acid, has also been used as an indicator for fruit juice authentication (Doner, 1985; Jamin *et al.*, 2005).

Other fruit juice components

Fruit juice contains lower levels of free amino acids compared to sugars and organic acids. Although the amount of free amino acids in orange juice is relatively high, typically around 0.3–0.4%, in other juices, such as apple juice, the level may be as low as 10 ppm (Fry, 1990; Lea, 1990). The amino acid profile differs among different fruit juices. For example, the most abundant amino acid in orange juice is proline, but in apple juice it is asparagine. The composition of amino acids in fruit juices is well documented (Fry, 1990; Lea, 1990). Amino acid analysis using column chromatography is a common method for the analysis of fruit juice.

Fruit juice is not a major source of dietary vitamins and minerals, although orange juice is rich in vitamin C. Table 13.2 shows the mineral content of some common fruit juices. The major inorganic ion is potassium.The mineral content may change during fruit juice processing and storage. For example, iron content may be higher in

Table 13.2 Mineral content (mg/L) of some fruit juices[a]

	Orange, fresh	Orange, canned	Orange[b]	Grapefruit, fresh	Apple juice[b]	Grape juice[b]
Na	8	21	8	8	29	33
Ca	113	83	92	92	71	96
Mg	113	113	104	125	29	104
Zn	0.50	0.71	0.50	0.50	0.29	0.54
Mn	0.15	0.15	0.15	0.21	1	4
K	2067	1817	1971	1667	1229	1392
P	175	146	1667	154	71	117
Fe	2	5	1	2	4	2.5
Cu	0.45	0.59	0.46	0.34	0.23	3

[a]Adapted from Pennington (1997).
[b]From concentrated/bottled.

canned juice than fresh juice. The mineral content of reconstituted juice may differ from fresh juice, since the water used for juice reconstitution may add some minerals to the product. The profile of minerals in fruit juice can be used for juice authentication (Fry, 1990). New juice based drinks are appearing on the market which have been fortified with calcium, specifically citrus juices. Also, vitamins, dietary essential fatty acids, phytochemicals, antioxidants, and proteins are added to increase the nutritional functionality of juices. Other ingredients, such as pigments, phenolics, and volatile compounds also contribute to the sensory and nutritional qualities of fruit juices. The level of endogenous antioxidants and bioactive compounds is a current area of interest and much work is being conducted to determine what the profile of juice pigments and antioxidants is and how this may be nutritionally beneficial. The juice making process may or may not result in a loss of phenolic antioxidants. Biological variability between cultivars of fruit, as well as length of time in storage, cultivation conditions, and extraction methods can all affect the level of bioactive components in juice.

Application of infrared technology to juice analyses

Infrared technology has been applied to almost every aspect of agricultural and food science for the past four decades. The wide application of infrared (IR) technology in the area of food science is greatly attributed to Karl Norris' efforts to develop a rapid method for moisture determination in cereal products (McClure, 2003). Norris successfully integrated chemometrics with the analysis of near-infrared (NIR) spectral data to predict moisture, and later proximate composition of grain products. This led to a breakthrough in the use of NIR for both qualitative and quantitative analysis. Absorptions in the NIR region are mainly comprised of overtones and combinations bands, which allow the use of much longer path lengths than in the mid-infrared (MIR) region, and thus significantly simplify sample preparation and spectral acquisition. Furthermore, the application of chemometrics to spectral data analysis makes it possible to simultaneously analyze multiple food components in a few minutes. Although, initially, the NIR region drew the most attention in food analysis, particularly for rapid and/or non-destructive analysis of low and intermediate moisture food, with the development of new sampling accessories for the MIR region, such as attenuated total reflectance (ATR), the use of MIR and chemometrics in food analysis is gaining increasing interest since spectral features in MIR are more distinct, making data interpretation easier.

Because of the numerous components in a food, most of the time it is difficult to identify the absorption bands caused by a particular compound of interest. Chemometric techniques, such as PLS regression (Geladi and Kowalski, 1986), which can make use of full spectral data, can be effectively applied to correlate spectral features with the concentration of an analyte. Unlike traditional IR techniques, most of the current research on food analysis using IR spectroscopy involves chemometric methods, in which a mathematical model is devised to correlate absorbance or transmittance values at different wavelengths with a known concentration of an analyte. This model is used to predict the concentration of analyte in samples of the same or similar food items with unknown composition. Although the approaches to acquire spectra, as well as the discrete chemical information obtained from the spectra are quite different between NIR and MIR, the chemometric techniques used for NIR and MIR are basically the same.

Since the 1960s, consistent efforts have been made to seek powerful mathematical methods to interpret spectral data information. Stepwise multiple linear regression, the oldest data analysis method, is still used widely for spectral data analysis. Partial least squares regression (PLS) has become one of the most popular methods since the late 1980s. More and more mathematical methods, such as artificial neural networks, are being applied to IR data calibration to solve both linear and non-linear changes in spectral features (Gestal *et al.*, 2004, 2005; Huang *et al.*, 2007a).

Infrared methods have been applied for both qualitative and quantitative analyses to ensure food safety and quality. The application areas can be classified as follows:

- Determination of food components. Food components can be macronutrients (e.g. protein, fat, and carbohydrate) and micronutrients or non-nutritional (e.g. pigments). Some of the components, such as salt and minerals, have no specific absorption bands in the IR region; however, since the presence of these components affects the shape and position of water bands or other molecular bands in the IR region, it is possible to detect and quantify these components in a food (Huang *et al.*, 2003a).
- Determination of important quality indicators or parameters. Quality indicators cover both chemical (e.g. pH and acidity) and physical characteristics (e.g. hardness and viscosity). Although IR spectral features are essentially tied to the chemical properties of a food, some physical properties of the food may be closely related (in either a linear or non-linear fashion) to its chemical properties that are detectable in the IR region, thus making it possible to predict

the physical properties of a material (Lundin *et al.*, 1998; Segtnan *et al.*, 2003; Huang *et al.*, 2003b, 2007b).

- Detection or determination of toxicants (e.g. pesticides and myotoxins) and biological contaminants (e.g. pathogens). The amount of a contaminant may be very small, such as 100 ppb aflatoxin, and beyond the detection limit of an IR spectrometer, but if the level of contaminant is highly correlated with some other IR-sensitive components (e.g. protein and fat) in the same food, then it may be possible to devise a reliable correlation between the two components. Often, it is possible to predict the level of a contaminant, or its presence or absence by using multivariate analytical methods (Pearson *et al.*, 2001).
- Grading, classification, or authentication of food. This typically requires determination of more than one quality indicator. For example, acidity, brix value, brix-acid ratio are required for the grading of orange juice. Since IR methods can simultaneously determine several ingredients, or directly predict a quality attribute (e.g. its grade or compliance with a commercial or market specification), these methods can significantly save time and cost for this type of application (Twomey *et al.*, 1995; Gestal *et al.*, 2004).

Since physical and chemical properties, as well as the quality standard for fruit juices are different from other food categories (e.g. meat products), fruit juice analysis using IR methods has its own characteristics. For example, penetration depth is less of an impediment to the use of spectral methods for the analysis of juices compared to solid foods that have a well-defined tissue structure where it is often difficult to obtain spectra that are representative of the bulk properties of the food.

Basic spectral features of fruit juices

Water

Water is the principal component in food, particularly in liquid foods such as fruit juice. Water molecules have strong absorption bands in the IR region. The presence of water in food affects the determination or detection of other components, which requires mathematical approaches to overcome the interference from water molecules.

Water bands are dominant in a typical IR spectrum of fruit juice (Figure 13.2). In Figure 13.2, three water bands around 3328 cm^{-1} (OH-stretching, $\upsilon_{1,3}$), 2115 cm^{-1} (combination band, $\upsilon_2 + \upsilon_L$), and 1634 cm^{-1} (OH-bending, υ_2) are the primary absorption bands in the mid-IR region for fruit juices (Libnau *et al.*, 1994). In addition, three small water bands can be found around 5200 cm^{-1} (1923 nm, $\upsilon_{1,3} + \upsilon_2$), 7000 cm^{-1} (1429 nm, first overtone of the OH-stretching band, $2\upsilon_{1,3}$), 8500 cm^{-1} (1176 nm, $2\upsilon_{1,3} + \upsilon_2$) in the NIR region, which are the combinations of OH-bending, OH-stretching and/or the first overtone of the OH-stretching (Libnau *et al.*, 1994). The strength of these absorption bands is very small compared to the three water bands in the MIR region, and normally should be observed with a NIR spectrometer. In an even shorter wavelength NIR region (700–1100 nm), in which absorptions are comprised of third and fourth overtones and combination bands with very

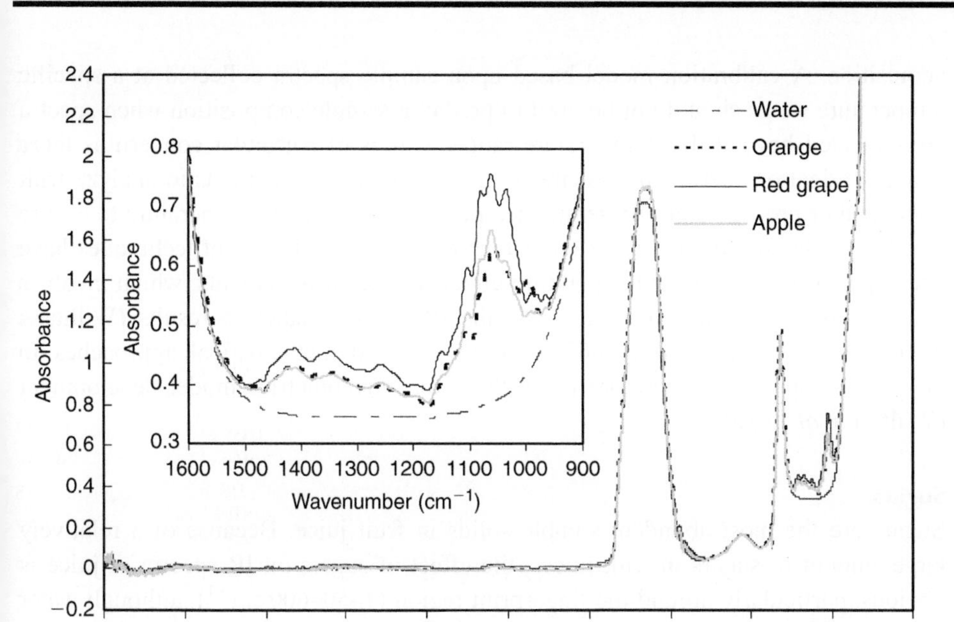

Figure 13.2 Fourier transform infrared spectra of water, orange juice, red grape juice, and apple juice. Spectra were recorded using an attenuated total reflectance through top plate at room temperature.

small absorptivity coefficients, water bands at about 970 nm, 840 nm and 750 nm are reported (McClure and Standfield, 2002; Huang *et al.*, 2007c).

The hydrogen bonds of water are highly temperature-dependent. Various studies have been conducted to study the hydrogen bond systems of water and their temperature-dependent behaviors. According to these studies, liquid water contains at least three components or classes with different structures, and the proportion of each class changes with temperature, which therefore affects the spectral features of water (Lin and Brown, 1992; Libnau *et al.*, 1994; McClure and Standfield, 2002). In both MIR and NIR regions, an increase in temperature results in an increase in the intensity of water bands due to the larger energy of hydrogen bonding (Finch and Lippincott, 1956). Temperature also affects the position of the water bands. Whether a water band shifts towards a lower or higher wavenumber as temperature increases, depends on the wavenumber at which the water band appears. In the MIR region, when temperature increases, the water band at 3328 cm^{-1} shifts to higher wavenumber, but those at 2115 cm^{-1} (combination band, $\upsilon_2 + \upsilon_L$), and 1634 cm^{-1} shift to lower wavenumber (Finch and Lippincott, 1956; Libnau *et al.*, 1994). In the NIR regions, three major bands at 5200 cm^{-1}, 7000 cm^{-1}, and 8500 cm^{-1} shift to higher wavenumber with an increase of temperature (Libnau *et al.*, 1994).

Because the spectral features of water, the dominant spectral features for juice, strongly depends upon temperature, the accuracy of using IR methods for fruit juice analysis can be affected by sample or environmental temperature fluctuations during measurements. It is important to keep the temperature consistent during spectral acquisition. A calibration model based upon sample spectra collected at a specific temperature range should not be used to predict a sample composition when spectra are collected beyond that temperature range. This is of particular concern as there is a great deal of interest in implementing non-invasive IR methods to analyze fruit juice held or processed at refrigerated temperatures or at room temperature to ensure that the foods are safe and meet regulatory requirements. Different techniques have been applied to correct the temperature effects on IR measurements, which involves simply applying a correction factor on the IR model prediction results (Williams *et al.*, 1982; Huang *et al.*, 2007c), or complicated mathematical approaches to pre-process spectral data before using them for chemometrics model development (Wülfert *et al.*, 2000).

Sugars

Sugars are the most abundant soluble solids in fruit juice. Because of a relatively large amount of sugars in fruit juices, the effect of sugars on IR spectra of juice is obvious, particularly around the fingerprint region (1450–600 cm^{-1}), although water bands remain the dominant features (Figures 13.2 and 13.3).

Figure 13.3 shows the Fourier transform infrared (FTIR) spectra of 10% (w/v) aqueous solutions of fructose, glucose, and sucrose from 10 000 to 600 cm^{-1}. The 10% sugar solutions were used to approximate the sugar content in fruit juice. The absorption bands around 1100–1000 cm^{-1} are mainly due to the C–H and C–O stretch vibrations, and the spectral signatures among sugars are somewhat

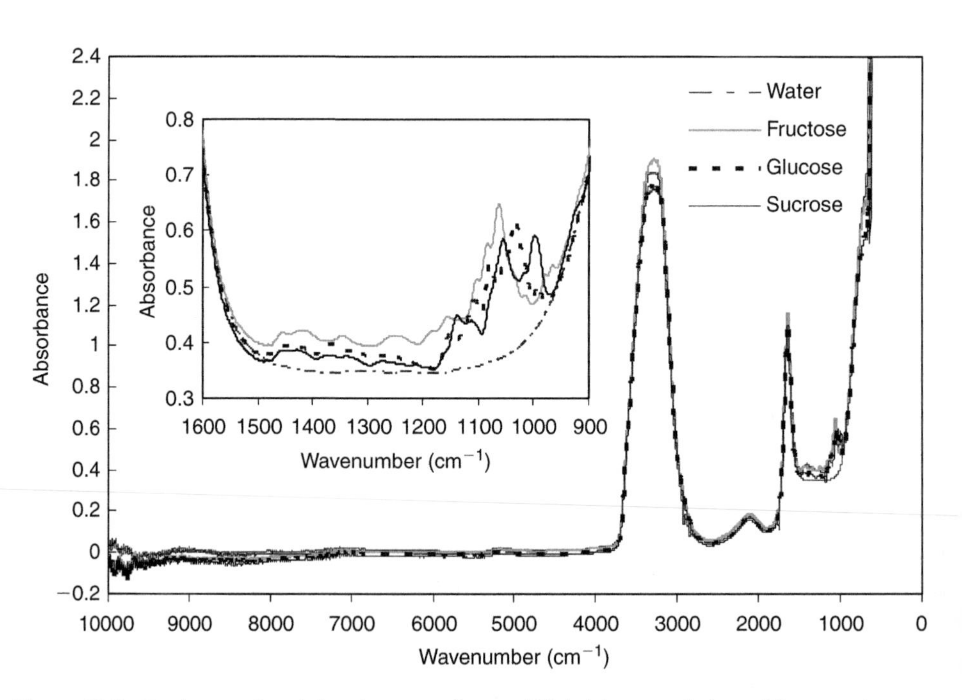

Figure 13.3 Fourier transform infrared spectra of water, 10% (w/v) water solutions of fructose, glucose and sucrose. Spectra were recorded using an attenuated total reflectance through top plate at room temperature.

different from one another (Max and Chapados, 2007). For example, aqueous glucose has a strong C–O stretch vibration at around 1030 cm^{-1}, while for fructose, the C–O stretch vibration band is around 1060 cm^{-1}, and for sucrose, around 1055 cm^{-1}. In addition, there are some low intensity bands around 1400 cm^{-1} caused by C–C–H and C–O–H deformation modes. Unlike other simple molecules, sugars have endocyclic and exocyclic C–O bonds, such as around 995 cm^{-1} (exocyclic) for sucrose, and around 1080 cm^{-1} (endocyclic) for glucose and fructose.

Sucrose, glucose, and fructose aqueous solutions have more than one hydrate form. Max and Chapados (2007) identified pentahydrate (sugar–$5H_2O$) and dihydrate (sugar–$2H_2O$) for D-glucose and sucrose, and pentahydrate and monohydrate (sugar–H_2O) for D-fructose using FTIR spectrophotometry with factor analysis. The existence of multiple hydrates further complicates the IR spectra of sugar solutions as well as fruit juices containing multiple sugars. This makes it difficult if not almost impossible to assign absorption bands in the fingerprint region for specific sugars in fruit juice.

Overall, sugar molecules in fruit juice not only have their own spectral signatures in the IR region, but also affect the shape and position of the water bands in this region. Sugars affect the shape and intensity of the absorption bands due to OH stretching vibration, such as at around 3300 cm^{-1}, although most of the time it is difficult to separate the OH stretch vibrations of water from those of sugars. Furthermore, the existence of sugar molecules changes the ratio of water components or classes (water contains at least three components with different molecular structures), and thus affects the water bands, such as the one at around 1640 cm^{-1} (Lin and Brown, 1992; McClure and Standfield, 2002).

Organic acids and other components

After sugars, organic acids are the major soluble solids in fruit juice. Organic acids in fruit juice contain C=O, C–O, O–H, C–H functional groups. The presence of these functional groups adds more spectral features to the water–sugar system. Since the amount of organic acids in fruit juice is relatively small (about 1%), the effects of these acids to the IR spectrum of a juice, most of the time, cannot be discerned by visual examination of the spectra. Nevertheless, these organic acids play an important role in analyzing fruit juice via IR methods, due to their spectral signatures as well as their effects on water or sugar bands in the IR region that can be ascertained from chemometric models. Furthermore, because of the important role of organic acids to the safety and quality control of fruit juices, the analysis of organic acids in fruit juice by IR methods remains an active area of research.

The concentrations of other components, such as minerals, vitamins, and polyphenols, tend to be very small in fruit juice. To eliminate the interference of the water band, small samples of juice are commonly applied to a filter first, then when the water has evaporated, sample spectra are acquired from the dehydrated sample. This technique is effective in MIR (FTIR) but is less so in the shorter wavelength NIR region. In addition, small amounts of components may have no absorption bands in the IR region (e.g. minerals), or even if they are absorptive (e.g. vitamin C), the effects of these components on the spectral features dominated by the water–sugars system are very limited. Detection of these components is largely due to their correlation with other molecules that can be directly detected in the IR region. This implies a relatively large error for analyzing small amounts of components or components with no absorption bands in the IR region by using IR techniques. Nevertheless, because it is time-consuming and/or costly to determine these components using current methods, the application of IR techniques is always attractive to both the research community and the food industry.

Analyses of composition and quality parameters

Sugars and organic acids are major soluble solids in fruit juices. The amount and composition of sugars and organic acids largely decide the sensory properties of fruit juices. On the other hand, total soluble solids (expressed as brix value), acids, and brix-acid ratio are the key quality indicators for fruit juices. Therefore, determination of total soluble solids, sugars, acidity, and organic acids of fruit juices are routine tests and represent the major areas of research applying IR technology to fruit juice analyses.

Routine analyses of total soluble solids and acidity of fruit juices are simple tasks, using a refractometer and a titration method, respectively. However, determinations of each individual sugar or organic acid is often a complicated task, normally requiring enzymatic analysis, or high-performance liquid chromatography (HPLC). The methodologies for either simple or complicated conventional tasks are similar for IR analysis, although it is more costly and time-consuming to develop a chemometric model for complicated tasks, since it is more difficult to obtain the reference values, such as organic acid profile vs. titratable acidity. Yet, once the models are built, they would be equally convenient to predict simple and complicated conventional tasks.

The study of Lanza and Li (1984) is among the earliest applications of IR technology in fruit juice analysis. Lanza and Li used NIR (1100–2500 nm) in the transmittance mode to predict the sugar content in 11 different types of fruit juices. Fruit juices were centrifuged to remove pulp before samples were transferred by pipette into a quartz transmission cell (path length = 2.2 mm) for NIR analysis. A step-forward multiple linear regression method was used to correlate the NIR spectral data with total sugar and individual sugars separately. This study indicated the potential of applying a NIR method for the analysis of total sugars in fruit juices in routine quality control, though it could not achieve ideal accuracy for individual sugar determination. With the improvement in IR instrumentation and development of more sophisticated mathematical models, the subsequent studies using NIR to determine total sugar and individual sugars in fruit juices or fruit juice model systems with minimal sample preparation or using dry extract of fruit juices gained acceptable accuracy and precision (Li *et al.*, 1996; Rambla *et al.*, 1997, 1998; Segtnan and Isaksson, 2000; Rodriguez-Saona *et al.*, 2001). Similarly, studies indicated that NIR methods have the potential for determination of acidity, organic acid profiles, pH, and even total calories of fruit juices (Li *et al.*, 1996; Moros *et al*,, 2005; Cen *et al.*, 2006; Chen *et al.*, 2006).

Since the strong absorption of water molecules in fruit juice makes it difficult for sampling in the MIR region, the spectrometers used for fruit juice analysis were exclusively dispersive NIR spectrometers until the late 1990s, when FT-NIR spectrometers and FTIR with attenuated total reflectance (ATR) became available and were applied to the analysis of fruit juices (Rambla *et al.*, 1998; Rodriguez-Saona *et al.*, 2001; Duarte *et al.*, 2002; Moros *et al.*, 2005). The application of FT-NIR spectrometers to fruit juices helps improve the quality of NIR spectra (Rodriguez-Saona *et al.*, 2001). On the other hand, the emergence of the ATR accessory has greatly simplified the sampling process when using MIR spectrometer, and allows the wider use of the MIR region for rapid and non-destructive analysis of both liquid and solid food samples. The ATR accessory is mainly made of a crystal with much higher refractive index than the sample. When an IR beam incides on an ATR plate, it causes total internal reflection. The internal reflectance creates an evanescent wave that extends into the sample for a few micrometers. The attenuated energy (due to the absorption by the sample in the IR region) from the evanescent wave is passed back to the IR beam and then to the detector. With the use of both NIR and IR techniques, we can foresee that the accuracy and precision of IR methods for compositional analysis of fruit juice will be much improved.

Detection and determination of biological contaminants

Both qualitative and quantitative analyses of biological contaminants, such as pathogens and spoilage bacteria, in foods are daunting tasks, because of the existence of numerous microorganisms with various growth patterns and complex biological responses to extrinsic factors, and also because of the complicated chemical and physical nature of a food. However, this complexity stimulates the interest in finding innovative approaches to replace the typically expensive and time-consuming microbial testing, such as microbial enumeration techniques which take days, or the intricate multi-step DNA-based methods which take several hours. Application of IR technology to directly detect microorganisms is a fairly recent development and is based on the pioneering study of Naumann and his colleagues (Naumann *et al.*, 1991).

Naumann *et al.* (1991) indicated that FTIR could be highly selective in differentiating microorganisms to the subspecies, strain, or even serotype levels. The possibility of using IR methods to characterize microorganisms is due to its ability to detect the small biochemical differences among the cellular components of microbes, such as polysaccharides, proteins, lipids, and nucleic acids, while these cell components depend upon the expression of the genomes of an organism. Naumann *et al.* (1991) transferred pure bacteria cells (about 10–60 μg) into a sample holder, dried them and then acquired IR spectra directly. Each spectrum was then analyzed to extract specific chemical information, and compared with a reference database for characterization purposes. By 1991, Naumann and his colleagues had built a database comprised of FTIR spectra for more than 300 bacterial strains.

The initial study involving the application of FTIR to characterize microorganisms used pure microbial cultures. Applications of FTIR techniques to detect microorganisms in foods basically follow the same methodology. The study of Lin *et al.*, (2005) is typical and involves the detection of bacteria in fruit juices. Lin *et al.* (2005) inoculated each of the eight selected strains of the spore forming spoilage

Alicyclobacillus strains into pasteurized apple juices separately, then incubated the juice at 43°C for 7 days until the cell numbers were about 10^6–10^8 CFU/mL. The bacterial cells were harvested by centrifuging, resuspending in 0.9% saline solution three times. The final bacteria saline solution was filtered through an aluminum oxide membrane. The membrane with bacteria was air-dried to obtain a bacteria cell film, and then placed on an ATR zinc selenide (ZnSe) crystal to acquire FTIR spectra. The spectra were smoothed, transformed to second derivative, and then analyzed with principal component analysis and soft independent modeling of class analogy. The study identified some important peaks that may distinguish one *Alicyclobacillus* strain distinct from the other in the fingerprint region, and revealed the possibility of using FTIR to detect *Alicyclobacillus* spoilage strains.

Other representative studies on the detection of bacteria in fruit juices included Yu *et al.* (2004), Rodriguez-Saona *et al.* (2004), Al-Qadiri *et al.* (2006), and Al-Holy *et al.* (2006). Yu *et al.* (2004) used FTIR with ATR to analyze apple juices contaminated with eight bacteria at different concentrations (10^3–10^8 CFU/mL). Contaminated apple juice was used directly for spectral acquisition. The studies showed that FTIR can differentiate apple juice contaminated with bacteria at a concentration level of 10^3 CFU/mL. Rodriguez-Saona *et al.* (2004) used FT-NIR instead of FTIR to differentiate pathogenic strains and apple juices contaminated with *E. coli* strains. Bacteria were concentrated on an aluminum oxide membrane for NIR analysis, and the study results were promising. Al-Qadiri *et al.* (2006) and Al-Holy *et al.* (2006) successfully applied FTIR to differentiate *E. coli* 0157:H7 from other bacteria in apple juice, using methodologies similar to those of Lin *et al.* (2005).

To apply IR technology for microorganism identification and/or determination, two basic conditions must be met. First, a purified or semi-purified single species, instead of a mixture of microorganisms, should be used for IR spectral acquisition, at least initially when sensitivities within a specific matrix are to be determined. Second, a database containing FTIR or FT-NIR spectra of microorganisms of interest should be available or should be developed. A spectrum should be collected under the same experimental conditions using the same instrumentation and setup as that used for the database spectra.

The greatest advantage of IR methods is that they require minimum sample preparation, but this may not be obvious for directly identifying bacteria in foods, since bacteria usually have to be extracted and purified from a food at least to some extent, before being subjected to IR analysis. Although some studies used IR methods to directly determine microorganisms in a food matrix with favorable results, more work needs to be done and the method may have practical limitations in the long term, unless the specific analysis needed for determining food safety in a particular food system is always tied to one or a few microorganisms.

Fortunately, directly identifying microorganisms is not the only approach to detect microbial contamination in foods. Most of the time, determination of the metabolites from microorganism growth is a viable alternative. For example, some *Alicyclobacillus* spp. produce guaiacol which causes off-flavor in fruit juice, so detection of guaiacol would indicate the contamination of *Alicyclobacillus* spp. Determination of a specific biochemical compound instead of the actual level of microorganisms is certainly an easier task, and can follow the routine for compositional analysis as discussed earlier in this chapter.

Authentication and classification of fruit juices

Commonly used methods for authentication

Falsification of juice composition not only affects the quality of fruit juices, it deceives consumers who pay more for a product with false claims on its ingredient statement. It is difficult to estimate the amount of adulterated fruit juices in the global food market, but there is no doubt that with an increased globalization, the impact of single events of food adulterations will affect a larger and wider population than ever. Fruit juice falsifications involve simple adulteration by adding water, sugar, and organic acids, or more complicated approaches by using fruit products, such as pulpwash, peel, and other fruit constituents (Fry *et al.*, 2005). Since the major components of fruit juices are water, sugars, and organic acids with lesser amounts of amino acids, vitamins, minerals, and phenolic compounds, detection and quantification of these components as well as determining their profile provide the basis for detection of adulteration. The soluble solid content is one of the two major quality indicators (the other is titratable acidity) for fruit juices, and is also used as an indicator to detect whether the juice has been diluted with water. Since the levels of soluble solids of fruit juices are regulated by government, international trade standard (e.g. codex standards) and industry specification, and at the same time the soluble solids of fruit juices are easy to measure, simply diluting fruit juices with water is not as common an adulteration practice as it has been in the past (Ashurst, 2005).

Detecting or determining whether unauthorized carbohydrates have been added is widely used for fruit juice authentication. Analytical methods for sugars vary from determining the stable oxygen and carbon isotope ratio of a selected sugar (Jamin *et al.*, 1997; Simpkins *et al.*, 1999; Antolovich *et al.*, 2001; Jamin *et al.*, 2003), to evaluating oligosaccharide profiles or sugar profiles of fructose, glucose, and sucrose (González *et al.*, 1999). Similarly, the addition of unauthorized organic acids in fruit juices can be detected or determined by stable isotope ratio and by HPLC organic acid profile (Doner, 1985; Jezek and Suhaj, 2001; Jamin *et al.*, 2005).

Measurement of minerals such as sodium, calcium, magnesium, and potassium is used together with other methods, or by itself, for fruit juice authentication. Recently, the evaluation of the polyphenol profile of fruit juice has also been used (Ooghe *et al.*, 1994; Mouly *et al.*, 1997).

To detect fruit juice adulteration, the test results of a juice need to be compared with that of authentic juice standards. This requires a database for the juices from various varieties and geographical origins, and it may become more difficult to obtain reliable standards as the sources of the juices become increasingly global.

Although some of the methods, such as determination of soluble solids and minerals, are relatively simple, other methods, such as stable isotope ratio or sugar and acid profile, are complicated and require expensive instruments, such as a mass spectrometer for stable isotope ratio and HPLC for sugar profiles. In addition, to combat sophisticated juice adulteration schemes, several different tests may be required to detect falsification of juice composition, which further increases the cost and complexity of the test.

Infrared methods for authentication and classification

Infrared spectroscopy methods are relatively simple and potentially less costly than the methods mentioned above. Being able to simultaneously determine multiple components in a few minutes fits the urgent need to combat sophisticated fruit juice adulteration. There are few studies on using IR spectroscopy methods for fruit juice authentication, and these studies are still at the laboratory stage, but results to date indicate the potential of applying IR techniques for fruit juice authentication in the near future.

With the limited reports available using IR spectroscopy for fruit juice authentication, the methods applied in these studies varied in every stage from sample preparation, selection of spectrometer, to chemometric techniques. Fruit juices were either directly used with no sample preparation, or with simple sample preparation such as filtration for IR analysis. Both NIR and FTIR involving reflectance and transmittance modes have been reported to detect adulterated fruit juices. The mathematical techniques used included principal component analysis, factorial discriminant analysis, linear discriminant and canonical variate analysis, neural networks, and PLS. A brief review of some representative studies follows.

Twomey *et al.* (1995) applied NIR (1100–2498 nm) with reflectance mode to detect concentrated orange juice adulteration that involved a total of 65 authentic concentrated orange juice samples and their adulterated samples with 5% or 10% addition of orange pulpwash, grapefruit, and sugar/acid mixture. The juices were vacuum-dried and ground to fine powder before analysis with a NIR spectrometer in a powder cell. The NIR spectral data were compressed to 20 principal components by principal component analysis, and then classified by factorial discriminant analysis. The method had about 90% accuracy with no false positive authentication.

Leon *et al.* (2005) applied NIR "transflectance" spectroscopy (400–2498 nm) and PLS to detect fresh apple juice adulterated with 10–40% of high fructose corn syrup (45% fructose and 55% glucose) or sugar solution (60% fructose, 25% glucose, and 15% sucrose). To acquire spectra, each juice sample was placed in a 0.2-mm-deep cell with a gold-plated backing plate to reflect light back. A total of 450 samples, including 150 authentic samples from 19 varieties, were used with an accuracy of over 90%.

Kelly and Downey (2005) used FTIR with ATR (800–4000 cm^{-1}) to detect sugar adulterants in apple juice. The study involved a total of 224 apple samples from 19 varieties, and 480 of their adulterated counterparts containing 10–40% solution of a mixture of sugars. Sample spectra and their second derivative spectra were compressed by using principal component analysis to eight principal components, respectively. A total of 16 principal components were then used for PLS regression and k-nearest neighbor (kNN) method to detect and quantify adulteration. FTIR could detect apple juices adulterated with sucrose at more than 80% accuracy, but there was greater difficulty on identifying apple juice adulterated with high fructose corn syrup (61–100% accuracy), or with a mixture of fructose, glucose, and sugar solution (76–97% accuracy), particularly at a low adulteration level. At 10% adulteration level, the PLS method used correctly classified 61% and 49% for apple juice adulterated with high fructose corn syrup and with a mixture of three single sugars, respectively, similar to the use of the kNN method (the corresponding data were 54% and 78% accurate).

Fruit juice classification or differentiation is to a large extent synonymous with authentication. Typical applications include classification of juice with different levels of adulteration, and differentiation of fresh-pressed juice from that from concentrated juices, which is in line with the application of NIR techniques for authentication.

Similar to other analytical techniques, the successful application of IR methods for fruit juice authentication requires a large database with spectral information of authentic juices from different varieties and geographic origins. This is further complicated by an increasing use of mixtures of fruit juice from different varieties or origins, and what is more, the increasing popularity of mixed fruit juices, for example, a tropical punch containing pineapple, orange, and banana juice.

The application of NIR or FTIR for relatively simple fruit juice authentication is very promising, but it is still difficult to detect complicated juice adulteration, such as only adding 10% of a mixture of sugar solution, using IR and any other techniques. The primary advantages of FTIR or NIR may be as a screening tool. However, most of the current studies using NIR or FTIR are non-invasive, which do not involve any sample preparation; while the use of other analysis methods, such as HPLC and stable isotope ratio, require complicated sample preparation, and yield similar results for complicated fruit juice adulteration. In addition, IR methods allow the use of full spectra that are composed of signals from all ingredients in a juice. These chemical signals together with more sophisticated mathematical approaches provide a potentially very powerful tool for the authentication of fruit juices.

Conclusions

Fruit juice products are an important food category, with a diversity of choices in the commercial market. Economic fraud is rampant in the juice trade, and because of the risk of juices being adulterated or misbranded, having analytical methods to authenticate juices is important. Infrared spectroscopy with multivariate analyses offers an approach for detecting juice adulteration, and is also a practical approach for online monitoring the safety and quality of various fruit juice products.

Infrared spectroscopy methods are relatively inexpensive to establish and relatively simple to use after analysts are properly trained. These spectroscopic methods can simultaneously determine multiple components within a few minutes. Analyses of full spectral data provide enormous potential for compositional analysis, classification, and authentication.

Currently, the major applications of IR technology for fruit juice analysis involve the determinations of components (e.g. sugar and acid content) and quality parameters (e.g. brix to acid ratio). It is also applied to detect biological contaminants in fruit juices, although this may have practical limitations for directly identifying microorganisms with sensitivities in the range of 100 CFU/mL. Finally, IR methods show great promise for combating fruit juice adulteration, since it would be difficult to fabricate a juice blend that would have the spectral features of a pure juice, which makes it possible to detect economic fraud. The successful applications of IR methods for fruit juice analyses very often requires that a large database with spectral information of juices from different varieties and geographic origins be included within a chemometric model to provide the greatest applicability.

Dedication

We would like to dedicate this chapter to the memory of our dear friend, colleague and long time collaborator, Dr David Mayes in recognition of his many important contributions to the field. Dr Mayes developed the instrumentation, chemometric algorithms and analytical software used by many of us in the USA for IR spectroscopic analysis.

References

Al-Holy MA, Lin M, Cavinato AG, Rasco BA (2006) The use of Fourier transform infrared spectroscopy to differentiate *Escherichia coli* 0157 : H7 from other bacteria inoculated into apple juice. *Food Microbiology*, **23**(2), 162–168.

Al-Qadiri HM, Lin MS, Cavinato AG, Rasco BA (2006) Fourier transform infrared spectroscopy, detection and identification of *Escherichia coli* 0157 : H7 and *Alicyclobacillus* strains in apple juice. *International Journal of Food Microbiology*, **111**(1), 73–80.

Antolovich M, Li X, and Robards K (2001) Detection of adulteration in Australian orange juices by stable carbon isotope ratio analysis (SCIRA). *Journal of Agricultural and Food Chemistry*, **49**(5), 2623–2626.

Ashurst PR (2005) Introduction. In: *Chemistry and Technology of Soft Drinks and Fruit Juices, 2nd edn* (Ashurst PR, ed.). Oxford: Blackwell Publishing.

Cen HY, He Y, Huang M (2006) Measurement of soluble solids contents and pH in orange juice using chemometrics and vis-NIRS. *Journal of Agricultural and Food Chemistry*, **54**(20), 7437–7443.

Chen JY, Zhang H, Matsunaga R (2006) Rapid determination of the main organic acid composition of raw Japanese apricot fruit juices using near-infrared spectroscopy. *Journal of Agricultural and Food Chemistry*, **54**(26), 9652–9657.

Doner LW (1985) Carbon isotope ratios in natural and synthetic citric acid as indicators of lemon juice adulteration. *Journal of Agricultural and Food Chemistry*, **33**, 770–772.

Duarte IF, Barros A, Delgadillo I, Almeida C, Gil AM (2002) Application of FTIR spectroscopy for the quantification of sugars in mango juice as a function of ripening. *Journal of Agricultural and Food Chemistry*, **50**(11), 3104–3111.

Finch JN, Lippincott E (1956) Hydrogen bond systems: temperature dependence of OH frequency shifts and OH band intensities. *Journal of Chemical Physics*, **24**, 908–909.

Food and Drug Administration (2002) Percentage juice declaration for foods purporting to be beverages that contain fruit or vegetable juice. 21 CFR 101.30.

Fry J (1990) Authentication of orange juice. In: *Production and Package of Non-carbonated Fruit Juices and Fruit Beverage* (Hicks D, ed.). Edinburgh: Blackie and Son Ltd, pp. 68–106.

Fry J, Martin GG, Lees MM (2005) Authentication of orange juice. In: *Production and Package of Non-carbonated Fruit Juices and Fruit Beverage* (Ashurst PR, ed.). New York: Aspen Publications, pp. 1–52.

Geladi P, Kowalski B (1986) Partial least-squares regression: A tutorial. *Analytica Chimica Acta*, **185**, 1–17.

Gestal M, Gómez Carracedo MP, Andrade JM, Dorado J, Fernández E, Prada D, Pazos A (2004) Classification of apple beverages using artificial neural networks with previous variable selection. *Analytica Chimica Acta*, **524**, 225–234.

Gestal M, Gómez Carracedo MP, Andrade JM, Dorado J, Fernández E, Prada D, Pazos A (2005) Selection of variables by genetic algorithms to classify apple beverages by artificial neural networks. *Applied Artificial Intelligence*, **19**(2), 181–198.

González J, Remaud G, Jamin E, Naulet N, Martin GG (1999) Specific natural isotope profile studied by isotope ratio mass spectrometry (SNIP-IRMS): C-13/C-12 ratios of fructose, glucose, and sucrose for improved detection of sugar addition to pineapple juices and concentrates. *Journal of Agricultural and Food Chemistry*, **47**(6), 2316–2321.

Huang Y, Cavinato AG, Mayes DM, Kangas LJ, Bledsoe GE, Rasco BA (2003a) Nondestructive determination of moisture and sodium chloride in cured Atlantic salmon (*Salmo salar*) (Teijin). *Journal of Food Science*, **68**, 482–486.

Huang Y, Tang J, Swanson BG, Cavinato AG, Lin M, Rasco BA (2003b) Near infrared spectroscopy: a new tool for studying physical and chemical properties of polysaccharide gels. *Carbohydrate Polymers*, **53**, 281–288.

Huang Y, Kangas LJ, Rasco BA (2007a) Application of artificial neural networks in food science. *Critical Reviews in Food Science & Nutrition*, **47**, 113–126.

Huang Y, Cavinato AG, Tang J, Swanson BG, Lin M, Rasco BA (2007b) Characterization of sol-gel transitions of food hydrocolloids with near infrared spectroscopy. *LWT-Food Science and Technology*, **40**, 1018–1026.

Huang Y, Cavinato AG, Mayes DM, Rasco BA (2007c) Influence of temperature on the measurement of NaCl content of aqueous solution by short-wavelength near infrared spectroscopy (SW-NIR). *Sensing and Instrumentation for Food Quality and Safety*, **1**, 91–97.

Jamin E, Gonzalez J, Remaud G, Naulet N, Martin GG, Weber D, Rossmann A, Schmidt HL (1997) Improved detection of sugar addition to apple juices and concentrates using internal standard C-13 IRMS. *Analytica Chimica Acta*, **347**(3), 359–368.

Jamin E, Guérin R, Rétif M, Lees M, Martin GJ (2003) Improved detection of added water in orange juice by simultaneous determination of the oxygen-18/oxygen-16 isotope ratios of water and ethanol derived from sugars. *Journal of Agricultural and Food Chemistry*, **51**(18), 5202–5206.

Jamin E, Martin F, Santamaria Fernandez R, Lees M (2005) Detection of exogenous citric acid in fruit juices by stable isotope ratio analysis. *Journal of Agricultural and Food Chemistry*, **53**(13), 5130–5133.

Jezek J, Suhaj M (2001) Application of capillary isotachophoresis for fruit juice authentication. *Journal of Chromatography A*, **916**(1–2), 185–189.

Kelly JD, Downey G (2005) Detection of sugar adulterants in apple juice using Fourier transform infrared spectroscopy and chemometrics. *Journal of Agricultural and Food Chemistry*, **53**(9), 3281–3286.

Lanza E, Li BW (1984) Application for near infrared spectroscopy for predicting the sugar content of fruit juices. *Journal of Food Science*, **49**(4), 995–998.

Lea AGH (1990) Apple juice. In: *Production and Package of Non-carbonated Fruit Juices and Fruit Beverage* (Hicks D, ed.). Edinburgh: Blackie and Son Ltd, pp. 182–225.

Leon L, Kelly JD, Downey G (2005) Detection of apple juice adulteration using near-infrared transflectance spectroscopy. *Applied Spectroscopy*, **59**(5), 593–599.

Li WJ, Goovaerts P, Meurens M (1996) Quantitative analysis of individual sugars and acids in orange juices by near-infrared spectroscopy of dry extract. *Journal of Agricultural and Food Chemistry*, **44**(8), 2252–2259.

Libnau FO, Kvalheim OM, Christy AA, Toft J (1994) Spectra of water in the near- and mid-infrared region. *Vibrational Spectroscopy*, **7**, 243–254.

Lin J, Brown CW (1992) Near-IR spectroscopic determination of NaCl in aqueous solution. *Applied Spectroscopy*, **46**(12), 1809–1815.

Lin MS, Al-Holy M, Chang SS, Huang Y, Cavinato AG, Kang D-H, Rasco BA (2005) Rapid discrimination of *Alicyclobacillus* strains in apple juice by Fourier transform infrared spectroscopy. *International Journal of Food Microbiology*, **105**(3), 369–376.

Lundin L, Stenlöf B, Hermansson A-M (1998) NIR spectra in relation to viscoelastic properties of mixtures of Na-κ-carrageenan, locust bean gum and casein. *Food Hydrocolloids*, **12**, 189–193.

McClure WF (2003) 204 years of near infrared technology: 1800–2003. *Journal of Near Infrared Spectroscopy*, **11**, 487–518.

McClure WF, Standfield DL (2002) Near-infrared spectroscopy of biomaterials. In: *Handbook of Vibrational Spectroscopy*. Chichester: John Wiley & Sons Ltd.

McLellan MR, Race EJ (1990) Grape juice processing. In: *Production and Package of Non-carbonated Fruit Juices and Fruit Beverage* (Hicks D, ed.). Edinburgh: Blackie and Son Ltd, pp. 226–242.

Max JJ, Chapados C (2007) Glucose and fructose hydrates in aqueous solution by IR spectroscopy. *Journal of Physical Chemistry A*, **111**(14), 2679–2689.

Moros J, Inon FA, Carrigues S, de la Guardia M (2005) Determination of the energetic value of fruit and milk-based beverages through partial-least-squares attenuated total reflectance-Fourier transform infrared spectrometry. *Analytica Chimica Acta*, **538**, 181–193.

Mouly PP, Gaydou EM, Faure R, Estienne JM (1997) Blood orange juice authentication using cinnamic acid Derivatives. Variety differentiations associated with flavanone glycoside content. *Journal of Agricultural and Food Chemistry*, **45**(2), 373–377.

Naumann D, Helm D, Labischinski H (1991) Microbiological characterizations by FT-IR spectroscopy. *Nature*, **351**(6321), 81–82.

Ooghe WC, Ooghe SJ, Detavernier CM, Huyghebaert A (1994) Characterization of orange juice (*Citrus sinensis*) by polymethoxylated flavones. *Journal of Agricultural and Food Chemistry*, **42**, 2191–2195.

Pearson TC, Wicklow DT, Maghirang EB, Xie F, Dowell FE (2001) Detecting aflatoxin in single corn kernels by transmittance and reflectance spectroscopy. *Transactions of the ASAE*, **44**(5), 1247–1254.

Pollack S, and Perez A (2006) *Fruit and Tree Nuts Situation and Outlook Yearbook.* Washington DC: United States Department of Agriculture. FTS-2006.

Rambla FJ, Garrigues S, de la Guardia M (1997) PLS-NIR determination of total sugar, glucose, fructose and sucrose in aqueous solutions of fruit juices. *Analytica Chimica Acta*, **344**(1–2), 41–53.

Rambla FJ, Garrigues S, Ferrer N, de la Guardia M (1998) Simple partial least squares-attenuated total reflectance Fourier transform infrared spectrometric method for the determination of sugars in fruit juices and soft drinks using aqueous standards. *Analyst*, **123**(2), 277–281.

Rodriguez-Saona LE, Fry FS, McLaughlin MA, Calvey EM (2001) Rapid analysis of sugars in fruit juices by FT-NIR spectroscopy. *Carbohydrate Research*, **336**(1), 63–74.

Rodriguez-Saona LE, Khambaty FM, Fry FS, Dubois J, Calvey EM (2004) Detection and identification of bacteria in a juice matrix with Fourier transform-near infrared spectroscopy and multivariate analysis. *Journal of Food Protection*, **67**(11), 2555–2559.

Segtnan VH, Isaksson T (2000) Evaluating near infrared techniques for quantitative analysis of carbohydrates in fruit juice model systems. *Journal of Near Infrared Spectroscopy*, **8**(2), 109–116.

Segtnan VH, Kvaal K, Rukke EO, Schüller RB, Isaksson T (2003) Rapid assessment of physico-chemical properties of gelatine using near infrared spectroscopy. *Food Hydrocolloids*, **17**, 585–592.

Simpkins WA, Patel G, Collins P, Harrison M, Goldberg D (1999) Oxygen isotope ratios of juice water in Australian oranges and concentrates. *Journal of Agricultural and Food Chemistry*, **47**(7), 2606–2612.

Twomey M, Downey G, McNulty PB (1995) The potential of NIR spectroscopy for the detection of the adulteration of orange juice. *Journal of the Science of Food and Agriculture*, **67**(1), 77–84.

USDA (1982) United States standards for grades of orange juice. *Federal Register* (47 FR 55455). Effective in January 10, 1983.

Williams PC, Norris KH, Zarowski WS (1982) Influence of temperature on estimation of protein and moisture in wheat by near-infrared reflectance. *Cereal Chemistry*, **59**, 473–477.

Wülfert F, Kok WT, Noord OE, Smilde AK (2000) Correction of temperature-induced spectral variation by continuous piecewise direct standardization. *Analytical Chemistry*, **72**(7), 1639–1644.

Yu CX, Irudayaraj J, Debroy C, Schmilovtich Z, Mizrach A (2004) Spectroscopic differentiation and quantification of microorganisms in apple juice. *Journal of Food Science*, **69**(7), S268–S272.

Wine and Beer

14

Daniel Cozzolino and Robert G Dambergs

Introduction

Spectroscopic techniques using the infrared wavelength region of the electromagnetic spectrum have been used in the food industry to monitor and assess the composition and quality value of foods produced. As in other food industries, the wine and beer industries have a clear need for simple, rapid, and cost-effective techniques for objectively evaluating the quality of grapes, wine and spirits. The use of near-infrared (NIR) spectroscopy in the wine industry dates back to some early work by Kaffka and Norris (1976). Their preliminary work was performed on a relatively small number of test samples prepared by standard addition of some of the main components of interest (viz. ethanol, fructose, and tartaric acid) to a red and a white wine, and analyzed in transmission using various path lengths (Kaffka and Norris,

Infrared Spectroscopy for Food Quality Analysis and Control
ISBN: 978-0-12-374136-3

1976). Although these samples represented alterations within the same two basic wine matrices, they allowed the identification of critical wavelengths that could be utilized for multiple linear regression (MLR) analysis.

Since the early work of Kaffka and Norris (1976) the main application of NIR spectroscopy in the beer and wine industries was the measurement of ethanol using filter instruments with two or three wavelengths (Osborne *et al.*, 1993; Cozzolino *et al.*, 2006a).

With the availability of new instruments and the development of software for chemometric/multivariate analysis that allows better interpretation of the spectra, new applications have been developed, such as online and process analysis (Gishen *et al.*, 2005; Cozzolino *et al.*, 2006a). This chapter aims to present the application of infrared, both NIR and mid-infrared (MIR), in different steps of wine and beer production.

Wine

Wine grapes

Grape composition at harvest is one of the most important factors that determine the future quality of the wine (Gishen *et al.*, 2005; Cozzolino *et al.*, 2006a; Dambergs *et al.*, 2007). The prediction of quality parameters in red grapes using NIR spectroscopy is usually conducted by scanning homogenized grape samples, whole grapes, or single berries using an NIR spectrophotometer (Gishen *et al.*, 2005; Cozzolino *et al.*, 2005, 2006a; Dambergs *et al.*, 2007). It has been reported that grape total anthocyanin concentration (color) is a good predictor of red wine composition and quality and is widely used by the Australian wine industry (Gishen *et al.*, 2000, 2005; Dambergs *et al.*, 2003, 2007). It has been demonstrated that total anthocyanins, TSS, and pH can be measured with partial least squares (PLS) regression using reflectance spectra of homogenates of red grape berries scanned over the wavelength range of 400–2500 nm (Dambergs *et al.*, 2003, 2006, 2007). With calibrations for total anthocyanins in red grapes, it has been observed that for large data sets incorporating many vintages, regions, and grape varieties, PLS calibrations show pronounced non-linearity (Dambergs *et al.*, 2003, 2006, 2007). The standard error of prediction (SEP) for total anthocyanins varied from 0.05 to 0.18 mg/g, and it increased with diverse sample sets in comparison to sample sets restricted on the basis of growing region and/or variety (Dambergs *et al.*, 2003, 2006, 2007). The prediction accuracy using calibrations derived from restricted sample sets approached that of the reference methods for total anthocyanins and pH. An alternative strategy to mitigate the effects of non-linearity on the NIR calibrations for total anthocyanins is to use LOCAL regression (Dambergs *et al.*, 2003, 2006, 2007).

The use of NIR spectroscopy has now been put into practise by several large Australian wine companies for determination of the concentration of total anthocyanins (color) in red grapes for payment purposes or for streaming grapes before crushing (Cozzolino *et al.*, 2006a). Similar NIR applications have been reported by private wineries and research groups in Chile, Spain, and Portugal (Arana et al. 2005; Jaren et al., 2001; Herrera et al., 2003). The possibility of simplifying the sample presentation (e.g. using whole grapes instead of homogenates) could dramatically increase sample throughput in the winery (Cozzolino et al., 2004a).

Investigations of whole grape berry presentation using a diode array spectrophotometer indicated that NIR may have potential for use at the weighbridge or for in-field analysis of total anthocyanins, TSS, and pH (Cozzolino et al., 2004a). The use of a diode array instrument to predict total acidity (measured by high-performance liquid chromatography as malic and tartaric acid) in a set of white grape varieties has also been reported by researchers in France (Chauchard et al., 2004). As well as measuring grape quality, there is also a need for objective measurements of negative quality parameters such as the degree of mold contamination, particularly with mechanically harvested grapes, where visual assessment can be difficult (Gishen et al., 2005; Cozzolino et al., 2006a).

Grape assessment for fungal infection at the weighbridge would normally be done by visual inspection, but this can be difficult with mechanically harvested fruit. The use of Vis/NIR spectroscopy was reported for the detection of powdery mildew (*Erysiphe necator*) in wine grapes (Gishen et al., 2005; Cozzolino et al., 2006a; Dambergs et al., 2007). The implication of this work is that it might be possible to discriminate infected fruit at the weighbridge to provide a "go/no-go" test to highlight suspect fruit for further detailed analysis to determine suitability for winemaking.

Wine composition

A large amount of the NIR work in the wine industry has concentrated on the measurement of ethanol (Osborne et al., 1993; Cozzolino et al., 2006a). Currently there are a number of dedicated NIR-based alcohol analyzers in the wine labs, and this technique has become a routine analysis method for alcohol content in wine. Ethanol has a strong NIR absorbance signal in alcoholic beverages (see Figure 14.1), usually second only to water, but accuracy and robustness of calibrations can be limited by matrix variations, particularly variations in sugar concentration (Kemeny

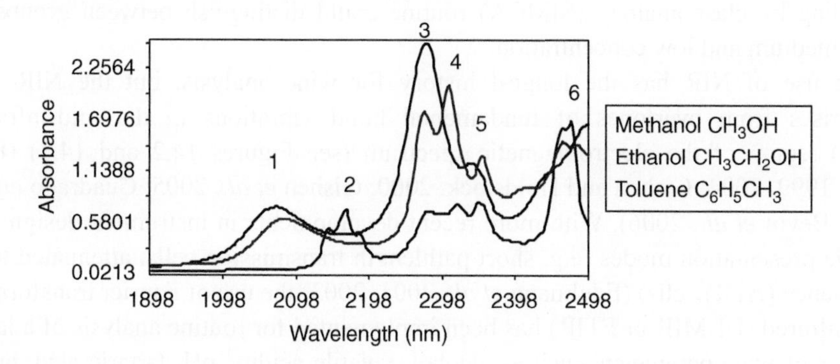

Figure 14.1 Spectra of ethanol, methanol and toluene in the NIR region between 1900 and 2500 nm. (1) OH stretch + deform, ROH; (2) aromatic CH stretch + CC stretch; (3) asymmetic CH stretch + deform, CH_3; (4) asymmetic CH stretch + deform, CH_2; (5) symmetric CH stretch + deform, CH_3; (6) aromatic CH stretch + deform.

et al., 1983; Davenel *et al.*, 1991; Dumoulin *et al.*, 1987). Medrano *et al.* (1995) examined the use of NIR spectroscopy and the use of a sample presentation method called dry extract system for infrared reflectance (DESIR) (Meurens *et al.*, 1987) to measure total phenolics, in addition to ethanol and sugar, in fortified wines.

More recently the use of NIR has been reported to measure several chemical parameters in wine such as volatile acidity, organic acids, malic acid, tartaric acid, lactic acid, reducing sugars, and sulfur dioxide (Urbano-Cuadrado *et al.*, 2004, 2005). However, only accurate determination of alcohol ($R^2 = 0.98$, SEP = 0.24% v/v), pH ($R^2 = 0.81$, SEP = 0.07), reducing sugars ($R^2 = 0.71$, SEP = 0.33 g/L), sulfur dioxide ($R^2 = 0.57$, SEP = 23.5 mg/L) and lactic acid ($R^2 = 0.81$, SEP = 0.41 mg/L) were obtained (Urbano-Cuadrado *et al.*, 2004, 2005). The comparison of NIR and Fourier transform-MIR (FT-MIR) to measure several wine parameters was also reported (Cuadrado *et al.*, 2005).

The use of NIR spectroscopy has also been reported to measure sodium (Na), potassium (K), magnesium (Mg), calcium (Ca), iron (Fe), and copper (Cu) in white and red wines (Sauvage *et al.*, 2002). Metal ions do not absorb in the NIR region, therefore the measurement of such elements is made indirectly, for example by interaction of metal ions with water, or other organic constituents (Sauvage *et al.*, 2002).

Sweet wines made from botrytized-grapes represent a complex matrix in that they have very high sugar, acid, and glycerol content in comparison with standard wines. Garcia-Jares and Medina (1997) examined a number of multivariate calibration routines such as MLR, stepwise regression (SWR), principal components regression (PCR), and PLS regression for the analysis of ethanol, glycerol, glucose, and fructose in botrytis-affected style wines, using NIR spectroscopy. Both PLS and SWR regression gave the best performance, in terms of the lowest SEP relative to mean values, but glycerol and glucose had high percentage errors with all calibration routines (Garcia-Jares and Medina, 1997).

Manley *et al.* (2001) attempted NIR calibrations to measure free amino nitrogen (FAN) in grape must and to monitor malolactic fermentation status of wines. Although calibrations could not accurately quantify the concentrations of the compounds of interest (malic acid, lactic acid, FAN), the development of a soft independent modeling by class analogy (SIMCA) routine could distinguish between groups of high, medium and low concentration.

The use of NIR has the longest history for wine analysis, but the NIR signal arises from overtones of fundamental bond vibrations in the mid-infrared (MIR) region of the electromagnetic spectrum (see Figures 14.2 and 14.3) (Patz *et al.*, 1999, 2004; Gishen and Holdstock, 2000; Gishen *et al.*, 2005; Cuadrado *et al.*, 2005; Bevin *et al.*, 2006). With more recent developments in instrument design and sample presentation modes (e.g. short pathlength transmission cells, attenuated total reflectance (ATR) cells) (Edelmann *et al.*, 2001, 2003) the use of Fourier transformed mid-infrared (FT-MIR or FTIR) has been implemented for routine analysis of a large number of wine parameters such as alcohol, volatile acidity, pH, tartaric acid, lactic acid, glucose plus fructose, acetic acid, glycerol, anthocyanins, polyphenols, and polysaccharides and reported by several authors (Patz *et al.*, 1999, 2004; Coimbra *et al.*, 2002; Kupina and Shrikhande, 2003; Mendes *et al.*, 2003; Cozzolino *et al.*,

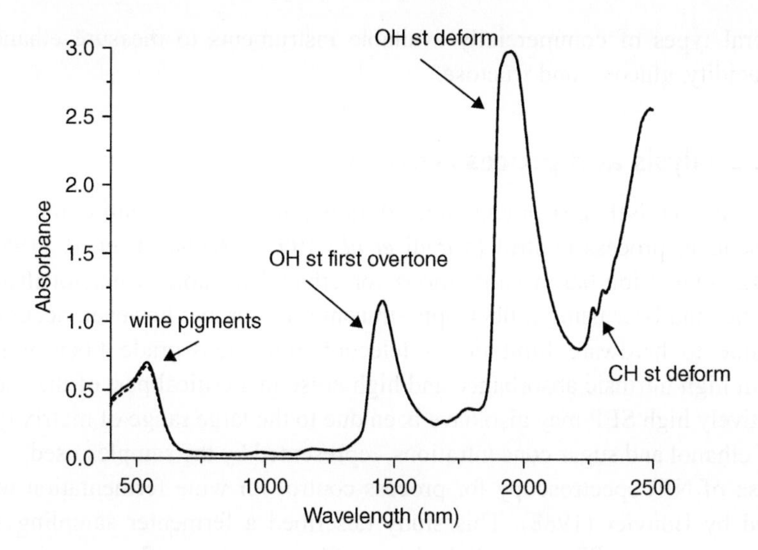

Figure 14.2 Visible and near-infrared spectra of red wine samples from different varieties (line = Shiraz, dotted line = Cabernet Sauvignon).

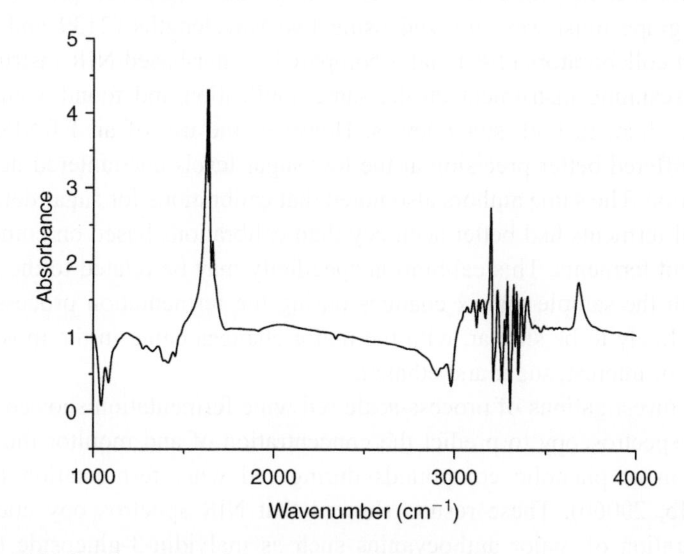

Figure 14.3 Mid-infrared spectrum of a wine sample analyzed in ATR mode.

2004b, 2006a; Moreira and Santos, 2004, Nieuwoudt *et al.*, 2004; Urbano-Cuadrado *et al.*, 2004, Cocciardi *et al.*, 2005; Lletí *et al.*, 2005, Sáiz-Abajo *et al.*, 2006; Versari *et al.*, 2006; Boulet *et al.*, 2007; Lachenmeier, 2007; Soriano *et al.*, 2007) and glycosylated compounds (G-G) in juice (Cynkar *et al.*, 2007). The application of FT-MIR rather than NIR in wine analysis was of special interest due to the presence of sharp and specific absorption bands for the wine constituents (McClure, 2003). Nowadays MIR spectroscopy is commonly used by the wine industry worldwide (Patz *et al.*, 1999, 2004; Kupina and Shrikhande, 2003; Gishen *et al.*, 2005; Cuadrado *et al.*, 2005; Bevin *et al.*, 2006; Versari *et al.*, 2006). Quality control labs use several types of commercially available instruments to measure ethanol, pH, volatile acidity, glucose, and fructose.

Online analysis and process control

Modern available NIR instrumentation offers opportunities for online measurement and subsequent process control (Varadi *et al.*, 1992). Buchanan *et al.* (1988) prepared PLS regression-based calibrations for ethanol in table wine, fortified wine, champagne, and beer using a fiber optic transmission probe; however, accuracy was limited due to hardware limitations—telecommunications-grade fiber optics was used, with high intrinsic absorbance and high noise in a critical part of the spectrum. The relatively high SEP may also have been due to the large range of matrix types, in terms of ethanol and sugar concentrations, represented by the samples used.

The use of NIR spectroscopy for process control for wine fermentation was also examined by Bouvier (1988). This study described a fermenter sampling system, with temperature equilibration, linked to a filter-based transflectance instrument. Wavelength selection was confirmed by examining spectra from a scanning instrument and calibrations suitable for determining sugar (glucose plus fructose) in fermenting grape must were utilized using two wavelengths (2139 and 2230 nm). Davenel and collaborators (1991) later compared a filter-based NIR instrument with an FT-NIR scanning instrument on the same application and found accuracy to be similar at medium to high sugar levels. However, the use of an FT-NIR scanning instrument offered better precision at the low sugar levels encountered near the end of fermentation. The same authors also noted that calibrations for sugar determination in individual ferments had better accuracy than calibrations based on combined data from different ferments. This calibration specificity may be related to the possibility that although the sample matrix changes during the fermentation process, the base spectrum is likely to be similar, with the major changes being in the most abundant constituents of interest, sugar and ethanol.

Recently, investigations of process-scale red wine fermentation showed the potential of NIR spectroscopy to predict the concentration of and monitor the extraction and evolution of phenolic compounds during red wine fermentation (Cozzolino *et al.*, 2004b, 2006b). These results showed that NIR spectroscopy could predict the concentration of major anthocyanins such as malvidin-3-glucoside ($R^2 = 0.91$ and standard error of cross-validation (SECV) = 28.0 mg/L), pigmented polymers ($R^2 = 0.87$ and SECV = 5.9 mg/L), and tannins ($R^2 = 0.83$ and SECV = 131.1 mg/L), in Cabernet Sauvignon and Shiraz wines during fermentation. Urtubia *et al.* (2004) reported the use of FT-MIR to monitor red wine fermentation. Zeaiter *et al.* (2006) tested the robustness of dynamic orthogonal projection as a method to maintain the robustness of calibration during the online monitoring of wine fermentation.

Wine quality grading

The ability accurately to assess wine quality is an important part of the winemaking process (Figure 14.4), particularly when allocating batches of wines to styles

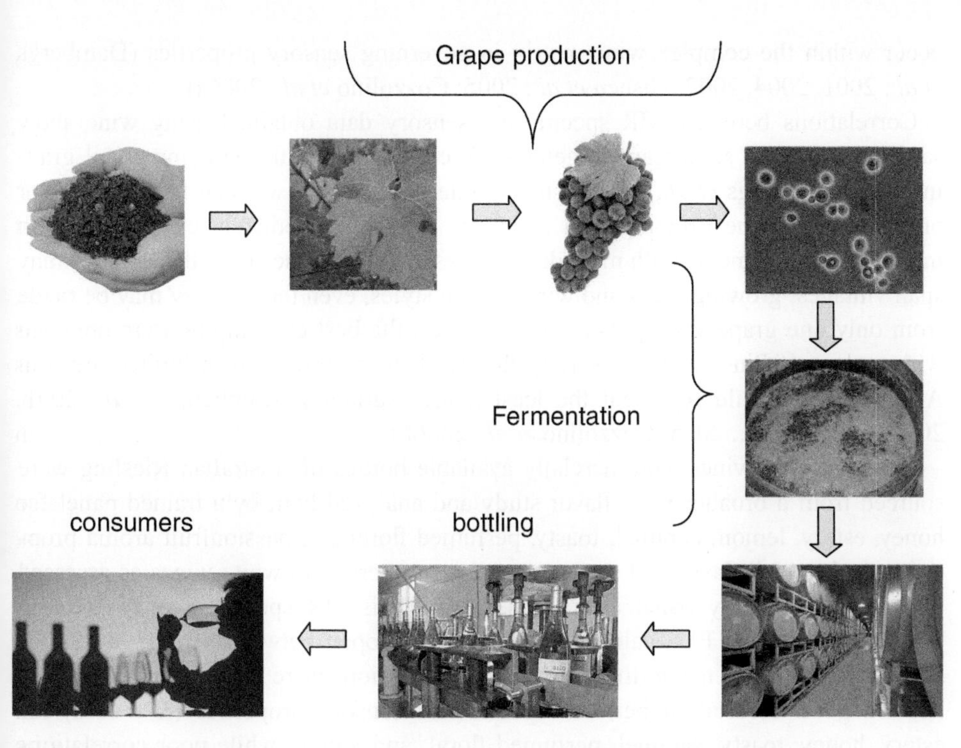

Figure 14.4 All parts of the winemaking process can be monitored by infrared spectroscopy.

determined by consumer requirements. Grape pricing is often determined by the quality category of the resulting wine—so-called "end use" payment (Gishen *et al.*, 2000). Wine quality, in terms of sensory characteristics, is normally a subjective measure, performed by experienced winemakers, wine competition judges, or wine tasting panellists. By nature, such assessments can be biased by individual preferences and may be subject to day-to-day variation. An objective quality grading method would therefore be of great assistance in the wine industry. Flavor compounds are often present in concentrations below the detection limit of NIR spectroscopy but the more abundant organic compounds offer potential for objective quality grading by this technique. It has also been demonstrated that wine quality rankings (as the score or allocation assigned to wines by sensory panels) for red and fortified wines could be discriminated by visible (Vis) and NIR spectroscopy (Gishen *et al.*, 2000; Dambergs *et al.*, 2001). Furthermore, it has been demonstrated that Vis/NIR spectroscopy can predict wine quality as judged by both commercial wine quality rankings and wine show scores (Gishen *et al.*, 2000; Dambergs *et al.*, 2001, 2007).

This application could provide a rapid assessment or pre-screening tool to allow preliminary blend allocation of large numbers of batches of wines prior to sensory assessment. Winemakers may be able to develop "profiles" for their blends as in-house NIR calibrations. NIR calibrations based on sensory scores will tend to be difficult to obtain due to variation between individual wine tasters and may not pick up compounds that are present at low concentrations, yet have strong sensory properties. Nevertheless, interpretation of spectral data may provide valuable insight into the more abundant parameters affecting wine quality and highlight the interactions that occur within the complex wine matrix in governing sensory properties (Dambergs *et al.*, 2001, 2004, 2007; Gishen *et al.*, 2005; Cozzolino *et al.*, 2006a).

Correlations between NIR spectra and sensory data obtained using wine show samples were less significant in general, in comparison with the commercial grading data (Dambergs *et al.*, 2001). The commercial samples were all from one major producer, from one growing area and were graded immediately ex-vintage, with minimal oak treatment. With most dry red classes in the wine show, the samples may span vintages, growing areas and winemaking styles, even though they may be made from only one grape variety. For dry red wines, the best calibrations were obtained with a class of Pinot Noir—a variety that tends to be produced in limited areas in Australia and would represent the least matrix variation (Dambergs *et al.*, 2001, 2004; Gishen *et al.*, 2005; Cozzolino *et al.*, 2006b).

Similar to red wines, commercially available bottles of Australian Riesling were sourced from a broader wine flavor study and analyzed both by a trained panel for honey, estery, lemon, caramel, toasty, perfumed floral and passionfruit aroma properties, and overall flavor and sweetness palate properties in white wines as assessed by a trained sensory panel and scanned using Vis/NIR spectroscopy (Cozzolino *et al.*, 2003, 2005). PLS calibration models developed between sensory attributes and Vis/NIR spectra using different wavelength regions were developed. The results showed good correlation between spectra and sensory properties ($R > 0.70$) for estery, honey, toasty, caramel, perfumed floral, and lemon, while poor correlations ($R <$ about 0.55) were found in most of the cases for passionfruit, sweetness and overall flavor, respectively.

Distillation

Grape spirit is produced by distillation of wine or wine/grape-derived process waste, and is used in the production of fortified wines. Methanol concentrations in grape marc, one of the major sources distillation raw materials, can be high due to the action of mold and bacteria in the raw product (Dambergs *et al.*, 2002). The methanol concentration in the final product must be minimized to comply with food regulations and operating continuous stills can be difficult without rapid methanol analysis to allow fine-tuning of the stills in a timely manner. In comparison to wine, the distillation process streams represent relatively simple matrices consisting of predominantly ethanol, water, and minor quantities of other volatile organic compounds. Two key analytes that are routinely monitored during the distillation process are ethanol and methanol, which have characteristic NIR spectra based on differences in relative concentrations of CH_3 groups, wavelength shifts for OH groups and a CH_2 group unique to ethanol (Figure 14.1). NIR calibrations have been developed for both compounds use PLS and MLR methods with transmission spectra of wine fortifying spirit using gas chromatography as the reference method. The PLS calibrations approached the accuracy of the reference methods, with an R^2 of 0.998 and a SECV of 0.06 g/L for methanol and an R^2 of 0.96 and SECV of 0.08% v/v for ethanol. The calibrations were very robust as indicated by high values for the ratio of the standard deviation of the reference data to the SEP of the calibration.

Despagne *et al.* (2000) compared PLS, locally weighted regression (LWR), and artificial neural network (ANN) in a model distillation system, where water content was monitored. Locally weighted regression outperformed PLS in that it could overcome concentration-related non-linearity but was prone to overfitting. ANN algorithms gave the best overall performance and with the appropriate training set, spanning a range of sample matrices, may represent the best calibration method for the distillation application. The applicability of NIR-based distillation process monitoring has also been demonstrated in a solvent recovery plant application separating, ethanol, methanol, ethyl acetate, acetone, toluene, dichloromethane, and water. The system used an instrument with acoustic-optic tunable filter technology (AOTF) to provide fast scanning over an 1100–2300 nm range, via an eight-channel multiplexer with fiber optics up to 75 m in length, allowing collection of transmission spectra at various key points of the distillation plant.

Another aspect of the distilled beverage industry is the blending and bottling of the final product. Rapid analysis is required for monitoring such operations and NIR-based analysis may offer some potential. However, considerable challenges might be expected with the large range of matrices likely to be encountered in most packaging plants, particularly in relation to ethanol and sugar concentration. Vandenberg *et al.* (1997) approached this problem with the analysis of spirit-based beverages that ranged in ethanol content from 20 to 40% v/v and in sugar content from 0.6 to 375 g/L. By using continuous spectral data rather than discrete wavelengths from a filter-based machine and using careful selection of wavelength regions simultaneously with the use of second derivative spectra and smoothing, they were able to prepare calibrations for ethanol, sugar, and density of sufficient accuracy for commercial use.

Yeast identification

Yeast identification is an important issue in process-scale (industrial) fermentations, where contamination with wild strains may introduce undesirable traits. As well as looking at yeast in various growth stages and after heat shock damage, Halasz *et al.* (1997) examined the possibility of discrimination between yeast strains with NIRS spectra of yeast slurries, prepared from yeast grown in synthetic media. Differences in protein profiles could be seen with electrophoresis methods and this was correlated with second derivative reflectance spectra at longer NIR wavelengths (2000–2500 nm). Sample sets were limited in size, but PCA plots clearly discriminated yeast growth phase yeast strain and could detect 10% cross-contamination of one strain with another.

In the last few years, both microbial and plant metabolite analysis has shifted from specific assays toward methods offering both high accuracy and sensitivity in highly complex mixtures of compounds. Large-scale metabolomic analysis is based on the use of gas chromatography mass spectroscopy (GC/MS) and liquid chromatography-MS (LC/MS) (Sweetlove *et al.*, 2004). Both NIR and FT-NIR spectroscopy have been examined to assess their suitability as a tool for yeast identification (Halasz *et al.*, 1997), yeast protein measurement (Majara *et al.*, 1998), and detection and identification of bacterial strains (Kansiz *et al.*, 1999; Rodriguez-Saona *et al.*,

2001; Irudayaraj *et al.*, 2002). The potential of NIR spectroscopy and multivariate analysis as a rapid screening technique to discriminate different yeast strains with particular metabolic profiles has been reported (Cozzolino *et al.*, 2007). The results showed that deletion strains were correctly classified as different from the wild-type laboratory strains and demonstrated the potential of combining NIR spectroscopy and multivariate techniques to enable the rapid selection of yeast strains with similar metabolic profiles. The use of FT-MIR was also explored as a rapid method of screening of the fermentation profiles of wine yeast (Nieuwoudt *et al.*, 2006).

Beer

Hops

Axcell *et al.* (1981a, 1981b) reported the use of NIR spectroscopy for the measurement of hop α-acids. Kiln-dried samples were ground and scanned in reflectance mode using an instrument with 19 filters covering the 1445–2348 nm wavelength range. Calibrations were prepared by stepwise multiple linear regression and the final calibrations used six filters over the 1680–2348 nm range. Samples were ground to varying degrees, and it was found that better reproducibility was obtained with more finely ground, sieved samples. The same study also demonstrated the feasibility of using NIR for measurement of residual moisture in the kiln-dried samples. Calibrations were also developed for hop β-acids and hop storage index.

Chandley (1993) achieved better accuracy with reflectance spectra of ground hops by using standard normal variate (SNV) and de-trend transformations to minimize the effects of particle size. The SEP values for NIR measurement of hop acids, moisture, and oil were lower than inter-laboratory errors for the reference methods and NIR analytical data correlated with commercial hop grading based on aroma and appearance. Garden and Freeman (1998) segregated samples from a large data set into high, medium, and low range for hop acids and storage index. For α-acids in particular, the reference data indicated three clusters of analyte concentration and it was found that the best performing calibrations were obtained with the low concentration samples. Segregation by analyte concentration did not improve performance of calibrations for β-acids and storage index.

Malt and wort analysis

Henry (1999) examined the ability of NIR spectroscopy to analyze β-glucans in barley. Standard preparations of glucan, starch, inulin, and various monosaccharides were used to identify spectral differences in the 1600–1800 nm region, where a three-wavelength calibration was used to predict β-glucan in ground barley, using an enzymatic procedure as the reference method. Halsey (1985, 1986, 1987a, 1987b, 1987c) scanned whole, malted grain with a sample transport module and prepared calibrations for moisture, total soluble nitrogen (TSN), FAN, friability, hot wort extract (HWE), total carbohydrates, fermentable sugars, and fermentability. The prediction error was high for the physical parameter, friability. The prediction error for FAN was higher than for TSN and it was suggested that the error with FAN calibrations was due to an unreliable reference method. Not withstanding prediction error, a comparison of repeatability of the NIR method and reference methods revealed equal if not better performance by the NIR method. Allison (1989) and co-workers (Allison *et al.*, 1978) used PCR to prepare calibrations for hot water extracts of unmalted barley flour, scanned over a range of 1100–2500 nm and used spectral information derived from the principal components to demonstrate a positive effect of starch on extract value and negative effects of β-glucan and protein content. PCA of spectra over the same wavelength range was also used to identify chemical changes occurring during the time course of the malting process.

In an attempt to improve β-glucan analysis by NIR spectroscopy, Czuchajowska *et al.* (1992) examined spectra of hull-less and covered barleys, meals of regular, high amylose and high amylo-pectin barleys, isolated starches and β-glucans. The 2000–2500 nm region showed the best discrimination and was used to prepare a calibration for β-glucan. This calibration was however affected by kernel size, hardness and protein content.

Garden and Freeman (1998) made an attempt at NIR analysis of diastatic power of malt (the potential for enzymatic conversion of starch to sugars). Whole malt samples were scanned and average spectra from three subsamples were used for calibration development. Calibrations for moisture and protein had low prediction error but the diastatic power calibration had high prediction error. The inability of NIR spectroscopy to predict diastatic power could be related to the fact that a diversity of factors may influence this parameter, some possibly related to compounds occurring at low concentrations. Halsey (1987c) evaluated the use of NIR spectroscopy for the analysis of HWE, total carbohydrates, fermentable sugars, fermentability, FAN, and TSN in malted and mashed barley. Samples were scanned by transmittance between 1100 and 2500 nm. To aid the choice of wavelengths for an MLR calibration, solutions of sugars, dextrins, amino acids, and polypeptides were also scanned. The precision of the reference and NIR methods were compared and other than for TSN the precision of the NIR method was adequate—in fact, in the case of sugars and carbohydrates, the precision of the NIR method exceeded the reference method. The SEP for HWE was relatively high and was influenced by the geographical origin of the samples.

The HWE reference method is relatively non-specific: better NIR results were obtained if the value was calculated from the NIR values for total carbohydrate and sugars, the compounds of interest. Sjoholm *et al.* (1996) investigated the use of NIR for analysis of HWE and fermentability. Calibrations for extract approached the precision of the reference method but although adequate, fermentability calibrations had higher error, particularly with varying sample matrix.

Beer composition

The potential for NIR analysis of a primary analyte in beer, alcohol (ethanol), was demonstrated by Coventry and Hunston (1984), using transflectance in the 1100–2500 nm range. Halsey (1987c) used an MLR model with two to three wavelengths to calibrate for alcohol and original gravity in finished beers. Better results were obtained with transmission compared with transflectance spectra. The original gravity calibration had high error, particularly with beers primed with low molecular weight sugars. It was pointed out that original gravity is an empirically derived measure and a good direct calibration may not be expected. Gallignani *et al.* (1993, 1994) examined the effect of maltose on ethanol calibration models and demonstrated a bias. A simple correction could be made by using a maltose solution or ethanol-free beer as the reference solution, resulting in good agreement of NIR-predicted alcohol with reference data. Chandley (1993) dried beer samples onto glass fiber filters and scanned in reflectance mode the dry extract system for infrared reflectance (DESIR) technique. Good calibrations were obtained for free amino nitrogen and total soluble nitrogen, but poor correlations of spectral data with bitterness values were observed. Similarly, Maudoux *et al.* (1997) scanned liquid beer samples in transmission mode and by reflectance with samples dried onto filters: better results for total nitrogen and polyphenols were obtained with the dried samples.

Although improvements in calibrations may be obtained by drying the samples, this must be balanced against the extra sample preparation complexity introduced and the fact that this technique limits the opportunity for simultaneous analysis of volatile compounds. Recently, the use of NIR, FT-MIR spectroscopy, and ATR cells were used to quantify ethanol and other compositional parameters in beer (Li *et al.*, 1999; Schropp *et al.* 2002; Kington and Jones, 2001; Engelhard *et al.*, 2004; Inon *et al.*, 2005, 2006; Llario *et al.*, 2006).

Process control

Inline and online monitoring and process control is seen as one of the potential strengths of NIR spectroscopy: this was demonstrated in an early application relating to blending of beer (Coventry and Hunston, 1984). A sampling system drew beer from a large diameter product line, temperature-equilibrated the product from the original −1°C to 20°C and applied pressure control to prevent out-gassing of carbon dioxide. The sample was passed through a pressure-resistant sample cell and scanned by transflectance to measure original gravity. A feedback system was set up to control the blending, based on the analytical data obtained from the online measurements. With this system, blending was achieved to well within specified tolerances. Petersen *et al.* (1992) prepared calibration models for online measurement of alcohol and original extract in beer (using a sampling system) and converted the calibration models for use with an inline probe with the aid of reference samples. Beer is relatively clear, so a transmission probe was used and scans collected over a 700–2100 nm wavelength range. Individual calibration models were prepared for specific beers and it was pointed out that although the equipment was simple to use and reliable, the calibration maintenance required expertise. An issue with a process-control instrument is the possibility of calibration drift when components of the probe or spectrometer are changed.

McDermott (1992) addressed this problem with an algorithm to match water reference spectra before and after changes. The system used a fiber optic probe with a sapphire rod and reflective tip separated by a 2.5 mm transmission gap, attached to a monochromator instrument scanning in the 1200–2400 nm range. Calibrations for ethanol, calories, and original gravity were prepared from first derivative spectra, using PLS regression. Fermentation monitoring is another potential application of NIR spectroscopy in the brewing process. Cavinato *et al.* (1990) described a system using a fiber optic reflectance probe placed in the side of fermentation vessel, feeding the back-scattered signal to a diode array spectrometer. The absorption band at 905 nm (third overtone CH stretch band from the methyl group) was used to prepare a calibration for ethanol, initially using ethanol and water, then ethanol, water and yeast mixtures. Ferments had the lowest signal-to-noise ratio, resulting in a calibration with low accuracy. Calibration problems may be encountered with ferments as they represent a shifting matrix, with progressive and large changes in turbidity, yeast content, sugar, and ethanol content.

Coventry (1994) increased the signal-to-noise ratio of the NIR spectra by using a double beam instrument with a reference cell. They demonstrated high precision ethanol and original gravity measurement in ferments with both an inline probe and a sampling system, using transmission spectra in the range of 750–3000 nm. This study demonstrated that short wavelengths (near the visible region) were prone to error produced by sample color and the long wavelengths were affected by sample temperature. The culmination of the early development work was a commercially available inline beer monitoring system. Calibration models for ethanol were insensitive to beer color, turbidity, and sugar concentration and were linear over the critical 0–8% ethanol range (Halsey, 1985; Gallignani *et al.*, 1993). Some more recent innovations have been the investigation of the use of ATR probes to produce spectra equivalent to those from short pathlength transmission probes, without the fouling issues often encountered in the crude beer mash (Inon *et al.*, 2005, 2006; Llario *et al.*, 2006).

Nørgaard *et al.* (2000) recognized the problems of covariance in fermentation systems, where a large number of dramatic matrix changes occur progressively and simultaneously. They discussed a model described as interval PLS (iPLS), for the systematic selection of optimal wavelengths to make a more robust calibration model with less factors and a lower SEP. Li *et al.* (1999) used an ANN algorithm to overcome non-linearity problems encountered with the wide range of sugar concentrations observed during fermentation and demonstrated significantly reduced SEP for sugar and ethanol measurement in comparison with a PLS model.

Yeast analysis and identification

A very early application of NIR spectroscopy in the beverage industry was in the monitoring of yeast concentration during beer brewing using a single wavelength (1000 nm) calibration for yeast count (Dambergs *et al.*, 2004). The calibration was non-linear but although it was derived from only one beer type was able to predict yeast counts in three other beer types in the normal working concentration range. This concept has been developed into a commercial inline monitoring device and has culminated in the development of an instrument that is unaffected by the presence of carbon dioxide bubbles, will discriminate yeast from smaller non-yeast particles, is unaffected by the turbidity of fermenting wort, and operates successfully in the high cell concentration range of the primary yeast inoculum (Dambergs *et al.*, 2004).

Mochaba *et al.* (1994) tested the ability of NIR spectroscopy to determine levels of glycogen, a major yeast storage carbohydrate. Yeast samples were microwave-dried and ground, in preparation for reflectance scanning. PLS calibrations were prepared from first derivative spectra: the SEP for glycogen was high, relative to the average value of samples measured, but the authors pointed out that the reference method also had relatively high error, and that the NIR method offered the advantage of speed. Better results were achieved by Moonsamy *et al.* (1995), who prepared NIR calibrations for glycogen and trehalose, the other major yeast storage carbohydrate. Spectra were acquired in reflectance mode over a more extended wavelength range and had more spectral data points than the previous study (Mochaba *et al.*, 1994). Calibrations for trehalose developed from scans of the dried samples had relatively high error, but it was found that scanning of wet yeast slurries gave better accuracy for trehalose and glycogen, implying that the drying method introduced errors. Scanning of slurries was also considered preferable since it would be more convenient for potential online monitoring applications.

Protein content of yeast is also an important physiological parameter and is used to determine payment in the sale of spent brewery yeast by-product. Majara *et al.* (1998) reported high accuracy of PLS calibrations for protein in dried yeast samples scanned in reflectance mode. The authors noted that to obtain a high degree of accuracy with the NIR method, care must be taken with the reference method to ensure complete and reproducible protein extraction. Yeast identification is an important issue in process-scale (industrial) fermentations, where contamination with wild strains may introduce undesirable traits.

Conclusions

Spectroscopy combined with multivariate methods has the potential to be a powerful tool for the assessment of beer and wine composition. Recent advances in chemometrics software and computing power have greatly enhanced the development of rapid analytical methods based on spectroscopic data and their subsequent application in a wide range of agricultural industries. Although the instrumentation may require a large capital outlay and can be reasonably complex to calibrate and maintain, but with more research, it is possible that more cost effective, simple instruments could be developed for general use by the beer and wine industry. Near-infrared (NIR) and mid-infrared (MIR) spectroscopy analytical techniques are beginning to gain acceptance in the beer and wine industry. As the technology of spectroscopic instrumentation and chemometrics advances further, the resulting spin-offs may further assist the industry in its quest to define and objectively measure product quality and to efficiently monitor the industrial process.

References

Allison MJ (1989) Areas of absorption relating to malt extract value in modified near infrared spectra of barley flour. *Journal of the Institute of Brewing*, **95**, 283–286.

Allison MJ, Cowe IA, McHale R (1978) The use of infra red reflectance for the rapid estimation of the soluble b-glucan content of barley. *Journal of the Institute of Brewing*, **84**, 153–155.

Arana C, Jaren C, Arazuri S (2005) Maturity, variety and origin determination in white grapes (*Vitis vinifera* L.) using near infrared reflectance technology. *Journal of Near Infrared Spectroscopy*, **13**, 349.

Axcell BC, Tulej R, Murray J (1981a) An ultra-fast system for hop analysis. I. The determination of alpha acids and moisture by near infrared reflectance spectroscopy. *Brewers Digest*, **56**, 18–19, 41.

Axcell BC, Tulej R, Murray J (1981b) An ultra-fast system fort hop analysis. II. The determination of beta acids and prediction of hop storage index by near infrared reflectance spectroscopy. *Brewers Digest*, **56**, 32–33.

Bevin CJ, Fergusson AJ, Perry WB, Janik LJ, Cozzolino D (2006) Development of a rapid "fingerprinting" system for wine authenticity by mid-infrared spectroscopy. *Journal of Agricultural and Food Chemistry*, **54**, 9713–9718.

Boulet JC, Williams P, Doco T (2007) A Fourier transform infrared spectroscopy study of wine polysaccharides. *Carbohydrate Polymers*, **69**, 79–85.

Bouvier JC (1988) Analys de automatique des sucres des mouts de raisin en fermentation par spectrometrie dans le proche infrarouge. *Sciences des Aliments*, **8**, 227–243.

Buchanan BR, Honigs DE, Lee CJ, Roth W (1988) Detection of ethanol in wines using optical-fiber measurements and near infrared analysis. *Applied Spectroscopy*, **42**, 1106–1111.

Cavinato AG, Mayes DM, Ge ZH, Callis JB (1990) Noninvasive method for monitoring ethanol in fermentation processes using fiber-optic near infrared spectroscopy. *Analytical Chemistry*, **62**, 1977–1982.

Chandley P (1993) The application of the DESIR technique to the analysis of beer. *Journal of Near Infrared Spectroscopy*, **1**, 133–139.

Chauchard F, Codgill R, Roussel S, Roger JM, Bellon-Meurel V (2004) Application of LS-SVM to non-linear phenomena in NIR spectroscopy: development of a robust and portable sensor for acidity prediction of grapes. *Chemometrics and Intelligent Laboratory System*, **71**, 141.

Cocciardi RA, Ismail AA, Sedman J (2005) Investigation of the potential utility of single-bounce attenuated total reflectance Fourier transform infrared spectroscopy in the analysis of distilled liquors and wines. *Journal of Agricultural and Food Chemistry, Chemometrics and Intelligent Laboratory System*, **53**, 2803–2809.

Coimbra MA, Goncalves F, Barros AS, Delgadillo I (2002) Fourier transform infrared spectroscopy and chemometric analysis of white wine polysaccharide extracts. *Journal of Agricultural and Food Chemistry*, **50**, 3405–3411.

Coventry AG (1994) In line measurement of alcohol and original gravity using fiber optic sensors. *Cerevisia and Biotechnology*, **19**, 48–52.

Coventry AG, Hunston MJ (1984) Applications of near-infrared spectroscopy to the analysis of beer samples. *Cereal Foods World*, **29**, 715, 717–718.

Cozzolino D, Smyth HE, Gishen M (2003) Feasibility study on the use of visible and near-infrared spectroscopy together with chemometrics to discriminate between commercial white wines of different varietal origins. *Journal of Agricultural and Food Chemistry*, **51**, 7703–7708.

Cozzolino D, Esler M, Dambergs RG, Cynkar WU, Boehm D, Francis IL, Gishen M (2004a). Prediction of colour and pH using a diode array spectrophotometer (400–1100 nm). *Journal of Near Infrared Spectroscopy*, **12**, 105–111.

Cozzolino D, Kwiatkowski M, Parker M, Gishen M, Dambergs RG, Cynkar W, Herderich M (2004b) Prediction of phenolic compounds in red wine by near infrared spectroscopy. *Analytica Chimica Acta*, **513**, 73–80.

Cozzolino D, Smyth HE, Lattey KA, Cynkar W, Janik L, Dambergs RG, Francis IL, Gishen M (2005) Relationship between sensory analysis and near infrared spectroscopy in Australian Riesling and Chardonnay wines. *Analytica Chimica Acta*, **539**, 341–348.

Cozzolino D, Dambergs RG, Janik L, Cynkar W, Gishen M (2006a) Analysis of grape and wine by near infrared spectroscopy—a review. *Journal of Near Infrared Spectroscopy*, **14**, 279–289.

Cozzolino D, Parker M, Dambergs RG, Herderich M, Gishen M (2006b) Chemometrics and visible-near infrared spectroscopic monitoring of red wine fermentation in a pilot scale. *Biotechnology and Bioengineering*, **95**, 1101–1107.

Cozzolino D, Flood L, Bellon J, Gishen M, De Barros Lopes M (2007) Combining near infrared spectroscopy and multivariate analysis: a tool to differentiate different strains of *Saccharomyces cerevisiae:* a metabolomic study. *Yeast*, **23**, 1089–1096.

Cuadrado MU, Luque de Castro MD, Perez Juan PM, Gómez-Nieto MA (2005) Comparison and joint use of near infrared spectroscopy and Fourier transform mid infrared spectroscopy for the determination of wine parameters. *Talanta*, **66**, 218–224.

Cynkar WU, Cozzolino D, Dambergs RG, Janik L, Gishen M (2007) Effect of variety, vintage and winery on the prediction of glycosylated compounds (G-G) in white grape juice by visible and near infrared spectroscopy. *Australian Journal of Grape and Wine Research*, **13**, 101–105.

Czuchajowska Z, Szczodrak J, Pomeranz Y (1992) Characterization and estimation of barley polysaccharides by near-infrared spectroscopy. I. Barleys, starches and β-d-glucans. *Cereal Chemistry*, **69**, 413–418.

Dambergs RG, Kambouris A, Schumacher N, Francis IL, Esler MB, Gishen M (2001) Wine quality grading by near infrared spectroscopy. *Proceedings of the 10th International Near Infrared Spectroscopy Conference.* Chichester: NIR Publications, pp. 187–189.

Dambergs RG, Kambouris A, Francis IL, Gishen M (2002) Rapid analysis of methanol in grape derived distillation products using near infrared transmission spectroscopy. *Journal of Agricultural and Food Chemistry*, **50**, 3079–3084.

Dambergs RG, Cozzolino D, Cynkar WU, Kambourious A, Francis IL, Gishen M, Høj P (2003) The use of near infrared reflectance for grape quality measurement. *The Australian and New Zealand Grapegrowers and Winemakers Journal*, **476**, 69–75.

Dambergs RG, Esler MB, Gishen M (2004) Application in analysis of beverages and brewing products. In: *Near Infrared Spectroscopy in Agriculture* (Roberts CA, Workman J, Reeves JB III, eds). Agronomy Monograph 44. Madison, WI: ASA, CSSA, and SSSA, pp. 465–486.

Dambergs RG, Cozzolino D, Cynkar WU, Janik L, Gishen M (2006) The determination of red grape quality parameters using the LOCAL algorithm. *Journal of Near Infrared Spectroscopy*, **14**, 71–79.

Dambergs RG, Cozzolino D, Francis, IL, Cynkar WU, Janik L, Gishen M (2007) Applications of visible and near infrared spectroscopy in the wine industry. In: *Proceedings of the 12th International NIR Conference* (Burling-Claridge GR, Holroyd SE, Sumner RMW, eds). Chichester: IM Publications, pp. 378–381.

Davenel A, Grenier P, Foch B, Bouvier JC, Verlaque P, Pourcin J (1991) Filter, Fourier transform infrared, and areometry, for following alcoholic fermentation in wines. *Journal of Food Science*, **56**, 1635–1638.

Despagne F, Massart DL, Chabot P (2000) Development of a robust calibration model for nonlinear in-line process data. *Analytical Chemistry*, **72**, 1657–1665.

Dumoulin ED, Azais BP, Guerain JT (1987) Determination of sugar and ethanol content in aqueous products of molasses distilleries by near infrared spectrophotometry. *Journal of Food Science*, **52**, 626–630.

Edelmann AD, Schuster KH, Lendl B (2001) Rapid method for the discrimination of red wine cultivars based on mid-infrared spectroscopy of phenolic wine extracts. *Journal of Agricultural and Food Chemistry*, **49**, 1139–1145.

Edelmann A, Diewok J, Baena JR, Lendl B (2003) High-performance liquid chromatography with diamond ATR-FTIR detection for the determination of carbohydrates, alcohols and organic acids in red wine. *Analytical and Bioanalytical Chemistry*, **376**, 92–97.

Engelhard S, Lohmannsroben HG, Schael F (2004) Quantifying ethanol content of beer using interpretive near-infrared spectroscopy. *Applied Spectroscopy*, **58**, 1205–1209.

Gallignani M, Garrigues S, Delaguardia M (1993) Direct determination of ethanol in all types of alcoholic beverages by near-infrared derivative spectrometry. *The Analyst*, **118**, 1167–1173.

Gallignani M, Garrigues S, de la Guardia M (1994) Derivative Fourier transform infrared spectrometric determination of ethanol in alcoholic beverages. *Analytica Chimica Acta*, **287**, 275–283.

Garcia-Jares CM, Medina B (1997) Application of multivariate calibration to the simultaneous routine determination of ethanol, glycerol, fructose glucose and total residual sugars in botryized-grape sweet wines by means of near-infrared reflectance spectroscopy. *Fresenius Journal of Analytical Chemistry*, **357**, 86–91.

Garden SW, Freeman PL (1998) Applications of near-infrared spectroscopy in malting: calibrations for analysis of green malt. *Journal of the American Society of Brewing Chemists*, **56**, 159–163.

Gishen M, Holdstock MG (2000) Preliminary evaluation of the performance of the Foss WineScan FT120 instrument for the simultaneous determination of several wine analyses. *Australian Grapegrower and Winemaker Journal*, **438**, 75–78, 81.

Gishen M, Dambergs RG, Kambouris A, Kwiatkowski M, Cynkar WU, Høj PB, Francis IL (2000) Application of near infrared spectroscopy for quality assessment of grapes, wine and spirits. In: *Proceedings of 9th International Conference on Near Infrared Spectroscopy. Verona, Italy.* Chichester: NIR Publications, pp. 917–920.

Gishen M, Dambergs RG, Cozzolino D (2005) Grape and wine analysis—enhancing the power of spectroscopy with chemometrics. A review of some applications in the Australian wine industry. *Australian Journal of Grape and Wine Research*, **11**, 296–305.

Halasz A, Hassan A, Toth A, Varadi M. (1997) NIR techniques in yeast identification. *Z. Lebensm. Unters. Forrsch A*, **204**, 72–74.

Halsey SA (1985) The use of transmission and transflectance near infrared spectroscopy for the analysis of beer. *Journal of the Institute of Brewing*, **91**, 306–312.

Halsey SA (1986) The application of transmission near infrared spectroscopy to the analysis of worts. *Journal of the Institute of Brewing*, **92**, 387–393.

Halsey SA (1987a) Near infrared reflectance analysis of whole hop cones. *Journal of the Institute of Brewing*, **93**, 399–404.

Halsey SA (1987b) Analysis of whole barley kernels using near infrared reflectance spectroscopy. *Journal of the Institute of Brewing*, **93**, 461–464.

Halsey SA (1987c) Rapid analysis of whole malt using near infrared reflectance spectroscopy. *Journal of the Institute of Brewing*, **93**, 407–412.

Henry CM (1999) Near-IR gets the job done. *Analytical Chemistry*, **71**, 625A–628A.

Herrera J, Guesalaga A, Agosin E (2003) Shortwave-near infrared spectroscopy for non-destructive determination of maturity of wine grapes. *Measurement Science and Technology*, **14**, 689–697.

Inon FA, Llario R, Garrigues S, de la Guardia M (2005) Development of a PLS based method for determination of the quality of beers by use of NIR: spectral ranges and sample-introduction considerations. *Analytical and Bioanalytical Chemistry*, **382**, 1549–1561.

Inon FA, Garrigues S, de la Guardia M (2006) Combination of mid- and near-infrared spectroscopy for the determination of the quality properties of beers. *Analytica Chimica Acta*, **571**, 167–174.

Irudayaraj J, Yang H, Sakhamuri S (2002) Differentiation and detection of microorganism using Fourier transform infrared photoacoustic spectroscopy. *Journal of Molecular Structure*, **606**, 181–188.

Jaren C, Ortuño JC, Arazuri S, Arana JI, Salvadores MC (2001) Sugar determination in grapes using NIR technology. *International Journal of Infrared and Millimeter Waves*, **22**, 1521.

Kaffka KJ, Norris KH (1976) Rapid instrumental analysis of composition of wine. *Acta Alimentaria*, **5**, 267–279.

Kansiz M, Heraud P, Wood B, Burden F, Beardall J, McNaughton D (1999) Fourier transform infrared micro spectroscopy and chemometrics as a tool for the discrimination of cyanobacterial strains. *Phytochemistry*, **52**, 407–417.

Kemeny G, Pokorny T, Forizs K, Leko L (1983) Use of the near infrared measurement technique in the wine industry (Hungarian). *Borgazdasag*, **31**, 127–132.

Kington LR, Jones TM (2001) Application for NIR analysis of beverages. In: *Handbook of Near-Infrared Analysis* (Burns TA, Ciurczak EW, eds). New York: Marcel Dekker, pp. 535–542.

Kupina SA, Shrikhande AJ (2003) Evaluation of a Fourier transform infrared instrument for rapid quality-control wine analyses. *American Journal of Enology and Viticulture*, **54**, 131–134.

Lachenmeier DW (2007) Rapid quality control of spirit drinks and beer using multivariate data analysis of Fourier transform infrared spectra. *Food Chemistry*, **101**, 825–832.

Li Y, Brown CW, Lo S-C (1999) Near infrared spectroscopic determination of alcohols-solving non-linearity with linear and non-linear methods. *Journal of Near Infrared Spectroscopy*, **7**, 55–62.

Llario R, Inon FA, Garrigues S, de la Guardia M (2006) Determination of quality parameters of beers by the use of attenuated total reflectance-Fourier transform infrared spectroscopy. *Talanta*, **69**, 469–480.

Lletí R, Meléndez E, Ortiz MC, Sarabia LA, Sánchez MS (2005) Outliers in partial least squares regression: Application to calibration of wine grade with mean infrared data. *Analytica Chimica Acta*, **544**, 60–70.

Majara M, Mochaba FM, O'Connor-Cox ESC, Axcell BC, Alexander A (1998) Yeast protein measurement using near-infrared reflectance spectroscopy. *Journal of the Institute of Brewing*, **104**, 143–146.

Manley M, van Zyl A, Wolf EEH (2001) The evaluation of the applicability of Fourier transform near-infrared (FT-NIR) spectroscopy in the measurement of analytical parameters in must and wine. *South African Journal of Oenology and Viticulture*, **22**, 93–100.

Maudoux M, Yan SH, Collin S (1997) Quantitative analysis of alcohol, real extract, original gravity, nitrogen and polyphenols in beers using NIR spectroscopy. In: *Proceedings of the International Conference on Near Infrared Spectroscopy, Essen, Germany*. Chichester: NIR Publications.

McClure WF (2003) 204 years of near infrared technology: 1800–2003. *Journal of Near Infrared Spectroscopy*, **11**, 487–518.

McDermott LP (1992) On-line blending control for beer production using near infrared spectroscopy. *Technical Quarterly, Master Brewers Association of the Americas*, **29**, 96–100.

Medrano R, Yan SH, Madoux M, Baeten V, Meurens M (1995) Wine analysis by NIR. *Leaping Ahead with Near Infrared Spectroscopy*. Victoria: Royal Australian Chemical Institute, pp. 303–306.

Mendes LS, Oliveira FCC, Suarez PAZ, Rubim JC (2003) Determination of ethanol in fuel ethanol and beverages by Fourier transform (FT)-near infrared and FT-Raman spectrometries. *Analytica Chimica Acta*, **493**, 219–231.

Meurens M, Van den Eynde P, Vanbelle M (1987) Fine analysis of liquids by NIR spectroscopy of dry extract on solid support (DESIR). In: *Near Infrared Diffuse Reflectance/Transmittance Spectroscopy* (Hollo J, Kaffka KJ, Gonczy JL, eds). Budapest: Akademiai Kiado, pp. 297–302.

Mochaba F, Torline P, Axcell BC (1994) A novel and rapid approach for the determination of glycogen in pitching yeasts. *Journal of the American Society of Brewing Chemists*, **52**, 145–147.

Moonsamy N, Mochaba F, O'Connor-Cox ESC, Axcell BC (1995) Rapid yeast trehalose measurement using near infrared reflectance spectroscopy. *Journal of the Institute of Brewing*, **101**, 203–206.

Moreira JL, Santos L (2004) Spectroscopic interferences in Fourier transform infrared wine analysis. *Analytica Chimica Acta*, **513**, 263–268.

Nieuwoudt HH, Prior BA, Pretorius IS, Manley M, Bauer FF (2004) Principal component analysis applied to Fourier transform infrared spectroscopy for the design of calibration sets for glycerol prediction models in wine and for the detection and classification of outlier samples. *Journal of Agricultural and Food Chemistry*, **52**, 3726–3735.

Nieuwoudt HH, Pretorius IS, Bauer FF, Nel DG, Prior BA (2006) Rapid screening of the fermentation profiles of wine yeasts by Fourier transform infrared spectroscopy *Journal of Microbiological Methods*, **67**, 248–256.

Nørgaard L, Saudland A, Wagner J, Nielsen JP, Munck L, Engelsen SB (2000) Interval partial least-squares regression (iPLS): A comparative chemometric study with an example from near-infrared spectroscopy. *Applied Spectroscopy*, **54**, 413–419.

Osborne BG, Fearn T, Hindle PH (1993) *Practical NIR Spectroscopy with Applications in Food and Beverage Analysis.* Harlow: Longman Scientific and Technical.

Patz C-D, David A, Thente K, Kurbelm P, Dietrich H (1999) Viticulture and enology. *Science*, **54**, 80.

Patz C-D, Blieke A, Ristow R, Dietrich H (2004) Application of FT-MIR spectrometry in wine analysis. *Analytica Chimica Acta*, **513**, 81–89.

Petersen PB, Andersen JK, Johansen JT, Roge EH (1992) In-line measurement of important beer variables with near-infrared spectroscopy. *EBC Symposium, Instrumentation and Measurement, Monograph XX, Copenhagen, Denmark.* Weihert-Druck GmbH: Darmstadt, pp. 56–72.

Rodriguez-Saona LE, Khambaty FM, Fry FS, Calvey EM (2001) Rapid detection and identification of bacterial strains by Fourier transform near infrared spectroscopy. *Journal of Agricultural and Food Chemistry*, **49**, 574–579.

Sáiz-Abajo MJ, González-Sáiz JM, Pizarro C (2006) Prediction of organic acids and other quality parameters of wine vinegar by near-infrared spectroscopy. A feasibility study. *Food Chemistry*, **99**, 615–621.

Sauvage L, Frank D, Stearne J, Milikan MB (2002) Trace metal studies of selected white wines and alternative approach. *Analytica Chemica Acta*, **458**, 223–230.

Schropp P, Bruder T, Forstner A (2002) Evaluation of the NIR method for measuring the alcohol content and other connected parameters in beer (Alcolyzer Beer). *Monatsschrift Fur Brauwissenschaft*, **55**, 212–216.

Sjoholm K, Tenhunen J, Tammisola J, Pietila K, Home S (1996) Determination of the fermentability and extract content of industrial worts by NIR. *Journal of the American Society of Brewing Chemists*, **54**, 135–140.

Soriano PM, Pérez-Juan A, Vicario JM, González Pérez-Coello MS (2007) Determination of anthocyanins in red wine using a newly developed method based on Fourier transform infrared spectroscopy. *Food Chemistry*, **104**, 1295–1303.

Sweetlove LJ, Last RL, Fernie AR (2004) Predictive metabolic engineering: a goal for systems biology. *Plant Physiology*, **132**, 420–425.

Urbano-Cuadrado M, Luque de Castro MD, Pérez-Juan PM, Garcia-Olmo J, Gómez-Nieto MA (2004) Near infrared reflectance, spectroscopy and multivariate analysis in enology—Determination or screening of fifteen parameters in different types of wines. *Analytica Chimica Acta*, **527**, 81–88.

Urbano-Cuadrado M, Luque de Castro MD, Pérez Juan PM, Gómez-Nieto MA (2005) Comparison and joint use of near infrared spectroscopy and Fourier transform mid infrared spectroscopy for the determination of wine parameters. *Talanta*, **66**, 218–224.

Urtubia A, Pérez-Correa JR, Meurens M, Agosin E (2004) Monitoring large scale wine fermentations with infrared spectroscopy. *Talanta*, **64**, 778–784.

Vandenberg FWJ, Vanosenbruggen WA, Smilde AK (1997) Process analytical chemistry in the distillation industry using near-infrared spectroscopy. *Process Control and Quality*, **9**, 51–57.

Varadi M, Toth A, Rezessy J (1992) Application of NIR in a fermentation process. In: *Making Light Work: Advances in Near Infrared Spectroscopy* (Murray I, Cowe IA, eds). Weinheim: VCH, pp. 382–386.

Versari A, Boulton RB, Parpinello GP (2006) Effect of spectral pre-processing methods on the evaluation of the color components of red wines using Fourier-transform infrared spectrometry. *Italian Journal of Food Science*, **18**, 423–431.

Zeaiter M, Roger JM, Bellon-Maurel V (2006) Dynamic orthogonal projection. A new method to maintain the on-line robustness of multivariate calibrations. Application to NIR-based monitoring of wine fermentations. *Chemometrics and Intelligent Laboratory Systems*, **80**, 227–235.

Eggs and Egg Products

15

Romdhane Karoui, Bart De Ketelaere, Bart Kemps, Flip Bamelis, Kristof Mertens, and Josse De Baerdemaeker

Introduction

By nature, avian eggs are vehicles for reproduction and form the basis of the incubation and poultry meat industry (Kemps, 2006). They are also an important food within the human diet. From a nutritional point of view, eggs are one of the most complete foods since they are rich in protein, lipid, and carbohydrates which are essential for a good diet; eggs also contain vitamins and mineral elements that are necessary for the development of young and elderly people. A large quantity of eggs are sold as intact eggs, but many are also used in the food industry. Indeed, egg white and egg yolk are extensively used as ingredients because of their unique functional properties, such as gelling and foaming. Foams are used in the food industry for manufacturing bread, cakes, crackers, ice creams, etc. Hen egg yolk has good emulsifying properties. The foaming and emulsifying properties of albumen and yolk are affected respectively by protein concentration, pH, ionic strength, among others.

The changes that occur in eggs during storage are many and complex and affect the functional properties of both egg yolk and egg albumen. These changes include: thinning

Infrared Spectroscopy for Food Quality Analysis and Control
ISBN: 978-0-12-374136-3

of albumen, increase of pH value, weakening and stretching of the vitelline membrane, and increase in water content of the yolk. Furthermore, Hardy (1995) and Stevens (1991) reported that eggs could provide active substances that can be used for therapeutic and diagnostic uses.

Several analytical techniques are available to quantify important constituents that are found in eggs and egg products such as proteins and amino acids, essential fats, sugars, vitamins, and minerals. These techniques include chromatographic and enzymatic techniques, atomic absorption spectroscopy, immunochemical assays, and mass spectrometry. Almost all the analytical methodologies that are based on these techniques require a number of manipulations to make sample properties suitable for the final quantitative measurements; these manipulations often require multiple steps, including treatment with chemicals and enzymes, isolation, thermal treatment, homogenization, filtration, etc. Although these analytical techniques have proven their validity under laboratory circumstances, their widespread industrial application for determining the nutritional quality of eggs and egg products is currently very limited.

Moreover, most of these techniques are characterized as relatively expensive, time-consuming, and labor-intensive and require highly skilled operators. As a result, there is a clear need for simple-in-use, rapid, non-destructive, and relatively low-cost analytical tools that can be utilized in both fundamental research and industrial applications. The present chapter provides the reader with an overview of the use of visible (Vis)/near-infrared (NIR) spectroscopy to assess the quality of egg and egg products.

Egg composition

The composition of an egg is depicted in Figure 15.1 and will be described briefly in the following paragraphs.

Eggshell

The eggshell has an average thickness of about 300 μm (Freeman and Vince, 1974) and accounts for about 9–12% of the total egg weight depending on egg size. It comprises about 94% calcium carbonate (in the shape of calcite crystals) with small amounts of magnesium carbonate, calcium, phosphate, and other organic matter including protein. These crystals are organized in pillars and form the palisade layer. In between this structure an organic matrix of proteins and collagen fibers are found. The organic matrix contains 0.63% of uronic acid and 0.48% of sialic acid as reported by Nakano *et al.* (2003). The same research group pointed out the presence of amino acids such as glycine, alanine, valanine, threonine, and lysine among others.

Shell strength is influenced by two factors: (1) the hen's diet, particularly its calcium, phosphorus, manganese, and vitamin D intake and (2) egg size, which increases as the hen ages, while the mass of shell material remains stable. Hence the shell is thinner on larger eggs.

Between 7000 and 17 000 tiny pores are distributed over the shell surface (Kemps, 2006). As the egg ages, moisture and carbon dioxide (CO_2) diffuses out and air diffuses in, inducing a growth in the air cell and a decrease in the net mass.

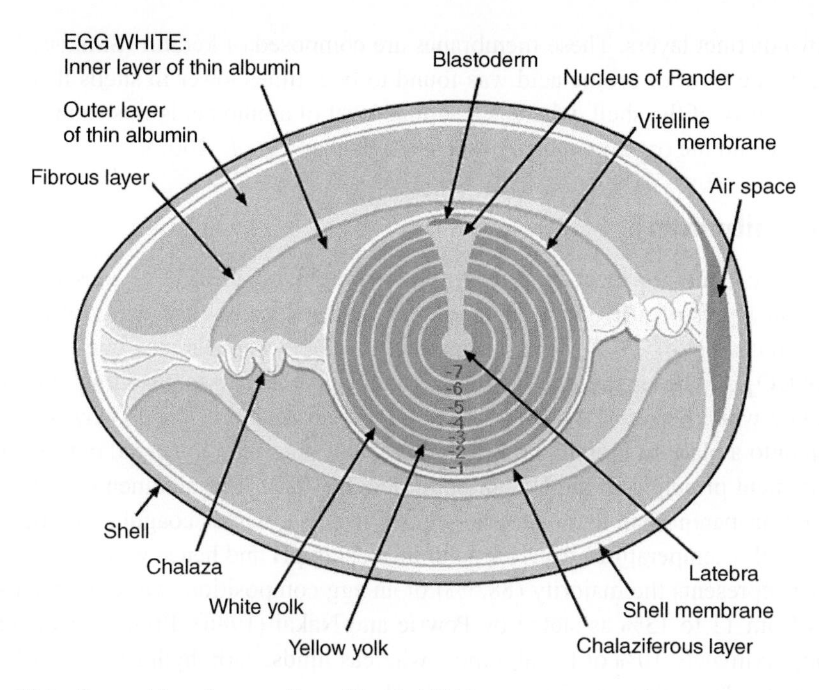

Figure 15.1 Composition of an egg. (From Allcroft, 1964.)

Brown-shelled eggs tend to be more expensive than white-shelled eggs. This could be explained by the fact that the former come from larger birds, which are more costly to feed. The brown color is mostly due to the presence of a substance called protoporphyrin, which is present at different concentrations in eggshell, even if eggs look white to the human eye (Solomon, 1976).

Cuticle

The shell is covered with a protective coating called the cuticle. By blocking the pores, the cuticle helps to preserve freshness and prevent microbial contamination of the egg contents. This is why, according to US advise, good-quality eggs (e.g. class A) should not be washed, which removes the cuticle, until immediately before they are to be used; this recommendation is under debate in Europe. As explained above, although some pigmentation can be found within the shell, the majority is present in the cuticle (Tullet, 1987). The cuticle was found to be composed of 85–87% protein, 3.5–4.4% carbohydrates, 2.5–3.5 fat, and 3.5% ash (Wedral *et al.*, 1974). The same research group reported the presence of amino acids such as aspartic and glutamic acids and glycine on one hand and carbohydrates such as glucose, galactose, mannose, and xylulose on the other hand.

Shell membranes

Between the shell and the albumen there are two shell membranes (outer and inner). The outer membrane consists of three layers of fibers, whereas the inner one contains only two distinct layers. These membranes are composed of keratin (Simkiss, 1958). Recently, the level of uronic acid was found to be 5 times lower in shells than in the organic matrix of the shell, whereas a similar level of amino acids was found in shell membrane and the organic matrix of the shell (Nakano *et al.*, 2003).

White (albumen)

Albumen accounts for most of an egg's liquid weight, about 67%. It consists of four opalescent layers of alternately thick and thin consistencies. The white of a freshly laid egg has a pH of 7.6–7.9 and an opalescent (cloudy) appearance due to the presence of CO_2. As the egg ages, CO_2 escapes and the pH increases, resulting in thinning of the egg white because of changes in protein conformation. This is why fresh eggs broken onto a plate sit up tall and firm, while older ones tend to spread out, being the measurement principle of the Haugh unit (Haugh, 1937). The albumen of older eggs is more transparent than that of fresher eggs. Fresh egg whites coagulate in the range 62–65°C; this temperatures decrease with increasing pH and hence with age.

Water represents the majority (88.5%) of an egg composition, while the total solid ranges from 11 to 13% as stated by Powrie and Nakai (1986). Protein levels represent approximately 10% of the albumen, whereas lipids, carbohydrates, and ash only represent 0.03%, 0.9%, and 0.5%, respectively.

The albumen of a newly laid egg consists of a gel (thick albumen) interposed between two liquid fractions called the outer thin and the inner thin albumen. The protein compositions of thick and thin albumen are similar except for the ovomucin content, which is 4 times higher in thick than in thin albumen (Brooks and Hale, 1959; Baliga *et al.*, 1968). However, discussion on albumen quality invariably centers on the characteristics of the thick albumen and there seems to be virtually no information available on the characteristics of thin albumen (Leeson and Caston, 1997).

Yolk

The yolk (yellow portion) represents approximately 33% of the liquid weight of an egg. It contains all of the fat in the egg and slightly less than half of the protein. With the exception of riboflavin and niacin, the yolk contains a higher proportion of the egg's vitamins than the white. All of the egg's vitamins A, D, and E are situated in the yolk. Egg yolks are one of the few foods naturally containing vitamin D. The yolk also contains more phosphorus, manganese, iron, iodine, copper, and calcium than the egg white, and it contains all of the zinc. Egg yolks have a pH of about 6.0, which stays relatively constant during aging, as there is no CO_2 loss.

Application of Vis/NIR in egg and egg products

Albumen quality

Freshness is the most important criterion to classify whole eggs, together with shell integrity. The most common quantitative parameters used to evaluate egg freshness are air cell height, which is affected by egg weight and storage relative humidity (Sauveur and De Reviers, 1988; Kessler *et al.*, 1990; Rossi *et al.*, 1995), and Haugh unit (HU) measurement. The HU is influenced by hen age (Eisen *et al.*, 1962; Sauveur and De Reviers, 1988; Silversides *et al.*, 1993; Silversides and Villeneuve, 1994). The characteristic of fresh eggs changes during aging, being influenced by both storage temperature and environmental conditions (Burley and Vadehra, 1989; Lucisano *et al.*, 1996; Rossi *et al.*, 2001). During storage, some well-known physical and chemical modifications caused by a loss of CO_2 from the egg through the pores in the shell are mainly the thinning of thick albumen (Kato *et al.*, 1981) and the increase of pH albumen (Hill and Hall, 1980). The pH of the albumen depends on the equilibrium between dissolved CO_2, bicarbonate ions, carbonate ions, and proteins.

Recently, new chemical indices that vary during the storage of eggs have been considered as descriptors of shell egg freshness (Rossi *et al.*, 1995). Among these chemical indices there is the determination of uridine and pyroglutamic acid concentration. Their increase in the albumen as well as in the yolk during the storage of eggs depends on the temperature at which eggs are stored. Furosine, ε-*N*-(2-furoylmethyl-L-lysine), an indicator of the Maillard reaction, was also used as an index that determine the shell egg freshness. Amadori products are formed during the early stages of the Maillard reaction between reducing sugars and proteins. The formation of these products decreases the nutritional value because it reduces the biological availability of lysine in the final product as shown in milk (Erbersdobler, 1986; Erbersdobler *et al.*, 1987). Since lysine and glucose are present in whole egg at a level of 0.82% and 0.34% respectively (Posati and Orr, 1976), the Maillard reaction could occur during the aging of shell eggs. In this sense, furosine, which is a product of Maillard reaction, was successfully used for the evaluation of egg freshness (Hidalgo *et al.*, 1995, 2006; Lucisano *et al.*, 1996; Rossi *et al.*, 2001). Hidalgo *et al.* (2006) confirmed the possibility of expressing shell egg freshness as equivalent egg age using furosine as a reference index. Indeed, several papers reported that furosine content showed high repeatability and low natural variability in fresh eggs and moreover, it is independent from egg weight, hen age, and storage relative humidity (Hidalgo *et al.*, 1995, 2006).

Although these methods provide useful information on the evaluation of egg freshness, they require sophisticated analytical equipment and skilled operators; they are also time-consuming and both necessitate the purchase and disposal of chemical reagents. The physicochemical methods are not effective enough to cover the growing demand for listing that the food industry must comply with. A number of non-invasive and non-destructive instrumental techniques such as optical methods in visual range could be used to fulfill these requirements. Infrared and fluorescence spectroscopic techniques have been used for the determination of egg freshness (Schmilovitch *et al.*, 2002; Bamelis, 2003; De Ketelaere *et al.*, 2004; Kemps *et al.*, 2006; Karoui *et al.*, 2006a, 2006b, 2007a, 2007b). These new analytical techniques are relatively low cost and can be applied in both fundamental research and in the factory as online sensors for monitoring egg products.

Norris (1996) investigated the usefulness of NIR spectroscopy in determining egg quality during storage. The authors reported some changes in the spectral data that occurred in eggs after lay. However, no relationship between these changes and the internal egg quality was given. One explanation could arise from the fact that eggs were stored for only few hours. In another study, Schmilovitch *et al.* (2002) used near-near-infrared spectroscopy (NNIR) in the transmittance mode (530–1130 nm) for the estimation of some quality parameters: storage days, pH, weight, air chamber size, relative loss of weight, and relative air chamber size. The authors applied partial least squares (PLS) regression to the NNIR data and the results obtained showed that days after hatching, air chamber size, weight loss, and pH could be predicted by NNIR with a determination coefficient (R^2) varying from 0.90 to 0.92. The authors suggested the development of a designated sensor to facilitate rapid testing of large egg samples, which could lead to improvement in the marketed produce and to the extension of egg shelf-life. The most critical point of this study is that these high correlations refer to group means rather than to individual eggs. In order to avoid this problem, Bamelis (2003) continued this work and used Vis/NIR to monitor quality changes in eggs stored at 18°C in an air-conditioned room over 21 days. Transmission spectra were acquired daily (except weekends) on egg samples, giving a total number of 16 spectra for each egg. The spectra were scanned between 500 and 880 nm, with an integration time of 250 ms. An increase and a decrease around 674 nm and 663 nm, respectively was observed when the storage time of eggs increased (Figure 15.2). However, a large variation between eggs presenting the same age was noticed, in agreement with the findings of Kemps *et al.* (2007). This variation was ascribed to the measurement errors, egg-dependent effects and different times.

Bamelis (2003) used another indicator for the evaluation of egg freshness and found that the 674 nm/663 nm ratios could be more suitable indicators for the determination of egg freshness. This work was continued by the same research group (Kemps *et al.*, 2006) by recording the transmission spectra of 600 eggs stored for 0, 2, 4, 6, 8, 10, 12, 14, 16, and 18 days at 18°C and relative humidity (RH) 55%;

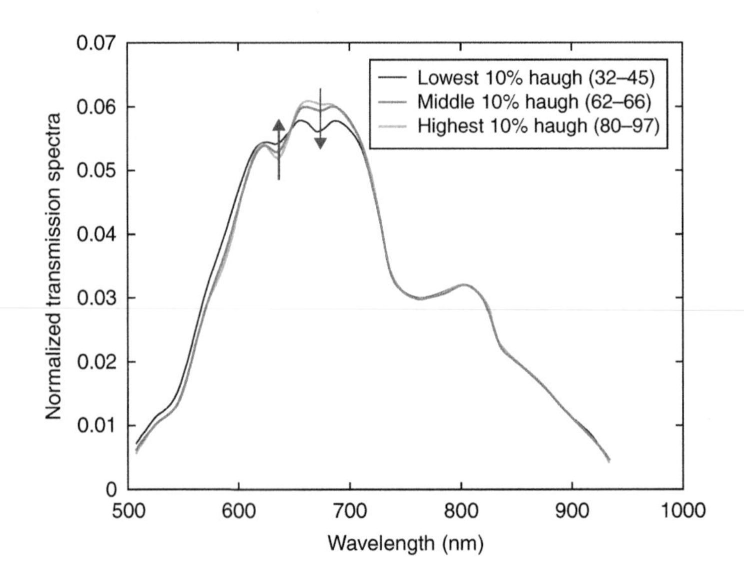

Figure 15.2 Changes in optical transmission during aging of eggs. An increase and a decrease around 674 nm and 663 nm, respectively can be observed when the storage time of eggs increases and Haugh units decrease.

the Vis/NIR spectra obtained on intact eggs showed large variation in proportional transmission values between eggs with a comparable albumen quality, confirming those obtained previously by Bamelis (2003).

In order to eliminate this problem, Kemps (2006) pointed out the necessity for spectral pre-treatment in order to assess the quality of albumen quality. Thus, multiplicative scatter correction (MSC) was applied for each spectrum recorded on intact eggs and the results obtained showed that the main changes in transmission spectra that occurred during storage of eggs was in the 500–750 nm spectral region. In addition, an increase around 630 nm and a decrease around 655 nm of the transmission values were observed during egg storage.

In order to assess the ability of Vis/NIR to determine egg quality non-destructively, PLS regression was applied (Kemps *et al.*, 2006). The results showed that the correlation coefficient between the measured and the predicted HU was 0.84 and 0.82 for the calibration and validation set, respectively. Better results were obtained for the pH since the correlation coefficient between the measured and the predicted pH was 0.87 and 0.86 for the calibration and validation set, respectively.

In order to interpret at the molecular level, regression coefficients of the PLS models of both pH and HU were studied (Kemps *et al.*, 2006). The most information was found between 570 and 750 nm. The researchers attributed these bands to the Maillard reaction with products inducing the formation of melanoidins (Burley and Vadehra, 1989), which absorbs visible light notably between 600 and 700 nm. However, the authors gave no more information about this phenomenon.

To determine the effect of the color of the eggshell, Kemps (2006) measured transmission and reflection spectra of brown and white eggs between 500 and 1150 nm. Considerably more light in the visible part of the spectrum passes through white-shelled eggs. This difference was attributed to the shell pigmentation, which is present at high levels in the case of brown-shelled eggs. Indeed, brown-shelled egg spectra showed a peak around 643 nm, which is caused by shell pigmentation. Above 750 nm, the transmission spectra of brown and white eggs showed a similar trend. Indeed, for both groups of eggs, peaks located at 760 and 960 nm were observed and were attributed to water (O–H) (Matcher *et al.*, 1994).

Recently, Liu *et al.* (2007) pointed out that the major changes in the spectra were in the range 400–520 nm and the transmittance decreased with increased storage time. A high correlation between Vis/NIR spectra, and HU and yolk coefficient was found. The main conclusion of this study is that chicken egg internal quality inspection could be investigated by Vis/NIR spectroscopy. However, in their experiment, only one kind of chicken egg in the spectral range 200–580 nm was investigated. Thus, more research is needed and should include controlled storage experiments (temperature, RH, etc.) for different strains and ages of hens, to obtain valid parameters for use in more generally used mathematical models.

Shell pigmentation

The eggshell color has no influence on the nutritional value of the egg, but it has an important impact on consumer preferences, making the color of the eggshell an important quality among other parameters (Wei and Bitgood, 1989). The color of an eggshell is determined by the presence of a pigment called protoporphyrin in the eggshell as pointed out by Lang and Wells (1987); another pigment called biliverdin also contributes to eggshell color (Kennedy and Vevers, 1973). Both pigments have different absorption peaks in the NIR. De Ketelaere *et al.* (2004) reported that for measuring shell color only reflected light and not transmitted light should be investigated, since pigments inside the egg may disturb transmitted light. Indeed, Wei and Bitgood (1989) determined eggshell color by measuring the reflected light at three specific wavelength bands, characteristic of red, green, and blue light. For table eggs, De Ketelaere *et al.* (2004) reported that shells must be strong enough to prevent failure during packing and/or transportation. Thus, the detection of eggs that have unfit shells should be removed for incubation, processing, and/or transportation.

Narushin *et al.* (2004) used mid-infrared spectroscopy (MIR) as a rapid technique for the determination of eggshell quality. The authors reported that MIR gave better results than egg size parameters. In addition, fracture force, maximal deformation, and shell stiffness were found to be predicted with comparable accuracy by both MIR and egg size parameters. Because the investigated correlations are only moderate, further research is needed to make the proposed technology useful for practical implementation.

Blood and meat spots

Blood and meat spots are considered to be the most common defects found in eggs (De Ketelaere *et al.*, 2004). These defects could influence the choice of the consumer in a range varying from 1% to nearly 100%. Eggs are usually given a lowered grade or declared inedible when a spot is detected in a whole egg by candling. However, the candling operation is an imperfect method of detecting blood and meat spots and could induce a monetary loss to the industry through consumer dissatisfaction with defective eggs. It could also result in a considerable loss to producers and handlers of eggs when non-defective eggs are mistakenly candled out (De Ketelaere *et al.*, 2004). Candling accuracy with respect to blood and meat spot detection reported in the literature indicates a range of accuracies varying from 20 to 90%.

Spectroscopic technique such as Vis/NIR could be used to avoid this problem, since De Ketelaere *et al.* (2004) showed that the presence of blood in the albumen produced absorption bands located at 415, 541, and 577 nm.

Brant *et al.* (1953) suggested that as the calciferous shell of the egg absorbs all transmitted light under 550 nm, only the 577 nm absorption peak could be used to detect the presence of blood in eggs. The above-mentioned authors showed a clear difference between eggs containing no blood and the others, since an accuracy of 99.7% was found.

De Ketelaere *et al.* (2004) continued this work and showed a difference between the transmission spectra of blanco and blood-containing eggs. The authors divided the transmission at 577 nm by the transmission at a reference wavelength in order to correct for eggshell thickness, egg size, and other non-hemoglobin-related characteristics. Gielen *et al.* (1979) suggested that a wavelength between 585 and 610 nm should be chosen. The ratio between the two transmission values is called the "blood value" and is used as an index. Unpublished results at the laboratory showed that a small amount of blood in the albumen could only be detected when part of it is diffused in the albumen, while very small bloodspots are often not accompanied by dispersed blood and, consequently, are hard to detect. On the other hand, a small amount of dispersed blood without the presence of a bloodspot cannot be seen with the human eye, but is recognized by the detecting mechanism and these are often classified as false rejects.

The detection success of blood spots in eggs is highly dependent on shell color. A higher detection rate of blood in white eggs could be achieved when compared with brown-colored shells. Indeed, the brown pigment of the eggshell (protoporphyrin) has optical properties that are closely related to those of hemoglobin. It shows a band located around 589 nm, which is very close to the absorption peak of hemoglobin (577 nm). This phenomenon makes the detection of blood in brown-shelled eggs difficult; even the use of a reference wavelength cannot solve this problem (Gielen *et al.*, 1979).

Hatching eggs

When an embryo develops in the egg, blood formation takes place from day 2 of development onwards (Romanoff, 1960; Bodemer, 1970). Initially, blood is formed on the surface of the yolk sac, close to the embryo on the upper side of the yolk (Romanoff, 1960) and not in the albumen.

The blood value mentioned, previously presents a means for monitoring the growth of the embryo, and had been thoroughly described by Bamelis *et al.* (2004), who followed individual eggs during incubation, and showed that the blood value suddenly decreases after about 72 h of incubation. Although blood is already formed more than 20 h before, it stays obscure for more than 20 h. Research has provided a plausible explanation of this finding. Since the initial formation of blood is situated at the surface of the yolk, it could be obscured by the yolk because of its high optical density (Williams and Norris, 1987).

After the second day of incubation, however, ion pumps become active and transport Na^+ ions from the albumen into the yolk sac (Babiker and Baggott, 1995), followed by a passive accumulation of water beneath the developing embryo (subembryonic fluid or SEF) (Adolph, 1967; Simkiss, 1980; Deeming *et al.*, 1987). Because of this movement of water into the yolk sac, it becomes optically transparent allowing the detection of blood attached to it.

The use of such non-destructive techniques is not only of great interest for the industry in detecting fertile eggs, but also for fundamental research. By monitoring a batch of eggs over a critical period, the exact time point at which any physiological event is happening (such as SEF initiation) can accurately be determined. These time points are more appropriate than the incubation time in determining the developmental phase of an embryo because growth rates vary. Since the technique is non-invasive, the development of the measured egg will not be disturbed enabling extra-embryonic growth factors during the very early stages of incubation to be linked with pre-incubation characteristics (e.g. storage time, age of the parent flock) and post-natal performance (Bamelis *et al.*, 2004).

Compositional analysis of egg products

Most analytical methods currently available for accurately determining moisture, fat, protein, and other major constituents in eggs and egg products are quite time-consuming. Therefore, the development of more rapid procedures is desirable. NIR reflectance, among other spectroscopic techniques, was used for the determination of moisture, fat, and protein in spray-dried whole egg (Wehling *et al.*, 1988). A standard error of performance of 0.15%, 0.20%, and 0.28% was obtained for moisture, protein, and fat, respectively, using a calibration based on three wavelengths. However, the authors suggested that it was necessary to use additional wavelengths to adequately measure these constituents in samples with particle size variability.

Liquid egg products are usually traded on a total solid content specification. In commercial egg breaking operation, there is often an imperfect separation of the yolk from the albumen; the total solids of these products and blends made from them will vary somewhat between batches. Therefore, in order to maintain consistent product quality, a rapid control procedure for this constituent would be advantageous. In this context, Osborne and Barrett (1984) used NIR transmission for the measurement of protein, total lipid, and total solid contents of liquid egg products. An R^2 of 0.96, 0.98, and 0.99 and residual standard deviation (RSD) of 0.37%, 0.66%, and 1.06% for protein, total lipid, and total solid contents, respectively, was obtained for NIR transmission compared with standard procedures. Although this feasibility study was performed on a relatively small number of samples ($n = 32$), the researchers suggested that NIR could be successfully applied to the determination of these parameters in liquid egg products.

In another study, NIR reflectance was used to assess its ability for the prediction of the physicochemical composition of freeze-dried egg yolk samples from laying hens fed with four different diets enriched with different sources of n-3 polyunsaturated fatty acids (Dalle Zotte *et al.*, 2006). NIR spectra were scanned between 1100 and 2498 nm on 365 yolk samples. The calibration results showed that NIR spectra could be used to predict their chemical composition, but also highlighted some limitations, which were related by the researchers to the measurement realized by reference method. The pH, cholesterol, and CIE color parameters were not successfully predicted. An explanation of the low accuracy of NIR in predicting color was attributed to the fact that the Vis region was not scanned, which suggested that color attributes could not be predicted from NIR spectra alone. The prediction of polyunsaturated fatty acid content was found to be accurate. By using partial least squares discriminant analysis (PLSDA), yolks from hens fed with the commercial diet and from hens fed commercial diet supplemented with marine origin were classified with 100% accuracy. From the obtained results, the researchers concluded that NIR could be used as a rapid screening of egg yolk samples originating from different feeding systems.

However, further studies should investigate the accuracy of NIR in the prediction of macronutrients (fat, protein, and ash) at highly accurate analytical values used in calibration. So far, egg product manufacturers are missing possibilities for fast raw material control because the usual refraction measurements do not allow for conclusions about the actual composition of egg products.

Büning-Pfaue *et al.* (2004) used NIR reflection spectra on liquid shell egg samples and sample mixtures from yolk, egg white, and water. By applying cluster analysis, no clear difference was obtained between the investigated samples. However, using PLS regression on the NIR spectra and physicochemical parameters, a good correlation was found for dry matter, crude protein, total fat, cholesterol, and lecithin phosphate contents since the R^2 was higher than 0.98, except for that of crude protein amount, which was equal to 0.79.

Thermal treatment could lead to a reduction in the nutritional value of the food as a result of the Maillard reaction, which makes amino compounds biologically unavailable. The F_{70}^{10} parameter is generally calculated from processing data in order to evaluate the thermal treatment to compare different time–temperature processing combinations and to calculate the effectiveness of the thermal process on pathogenic and spoilage organisms. The temperature in the center of the product that corresponds to each time interval must be obtained from the penetration heat curve, and a F_{70}^{10} value defined by the Bigelow law calculated. In general, the F_{70}^{10} is calculated from processing data in order to: (1) evaluate thermal treatment, (2) compare different time–temperature processing combinations, and (3) calculate the effectiveness of the thermal process on pathogenic and spoilage organisms.

Recently, NIR spectroscopy was used to evaluate the thermal treatment of fresh egg pasta (Zardetto, 2005). Eighty-seven fresh egg pasta samples were scanned by reflectance spectroscopy in the range 1000–2500 nm. The models predicted the F_{70}^{10} values with a standard error of prediction (SEP) of 0.16 and an R of 0.91. The researchers concluded that NIR spectroscopy could be used as a rapid tool for the determination for this parameter. Also, the examination of loading vectors suggested that NIR could be used to monitor physical changes that occurred during heat treatment.

Nowadays, there is an increasing consumer demand for high-quality and microbiologically safe foods. Non-thermal food-processing techniques such as ultra-high hydrostatic pressure and gamma irradiation leave the remaining sensory and nutritional qualities unaffected. In this sector, NIR was used to investigate the properties of egg white pasteurized by ultra-high hydrostatic pressure and gamma irradiation by using both NIR and chemo-sensor array sensor signal response (Seregély *et al.*, 2006). The researchers applied multivariate statistical analysis in order to test if there was a difference between the two techniques used for the pasteurization of egg white. Based on the presented results, they concluded that irradiation causes more drastic changes in the volatiles compounds and in the NIR properties than the ultra-high hydrostatic pressure. However, no explanation about the chemical compounds that are involved in this difference was given.

Conclusion

It is clear that Vis/NIR spectroscopy recorded on egg and egg products contains valuable information about thc quality of these products. The Vis/NIR spectra have shown to provide information about the quality of egg and egg products, including the detection of blood and meat spots and hatching as well as the composition of egg products. The present chapter has revealed the large amount of research in the field that has occurred within the past decade facilitated by the widespread use of chemometric tools. The increasing research activities can, it is hoped, address some of the challenges of Vis/NIR measurements of intact eggs and further explore the chemical systems and causality, which in many cases are not fully understood, as indicated by the tentative assignments of several peaks in egg studies.

Vis/NIR spectroscopy may be suited for online measurements because it is multidimensional, selective, and sensitive and because industrial online sensors are highly feasible. It is therefore expected that in the coming years, Vis/NIR spectroscopy combined with chemometric tools will be a reliable tool for understanding the basics of the quality of egg products.

References

Adolph EF (1967) Ontogeny of volume regulations in embryonic extra cellular fluids. *Quarterly Review of Biology*, **42**, 1–39.

Allcroft WM (1964) *Incubation and Hatchery Practice.* London: Her Majesty's Stationery Office.

Babiker EM, Baggott GK (1995) The role of ion transport in the formation of subembryonic fluid by the embryo of the Japanese quail. *British Poultry Science*, **36**, 371–383.

Baliga BR, Kadkol SB, Lahiry NL (1968) Changes in ovomucin concentration during thinning of thick white in eggs. Technical Report, Central Food Technological Research Institute, Mysore, India, 3pp.

Bamelis F (2003) Non invasive assessment of eggshell conductance and different developmental stages during incubation of eggs. PhD thesis, Catholic University of Leuven, Leuven, Belgium.

Bamelis F, Kemps B, Mertens K, Tona K, De Ketelaere B, Decuypere E, De Baerdemaeker J (2004) Non destructive measurements on eggs during incubation. *Avian and Poultry Biology Reviews*, **15**(3–4), 150–159.

Bodemer CW (1970) *Modern Embryology.* London: Holt, Rinehart and Winston.

Brant AW, Dull GG, Renfore WT, Kays SJ (1953) A spectrophotometric method for detecting blood in white shelled eggs. *Poultry Science*, **32**, 357–363.

Brooks J, Hale HP (1959) The mechanical properties of the thick white of the hen's egg. *Biochimica and Biophysica Acta*, **32**, 237–250.

Büning Pfaue H, Mielke K, Wambold C (2004) Near infrared spectrometric analysis of egg products. In: *Near Infrared Spectroscopy: Proceedings of the 11th International*

Conference (Davies AMC, Garrido-Varo A, eds). Chichester: NIR Publications, pp. 627–630.

Burley RW, Vadehra DV (1989) The albumen chemistry. In: *The Avian Egg. Chemistry and Biology* (Burley RW, Vadehra D, eds). New York: John Wiley and Sons, pp. 65–128.

Dalle Zotte A, Berzaghi P, Jansson LM, Andrighetto I (2006) The use of near-infrared reflectance spectroscopy (NIRS) in the prediction of chemical composition of freeze-dried egg yolk and discrimination between different n-3 PUFA feeding sources. *Animal Feed Science and Technology*, **128**, 108–121.

Deeming DC, Rowlett K, Simkiss K (1987) Physical influences on embryo development. *Journal of Experimental Zoology, Supplement*, **1**, 341–345.

De Ketelaere B, Bamelis F, Kemps B, Decuypere E, De Baerdemaeker J (2004) Non-destructive measurements of egg quality. *Worlds Poultry Science Journal*, **60**, 289–302.

Eisen EJ, Bohren BB, McKean HE (1962) The Haugh Unit as a measure of egg albumen quality. *Poultry Science*, **41**, 1461–1468.

Erbersdobler HF (1986) Twenty years of furosine, better knowledge about the biological significance of Maillard reaction in food and nutrition. In: *Aminocarbonyl Reaction in Food and Biological Systems* (Fijumaki M, Namiki M, Kato H, eds). Tokyo: Elsevier Science, pp. 481–491.

Erbersdobler HF, Dehn B, Nangpal A, Reuter H (1987) Determination of furosine in heated milks as a measure of heat intensity during processing. *Journal of Dairy Research*, **54**, 147–151.

Freeman BM, Vince MA (1974) *Development of the Avian Embryo.* London: Chapman and Hall.

Gielen RMAM, De Jong LP, Kerkvliet HMM (1979) Electro-optical blood-spot detection in intact eggs. *IEEE Transactions on Instrumentation and Measurements*, **IM-28**, 177–183.

Hardy CT (1995) Egg fluids and cells of the chorioallantoic membrane of embryonated chicken eggs can select different variants of influenza A (H3N2) viruses. *Virology*, **211**, 302–305.

Haugh RR (1937) The Haugh Unit for measuring egg quality. *Egg Poultry Magazine*, **43**, 552–555, 572–573.

Hidalgo A, Rossi M, Pompei C (1995) Furosine as a freshness parameter of shell eggs. *Journal of Agricultural and Food Chemistry*, **43**, 1673–1677.

Hidalgo A, Rossi M, Pompei C (2006) Estimation of equivalent egg age through furosine analysis. *Food Chemistry*, **94**, 608–612.

Hill AT, Hall JW (1980) Effects of various combinations of oil spraying, washing, sanitizing, storage time, strain, and age upon albumen quality changes in storage and minimum sample sizes required for their measurement. *Poultry Science*, **59**, 2237–2242.

Karoui R, Kemps B, Bamelis F, De Ketelaere B, Mertens K, Schoonheydt R, Decuypere E, De Baerdemaeker J (2006a) Development of a rapid method based on front face fluorescence spectroscopy for the monitoring of egg freshness: 1 evolution of thick and thin albumens. *European Food Research and Technology*, **223**, 303–312.

Karoui R, Kemps B, Bamelis F, De Ketelaere B, Mertens K, Schoonheydt R, Decuypere E, De Baerdemaeker J (2006b) Development of a rapid method based on front face fluorescence spectroscopy for the monitoring of egg freshness: 2 evolution of yolk. *European Food Research and Technology*, **223**, 180–188.

Karoui R, Schoonheydt R, Decuypere E, Nicolaï B, De Baerdemaeker J (2007a) Front face fluorescence spectroscopy as a tool for the assessment of egg freshness during storage at a temperature of 12.2°C and relative humidity of 87%. *Analytical Chimica Acta*, **582**, 83–91.

Karoui R, Schoonheydt R, Decuypere E, Nicolaï B, De Baerdemaeker J (2007b) Front-face fluorescence spectroscopy as a tool for the assessment of egg freshness during storage under modified atmosphere. *Food Chemistry,* in press.

Kato A, Ogata S, Matsudomi N, Kobayashi K (1981) Comparative study of aggregated and disaggregated ovomucin during egg white thinning. *Journal of Agricultural and Food Chemistry*, **29**, 821–823.

Kemps B (2006) VIS/NIR spectroscopy for the assessment of internal egg quality. PhD thesis, Division of Mechatronics, Biostatistics and Sensors, Department of Biosystems, Catholic University of Leuven, Leuven, Belgium.

Kemps B, Bamelis F, De Ketelaere B, Mertens K, Kamers B, Tona K, Decuypere E, De Baerdemaeker J (2006) Visible transmission spectroscopy for egg freshness. *Journal of the Science of Food and Agriculture*, **86**, 1399–1406.

Kemps B, De Ketelaere B, Bamelis F, Mertens K, Decuypere E, De Baerdemaeker J, Schwägele F (2007) Albumen freshness assessment by combining visible near-infrared transmission and low-resolution proton nuclear resonance spectroscopy. *Poultry Science*, **86**, 752–759.

Kennedy GY, Vevers HG (1973) Eggshell pigments of the Araucana fowl. *Comparative Biochemistry and Physiology*, **44B**, 11–25.

Kessler C, Sinell HJ, Wiegner J (1990) Beurteilung des Frischezustandes von Hühnereiern in Abhängigkeit von der Gewichtsklasse. *Archiv fur Lebensmittelhygiene*, **41**, 81–85.

Lang MR, Wells JW (1987) A review of eggshell pigmentation. *Worlds Poultry Science Journal*, **43**, 238–246.

Leeson S, Caston LJ (1997) A problem with the characteristic of thin albumen in laying hens. *Poultry Science*, **76**, 1332–1336.

Liu Y, Ying Y, Ouyang A, Li Y (2007) Measurement of internal quality in chicken eggs using visible transmittance spectroscopy technology. *Food Control*, **18**, 18–22.

Lucisano M, Hidalgo A, Comelli EM, Rossi M (1996) Evolution of chemical and physical albumen characteristics during the storage of shell eggs. *Journal of Agricultural and Food Chemistry*, **44**, 1235–1240.

Matcher SJ, Cope M, Delpy DT (1994) Use of the water absorption spectrum to quantify tissue chromophore concentration changes in near infrared spectroscopy. *Physical Medical Biology Journal*, **39**, 177–196.

Nakano T, Ikawa NI, Ozimek L (2003) Chemical composition of chicken eggshell and shell membranes. *Poultry Science*, **82**, 510–514.

Narushin VG, van Kempen TA, Wineland MJ, Christensen VL (2004) Comparing infrared spectroscopy and egg size measurements for predicting eggshell quality. *Biosystems Engineering*, **87**, 367–373.

Norris KH (1996) History of NIR. *Journal of Near Infrared Spectroscopy*, **4**, 31–37.

Osborne BG, Barrett GM (1984) Compositional analysis of liquid egg products using infrared transmission spectroscopy. *Journal of Food Technology*, **19**, 349–353.

Posati LP, Orr ML (1976) *Composition of Foods: Dairy and Egg Products raw-processed–prepared; Handbook 8–1.* Washington, DC: US Department of Agriculture.

Powrie W, Nakai S (1986) The chemistry of egg and egg products. In: *Egg Science and Technology* (Williams JS, Owen JC, eds). Westport: AVI Publishing Co., pp. 97–139.

Romanoff AL (1960) *The Avian Embryo.* New York: Macmillan.

Rossi M, Pompei C, Hidalgo A (1995) Freshness criteria based on physical and chemical modifications occurring in eggs during aging. *Italian Journal of Food Science*, **7**, 147–156.

Rossi M, Hidalgo A, Pompei C (2001) Reaction between albumen and 3,3', 5,5-tetramethylbenzidine as a method to evaluate egg freshness. *Journal of Agriculture and Food Chemistry*, **49**, 3522–3526.

Sauveur B, De Reviers M (1988) Egg quality. In: *Reproduction des volailles et production d'œufs* (Sauveur B, De Reviers M, eds). Paris: INRA, pp. 377–436.

Schmilovitch Z, Hoffman A, Egozi H, Klein E (2002) Determination of egg freshness by NNIRS (near-near infrared spectroscopy). In: *Proceedings of Agricultural Engineering Conference* (Paper number 02-AP-023). Budapest, Hungary.

Seregély Z, Farkas J, Tuboly E, Dalmadi I (2006) Investigating the properties of egg white pasteurised by ultra-high hydrostatic pressure and gamma irradiation by evaluating their NIR spectra and chemosensor array sensor signal responses using different methods of qualitative analysis. *Chemometrics and Intelligent Laboratory Systems*, **82**, 115–121.

Silversides FG, Twizeyimana F, Villeneuve P (1993) Research note: A study relating to the validity of the Haugh Unit correction for egg weight in fresh eggs. *Poultry Science*, **72**, 760–764.

Silversides FG, Villeneuve P (1994) Is the Haugh Unit correction for egg weight valid for eggs stored at room temperature? *Poultry Science*, **73**, 50–55.

Simkiss K (1958) The structure of eggshell with particular reference to the hen. PhD thesis, University of Reading, Reading.

Simkiss K (1980) Water and ionic fluxes inside the egg. *American Zoologist*, **20**, 385–393.

Solomon SE (1976) The thinning of thick egg white—an ultrastructural and histochemical evaluation. *Anatomia, Histologica Embryologic*, **5**, 90–104.

Stevens L (1991) Mini-review: egg white proteins. *Comparative Biochemistry and Physiology*, **100**, 1–9.

Tullet SG (1987) Egg shell formation and quality. In: *Egg Quality—Current Problems and Recent Advances* (Wells RG, Belyavin CG, eds). London: Butterworths, pp. 123–146.

Wedral EM, Vadehra DV, Baker RC (1974) Chemical composition of the cuticule and inner and outer membranes from egg of Gallus. *Comparative Biochemistry and Physiology*, **47**, 631–640.

Wehling RL, Pierce MM, Froning GW (1988) Determination of moisture, fat and protein in spray-dried whole egg by near infrared reflectance spectroscopy. *Journal of Food Science*, **53**, 1356–1359.

Wei R, Bitgood JJ (1989) A new objective measurement of eggshell color. 1. A test for potential usefulness of two color measuring devices. *Poultry Science*, **69**, 1175–1780.

Williams P, Norris K (1987) *Near-infrared Technology in Agricultural Food Industries.* St. Paul, MN: American Association of Cereal Chemists, Inc.

Zardetto S (2005) Potential applications of near infrared spectroscopy for evaluating thermal treatments of fresh egg pasta. *Food Control*, **16**, 249–256.

Index

Infrared Spectroscopy for Food Quality Analysis and Control
ISBN: 978-0-12-374136-3

Printed and bound by CPI Group (UK) Ltd, Croydon, CR0 4YY
08/05/2025
01864824-0001